Acid-Base Balance and Nitrogen Excretion in Invertebrates

Dirk Weihrauch • Michael O'Donnell

Editors

Acid-Base Balance and Nitrogen Excretion in Invertebrates

Mechanisms and Strategies in Various Invertebrate Groups with Considerations of Challenges Caused by Ocean Acidification

 Springer

Editors
Dirk Weihrauch
University of Manitoba
Winnipeg
Manitoba
Canada

Michael O'Donnell
Department of Biology
McMaster University
Hamilton
Ontario
Canada

ISBN 978-3-319-39615-6   ISBN 978-3-319-39617-0   (eBook)
DOI 10.1007/978-3-319-39617-0

Library of Congress Control Number: 2016959038

Printed on acid-free paper

This Springer imprint is published by Springer Nature
The registered company is Springer International Publishing AG
The registered company is Gewerbestrasse 11, 6330 Cham, Switzerland

Preface

The Importance of Invertebrates

Invertebrates rule the world in terms of number of species, number of individuals, and biomass. Over 95 % of species are invertebrates, and this diversity may be a consequence of the small size of most invertebrates (Wilson 1987). Their niches are thus also small, so the environment can be divided into multiple domains in which specialists can coexist. Biodiversity may also follow from their greater antiquity, in that invertebrates have had more time for exploration of the environment and evolution of specialist traits. Invertebrates are also the most common animals in terms of number of individuals; copepods and nematodes are the most common multicellular animals in marine and terrestrial environments, respectively. Invertebrates also rule the earth in terms of body mass; a typical hectare of Brazilian rain forest may contain 200 kg dry weight of animal body tissue, of which 93 % consists of invertebrates. Recent estimates suggest that the world holds 10^{18} insects, equivalent to more than 200 million insects for each human on the planet and 300 pounds of insects for every pound of humans. Their numbers and biomass have important implications for the earth's ecosystems. For example, the principal consumers of vegetation in Central and South America are not deer, rodents, or birds, but leaf-cutter ants.

Invertebrates also play dominant roles in ecosystem services, which are the benefits provided by ecosystems for humankind and which include supporting, provisioning, and regulating services (Prather et al. 2013). Invertebrates provide supporting services for primary production directly through pollination and seed dispersal and indirectly through a cascades and nutrient cycling. Insect pollination is required for ~75 % of all the world's food crops and is estimated to be worth ~10 % of the economic value of the world's entire food supply. Globally, pollinators appear to be strongly declining in both abundance and diversity. The loss of pollinator species may result in reduced seed set, reduced fitness, and lower population viability of flowering plants. Most primary production eventually enters detrital food webs in which the dominant consumers and nutrient cyclers are invertebrates. Soil invertebrates break down soil organic matter and recycle nutrients in terrestrial ecosystems. Earthworms, for example, burrow and facilitate water flow and storage, soil aeration, and root development. As a consequence, changes in the decomposer fauna alter the physical and chemical environment, with consequent "bottom-up" effects on primary producers and higher trophic levels. Reef-building corals also provide supporting services by producing structures that serve as habitat for most coastal fish species.

Provisioning services refer to goods obtained from ecosystems and include food, natural products, and pharmaceuticals obtained from insects, shrimp, crabs, scallops, oysters, lobsters, corals, sea cucumbers, and many others.

Regulating services are those that regulate ecosystem processes or maintain ecosystem structure. Invertebrates have important impacts on water quality, stabilization of food webs, and regulation of diseases and pests or invasive species. In shallow water, bivalves (i.e., mussels and oysters) may filter 10–100 % of the water column; in the case of the invasive zebra mussel (*Dreissena polymorpha*) in the Great Lakes, excessive water filtration has reduced plankton levels and thus altered food webs. Invertebrates are important not only as hosts for parasites and pathogens but also as predators and parasitoids that regulate many parasites and disease vectors. Human diseases transmitted by invertebrate vectors include malaria, dengue, onchocerciasis, and Lyme disease. Schistosomes and nematodes may cause disease directly. On the other hand, invertebrates play important roles in biocontrol of crop-feeding insects and disease vectors through parasitism, direct predation, or transmission of viruses, bacteria, and toxins.

For scientific, economic, and ethical reasons, invertebrates also play increasingly important roles as model organisms in fields such as genetics, developmental biology, toxicology, and physiology. Insects (*Drosophila*), nematodes (*C. elegans*), and protozoans (*Tetrahymena*) have long been used in studies of genetics and developmental biology. The flatworm *Schmidtea mediterranea* has been useful in studies of regeneration, and the crustacean *Daphnia* is widely used in studies of toxicology.

This book had its genesis, in part, on the authors' concerns that anthropogenic activities, especially the increasing levels of carbon dioxide in the earth's atmosphere resulting from the burning of fossil fuels, pose enormous challenges for most species. In particular, the loss of species abundance and diversity is troubling. For invertebrates, 67% of monitored populations show a 45% decline in mean abundance (Dirzo et al. 2014). For aquatic organisms, reductions in water pH as a consequence of elevations in ambient carbon dioxide levels create challenges on shell deposition in crustaceans, mollusks, and corals. However, the lower pH may facilitate the elimination of ammonium as a nitrogenous waste. Too much nitrogen in soils following application of fertilizer may lead to depletion of other important minerals such as calcium, phosphorus, and magnesium. Nitrogen-polluted air, resulting from release of nitrates from automobiles and industrial plants, also results in this acidification of the soil when acid rain falls.

This book summarizes therefore the most current views on acid–base homeostasis and nitrogen excretion strategies as well as the effects of increasing environmental CO_2 levels in a broad range of phyla and subphyla, including reef-building cnidarians, planarians, nematodes, leeches, echinoderms, aquatic and terrestrial crustaceans, cephalopods, and insects. The book also describes regulatory mechanisms in key species like *Caenorhabditis elegans* (genetic model system), *Schmidtea mediterranea* (regenerative model system), *Strongylocentrotus droebachiensis* (developmental model system), *Carcinus maenas* (physiologic model system, aggressive invasive species), *Sepioteuthis lessoniana* (commercially relevant cephalopod), and *Aedes aegypti* and *Rhodnius prolixus* (disease vectors). Additionally, a chapter is dedicated to elaborate on the physiological importance of the Na^+/K^+-ATPase, which is essential for ammonia excretion and acid–base regulatory processes.

Dirk Weihrauch
Winnipeg, MA, Canada

Michael O'Donnell
Hamilton, ON, Canada

References

Dirzo R, Young HS, Galetti M, Ceballos G, Isaac NJ, Collen B (2014) Defaunation in the Anthropocene. Science 345(6195):401–406

Prather CM, Pelini SL, Laws A, Rivest E, Woltz M, Bloch CP, Del Toro I, Ho CK, Kominoski J, Newbold T (2013) Invertebrates, ecosystem services and climate change. Biol Rev 88(2):327–348

Wilson EO (1987) The little things that run the world (the importance and conservation of invertebrates). Conser Biol 1:344–346

Contents

Contributors

Aida Adlimoghaddam Department of Biological Sciences, University of Manitoba, Winnipeg, MB, Canada

Katie L. Barott Scripps Institution of Oceanography, University of California San Diego, La Jolla, CA, USA

Megan E. Barron Scripps Institution of Oceanography, University of California San Diego, La Jolla, CA, USA

Dimitri D. Deheyn Scripps Institution of Oceanography, University of California San Diego, La Jolla, CA, USA

Andrew Donini Department of Biology, York University, Toronto, ON, Canada

Sandra Fehsenfeld Department of Zoology, University of British Columbia, Vancouver, BC, Canada

Daniela P. Garçon Universidade Federal do Triângulo Mineiro, Campus de Iturama, Iturama, MG, Brazil

Caitlin G. Howe Department of Environmental Health Sciences, Mailman School of Public Health, Columbia University, New York, NY, USA

Marian Y. Hu Institute of Physiology, Christian-Albrechts-University of Kiel, Kiel, Germany

David I. Kline Scripps Institution of Oceanography, University of California San Diego, La Jolla, CA, USA
Smithsonian Tropical Research Institute, Balboa, Panama

Francisco A. Leone Departamento de Química, Ciências e Letras de Ribeirão Preto, Universidade de São Paulo, Ribeirao Preto, Sao Paulo, Brazil

Lauren B. Linsmayer Scripps Institution of Oceanography, University of California San Diego, La Jolla, CA, USA

Stuart M. Linton School of Life and Environmental Sciences, Deakin University, Waurn Ponds, VIC, Australia

Malson N. Lucena Departamento de Química, Ciências e Letras de Ribeirão Preto, Universidade de São Paulo, Ribeirão Preto, Ribeirao Preto, Sao Paulo, Brazil

Philip G.D. Matthews Department of Zoology, University of British Columbia, Vancouver, BC, Canada

John C. McNamara Departamento de Biologia, Faculdade de Filosofia, Ciências e Letras de Ribeirão Preto, Universidade de São Paulo, Ribeirão Preto, Sao Paulo, Brazil
Centro de Biologia Marinha, Universidade de São Paulo, Ribeirão Preto, Sao Paulo, Brazil

David F. Moffett School of Biological Sciences, Washington State University, Pullman, WA, USA

M.J. O'Donnell Department of Biology, McMaster University, Hamilton, ON, Canada

Horst Onken Department of Biological Sciences, Wagner College, Staten Island, NY, USA

Marcelo R. Pinto Departamento de Química, Ciências e Letras de Ribeirão Preto, Universidade de São Paulo, Ribeirão Preto, Ribeirão Preto, Sao Paulo, Brazil

Alex R. Quijada-Rodriguez Department of Biological Sciences, University of Manitoba, Winnipeg, MB, Canada

Meike Stumpp Department of Zoophysiology, Zoological Institute at Christian-Albrechts-University Kiel, Kiel, Germany

Martin Tresguerres Scripps Institution of Oceanography, University of California San Diego, La Jolla, CA, USA

Yung-Che Tseng Marine Research Station, Institute of Cellular and Organismic Biology, Academia Sinica, Taipei, Taiwan

Dirk Weihrauch Department of Biological Sciences, Institute of Cellular and Organismic Biology, Academia Sinica, Taipei, Taiwan

Jonathan C. Wright Department of Biology, Pomona College, Claremont, CA, USA

Nitrogen Excretion in Aquatic Crustaceans

Dirk Weihrauch, Sandra Fehsenfeld, and Alex Quijada-Rodriguez

© Springer International Publishing Switzerland 2017
D. Weihrauch, M. O'Donnell (eds.), *Acid-Base Balance and Nitrogen Excretion in Invertebrates*,
DOI 10.1007/978-3-319-39617-0_1

1.1   Summary

This chapter summarizes the current knowledge of processes involved in nitrogen excretion in aquatic crustaceans with focus on the species-rich (up to 10,000 species) infraorder Brachyura, the true crabs (Martin and Davis 2001). Besides the introduction that briefly covers pathways of the synthesis of nitrogenous waste products and the toxicity of ammonia, organs involved in the excretory processes will be introduced, such as the antennal gland and the gills. More emphasis will be given toward the gills and their capability to actively excrete ammonia in different haline species including the marine crabs *Cancer pagurus* and *Metacarcinus magister*, brackish water living *Carcinus maenas*, as well as freshwater dwelling *Eriocheir sinensis*. In more detail this chapter reviews the branchial ammonia excretion mechanisms in the green shore crab *C. maenas*, which is summarized in a working model at the end. Here the potential roles of the Na^+/K^+-ATPase, K^+-channels, Rh-proteins, V-type H^+-ATPase, the microtubule network, Na^+/H^+ exchangers (NHEs), ammonia transporters (AMTs), and aquaporins (AQPs) will be discussed. Besides cited literature, this chapter contains original data as well as cross-references to corresponding chapters within this book.

1.2   Introduction

1.2.1 Synthesis of Nitrogenous Waste Products in Aquatic Crustaceans

As their primary nitrogenous waste product, aquatic crustaceans excrete ammonia (in this chapter NH_3 refers to molecular ammonia, NH_4^+ to ammonium ions, and ammonia to the sum of both), originating from protein catabolism, as seen in other aquatic invertebrates and fish (Claybrook 1983; Larsen et al. 2014). The majority of ammonia derives from the deamination of glutamine, glutamate, serine, and asparagine by the enzymes glutaminase, glutamate dehydrogenase, serine dehydrogenase, and asparaginase, respectively (Greenaway 1991; Krishnamoorthy and Srihari 1973). A summary of the synthesis of nitrogenous waste products in crustaceans is depicted in ◘ Fig. 1.1.

Additionally, some ammonia is produced via the uricolytic pathway due to the metabolism of nucleic acids. Compared to the predominant production of NH_3 from amino acids, ammonia originating from this pathway is considered being small (Hartenstein 1970; Schoffeniels and Gilles 1970; Weihrauch et al. 2004b).

In addition, the production of ammonia via the purine nucleotide cycle has been discussed (Weihrauch et al. 2004b). The purine nucleotide cycle is involved in the synthesis of nucleic acids and coupled to various transaminase reactions. At the start of the cycle, inosine monophosphate (IMP) and aspartic acid, an amino acid synthesized from oxaloacetate, and the α-amino group derived from the deamination of glutamic acid, form adenylosuccinate, which is a precursor of adenosine monophosphate (AMP). After the catalytic deamination of AMP by AMP deaminase, ammonia is liberated and IMP enters the cycle again. The existence of this pathway has further been supported by the analysis of expressed sequence tags (ESTs) from tissues of lobster *Homarus americanus* and the green shore crab *Carcinus maenas* (Towle and Smith 2006) verifying the abundance of mRNA coding for adenylosuccinate synthase. Adenylosuccinate synthase is an enzyme important for the initial conversion of inosine monophosphate to adenylosuccinate

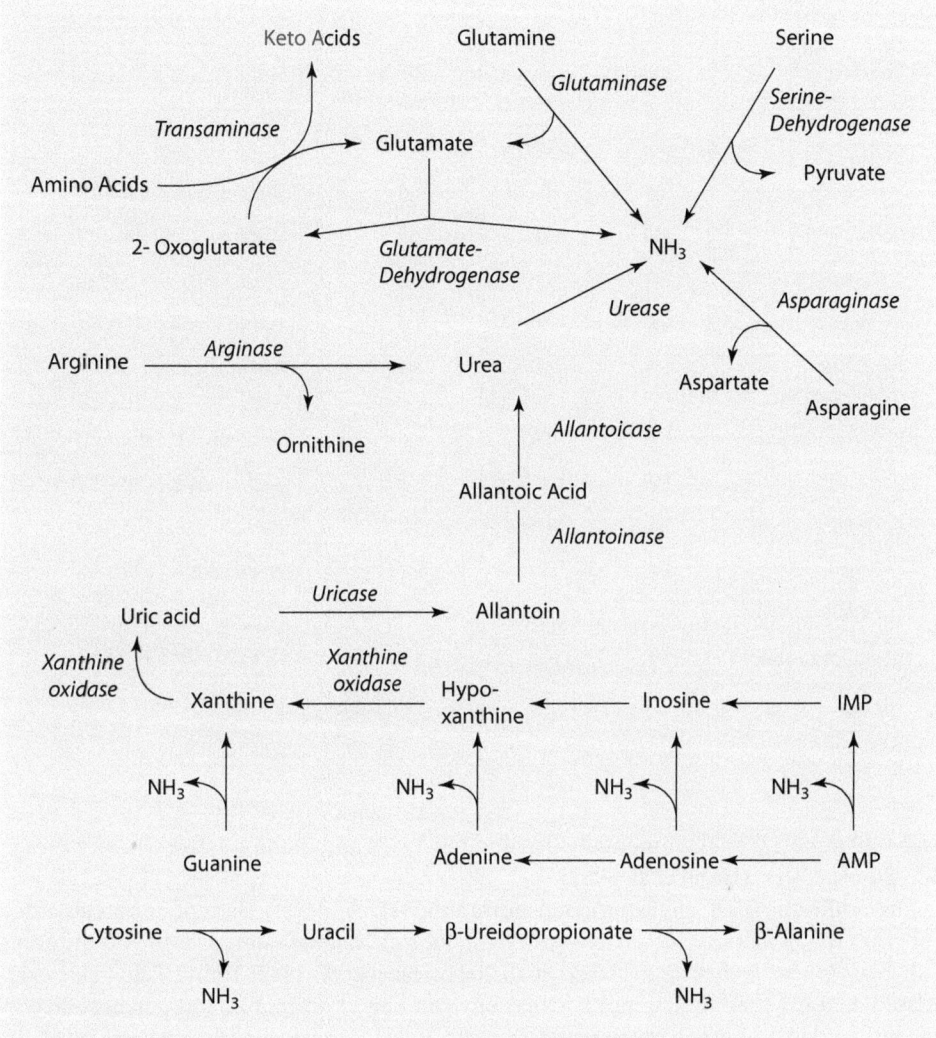

☐ **Fig. 1.1** Metabolic formation of nitrogen excretion products in Crustacea. The expression of xanthine oxidase in crustaceans is unclear to date, which is indicated by "?" within the figure (The image is taken with permission from Weihrauch et al. **2004b**) and modified after Claybrook (1983))

(Lowenstein 1972). The overall importance of the purine nucleotide cycle in ammonia-genesis in crustaceans is, however, not clear to date. For details on the nitrogen metabolism of terrestrial crabs, the reader is referred to ► Chap. 2.

1.2.2 Toxicity of Ammonia in Crustaceans

In crustaceans, toxic effects have been observed and for a number of aquatic crustaceans, ammonia is lethal at relatively low doses. In *Penaeus paulensis*, the Sao Paulo shrimp, $LC50_{96h}$ was determined to 0.307 mmol·l^{-1} total ammonia and 19 µmol·l^{-1} NH_3 (Ostrensky et al. 1992). In the red tail prawn *Penaeus penicillatus*, $LC50_{96h}$ was 58 µmol·l^{-1} NH_3

◘ **Table 1.1** The concentration of ammonia in the hemolymph of selected decapod crustaceans

Species	Salinity	Ammonia ($\mu mol \cdot l^{-1}$)	Urea ($\mu mol \cdot l^{-1}$)	Source
Brachyura				
Maia squinado	SW	50–70	nd	Durand et al. (2000)
Necora puber	SW	120	nd	Durand and Regnault (1998)
Cancer pagurus	SW	80	38	Weihrauch et al. (1999b)
Metacarcinus magister	SW	180	nd	Martin et al. (2011)
Portunus trituberculatus	SW	120	80	Ren et al. (2015)
Callinectes sapidus	BW	390	nd	Cameron and Batterton (1978)
Carcinus maenas	SW	83	78	Weihrauch et al. (1999a)
Carcinus maenas	BW	99	1029	Weihrauch et al. (1999b)
Eriocheir sinensis	FW	160	824	Weihrauch et al. (1999b)
Other crustaceans				
Cherax destructor	FW	100	nd	Fellows and Hird (1979)

SW seawater, *BW* brackish water, *FW* freshwater, *nd* no data available

and 1.39 mmol·l^{-1} total ammonia (Chen and Lin 1992) and in the crayfish *Orconectes nais* 186 µmol·l^{-1} NH$_3$ (Hazel et al. 1982).

In addition, high environmental ammonia (HEA) levels disrupt ionoregulatory function in low salinity acclimated American lobster *Homarus americanus* and the crayfish *Pacifastacus leniusculus* (Harris et al. 2001, Hazel et al. 1982; Young-Lai et al. 1991). Also, when exposed to 1 mmol l^{-1} total environmental ammonia, ion permeability in the gill epithelia of green shore crabs *Carcinus maenas* increased (Spargaaren 1990). In the Dungeness crab *Metacarcinus magister*, a chronic exposure of 2 weeks to the same doses caused an increase of the hemolymph ammonia from 179 µmol l^{-1} to nearly 1 mmol l^{-1}, accompanied with a total loss of the capability to actively excrete ammonia via the gills. In fact, ammonia excretion in these animals ceased completely, with a change of excretion to a so far unknown nitrogenous waste product (Martin et al. 2011). In the shrimp *Penaeus stylirostris*, the total number of immune active hemocytes was reduced by ca. 50 % when exposed to elevated environmental ammonia levels (Le Moullac and Haffner 2000).

Accordingly, due to the toxic effects, hemolymph ammonia concentrations in aquatic crustaceans are usually low, ranging between 50 and 400 µmol·l^{-1}, as seen in ◘ Table 1.1. for selected aquatic crustacean species. As described in ► Chap. 2, semiterrestrial and terrestrial crustaceans exhibit usually a higher tolerance for ammonia in their body fluids with values detected up to 2 mmol·l^{-1} (Weihrauch et al. 2004b).

1.2.3 Habitat and Physiological Adaptations

Crustaceans have conquered all aquatic habitats including freshwater, brackish water environments as found in estuaries and the Baltic Sea, fully marine water bodies, but also high-saline environments, such as the Mediterranean Sea, and, e.g., in the case of the brine shrimp *Artemia salina*, extreme high-saline ($\geq$250 ppt.) habitats (Cole and Brown 1967). Additionally, a number of crustacean species inhabit semiterrestrial and even fully terrestrial environments. Specific adaptations and nitrogen excretion strategies of those animals are covered in ▶ Chap. 2.

According to the salinity of the habitat, the ion conductance of the ion and gas-transporting epithelia (e.g., gills) can vary dramatically. While most marine crustaceans are osmoconformers, keeping their body fluid osmolality close to the osmolality of the environment, brackish water and freshwater species are hyper-regulators, maintaining a higher osmolality in their hemolymph compared to the exterior medium (Larsen et al. 2014; Mantel and Farmer 1983; Weihrauch and O'Donnell 2015). Marine species usually exhibit very leaky gill epithelia. For instance, the gill epithelium of the edible crab *Cancer pagurus* exhibits a conductance of ca. 250–280 mS cm^{-2}. In brackish water species such as the green shore crab *Carcinus maenas* (acclimated to 10 ‰ S.) or the isopod *Idotea baltica* (acclimated to 20 ‰ S.), the conductance of the gills and pleopods is much lower, assessed to approximately 40–60 mS cm^{-2} and 80 mS cm^{-2}, respectively. In freshwater species like the Chinese mitten crab *Eriocheir sinensis* or the red freshwater crab *Dilocarcinus pagei*, the branchial ion permeability is very low (below 5 mS cm^{-2}), minimizing thereby dangerous passive ion losses into the highly dilute environment via the paracellular route (Postel et al. 2000; Weihrauch et al. 1999b, 2004a).

Besides the environmental salinity dictating the ion permeability of the body surfaces and thereby the paracellular pathway for nitrogenous compounds of an animal inhabiting these environments, also the respective lifestyle demands particular transport characteristics to ensure sufficient excretion of metabolically produced ammonia. The majority of crustaceans exhibiting a pelagic lifestyle, colonizing the open water column, where ammonia concentrations are usually below 5 µmol l^{-1}, and thus ten to forty times lower compared to ammonia concentrations usually found in the body fluids of aquatic crustaceans (Weihrauch et al. 2004a). However, benthic-living species may face higher environmental ammonia concentrations. For instance, studies on the interstitial pore water in the North Sea showed that ammonia concentrations can already be quite high here with values between 100 and 300 µmol l^{-1} near the surface of the sediment, but can reach extreme concentrations of up to 2.8 mmol l^{-1} in 4–9 cm sediment depth due to the lack of ammonia assimilation by phytoplankton and a reduced bacterial nitrification (Enoksson and Samuelsson 1987; Lohse et al. 1993). As described below, gills of a variety of crab species are capable to excrete ammonia in an active mode against inwardly directed gradients, counteracting thereby these challenging high environmental ammonia concentrations. In fact, the capability of the gills to actively excrete ammonia is an important physiological adaptation, e.g., benthic-living brachyuran crabs are often feeding on carrion and do not hesitate to enter high ammonia environments in order to feed (Weihrauch, personal observation). In addition, during low tide and in the winter months, many decapod crabs exhibit a burrowing behavior (Bellwood 2002; McGaw 2004). During these time periods, where water exchange is limited but metabolic ammonia production continues, animals likely encounter elevated environmental ammonia concentrations in their burrows that might exceed the ammonia levels of their body fluids, and active excretion mechanisms are required to counteract potential influxes.

1.2.4 Hemolymph Ammonia and Urea Concentrations in Aquatic Crustaceans

Listed in ◘ Table 1.1, hemolymph ammonia concentrations in aquatic environments are fairly low and ranging mostly between 100 and 200 $\mu mol\ l^{-1}$. Hemolymph urea concentrations vary and seem to be dependent on the environmental salinity. While urea concentrations are rather low in marine species, fairly high values were detected in osmoregulating species.

It is to note that no urea was excreted or transformed into other molecules, when isolated posterior gills from brackish water-acclimated *C. maenas* were perfused with 600 $\mu mol\ l^{-1}$ urea (Weihrauch et al. 1999a, b). The purpose of retaining urea within the body fluids in osmoregulating crabs is unknown to date but likely does not play a significant role in the detoxification of elevated hemolymph ammonia levels since, at least in *M. magister*, urea concentration ($467.2 \pm 33.5\ \mu mol\ l^{-1}$) increased only by ca. 30 % when animals were chronically exposed to 1 $mmol\ l^{-1}$ ammonia. Under these conditions the animals fully ceased the excretion of ammonia and urea, while hemolymph ammonia concentrations raised to new steady-state levels to nearly 900 $\mu mol\ l^{-1}$ (Martin et al. 2011).

1.3 Tissues Involved in the Excretion of Nitrogenous Waste Products

1.3.1 Excretion Rates of Ammonia and Urea in Aquatic Crustaceans

As depicted in ◘ Table 1.2, in aquatic crabs the majority of the nitrogenous waste is excreted in the form of ammonia, with a lower amount of nitrogen being excreted as urea.

1.3.2 Antennal/Maxillary Glands

The antennal and maxillary glands feature a mammalian nephron-like ultrafiltration with modifications of ion composition of the primary filtrate in the downstream sections of the gland, namely, the proximal tubular region, distal tubular region, end tubular region, and bladder (Freire et al. 2008; Tsai and Lin 2014). As described for the crayfish *Pacifastacus* sp. and *Orconectes virilis*, molecules of a molecular mass less than 20 kDa are filtered freely into coelomic end sac, the first part of the gland. The structure and organization of the "internal excretory organs" such as the antennal and maxillary glands have been described in detail by Freire and others (Freire et al. 2008). In aquatic crustaceans, the excretion of nitrogenous waste via the antennal/maxillary glands was considerate being small. For instance, in the low-salinity-acclimated blue crab *Callinectes*, the actual urinary excretion rate accounted to only 1–2 % of the animals' total ammonia excretion rate (Cameron and Batterton 1978).

◻ **Table 1.2** Ammonia and urea excretion rates in selected aquatic crustaceans

Species	Salinity	Ammonia (nmol·l gFW^{-1} h^{-1})	Urea (nmol·l gFW^{-1} h^{-1})	Source
Brachyura				
Maia squinado (winter)	SW	46	1.7	Durand et al. (2000)
Necora puber	SW	552	14	Durand and Regnault (1998)
Cancer pagurus	SW	336	41	Weihrauch et al. (1999b)
Cancer antennarius	SW	142	nd	Hunter and Kirschner (1986)
Cancer antennarius	BW	374	nd	Hunter and Kirschner (1986)
Metacarcinus magister	SW	367	15	Martin et al. (2011)
Callinectes sapidus	BW	540	nd	Cameron and Batterton (1978)
Carcinus maenas	SW	187	13	Weihrauch et al. (1999a)
Carcinus maenas	BW	136	24	Weihrauch et al. (1999b)
Eriocheir sinensis	FW	123	13	Weihrauch et al. (1999b)
Other crustaceans				
Astacus leptodactylus	FW	67	nd	Ehrenfeld (1974)
Penaeus monodon	SW	63	nd	Chen and Cheng (1995)

SW seawater, *BW* brackish water, *FW* freshwater, *nd* no data available

On the other hand, other more recent observation indicates that the potential function of the antennal gland for the excretion of nitrogenous waste in decapod crabs should not be underestimated. An analysis of the urine of various crab species revealed that while ammonia concentrations in *C. pagurus* (seawater) and *E. sinensis* (freshwater) were found to be very similar compared to their respective hemolymph ammonia concentrations, urine in brackish water-acclimated *C. maenas* contained ca. 300–400 µmol l^{-1}, approximately three times the value measured in the hemolymph. This indicates a modification of the primary urine within the nephron, as more pronounced observed in semiterrestrial and terrestrial crabs. For instance, in the ghost crab *Ocypode quadrata*, ammonia concentrations measured in the urine were with 116 mmol l^{-1} exceptional high and roughly 130 times greater the values measured in the hemolymph (DeVries et al. 1994). For more information, please also refer to ▶ Chap. 2.

In addition to ammonia, also urea was detected in the urine of crabs. While in osmo-conforming *C. pagurus* urea concentrations were with 25 $\mu mol\cdot l^{-1}$ similar to the respective hemolymph value ($\rightarrow$ no modification of the primary urine during the passage through the antennal gland), in osmoconforming green crabs *C. maenas*, the urine contained 388 $\mu mol\cdot l^{-1}$ urea, roughly five times as much, compared to values found in the hemo-lymph. In contrast, in hyper-regulating *C. maenas* and *E. sinensis*, urea concentrations were with 125 and 199 $\mu mol\ l^{-1}$ much lower compared to the respective hemolymph val-ues, indicating that in these species freely filtrated urea is actively reabsorbed within the nephron (Weihrauch et al. 1999a, b). The observation is in line with the notion that osmo-regulating crabs retain, for a so far unknown reason, a certain high amount of urea within their body fluids as described above ($\blacktriangleright$ Sect. 1.3.1).

Besides the process of ultrafiltration occurring in the first section of the antennal gland, the coelomic end sac, mechanisms involved in the adjustment of ammonia and urea concentrations along the nephron have not investigated to date. Studies by Tsai and others on the antennal gland of the semiterrestrial crab *Ocypode stimpsoni* showed abun-dance of the basolateral Na^+/K^+-ATPase and an apical V-ATPase in labyrinthine and end-labyrinthine cells in all sections of the nephron (Tsai and Lin 2014). As described in more detail below, both pumps are usually directly involved in energizing trans-epithelial ammonia transport processes in excretory tissues (Larsen et al. 2014; Weihrauch et al. 2004b; Weihrauch and O'Donnell 2015).

1.3.3 Gills

Gills of crustaceans are multifunctional organs involved in gas exchange, acid-base balance, osmoregulation, ammonia excretion, and Ca^{2+} transport (Freire et al. 2008; Henry et al. 2012). In decapod crustaceans three types of gills can be distinguished: (a) dendrobran-chiae, found, e.g., in penaeoid and sergestoid shrimps, (b) trichobranchiae, found, e.g., in crayfish and lobster, and (c) phyllobranchiae, the gills of the Brachyura, some Anomura, and the caridean shrimp (Freire et al. 2008) (Taylor and Taylor 1992). The single-layered gill epithelium is covered with a thin, ion-selective cuticle as illustrated in $\square$ Fig. 1.2. In brackish water-acclimated *C. maenas*, for instance, the isolated cuticle exhibits a slightly greater conductance for NH_4^+ (683 ± 165 mS cm^{-2}) than for Na^+ (583 ± 71 mS cm^{-2}). Cau-tion must be taken when interpreting data deriving from flux experiments employing amiloride, since this drug inhibits the cation conductance of the isolated cuticle in a dose-dependent manner (Lignon 1987; Onken and Riestenpatt 2002; Weihrauch et al. 2002).

Most studies regarding ion transport, acid-base regulation, and ammonia fluxes have been performed on the phyllobranchiate gills of brachyuran crabs due their relatively large size that allows for its perfusion with a hemolymph-like saline. Additionally, splitting a single gill lamellar into two half-lamellae (which consists only of the cuticle-covered epithelium layer) allows for tracer flux studies as well as electrophysiological experiments in micro-Ussing chambers.

Although not studied in this regard, one can expect that other gill-like structures, such as the endopodites from isopods, are also responsible for excretory processes. For instance, in the euryhaline isopod *Idotea baltica*, the NaCl transport mechanism in the endopodites was found to be very similar to that identified in the osmoregulating posterior gills of *C. maenas* (Postel 2001). As described by Weihrauch and O'Donnell, a number of transporters involved in the ionoregulatory machinery play also a crucial role in trans-epithelial ammonia excretion

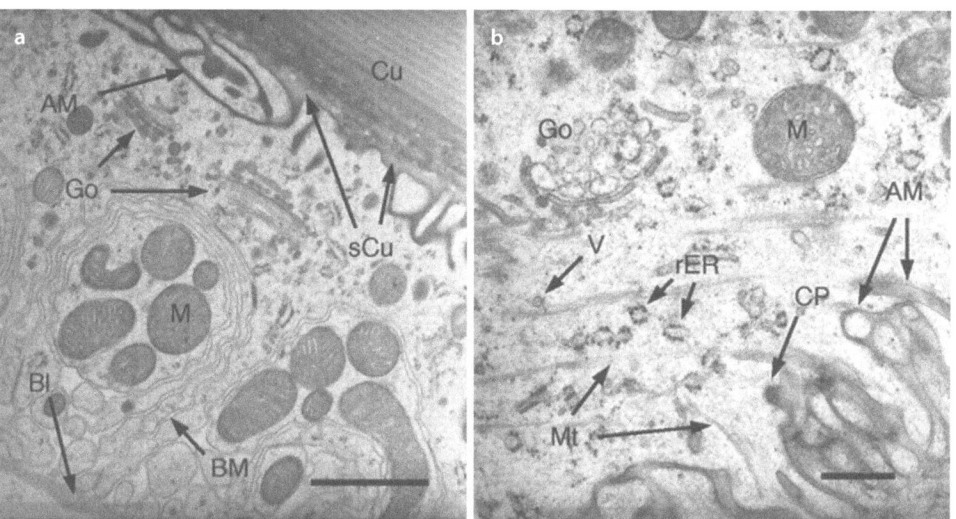

▢ Fig. 1.2 Electron micrographs of the posterior gill epithelial cells of brackish water-acclimated shore crab *Carcinus maenas*. *AM* apical membrane, *Bl* basal lamina, *BM* basolateral membrane, *CP* clathrin-coated pit, *Cu* cuticle, *Go* Golgi apparatus, *M* mitochondria, *Mt* microtubules, *rER* rough endoplasmic reticulum, *sCu* subcuticular space, *V* vesicle. Scale bars, 1 μm (**a**); 0.5 μ m (**b**) (From Weihrauch et al. (2002))

(Weihrauch and O'Donnell 2015). The best example for this is the Na^+/K^+-ATPase, one of the key transporter responsible for energizing active NaCl transport processes in invertebrates and vertebrates alike (Larsen et al. 2014), but equally important as a NH_4^+ pump (▸ see Chap. 3). This pump is sometimes also referred to as a NH_4^+/K^+-ATPase. Accordingly, other osmoregulatory tissues identified in crustacean, such as the Na^+/K^+-ATPase-rich Crusalis organ (Johnson et al. 2014), a structure located in the swimming legs of freshwater-invading copepod *Eurytemora affinis*, might also be responsible for the animals' ammonia excretion.

1.4 Branchial Ammonia Excretion

1.4.1 Capacity of Active Ammonia Excretion in Gills of Different Haline Crab Species

The gills of aquatic brachyurans are metabolically active organs. This becomes obvious when comparing ammonia excretion rates detected from intact animals with metabolically released ammonia from the branchial tissues. Regardless whether gills from seawater *Cancer* species, brackish water *C. maenas*, or freshwater *E. sinensis* crabs were examined, the branchial tissues generated and released roughly 30–40 times more ammonia when compared to the excretion rates of the respective intact animals. Of that branchially produced ammonia, the vast majority (65–93 %) is transported toward the environment by both the osmoregulatory active posterior gills and the (respiratory) anterior gills (Martin et al. 2011; Weihrauch et al. 1999b).

In addition to the release of metabolic ammonia, all anterior and posterior gills from various haline brachyurans investigated so far in this respect were capable to actively excrete ammonia against a four- to eightfold inwardly directed gradient (◘ Fig. 1.3). When gills of the marine edible crab *C. pagurus*, for example, were perfused and bathed symmetrically with 100 µmol l⁻¹ NH₄Cl, approximately 29 % (posterior gills) to 36 % (anterior gills) of the ammonia was actively eliminated from the perfusion solution during a single run through the gill (Weihrauch 1999). In the Dungeness crab *M. magister*, gills perfused with 100 µmol l⁻¹ NH₄Cl were able to maintain a steady excretion rate even against a 16-fold inwardly directed gradient (Martin et al. 2011). However, when Dungeness crabs were exposed for 14 days to 1 mmol l⁻¹ NH₄Cl, the gills lost the capability entirely to mediate an active ammonia excretion. Interestingly, in hyper-regulating green shore crabs, lower active ammonia excretion rates were detected in the osmoregulatory more active posterior gills compared to the far lesser active anterior gills as shown in ◘ Figs. 1.3 and 1.6 (Weihrauch et al. 1998). These results clearly indicate that osmoregulatory processes are not directly linked to ammonia excretion and that excretion of toxic ammonia must proceed somewhat independently of other physiological processes. This is supported also by experiments on perfused *C. maenas* gills, showing that the ammonia excretion rates are reduced by only 24 %, when all Cl⁻ ions were omitted from the experimental solutions (Weihrauch et al. 1998). This contrasts terrestrial species where ammonia excretion and sodium reabsorption, and hence osmoregulation are linked (► see Chap. 2).

1.4.2 Mechanism of Branchial Ammonia Excretion in Decapod Crabs

Due to their organization, the phyllobranchiate gills from decapod crabs are ideal tissues to investigate ion fluxes (Freire et al. 2008). The gills offer a large surface area (i.e., 247 cm²/ gFW for *C. maenas* (Riestenpatt 1995)) and can easily be perfused with hemolymph-like solutions. For certain model organisms, such as the Chinese mitten crab *E. sinensis*, the red freshwater crab *Dilocarcinus pagei*, *Neohelice (Chasmagnathus) granulata*, the blue crab *Callinectes sapidus*, or the green shore crab *C. maenas*, we have a fairly good understanding of the branchial mechanism for osmoregulatory NaCl transport (Freire et al. 2008; Henry et al. 2012; Larsen et al. 2014; Weihrauch et al. 2004a). The brachial mechanism for ammonia excretion in aquatic crustaceans, however, has been studied only in a few species and in greater detail only in the green shore crab *C. maenas* and to some extent also in the Dungeness crab *M. magister* and *E. sinensis*. ◘ Table 1.3 summarizes the effect of various pharmacological reagents on the branchial ammonia excretion in *C. maenas*.

Role of the Na⁺/K⁺-ATPase

As described in detail in ► Chap. 3, the Na⁺/K⁺-ATPase plays a major role in branchial ammonia transport. Not only does this pump, localized to the basolateral membrane of the gill epithelia (Henry et al. 2012; Towle et al. 2001), accept NH₄⁺ as a substrate, and thereby actively transport hemolymph ammonia into the epithelial cells (Lucu et al. 1989; Masui et al. 2009; Weihrauch et al. 1999b), but it also creates low cytoplasmatic [Na⁺] and generates a negative cell potential of predicted −55 mV (Riestenpatt 1995). Blocking this pump with basolateral applied ouabain, a specific inhibitor of the Na⁺/K⁺-ATPase (Skou 1965), reduced the gradient driven (200 µmol l⁻¹ NH₄Cl in the hemolymph vs. 0 µmol l⁻¹ NH₄Cl in the bathing solution) and active (symmetrically applied 100 µmol l⁻¹ NH₄Cl) branchial ammonia excretion by 52 % and 60 %, respectively (Weihrauch et al. 1998). Also

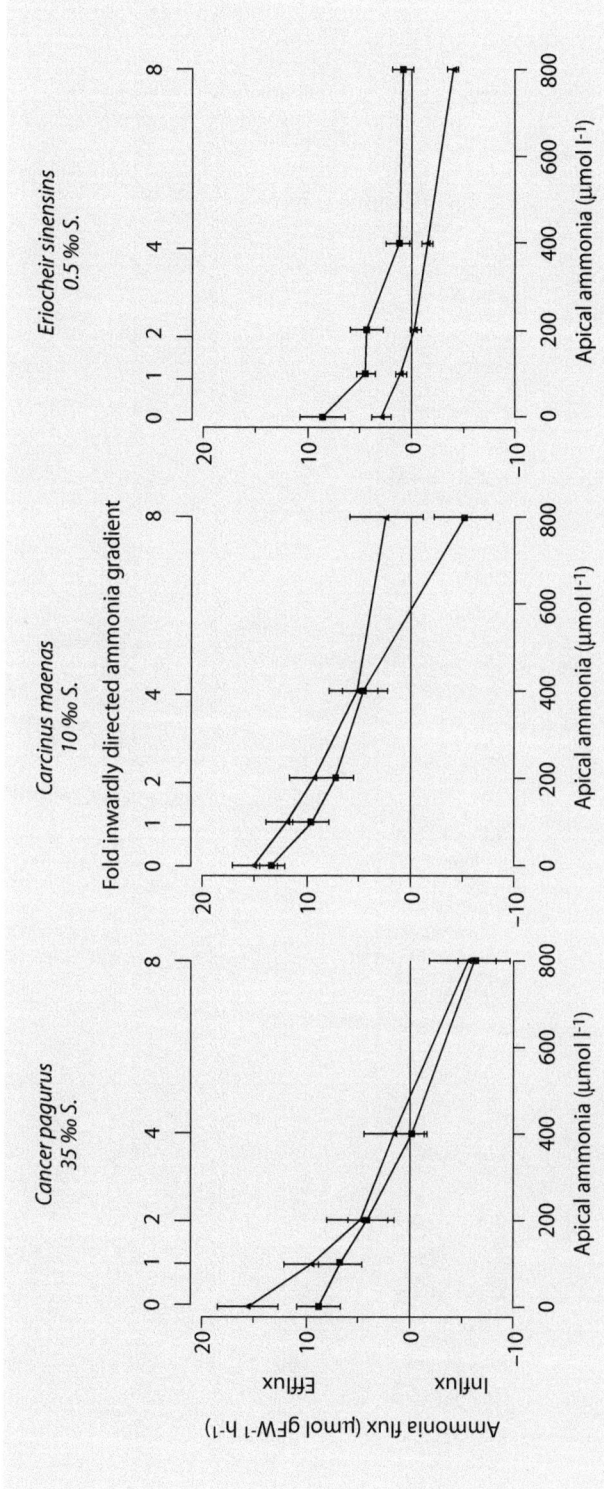

◘ **Fig. 1.3** Branchial ammonia fluxes in anterior (▲) and posterior (■) gills of osmoconforming *C. pagurus* and osmoregulating *C. maenas* and *E. sinensis*. Gills were perfused with hemolymph-like salines enriched with 100 μmol l⁻¹ NH₄Cl and immersed in various environmental ammonia concentrations. Data represent means ± SEM (Modified after Weihrauch et al. 1999a)

◘ **Table 1.3**   Percentage inhibition of the branchial ammonia excretion in posterior gills of the hyper-regulating green shore crab

Manipulation	[Inhibitor] (mmol l^{-1})	Target	Gradient (micromo l^{-1}) Basal/Apical	Reduction of efflux (%)	Source
Dinitrophenol[b]	0.5	H$^+$-gradient driven mechanisms	200:0	55[c]	Weihrauch et al. (1998)
Acetazolamide[b]	1	Carbonic anhydrase	200:0	18	Weihrauch et al. (1998)
Omission of Cl$^-$[a, b]	–	Cl-dependent transports	200:0	24[c]	Weihrauch et al. (1998)
Cs$^+$[a]	10	K$^+$-channels	200:0	12	Weihrauch et al. (1998)
Ouabain[b]	5	Na$^+$/K$^+$-ATPase	200:0	52[c]	Weihrauch et al. (1998)
Amiloride[a]	0.1	Na$^+$-channels/ NHEs	200:0	55[c]	Weihrauch (1999)
Omission of Na$^+$[a,b]	–	Na$^+$-dependent transports	100:100	57[c]	Weihrauch (1999)
Ouabain[b]	5	Na$^+$/K$^+$-ATPase	100:100	60[c]	Weihrauch (1999)
Cs$^+$[b]	10	K$^+$-channels	100:100	58[c]	Weihrauch (1999)
Ouabain[b]/Cs$^+$[b]	5/10	Na$^+$/K$^+$-ATPase/ K$^+$-channels	100:100	86[c]	Weihrauch (1999)
ZD7288[b]	0.01	hyperpolarization-activated cyclic nucleotide-gated K$^+$-channel	130:0	40[c]	Fehsenfeld and Weihrauch (2016)
Ba^{2+}[b]	12	K$^+$-channels	130:0	70[c]	Fehsenfeld and Weihrauch (2016)
Bafilomycin[a,b]	0.001	V-ATPase	100:100	66[c]	Weihrauch et al. (2002)
Bafilomycin[a,b]/ Ouabain[b]	0.001/5	V-ATPase/Na$^+$/ K$^+$-ATPase	100:100	95[c]	Weihrauch et al. (2002)
Colchicine[b]	0.2	Microtubule network	200:0	74[c]	Weihrauch et al. (2002)
Colchicine[b]	0.2	Microtubule network	100:100	92[c]	Weihrauch et al. (2002)
Thiabendazole[b]	0.2	Microtubule network	100:100	100[c]	Weihrauch et al. (2002)

Table 1.3 (continued)

Manipulation	[Inhibitor] (mmol l⁻¹)	Target	Gradient (micromo l⁻¹) Basal/Apical	Reduction of efflux (%)	Source
Taxol[b]	0.01	Microtubule network	100:100	77[c]	Weihrauch et al. (2002)
Cytochalasin D[b]	0.005	Actin filaments	100:100	18	Weihrauch et al. (2002)

Carcinus maenas, [a]apical application, [b]basolateral application, [c]denotes significant differences from control values

in the osmoconforming edible crab *Cancer pagurus*, active branchial ammonia excretion was inhibited by ouabain, here however to 100 % (Weihrauch et al. 1999b). Further, 6 h after exposure to elevated environmental ammonia concentrations (ca. 20–100 µmol l⁻¹), branchial Na^+/K^+-ATPase activity increased significantly in the swimming crab *P. trituberculatus* (Ren et al. 2015).

Role of K⁺-Channels

Pharmacological experiments indicated further that in osmoregulating *C. maenas* hemolymph ammonia (NH_4^+) additionally enters the epithelial cells via basolateral Cs^+- and Ba^{2+}-sensitive pathways, likely K^+-channels (Larsen et al. 2014; Weihrauch et al. 1999b), driven by the negative cell potential. Accordingly, a study by Ren and coworkers demonstrated elevated mRNA expression levels of a branchial expressed K^+-channel 6 h after exposure to ca. 20 and 100 µmol l⁻¹ ammonia (Ren et al. 2015). Additionally, a hyperpolarization-activated cyclic nucleotide-gated potassium channel (HCN) has been shown to play a role in basolateral NH_4^+ transport. An mRNA expression analysis revealed that HCN is expressed in the gills of *C. maenas*, and branchial ammonia excretion was inhibited by approximately 40 % applying the HCN-specific inhibitor ZD7288 (Fehsenfeld and Weihrauch 2016). Functional expression analysis of rat HCN2 in frog oocytes has shown that this transporter does not only mediate the transport of K^+ but also of NH_4^+, although to a lesser extent (Carrisoza-Gaytan et al. 2011). Due to the action of the Na^+/K^+-ATPase (see above) and the NH_4^+ permeable pathways within the basolateral membrane of the epithelium, cytoplasmatic NH_4^+ concentrations are likely very high, possibly between 1 and 2 mmol l⁻¹. Interestingly, basolateral application of Cs^+ on the isolated perfused gill in osmoconforming *C. pagurus* showed no significant inhibition in the ammonia excretion, indicating that different transport mechanisms can be found in different haline crab species.

Role of Rh-Proteins

Gaseous ammonia, NH_3, might enter the cytoplasm together with CO_2 from the hemolymph via branchially expressed Rh-like proteins (RhCM) (Weihrauch et al. 2004b). Rh-glycoproteins are highly conserved proteins within the animal kingdom and have strongly been suggested to be dual gas channels, allowing the passage of NH_3 and CO_2 (Endeward et al. 2008; Musa-Aziz et al. 2009; Perry et al. 2010). In crustaceans, Rh-proteins belong to the "primitive" Rh-protein cluster 1, RhP1 (Huang and Peng 2005) (◘ Fig. 1.6), while in mammals three major Rh-protein isoforms exist (Rhag, Rhbg, Rhcg); only one isoform has so far been identified in decapod crustaceans. In the marine Dungeness crab *M. magister*, this crustacean Rh-like protein (RhMM) was particularly highly expressed in the gill tissues (Martin et al. 2011). At least for vertebrate epithelia, such as the intestine, renal tissues, gills, and skin, high abundance of Rh-protein as well as basolateral localization was confirmed for usually highly expressed Rhbg, while vertebrate Rhcg shows usually much lower expression levels (Larsen et al. 2014). Also for crustacean gills, basolateral localization of the (sole) Rh-protein was assumed (Ren et al. 2015). However, a localization in membrane intracellular vesicles would also be plausible, as Rh-proteins could promote here an ammonia-trapping mechanism. A participation of Rh-protein in epithelial ammonia transport for invertebrates was first suggested by Weihrauch and coworkers in 2002 for the gills of *C. maenas* (Weihrauch et al. 2002) (see the following paragraph), but later also for the insect midgut and the hypodermis of *C. elegans* (Adlimoghaddam et al. 2015; Weihrauch 2006) (▶ see also Chaps. 4 and 5).

Role of the V-Type H$^+$-ATPase and the Microtubule Network

Cytoplasmic ammonia, either imported from the hemolymph or produced by the brachial epithelium itself, is believed to be at least partially trapped as NH$_4^+$ in acidified vesicles. A strong abundance of an intracellular V-type H$^+$-ATPase (◘ see Fig. 1.4) (Weihrauch et al. 2001), as well as the fact that the active and gradient driven ammonia excretion in *C. maenas* can nearly completely be blocked by inhibitors targeting the microtubule network, such as colchicine, taxol, and thiabendazole (◘ Fig. 1.5), strongly indicates that NH$_4^+$-laden vesicles are transported to the apical membrane, where NH$_4^+$ is then released by exocytosis (Weihrauch et al. 2002). In addition, perfusion experiments of gills of brackish water-acclimated green shore crabs showed impressively that the V-type H$^+$-ATPase plays indeed a major role in branchial ammonia excretion: After blockage of the pump by bafilomycin A$_1$, active ammonia excretion was reduced by 66 %.

In addition, exocytotic excretion of ammonia was further supported by studies from Ren and coworkers, as exposure to increased environmental ammonia led to increased mRNA expression levels of the branchial V-type H$^+$-ATPase and a vesicle-associated membrane protein (VAMP). VAMPs have known to be involved in exocytotic processes (Edelmann et al. 1995).

The suggested vesicular ammonia excretion mechanism is advantageous for marine/brackish water-inhabiting species since the process is independent of environmental ammonia concentrations and pH regimes. As seen in ◘ Fig. 1.5, also active ammonia excretion across the gills of the freshwater-acclimated Chinese mitten crab *E. sinensis* can be blocked by colchicine, although to a much lesser extent. In general, for animals inhabiting freshwater environments with a low buffer capacity, such as fish, amphibians, planarians, and leeches, evidence has been provided that ammonia is excreted in the environmental facing excretory organs (e.g., gills, skin) over the apical membrane of the respective epithelia (Cruz et al. 2013; Quijada-Rodriguez et al. 2015; Weihrauch et al. 2012, 2009). Accordingly, it is assumed that also in *E. sinensis* and other freshwater crustaceans, the majority of ammonia is directly excreted over the apical membrane of the transport epithelia (▶ refer also to Chap. 5). This might be accomplished via an ammonia-trapping (also called acid trapping) mechanism. For this mechanism an apically localized V-type H$^+$-ATPase – in *E. sinensis* this pump is localized to the pillar cells (Freire et al. 2008) – generates a partial pressure gradient for NH$_3$ (ΔP_{NH3}) by acidifying the subcuticular space

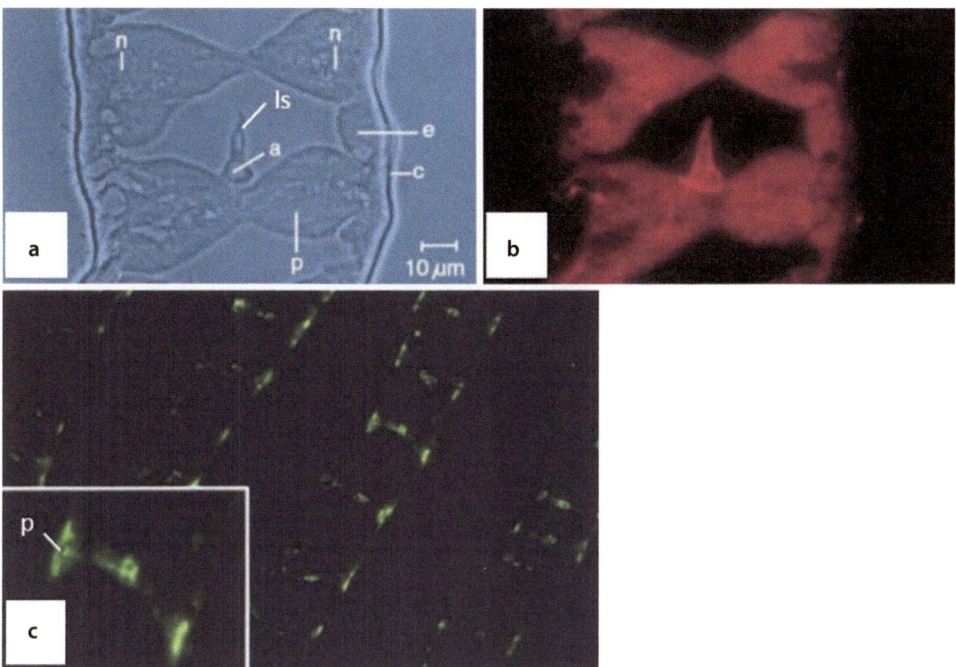

☐ **Fig. 1.4** Branchial localization of the V-type H+-ATPase (subunit B) in the posterior gills of brackish water-acclimated *Carcinus maenas*. Phase contrast micrograph of the branchial epithelium of *C. maenas* (**a**). Imunocytochemical staining for the subunit B of the V-type- H+-ATPase in *C. maenas* (**b**) and *Eriocheir sinensis* (**c**). In *E. sinensis* V-type H+-ATPase is exclusively present in the pillar cells. *p* pillar cells, *c* cuticle, *a* arteriole, *ls* lamellar septum, *e* epithelial cell, *n* nucleus ((**a, b,** taken with permission from (Weihrauch et al. 2001)) and freshwater-acclimated *Eriocheir sinensis* (**c,** taken with permission from (Freire et al. 2008)))

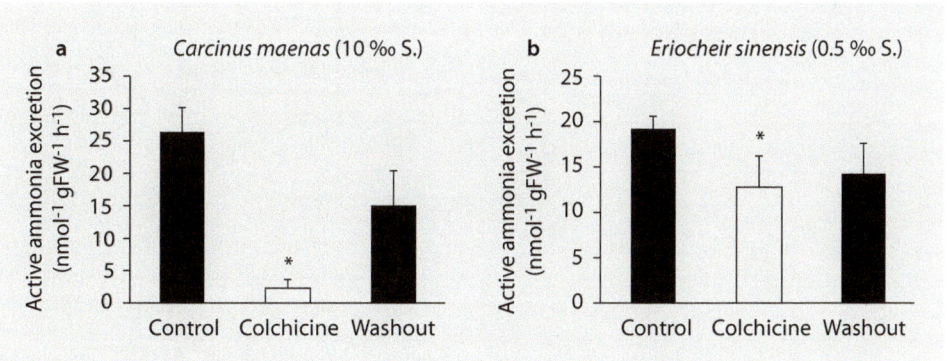

☐ **Fig. 1.5** Microtubule-dependent branchial ammonia excretion in two osmoregulating crabs, *Carcinus maenas* acclimated to 10‰ S. and *Eriocheir sinensis* acclimated to 5‰ S. (**a**) Gills from *C. maenas* were perfused and bathed with 100 µmol l^{-1} NH_4Cl (Values were taken from (Weihrauch et al. 2002)). (**b**) Gills from *E. sinensis* were perfused with 100 µmol l^{-1} NH_4Cl and bathed in 150 µmol l^{-1} NH_4Cl. Data represent means ± SEM of four observations. *indicates statistical significance ($P < 0.05$)

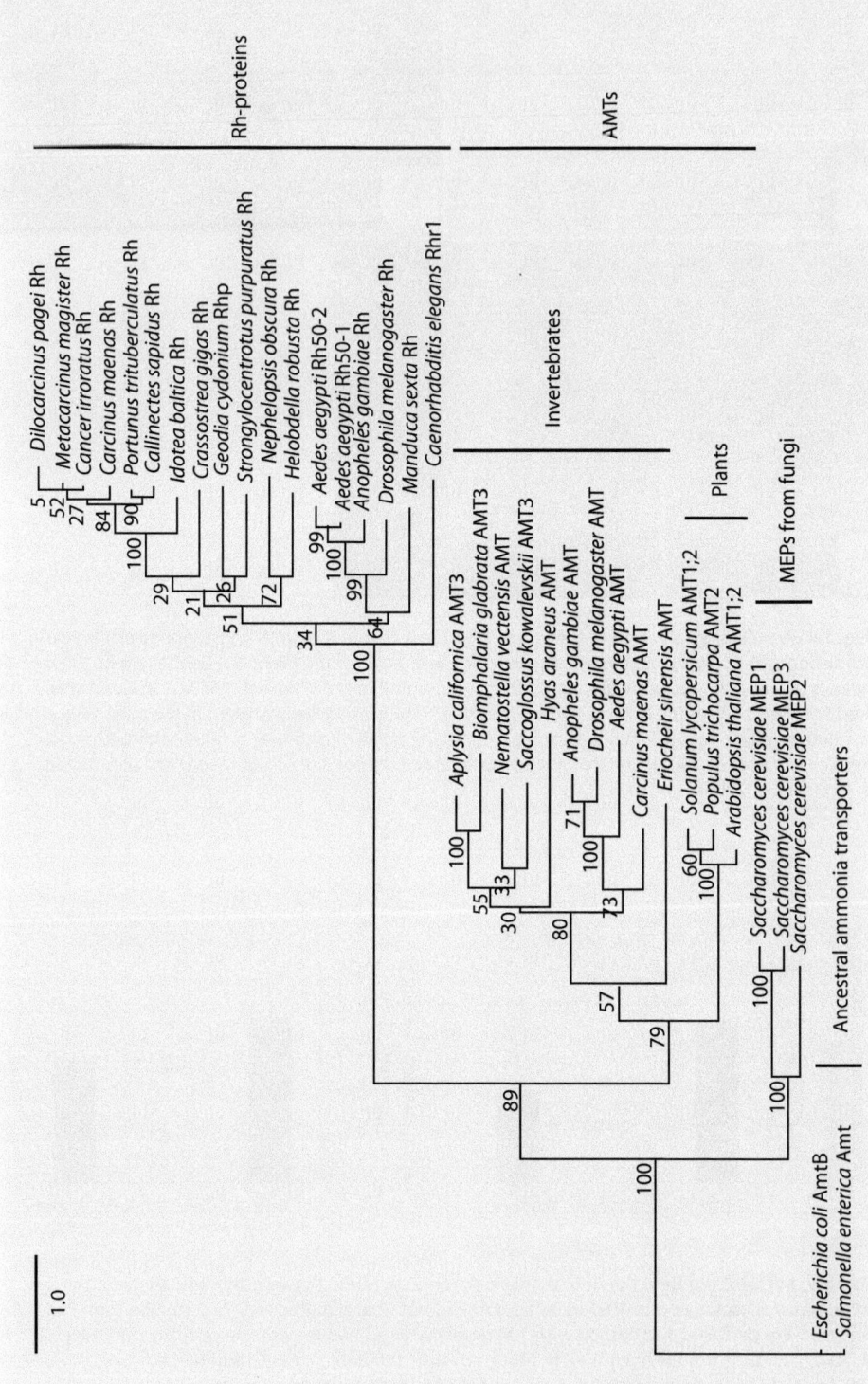

◻ **Fig. 1.6** Maximum likelihood tree of invertebrate Rh-proteins and AMTs. Amino acid sequences were aligned by muscle alignment on Geneious 7 (Kearse et al. 2012). Analysis of aligned sequences was performed with Geneious using PhyML maximum likelihood method with the LG model (Guindon et al. 2010). Numbers beside branches represent bootstrap values from 500 replicates. The tree branches are drawn to scale, with the scale bar representing the number of amino acid substitutions per site. GenBank accession numbers are given in brackets behind the species name. *Escherichia coli* AMTB (1U77_A); *Salmonella enterica* AMT (WP_049281994.1); *Aedes aegypti* AMT1 (AAY63898.1); *Anopheles gambiae* AMT (XP_318439.3); *Aplysia californica* AMT3 (XP_012946834.1); *Arabidopsis thaliana* AMT1;2 (NP_176658.1); *Biomphalaria glabrata* AMT3 (XP_013085164.1); *Carcinus maenas* AMT (PRJNA255867); *Drosophila melanogaster* AMT (PRJNA255867); *Eriocheir sinensis* AMT (FG984314.1); *Hyas araneus* AMT (PRJEB1422); *Nematostella vectensis* AMT (XP_001639101.1); *Populus trichocarpa* AMT2 (XP_002325790.2); *Saccoglossus kowalevskii* AMT3 (XP_006819255.1); *Solanum lycopersicum* AMT1;2 (O04161.1); *Aedes aegypti* Rh50-1 (AAX19513.1); *Aedes aegypti* Rh50-2 (AY926464); *Anopheles gambiae* Rh (AAP47144.1); *Caenorhabditis elegans* Rhr1 (AAF97864.1); *Callinectes sapidus* Rh (AAM21147.1); *Cancer irroratus* Rh (AAM21148.1); *Carcinus maenas* Rh (AAK50057.2); *Crassostrea gigas* Rh type B (EKC21768.1); *Dilocarcinus pagei* Rh (AAM21149.1); *Drosophila melanogaster* Rh (AF193812); *Geodia cydonium* Rhp1 (CAA73029.1); *Helobdella robusta* Rh (XP_009016010.1); *Idotea baltica* Rh (AAM21150.1); *Metacarcinus magister* Rh (AEA41159.1); *Manduca sexta* Rh (ABI20766.1); *Nephelopsis obscura* Rh (KM923907); *Portunus trituberculatus* Rh (AHY27545.2); *Strongylocentrotus purpuratus* Rh type B (XP_789738.3); *Saccharomyces cerevisiae* MEP1 (NP_011636.3); *Saccharomyces cerevisiae* MEP2 (CAA96025.1); *Saccharomyces cerevisiae* MEP3 (EGA76487.1) and lowering thereby subcuticular [NH_3], by protonating the gas into NH_4^+. Cytoplasmic NH_3 then follows this gradient via a putatively apically localized Rh-protein into the subcuticular space. From here NH_4^+ is released via cation permeable structures of the cuticle into the environment. As long as the subcuticular space has a lower pH compared to the cytoplasm, this process continues. It is however also possible that NH_4^+ is directly excreted through apical NH_4^+-channels (AMT, see paragraph below) or a combination of both processes. Apical Cs^+-sensitive K^+-channels as identified in the gills of *C. maenas* (Riestenpatt et al. 1996) seem not to be involved in apical ammonia (NH_4^+) transport, as application of Cs^+, a rather nonspecific K^+-channel blocker, had no significant effect on the ammonia efflux in the green crab (Weihrauch et al. 1998).

Role of Na$^+$/H$^+$ Exchangers (NHEs)

For ammonia release out of the cytoplasm into the environment, an apical Na^+/H^+ exchanger (NHE) was discussed to play a role, due to the fact that branchial ammonia excretion was partially blocked in *C. maenas* and *Cancer pagurus* by apically applied amiloride, a rather nonspecific inhibitor of NHEs (Kleyman and Cragoe 1988; Lucu et al. 1989; Weihrauch et al. 1999a). Also in intact marine *Cancer antennarius* and *Petrolisthes cinctipes* crabs and in the freshwater crayfish *Astacus leptodactylus*, ammonia excretion was reduced after application of amiloride (Ehrenfeld 1974; Hunter and Kirschner 1986). Support for this hypothesis was further gained by studies showing that the mammalian NHE-3 seems to be involved in renal ammonia excretion as experiments employing brush border vesicles isolated from the proximal tubule indicated a Na^+/NH_4^+ exchange (Kinsella and Aronson 1981). Also, in the gills of *C. maenas*, an NHE (likely electrogenic) was identified on the molecular basis (Towle et al. 1997) and it cannot be excluded that this transporter accepts NH_4^+ at the cation or proton-binding side. All physiological experiments involving the use of amiloride, however, must be considered with caution, as the conductance (G_{cut}) of the isolated cuticle for Na^+ and NH_4^+ in *C. maenas* can be blocked in a dose-dependent manner (K_{ami} for $NH_4^+ = 20.4$ mmol l^{-1}) by this drug (Onken and Riestenpatt 2002; Weihrauch et al. 2002). Also, in the marine crabs *M. magister* and *P. trituberculatus*, branchial mRNA expression levels of NHE were downregulated upon a 2-day exposure to increased environmental ammonia levels. During this time point, *M. magister* maintained considerately lower hemolymph ammonia concentrations (ca. 500 µmol l^{-1}) compared to the high environmental levels (1 mmol l^{-1}). These data imply that NHEs are likely not directly involved in branchial ammonia excretion. NHEs also in crustacean gills might be localized to all membranes, apical, basolateral, but also to membranes of intracellular compartments (e.g., exocytotic vesicles) as describe, e.g., for the mammalian NHE and the NHX transporters in the nematode *Caenorhabditis elegans* (Nehrke and Melvin 2002; Orlowski and Grinstein 2004; Zachos et al. 2005). Clearly, their particular role in branchial ammonia excretion processes requires further investigations.

Role of Ammonia Transporters (AMTs)

Increasing numbers of genome and transcriptome projects revealed that invertebrates, but not vertebrates, do express AMTs, transporters typically found in roots of plants, here being responsible for the uptake of NH_4^+ from the soil (Ludewig et al. 2002).

In a recent study, an evidence was provided indicating that also AMTs in mosquitos promote NH_4^+ transport (Pitts et al. 2014) (▶ see also Chap. 4). Also in the genomes of crustaceans, AMTs have been identified, e.g., for *C. maenas*, *E. sinensis*, and the great spider crab *Hyas araneus*, which group together with the insect AMTs but also the plant

AMTs (■ see Fig. 1.6). At this moment, however, these transporters have not been tested for their transport specificity and tissue and cellular localization. It is conceivable that in crustaceans AMTs are expressed in the ammonia-transporting tissues such as the antennal glands and the gills. In both tissues an apical localization is assumed, since here it would allow the direct exit of cytoplasmic NH_4^+ that has been pumped from the hemolymph into this compartment by the basolateral Na^+/K^+-ATPase.

Role of Aquaporins (AQPs)

Aquaporins (AQPs) belong to the larger family of major intrinsic proteins (MIP). These proteins facilitate first and foremost the transport of water in and out of the cell (Knepper 1994). However, for some members of this family, particularly the mammalian AQP3, AQP7, AQP8, and AQP9, ammonia transport capabilities have been confirmed at least when expressed in *Xenopus* oocytes (Holm et al. 2005; Litman et al. 2009; Saparov et al. 2007). Also in crustaceans AQPs have now been identified, but not yet tested for their transport characteristics. However, as seen in ■ Fig. 1.7, employing a partially published sequence (GenBank Accession number: DV467482.1) of a AQP expressed in the branchial tissue of *C. maenas* (CmAQP), quantitative PCR revealed that mRNA abundance of CmAQP correlated with the active ammonia excretion rates observed in the anterior and posterior gills of seawater and brackish water-acclimated green crabs. In fact, when crabs were exposed for 14 days to 1 mmol l^{-1}, a significant upregulation of the mRNA expression of the transporter was observed in the posterior gills of seawater-acclimated *C. maenas* (data not shown). It is to assume that also AQPs from crustaceans are primarily responsible for the epithelial transport of water; however, the obtained data imply that the function of the branchial-expressed AQP and the active ammonia transport

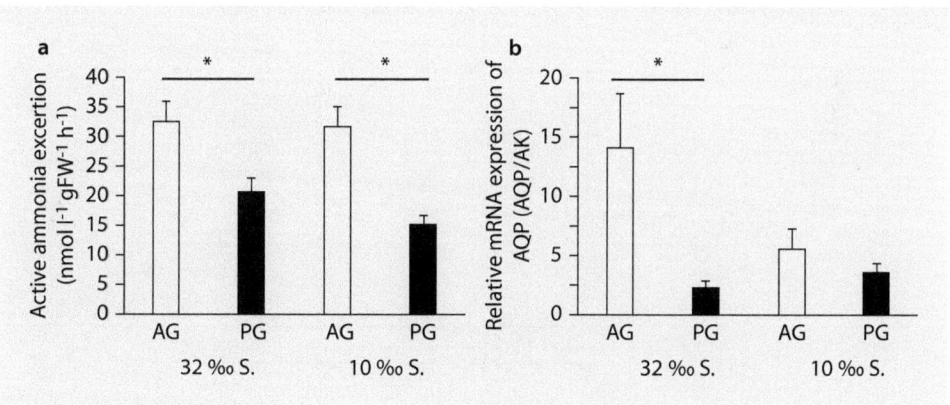

■ **Fig. 1.7** Active ammonia excretion rates (**a**) and mRNA expression levels of CmAQP (**b**) in the anterior and posterior gills of seawater (32‰ S.) and brackish water (10‰ S.)-acclimated *C. maenas* crabs. (**a**), gills were perfused with hemolymph-like salines containing 100 µmol l^{-1} NH_4Cl. The external bath was adjusted to the corresponding acclimation salinity and enriched also with 100 µmol l^{-1} NH_4Cl. Ammonia was measured according to Weihrauch et al. (1998). Data are represented as the mean ± SEM. (n = 4–9). * indicates statistical significance between groups ($P < 0.05$). (**b**) Relative branchial mRNA expression of CmAQP. AQP mRNA expression levels were standardized to the mRNA expression levels of arginine kinase (AK) in the respective samples. Data are represented as the mean ± SEM. (n = 4–9). * indicates statistical **significance between** groups ($P < 0.05$). *AG* anterior gills, *PG* posterior gills

might somehow be coupled. Future studies must clarify their particular role in the animal's physiology.

Finally, the paracellular passage of ammonia cannot be excluded, which would greatly depend on the "tightness" of the respective gill epithelia and the trans-epithelial ammonia gradients.

1.5   Conclusion and Summarizing Working Model of Branchial Ammonia Excretion Proposed for the Green Shore Crab *Carcinus maenas*

The described model below (◘ Fig. 1.8) is based on the data collected over the last 25 years, which have been acquired employing a number of techniques including gill perfusion studies (◘ see Table 1.3), Using chamber experiments (electrophysiology on the split gill lamella), immunohistochemistry, electron microscopy, and molecular techniques. As a

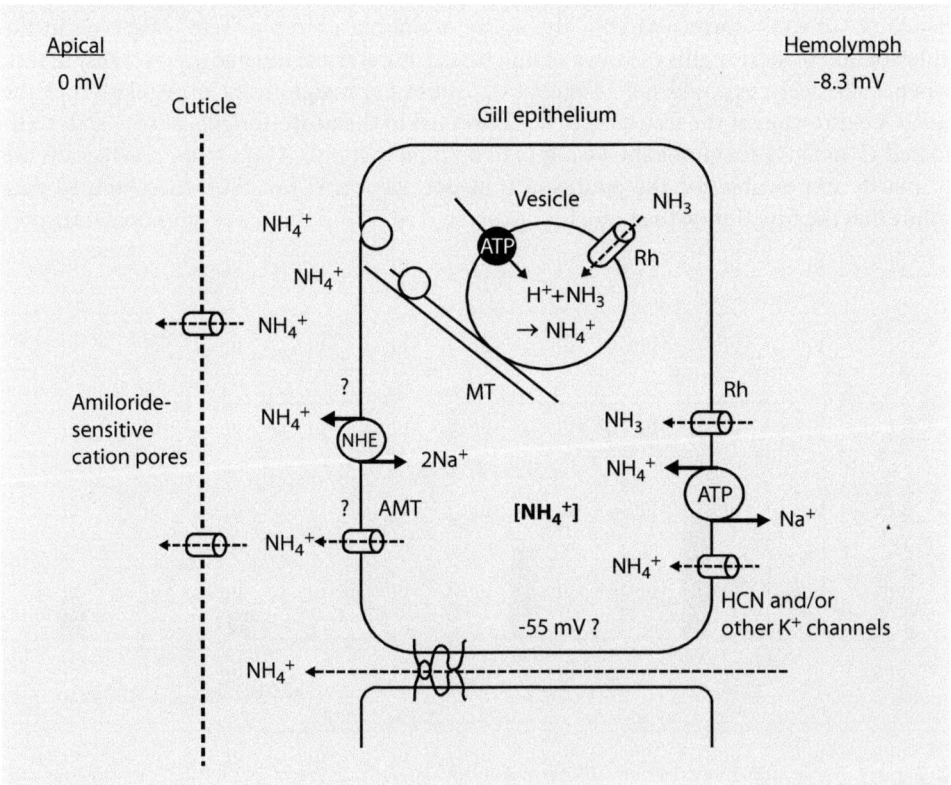

◘ **Fig. 1.8** Hypothetical model of the ammonia transport over gills of brackish water (10‰ S.)-acclimated green shore crabs *Carcinus maenas*. Transport processes are explained in the text above. AMT, NH_4^+ channel; HCN, hyperpolarization-activated cyclic nucleotide-gated potassium channel; MT, microtubule; and Rh, Rhesus protein. ? indicates unknown but speculated presence of apical NHEs and AMTs. Within the cuticle amiloride-sensitive NH_4^+, permeable pores are indicted. Millivolt values for the cytoplasm and hemolymph (rel. to the environment) were taken from Riestenpatt (1995)

result, the ammonia excretion mechanism functioning in the branchial epithelium of the green shore crab *C. maenas* as depicted in an updated working model is likely the best understood mechanism for crustaceans to date. Additional information regarding branchial acid-base regulatory mechanisms, which are believed to be closely coupled to the branchial ammonia excretion process, is provided in ▶ Chap. 6 for decapod crustaceans and specifically for *C. maenas*. Current and former research demonstrated that crustaceans, which populate now our planet for about 500 million years, and here particularly species belonging to the Brachyura (decapod crabs) with their readily perfusable gills and single cell-layered gill epithelia, provide excellent model systems to investigate complex transport processes, such as osmoregulatory NaCl uptake, acid-base regulation, and ammonia excretion.

References

Adlimoghaddam A, Boeckstaens M, Marini AM, Treberg JR, Brassinga AK, Weihrauch D (2015) Ammonia excretion in *Caenorhabditis elegans*: mechanism and evidence of ammonia transport of the Rh-protein CeRhr-1. J Exp Biol 218:675–683

Bellwood O (2002) The occurrence, mechanics and significance of burying behaviour in crabs (Crustacea: Brachyura). J Nat Hist 36:1223–1238

Cameron JN, Batterton CV (1978) Antennal gland function in the freshwater crab *Callinectes sapidus*: water, electrolyte acid-base and ammonia excretion. J Com Physiol 123:143–148

Carrisoza-Gaytan R, Rangel C, Salvador C, Saldana-Meyer R, Escalona C, Satlin LM, Liu W, Zavilowitz B, Trujillo J, Bobadilla NA et al (2011) The hyperpolarization-activated cyclic nucleotide-gated HCN2 channel transports ammonium in the distal nephron. Kidney Int 80:832–840

Chen J-C, Cheng S-Y (1995) Hemolymph oxygen content, oxyhemocyanin, protein levels and ammonia excretion in the shrimp *Penaeus monodon* exposed to ambient nitrite. J Com Physiol B 164:530–535

Chen JC, Lin CY (1992) Lethal effects of ammonia on *Penaeus chinensis* Osbeck juveniles at different salinity levels. J Exp Mar Biol Ecol 156(1):139–148

Claybrook DL (1983) The biology of crustacea. In: Mantel LH (ed) In Internal anatomy and physiological regulation. Academic Press, London, pp 163–213

Cole GA, Brown RJ (1967) The chemistry of artemia habitats. Ecology 48:858–861

Cruz MJ, Sourial MM, Treberg JR, Fehsenfeld S, Adlimoghaddam A, Weihrauch D (2013) Cutaneous nitrogen excretion in the African clawed frog *Xenopus laevis*: effects of high environmental ammonia (HEA). Aquat Toxicol 136–137:1–12

DeVries MC, Wolcott DL, Holliday CW (1994) High ammonia and low ph in the urine of the ghost crab, *Ocypode quadrata*. Biol Bull 186:342–348

Durand F, Regnault M (1998) Nitrogen metabolism of two portunid crabs, *Carcinus maenas* and *Necora puber*, during prolonged air exposure and subsequent recovery: a comparative study. J Exp Biol 201(Pt 17):2515–2528

Durand F, Devillers N, Lallier FH, Regnault M (2000) Nitrogen excretion and changes in blood components during emersion of the subtidal spider crab *Maia squinado* (L.). Comp Biochem Physiol A Mol Integr Physiol 127:259–271

Edelmann L, Hanson PI, Chapman ER, Jahn R (1995) Synaptobrevin binding to synaptophysin: a potential mechanism for controlling the exocytotic fusion machine. EMBO J 14:224–231

Ehrenfeld J (1974) Aspects of ionic transport mechanisms in crayfish *Astacus leptodactylus*. J Exp Biol 61:57–70

Endeward V, Cartron JP, Ripoche P, Gros G (2008) RhAG protein of the Rhesus complex is a CO_2 channel in the human red cell membrane. FASEB J 22:64–73

Enoksson V, Samuelsson M-O (1987) Nitrification and dissimilatory ammonium production and the effects on nitrogen flux over the sediment-water interface in bioturbated coastal sediments. Mar Ecol Prog Ser 36:181–189

Fehsenfeld S, Weihrauch D (2016) The role of an ancestral hyperpolarization-activated cyclic nucleotide-gated K^+ channel in branchial acid-base regulation in the green crab, *Carcinus maenas*. J Exp Biol 219:887–896

Fellows FCI, Hird FJR (1979) Nitrogen metabolism and excretion in the freshwater crayfish *Cherax destructor*. Com Biochem Physiol 64B:235–238

Freire CA, Onken H, McNamara JC (2008) A structure-function analysis of ion transport in crustacean gills and excretory organs. Comp Biochem Physiol A Mol Integr Physiol 151:272–304

Greenaway P (1991) Nitrogenous excretion in aquatic and terrestrial Crustacea. Mem Qld Mas 31:215–227

Guindon S, Dufayard JF, Lefort V, Anisimova M, Hordijk W, Gascuel O (2010) New algorithms and methods to estimate maximum-likelihood phylogenies: assessing the performance of PhyML 3.0. Syst Biol 59:307–321

Harris RR, Coley S, Collins S, McCabe R (2001) Ammonia uptake and its effects on ionoregulation in the freshwater crayfish *Pacifastacus leniusculus* (Dana). J Comp Physiol B 171:681–693

Hartenstein R (1970) Nitrogen metabolism in non-insect arthropods. In: Campbell JW (ed) In comparative biochemistry of nitrogen metabolism. Academic Press, New York, pp 299–385

Hazel RH, Burkhead CE, Huggins DG (1982) Development of water quality criteria for ammonia and total residual chlorine for the protection of aquatic life in two Johnson County, Kansas Streams. In: Pearson JG, Foster RB, Bishop WE (ed) Proceedings of annual symposium on aquatic toxicology, 5th edn. Philadelphia, pp 381–388

Henry RP, Lucu C, Onken H, Weihrauch D (2012) Multiple functions of the crustacean gill: osmotic/ionic regulation, acid-base balance, ammonia excretion, and bioaccumulation of toxic metals. Front Physiol 3:431

Holm LM, Jahn TP, Moller AL, Schjoerring JK, Ferri D, Klaerke DA, Zeuthen T (2005) NH3 and NH_4^+ permeability in aquaporin-expressing *Xenopus* oocytes. Pflugers Arch 450:415–428

Huang CH, Peng J (2005) Evolutionary conservation and diversification of Rh family genes and proteins. Proc Natl Acad Sci U S A 102:15512–15517

Hunter KC, Kirschner LB (1986) Sodium absorption coupled to ammonia excretion in osmoconforming marine invertebrates. Am J Physiol 251:R957–R962

Johnson KE, Perreau L, Charmantier G, Charmantier-Daures M, Lee CE (2014) Without gills: localization of osmoregulatory function in the copepod *Eurytemora affinis*. Physiol Biochem Zool 87:310–324

Kearse M, Moir R, Wilson A, Stones-Havas S, Cheung M, Sturrock S, Buxton S, Cooper A, Markowitz S, Duran C et al (2012) Geneious Basic: an integrated and extendable desktop software platform for the organization and analysis of sequence data. Bioinformatics 28:1647–1649

Kinsella JL, Aronson PS (1981) Interaction of NH_4^+ and Li^+ with the renal microvillus membrane Na^+-H^+ exchanger. Am J Physiol 241:C220–C226

Kleyman TR, Cragoe EJ Jr (1988) Amiloride and its analogs as tools in the study of ion transport. J Membr Biol 105:1–21

Knepper MA (1994) The aquaporin family of molecular water channels. Proc Natl Acad Sci U S A 91:6255–6258

Krishnamoorthy RV, Srihari K (1973) Changes in the patterns of the freshwater field crab *Paratelphusa hydrodromous* upon adaptation to higher salinities. Mar Biol 21:341–348

Larsen EH, Deaton LE, Onken H, O'Donnell M, Grosell M, Dantzler WH, Weihrauch D (2014) Osmoregulation and excretion. Compr Physiol 4:405–573

Le Moullac G, Haffner P (2000) Environmental factors affecting immune responses in Crustacea. Aquaculture 191:121–131

Lignon JM (1987) Ionic permeabilities of the isolated gill cuticle of the shore crab *Carcinus maenas*. J Exp Biol 131:159–174

Litman T, Sogaard R, Zeuthen T (2009) Ammonia and urea permeability of mammalian aquaporins. Handb Exp Pharmacol 190:327–358

Lohse L, Malschaert JFP, Slomp CP, Helder W, Van Raaphorst W (1993) Nitrogen cycling in North Sea sediments: Interaction of denitrification and nitrification in offshore and coastal areas. Mar Ecol Prog Ser 101:283–296

Lowenstein JM (1972) Ammonia production in muscle and other tissues: the purine nucleotide cycle. Physiol Rev 52:382–414

Lucu C, Devescovi M, Siebers D (1989) Do amiloride and ouabain affect ammonia fluxes in perfused *Carcinus* gill epithelia? J Exp Zool 249:1–5

Ludewig U, von Wiren N, Frommer WB (2002) Uniport of NH_4^+ by the root hair plasma membrane ammonium transporter LeAMT1;1. J Biol Chem 277:13548–13555

Mantel LH, Farmer IL (1983) Osmotic and ionic regulation. In: Mantel BA (ed) The biology of crustacea. Academic Press, London, pp 54–126

Martin JW, and Davis GE (2001) An updated classification of the recent crustacea. Natural History Museum of Los Angeles County, Science Series 39. Los Angeles, California. ISSN 1-891276-27-1

Martin M, Fehsenfeld S, Sourial MM, Weihrauch D (2011) Effects of high environmental ammonia on branchial ammonia excretion rates and tissue Rh-protein mRNA expression levels in seawater acclimated Dungeness crab *Metacarcinus magister*. Comp Biochem Physiol A Mol Integr Physiol 160:267–277

Masui DC, Mantelatto FL, McNamara JC, Furriel RP, Leone FA (2009) Na+, K+-ATPase activity in gill microsomes from the blue crab, *Callinectes danae*, acclimated to low salinity: novel perspectives on ammonia excretion. Comp Biochem Physiol A Mol Integr Physiol 153:141–148

McGaw IJ (2004) Ventilatory and cardiovascular modulation associated with burying behaviour in two sympatric crab species, *Cancer magister* and *Cancer productus*. J Exp Mar Biol Ecol 303:47–63

Musa-Aziz R, Chen LM, Pelletier MF, Boron WF (2009) Relative CO2/NH3 selectivities of AQP1, AQP4, AQP5, AmtB, and RhAG. Proc Natl Acad Sci U S A 106:5406–5411

Nehrke K, Melvin JE (2002) The NHX family of Na+-H+ exchangers in *Caenorhabditis elegans*. J Biol Chem 277:29036–29044

Onken H, Riestenpatt S (2002) Ion transport across posterior gills of hyperosmoregulating shore crabs (*Carcinus maenas*): amiloride blocks the cuticular Na(+) conductance and induces current-noise. J Exp Biol 205:523–531

Orlowski J, Grinstein S (2004) Diversity of the mammalian sodium/proton exchanger SLC9 gene family. Pflugers Arch 447:549–565

Ostrensky A, Marchiori MA, Poersch LH (1992) Aquatic toxicity of ammonia in the metamorphosis of postlarvae *Penaeus paulensis* Perez-Farfante. An Acad Bras Cienc 64(4):383–389

Perry SF, Braun MH, Noland M, Dawdy J, Walsh PJ (2010) Do zebrafish Rh proteins act as dual ammonia-CO2 channels? J Exp Zool A Ecol Genet Physiol 313:618–621

Pitts RJ, Derryberry SL Jr, Pulous FE, Zwiebel LJ (2014) Antennal-expressed ammonium transporters in the malaria vector mosquito *Anopheles gambiae*. PLoS One 9:e111858

Postel U (2001) Physiologie der Pleopoden mariner Asseln (Crustacea, Isopoda). VWFS (Akademische Abhandlungen zur Biologie), Berlin

Postel U, Becker W, Brandt A, Luck-Kopp S, Riestenpatt S, Weihrauch D, Siebers D (2000) Active osmoregulatory ion uptake across the pleopods of the isopod *Idotea baltica* (Pallas): electrophysiological measurements on isolated split endo- and exopodites mounted in a micro-ussing chamber. J Exp Biol 203(Pt 7):1141–1152

Quijada-Rodriguez AR, Treberg JR, Weihrauch D (2015) Mechanism of ammonia excretion in the freshwater leech *Nephelopsis obscura*: characterization of a primitive Rh protein and effects of high environmental ammonia. Am J Physiol Regul Integr Comp Physiol Ajpregu 00482:2014

Ren Q, Pan L, Zhao Q, Si L (2015) Ammonia and urea excretion in the swimming crab *Portunus trituberculatus* exposed to elevated ambient ammonia-N. Comp Biochem Physiol A Mol Integr Physiol 187:48–54

Riestenpatt S (1995) Die osmoregulatorische NaCl-Aufnahme über die Kiemen decapoder Crustaceen (Crustacea, Decapoda). In: Akademische Abhandlungen zur Biologie. Verlag für Wissenschaft und Forschung, Berlin, pp 1–131

Riestenpatt S, Onken H, Siebers D (1996) Active absorption of Na+ and Cl− across the gill epithelium of the shore crab *Carcinus maenas*: voltage-clamp and ion-flux studies. J Exp Biol 199:1545–1554

Saparov SM, Liu K, Agre P, Pohl P (2007) Fast and selective ammonia transport by aquaporin-8. J Biol Chem 282:5296–5301

Schoffeniels E, Gilles R (1970) Nitrogenous constituents and nitrogen metabolism in arthropods. In: Florkin M, Sheer B (eds) Chemical zoology. Academic Press, New York, pp 199–227

Skou JC (1965) Enzymatic Basis For Active Transport Of Na+ And K+ Across Cell Membrane. Physiol Rev 45:596–617

Spargaaren DH (1990) The effect of environmental ammonia concentrations on the ion-exchange of shore crabs, *Carcinus maenas* (L.). Com Biochem Physiol 97C:87–91

Taylor HH, Taylor EW (1992) Gills and lungs: the exchange of gases and ions. In: Harrison WF, Humes AG (eds) Microscopic anatomy of invertebrates. Wiley-Liss, New York, pp 203–293

Towle DW, Smith CM (2006) Gene discovery in *Carcinus maenas* and *Homarus americanus* via expressed sequence tags. Integr Comp Biol 46:912–918

Towle DW, Rushton ME, Heidysch D, Magnani JJ, Rose MJ, Amstutz A, Jordan MK, Shearer DW, Wu WS (1997) Sodium/proton antiporter in the euryhaline crab *Carcinus maenas*: molecular cloning, expression and tissue distribution. J Exp Biol 200(Pt 6):1003–1014

Towle DW, Paulsen RS, Weihrauch D, Kordylewski M, Salvador C, Lignot JH, Spanings-Pierrot C (2001) Na$^{(+)}$+K$^{(+)}$-ATPase in gills of the blue crab *Callinectes sapidus*: cDNA sequencing and salinity-related expression of alpha-subunit mRNA and protein. J Exp Biol 204:4005–4012

Tsai JR, Lin HC (2014) Functional anatomy and ion regulatory mechanisms of the antennal gland in a semi-terrestrial crab, *Ocypode stimpsoni*. Biol Open 3:409–417

Weihrauch D (1999) Zur Stickstoff-Exkretion aquatischer Brachyuren: *Carcinus maenas* (Linnaeus 1758, Decapoda, Portunidae), *Cancer pagurus* Linnaeus 1758 (Decapoda, Cancridae) und *Eriocheir sinensis* H. Milne Edwards 1853 (Decapoda, Grapsidae). Verlag fuer Wissenschaft und Forschung, Berlin

Weihrauch D (2006) Active ammonia absorption in the midgut of the Tobacco hornworm *Manduca sexta* L.: Transport studies and mRNA expression analysis of a Rhesus-like ammonia transporter. Insect Biochem Mol Biol 36:808–821

Weihrauch D, O'Donnell MJ (2015) Links between Osmoregulation and Nitrogen-Excretion in Insects and Crustaceans. Integr Comp Biol 55(5):816–829

Weihrauch D, Becker W, Postel U, Riestenpatt S, Siebers D (1998) Active excretion of ammonia across the gills of the shore crab *Carcinus maenas* and its relation to osmoregulatory ion uptake. J Comp Physiol B 168:364–376

Weihrauch D, Becker W, Postel U, Luck-Kopp S, Siebers D (1999a) Potential of active excretion of ammonia in three different haline species of crabs. J Comp Physiol B 169:25–37

Weihrauch D, Siebers D, Towle D (1999b) High levels of urea maintained in the hemolymph of the ammoniotelic euryhaline crabs *Carcinus maenas* and *Eriocheir sinensis*. In: Hochachka PW, Mommsen TP (eds) Fifth International Congress of Comparative Physiology and Biochemistry. Elsevier, Calgary/Alberta, p S81

Weihrauch D, Ziegler A, Siebers D, Towle DW (2001) Molecular characterization of V-type H$^{(+)}$-ATPase (B-subunit) in gills of euryhaline crabs and its physiological role in osmoregulatory ion uptake. J Exp Biol 204:25–37

Weihrauch D, Ziegler A, Siebers D, Towle DW (2002) Active ammonia excretion across the gills of the green shore crab *Carcinus maenas*: participation of Na$^{(+)}$/K$^{(+)}$-ATPase, V-type H$^{(+)}$-ATPase and functional microtubules. J Exp Biol 205:2765–2775

Weihrauch D, McNamara JC, Towle DW, Onken H (2004a) Ion-motive ATPases and active, transbranchial NaCl uptake in the red freshwater crab, *Dilocarcinus pagei* (Decapoda, Trichodactylidae). J Exp Biol 207:4623–4631

Weihrauch D, Morris S, Towle DW (2004b) Ammonia excretion in aquatic and terrestrial crabs. J Exp Biol 207:4491–4504

Weihrauch D, Wilkie MP, Walsh PJ (2009) Ammonia and urea transporters in gills of fish and aquatic crustaceans. J Exp Biol 212:1716–1730

Weihrauch D, Chan AC, Meyer H, Doring C, Sourial MM, O'Donnell MJ (2012) Ammonia excretion in the freshwater planarian *Schmidtea mediterranea*. J Exp Biol 215:3242–3253

Young-Lai WW, Charmantier-Daures M, Charmantier G (1991) Effect of ammonia on survival and osmoregulation in different life stages of the lobster *Homarus americanus*. Mar Biol 110:293–300

Zachos NC, Tse M, Donowitz M (2005) Molecular physiology of intestinal Na$^+$/H$^+$ exchange. Annu Rev Physiol 67:411–443

Nitrogenous Waste Metabolism Within Terrestrial Crustacea, with Special Reference to Purine Deposits and Their Metabolism

Stuart M. Linton, Jonathan C. Wright, and Caitlin G. Howe

© Springer International Publishing Switzerland 2017
D. Weihrauch, M. O'Donnell (eds.), *Acid-Base Balance and Nitrogen Excretion in Invertebrates*,
DOI 10.1007/978-3-319-39617-0_2

2.1 Introduction and Overview

Ammonia is the nitrogenous waste product resulting from the catabolism of proteins and amino acids. There is no metabolic cost associated with its production, and thus it is metabolically cheap. However, to avoid toxicity, the ammonia concentration within body fluids must be kept low. In aquatic gill breathers, ammonia is typically excreted into the surrounding water across the branchial surface before it builds up to toxic levels. Animals colonising land may not have access to large volumes of water in which to excrete ammonia, and in these species waste nitrogen is typically incorporated into less toxic compounds such as amino acids, urea and purines (uric acid and guanine). Amino acids may be temporarily stored and catabolised to release ammonia, which is then excreted. Urea and purines can be excreted as is. Urea is highly soluble in water and therefore needs to be excreted in solution, while purines have very low solubility and can be precipitated and excreted in semi-solid form. Purines in many ways are an ideal nitrogenous waste product; their very low solubility and high nitrogen content (four to five nitrogen atoms per molecule) allow for the elimination of waste nitrogen with minimal water.

Decapod crustaceans are comparatively recent terrestrial colonists, with the earliest groups probably radiating onto land approximately 50 mya (Hartnoll 1988; Little 1983). Terrestrial isopods (suborder Oniscidea) may have colonised land even earlier, perhaps 290–354 mya (Broly et al. 2013). Both taxa display different levels of terrestriality which can be categorised into four main groups: essentially aquatic crustaceans which are resident sub-tidally but are capable of surviving brief emersion during which they may or may not be active (T_1 species); amphibious species which require significant access to water but which spend substantial periods out of water (T_2 species); amphibious species which are resident and principally active on land, but which require regular or at least periodic immersion in standing water (often in burrows) (T_3 species); and terrestrial species which do not require immersion (T_4 species) (Hartnoll 1988). The majority of terrestrial crustaceans, from T_1 to T_4 species, have retained ammonotely, excreting ammonia either in solution or as a gas. Only one terrestrial crustacean, the robber crab, *Birgus latro*, is purinotelic and thus parallels the excretory strategy exploited by the majority of insects and arachnids.

Previous reviews (Greenaway 1991; O'Donnell and Wright 1995; Weihrauch et al. 2004) have described the physiological mechanisms of ammonotely by amphibious crabs, isopods, grapsid and gecarcinid crabs and purinotely in *Birgus latro*. However, these reviews have not discussed the detoxification of nitrogenous wastes between nitrogenous excretion bouts with reference to different levels of terrestriality, and this review will attempt to do so. There will be a particular focus on the morphology, biochemistry and metabolism of the purine deposits that are frequently observed in terrestrial Crustacea. It will review evidence that the purine deposits are not a nitrogen reserve, as previously thought, but may represent storage excretion. By comparing the nitrogenous waste metabolism between species with different levels of terrestriality, the review will attempt to hypothesise the steps required to evolve purinotely and how terrestrial crustaceans may display these steps.

2.2   Nitrogenous Excretion in Terrestrial Crustaceans

2.2.1 Ammonia Excretion by Amphibious T_{1-3} Species

Amphibious crabs (T_{1-3} species) that make terrestrial excursions onto land for various periods have a common nitrogenous excretion strategy. During their terrestrial excursions, these species do not excrete significant amounts of nitrogenous wastes. For example, the amphibious crabs *Carcinus maenas* (T_1), *Potamonautes warreni* (T_2) and *Discoplax celeste* (T_3) (described as *Cardisoma hirtipes* in the original literature (Ng and Davie 2012)) during respective emersion periods of 72 h, 96 h and 9 days had very low ammonia excretion rates of 24 ± 7 nmol.g^{-1} wet mass.h^{-1}, 0.95 ± 0.26 and 0.2 ± 0.1 nmol. g$^{-1} \cdot$h^{-1}, respectively (Wood et al. 1986; Dela-Cruz and Morris 1997a; Durand and Regnault 1998; Morris and Van Aardt 1998). When these species are emersed to prevent aqueous nitrogen excretion, waste nitrogen is temporarily stored (discussed later), and there is evidence in some species that protein catabolism is suppressed (Durand and Regnault 1998). Upon re-immersion, their gills excrete ammonia into the water (Linton and Greenaway 1995; Dela-Cruz and Morris 1997a; Durand and Regnault 1998; Morris and Van Aardt 1998). Amphibious species, such as *Carcinus maenas*, *Potamonautes warreni* and *Discoplax celeste*, excrete a large pulse of ammonia (excretion rates up to 4.9 ± 0.9 µmol.g$^{-1} \cdot$h^{-1}) within 1–5 min of re-immersion (Dela-Cruz and Morris 1997a; Durand and Regnault 1998; Morris and Van Aardt 1998). In *Discoplax celeste* and *Potamonautes warreni* at least, it is unlikely that ammonia is accumulated within the branchial chambers during the period of emersion. *Discoplax celeste* was emersed for 9 days, before re-immersion and measurement of ammonia excretion (Dela-Cruz and Morris 1997a). Similarly, *Potamonautes warreni* was held in air for 4 days before being submersed (Morris and Van Aardt 1998). These periods are too long for the crabs to hold significant amounts of fluid within the branchial chamber. Also, the large amounts of ammonia excreted (1.1 ± 0.3 µmol.g$^{-1} \cdot$h^{-1} for *Discoplax celeste* and 4.9 ± 0.9 µmol.g$^{-1} \cdot$h^{-1} for *Potamonautes warreni*) after re-immersion are likely to be toxic (Dela-Cruz and Morris 1997a; Morris and Van Aardt 1998). Such pulsatile excretion implies that the stored nitrogenous wastes are readily mobilised and that the transporters and enzymes involved in remobilisation and secretion of ammonia are always present. The potential transport mechanism of ammonia across the gills may be similar to that proposed for *Carcinus maenas*; it may involve active or facilitated basolateral transport (via the sodium potassium ATPase and/or ammonium channels) followed by diffusion of ammonia into acidified vesicles that are moved to the apical membrane via a microtubular network (for further details ▶ see Chap. 1, Weihrauch et al.). Ammonia excretion by *Carcinus maenas* remains elevated for up to 24 h, while ammonia excretion by *Potamonautes warreni* and *Discoplax celeste* returns to control levels within an hour (Dela-Cruz and Morris 1997a; Durand and Regnault 1998; Morris and Van Aardt 1998). Similarly, the amphibious gecarcinid, *Cardisoma carnifex*, does not excrete significant amounts of ammonia during terrestrial excursions. Ammonia is excreted into the permanent water at the base of the burrow upon return (Wood et al. 1986). The physiological requirement of excreting ammonia into water may limit the terrestrial excursions of the amphibious species and along with reproduction and development may represent one of the last ties to water to be severed for the colonisation of land (Dela-Cruz and Morris 1997b; Morris and Van Aardt 1998).

The nitrogenous excretion strategy of the amphibious Australian arid zone crab, *Austrothelphusa transversa* (née *Holthuisana* (Davie (2002)) (T_3 species), is different. This species is found in the flood plains of arid and semiarid Australia (Bishop 1963). Crabs may experience up to a year or more without water, with droughts of up to 3–5 years not uncommon. In that time, *Austrothelphusa transversa* aestivates within a dry burrow (MacMillen and Greenaway 1978; Greenaway 1984). It emerges after infrequent rain and/or flooding to reproduce and feed (Greenaway 1984). During emersion and like other amphibious species, there is no significant excretion of nitrogenous wastes (Linton and Greenaway 1995). Upon immersion, ammonia is excreted into water via the gills (Linton and Greenaway 1995). However, in contrast to the species described above, there is no initial large burst of ammonia excretion. Instead, ammonia excretion begins after approximately half an hour (S Linton personal observation) and remains elevated for up to 3 h following re-immersion (Linton and Greenaway 1995). Remobilisation of stored wastes is thus delayed, possibly due to the requisite synthesis and mobilisation of enzymes and transporters required. In the crab's natural environment, availability of water for re-immersion is unpredictable, and the delay in expression of enzymes and transporters would mean that *Austrothelphusa transversa* does not waste resources synthesising proteins it does not require. The form in which waste nitrogen is stored during aestivation remains to be determined.

Like amphibious decapods, amphibious and intertidal isopods (*Ligia beaudiana, Ligia occidentalis, Alloniscus perconvexus* and *Tylos punctatus*) excrete ammonia into the external seawater (Wieser 1972a; Nakamura and Wright 2013), though *Tylos punctatus* and *Alloniscus perconvexus* are also capable of volatilisation of ammonia gas. Like the terrestrial species, ammonia excretion is periodic and constrained by water availability. Depending on the species, the excretory pattern may follow either a tidal or a diel cycle. Thus, *Ligia occidentalis* excretes ammonia during immersion at high tide, while *Alloniscus perconvexus* and *Tylos punctatus* excrete ammonia diurnally while in their sand burrows (Nakamura and Wright 2013).

2.2.2 Ammonia Excretion by Terrestrial T_4 Crustaceans

Grapsidae, Isopods and Amphipods: Ammonia Gas Excretion

Two groups of terrestrial crustaceans, the grapsid crabs, *Geograpsus grayi* and *Geograpsus crinipes* and isopods, excrete their nitrogenous wastes as ammonia gas. Although they excrete a common product, the mechanism by which they do this is different. A probable common feature is that in the secretion of ammonia, ammonium bicarbonate is exchanged for sodium chloride.

The carnivorous grapsid crab, *Geograpsus grayi* (T_4 species) excretes ammonia gas in bouts lasting from 3 h to 3 days, with a maximal excretion rate of up to 220 ± 20 nmol.g^{-1}·h^{-1} (Varley and Greenaway 1994). Excretion of nitrogenous wastes begins when the urine produced by the antennal gland of *Geograpsus grayi* is passed into the branchial chamber for reabsorption of salts by the gills. During this reprocessing, ammonium is exchanged for sodium, and the ammonia concentration of the branchial fluid rises to 80.6 ± 11.3 mmol.L^{-1} (mean $\pm$ sem) (Varley and Greenaway 1994). There is also an alkalinisation of the branchial fluid, from 7.66 ± 0.02 (pH of haemolymph and possibly urine) to 8.06 ± 0.05 (Varley and Greenaway 1994). This is achieved by the secretion of the base, bicarbonate which is exchanged for chloride via a chloride-bicarbonate exchanger (Varley and Greenaway 1994). Given the high pKa of ammonium (9.24 at 25 °C; Aylward and Findlay 1971), this increase in pH causes the concentration of ammonia base (NH_3) to rise 2.5-fold (◻ Fig. 2.1). It is then free to diffuse away as ammonia gas (Greenaway and Nakamura 1991; Varley and Greenaway 1994). Gaseous ammonia excretion also occurs in a closely related species, *Geograpsus crinipes* (S Linton personal observation). *Geograpsus crinipes* (T_3 species) lives near the coast and makes frequent excursions back to sea water; thus, it is not as terrestrial as *Geograpsus grayi*. Hence, excretion of gaseous ammonia may be common to the family Grapsidae and may have evolved in a common amphibious ancestor.

Terrestrial isopods (*Porcellio scaber, Armadillidium vulgare* and *Oniscus asellus*) (T_4 species) excrete gaseous ammonia in bouts lasting from 40 min to 2 h. Ammonia is excreted diurnally when the animal is inactive in a humid environment (Dresel and Moyle 1950; Wieser et al. 1969; Wieser and Schweizer 1970; Kirby and Harbaugh 1974; Wright and O'Donnell 1993); the mean diurnal excretion rate for *A. vulgare* is 210 nmol. g^{-1} h^{-1} (Wright and Peña-Peralta (2005). In humidities greater than 86 % (critical humidity), *Porcellio scaber* secretes a pleon fluid with a high NaCl concentration (up to 4.1 mol.L^{-1} or 8.2 Osmol kg^{-1}) for water vapour absorption (Wright and Machin 1990; Wright and O'Donnell 1992). Ammonium is secreted into this pleon fluid, possibly in exchange for sodium (Wright and O'Donnell 1993). The mean ammonium concentration of the pleon fluid during excretory bouts ranges between 14.2 ± 2.9 and 25.6 ± 7.0 mmol.L^{-1}. During excretory bouts, the ammonia concentration within the haemolymph also undergoes an episodic increase from approximately 1.5 mmol.L^{-1} to 6.1 ± 1.3 mmol.L^{-1}; it frequently reaches 20 mmol.L^{-1} and may exceed 60 mmol.L^{-1} (Wright and O'Donnell 1993). The pH of the pleon fluid is similar to that of the haemolymph (pH = 7.6 ± 0.04), and thus there is no evidence of increased volatilisation through alkalinisation (Wright and O'Donnell 1993). Beating of the pleopods circulates air over the pleon fluid surface to facilitate convective removal of NH_3 and sustain a partial pressure gradient from the pleon fluid into the overlying air (Wright and O'Donnell 1993). Diffusion of gaseous ammonia from the pleon fluid would drive the ammonium/ammonia equilibrium to the right and thus help to sustain the production of ammonia (◻ Fig. 2.1).

Dissociation of ammonium and volatilization of the ammonia base poses a problem, however, in that accumulating protons will decrease the pleon fluid pH and in turn reduce the equilibrium concentration (and partial pressure) of NH_3 as predicted by the disassociation equation (◻ Fig. 2.1). Although terrestrial isopods appear not to increase the pH of the pleon fluid during ammonia volatilization, fluid acidification must be prevented. Wright and O'Donnell (1993) hypothesise that they use a similar process to the grapsid crabs, transporting bicarbonate into the pleon fluid via a chloride-bicarbonate exchanger in order to buffer protons and excrete the acid equivalents as carbon dioxide gas

◻ **Fig. 2.1** Acid base equilibria and disassociation constants (at 25 °C) of ammonium and carbonic acid (Aylward and Findlay 1971)

$$NH_4^+ \;\underset{}{\overset{pK_a = 9.24}{\rightleftharpoons}}\; NH_3 + H^+$$

$$H^+ + HCO_3^- \;\underset{}{\overset{pK_a = 6.35}{\rightleftharpoons}}\; H_2CO_3 \;\rightleftharpoons\; CO_2 + H_2O$$

(■ Fig. 2.1). The efficient conversion of carbonic acid to CO_2 in pleon fluid could employ a membrane-bound carbonic anhydrase. Carbonic anhydrase activity has been measured using pH stat assays (Henry and Cameron 1982) in three species of terrestrial isopods (*Ligia occidentalis*, *Ligidium lapetum* and *Armadillidium vulgare*) (Godlove et al. unpublished data 2003). In the two volatilising species (*Armadillidium vulgare* and *Ligidium lapetum*), high carbonic anhydrase activities were inhibited by acetazolamide, hence providing experimental support for this model (■ Fig. 2.2) (Godlove et al. unpublished data 2003).

Ammonia excretion and volatilisation of ammonia gas occurs in both terrestrial and amphibious isopods, though not in the basal intertidal genus *Ligia*. It may be an ancestral trait in the Holverticata (Oniscidea excluding the Ligiidae) and have evolved independently in the terrestrial genus *Ligidium* (Nakamura and Wright 2013).

Like isopods, amphipods are again primarily ammonotelic (Spicer et al. 1987), and this includes the supralittoral, sand-burrowing beachhopper *Orchestia gammarellus*

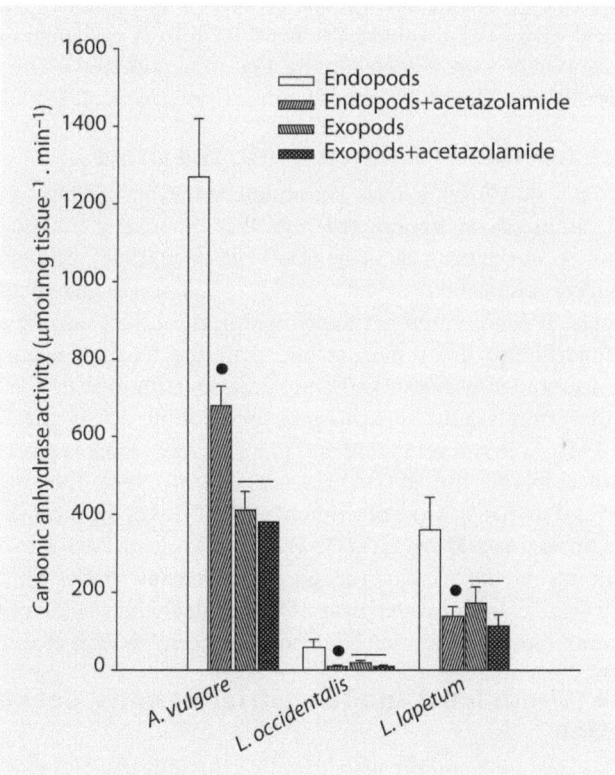

■ **Fig. 2.2** Carbonic anhydrase activity in pleopodal endopodites and exopodites of three species of Oniscidea (*Armadillidium vulgare*, *Ligia occidentalis* and *Ligidium lapetum*) (Godlove et al., 2003 unpublished data). Most carbonic anhydrase activity is concentrated in the endopodites and is significantly inhibited by 10 mM acetazolamide. Closed circles indicate that the means for the carbonic anhydrase activity within the endopods treated with acetazolamide were significantly different from those of the uninhibited controls ($p < 0.05$ Student's t-test). Acetazolamide did not inhibit carbonic anhydrase activity in the exopodites. Horizontal bars above the means indicate that the groups showed no significant difference ($p > 0.05$ Student's t-test). *Ligia occidentalis* excretes aqueous ammonia (into seawater) and has markedly lower carbonic anhydrase activity than that of *Ligidium lapetum* and *Armadillidium vulgare* which liberate volatile ammonia gas. mean ± SEM ($n = 5$–12)

(Talitridae) (Dresel and Moyle 1950). However, nitrogen excretion has yet to be studied in any of the fully terrestrial (T$_4$) talitrids. *Orchestia gammarellus* has been suggested to be able to excrete gaseous ammonia (unpublished work of A.C. Taylor and A.J. Wilkie cited by Spicer et al. (1987)), which would represent an elegant example of convergent evolution with the terrestrial isopods (Oniscidea). Further work in this area would be valuable.

Gecarcinidae: Excretion of Ammonia into the Urine During Its Reprocessing in the Branchial Chamber

The gecarcinid land crabs (*Gecarcoidea natalis*, *Gecarcinus lateralis*, and *Cardisoma guanhumi*) excrete ammonia in aqueous form rather than as a gas. In these species, urine produced by the antennal gland has a low ammonium concentration (0.36 ± 0.08 mmol.L^{-1}, 0.52 ± 0.53 mmol.L^{-1} and 0.53 ± 0.49 mmol.L^{-1}, respectively) (Greenaway and Nakamura 1991; Wolcott 1991). Ammonia levels increase 20- to 30-fold to approximately 10 mmol.L^{-1} when the urine is reprocessed in the branchial chamber (respective ammonia concentrations for *Gecarcoidea natalis*, *Gecarcinus lateralis* and *Cardisoma guanhumi* of 10.82 ± 7.79 mmol. L^{-1}, 9.94 ± 10.2 and 6.46 ± 10.14 mmol.L^{-1}); again sodium is exchanged for ammonium during its reuptake (Greenaway and Nakamura 1991; Wolcott 1991). The modified urine, called 'P', is then released (Greenaway and Nakamura 1991; Wolcott 1991).

Ocypodidae: Excretion of Ammonia into the Urine

Ghost crabs (*Ocypode quadrata*) are also ammonotelic but, in contrast to the grapsid and gecarcinid crabs, ammonia is transported into the antennal gland duct and excreted in high concentration in the primary urine (De Vries et al. 1994). The primary urine has an ammonia concentration of 116 ± 34.9 mmol.L^{-1}, 134 times that of the haemolymph (0.86 ± 0.39 mmol.L^{-1}). Ammonium is trapped within the urine, since at a pH of 5.5, the urine is approximately 100 times more acidic than the haemolymph (pH = 7.58). The lower sodium concentration of the urine (approximately 60 % of that of the haemolymph) again suggests that ammonium is exchanged for sodium across the duct epithelium (De Vries et al. 1994). As in the gecarcinid and grapsid crabs, urine is passed into the branchial chambers where it is alkalinised (De Vries and Wolcott 1993). The resulting increase in the concentration and partial pressure of ammonia (NH$_3$) drives the production of ammonia gas which then diffuses away (◘ Fig. 2.1) (De Vries and Wolcott 1993; De Vries et al. 1994). Secretion of ammonia into the urine is probably common to all species within Ocypodoidea since the primary urine of both *Uca pugilator* (Ocypodidae) and *Ucides cordatus* (Ucididae) contains high ammonium concentrations (Green et al. 1959; De Vries et al. 1994).

Coenobitidae (*Birgus latro* and Terrestrial Hermits, *Coenobita*): Purine Excretion

The robber or coconut crab, *Birgus latro* (family Coenobitidae), is the only terrestrial crustacean that is purinotelic. Its main nitrogenous waste product is a white purine pellet that is produced separately to the brown faecal pellet. The pellet consists of 60 % uric acid and 40 % guanine (Greenaway and Morris 1989; Linton et al. 2005). It is composed of spherules (1.6 ± 0.6 µm in diameter) and is covered by a peritrophic membrane (Dillaman et al. 1999) (◘ Fig. 2.3a, b). Thus, the pattern of nitrogenous excretion in this species is convergent with that of other highly terrestrial arthropods (e.g. insects and chelicerates), and it is the only reported excretion of guanine outside of the chelicerates (Little 1983).

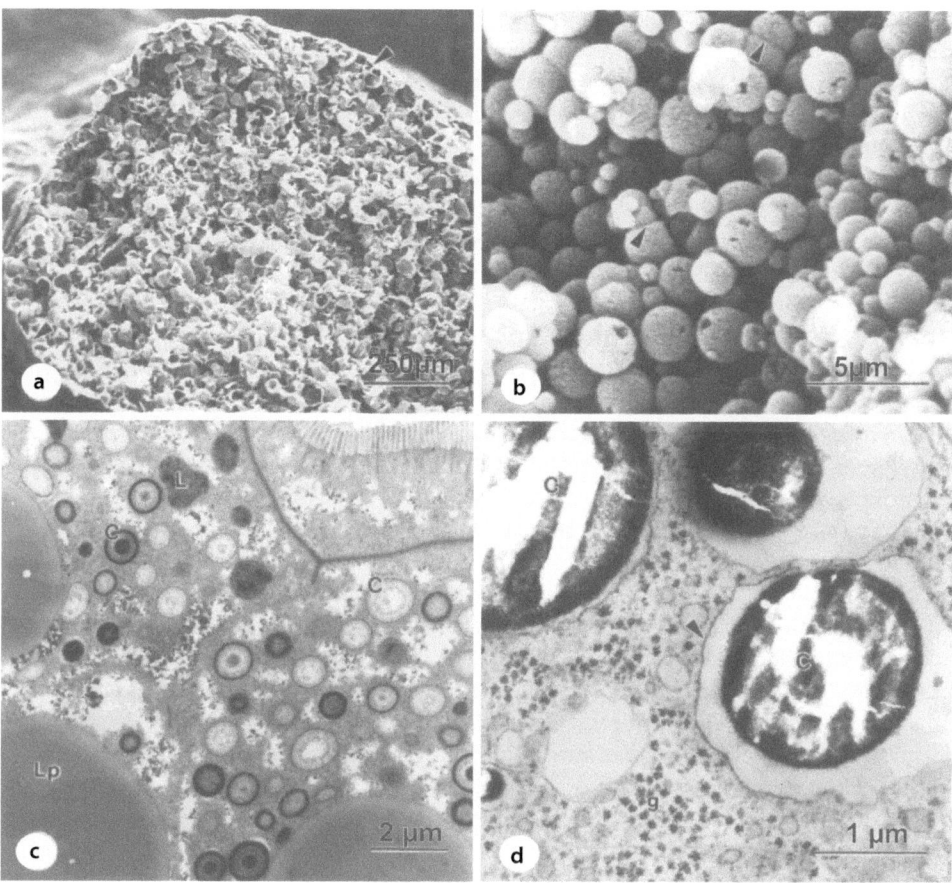

◻ Fig. 2.3 Scanning electron micrographs of the white faecal pellet produced by *Birgus latro* and containing uric acid and guanine (**a, b**). Faecal pellet (**a**) is bound by a peritrophic membrane (*black arrow head*) (**a**) and contains numerous purine spherules (**b**). Electron micrographs of the reserve inclusion cells (R cells) from the midgut gland of *Birgus latro* (**c, d**). (**c**) Cytoplasm contains numerous concretions with electron dense lamellae. (**d**) Membrane-bound vesicles containing electron dense purine concretions. Purine crystal has been damaged during sectioning (Figure reproduced with permission from Dillaman et al. (1999))

The purine is synthesised de novo in reserve inclusions cells (R cells) within the midgut gland. The gland cells possess high xanthine dehydrogenase activity and contain spherical membrane-bound purine concretions (◻ Fig. 2.3c, d) (Greenaway and Morris 1989; Dillaman et al. 1999). The purine spherules are probably exocytosed into the lumen of the midgut gland tubules. Secretion appears to be coordinated and is probably under hormonal control. It is accompanied by peristaltic contractions which push the urate spherules into the midgut. A peritrophic membrane is then secreted around the pellet as it is passed into the hindgut and before it is voided with the faeces (Dall and Moriarty 1983; Dillaman et al. 1999).

The main nitrogenous excretory product of the closely related terrestrial hermit crabs, *Coenobita* sp., is unknown. The terrestrial hermit crab *Coenobita brevimanus* (the species most closely related to *Birgus latro*) is not purinotelic as it does not produce a white purine pellet, and its faeces contains little purine (Linton et al. 2005). However, like *Birgus latro*, *Coenobita brevimanus* does possess purine deposits of urate within its body (Linton et al. 2005). Purine deposits in *Birgus latro* are a mixture of urate and guanine (Linton et al. 2005). The total purine concentration in the midgut gland of *Birgus latro* comprises 65.7 ± 12.0 ($n = 5$) μmol. urate g^{-1} tissue plus 26.1 ± 3.9 ($n = 5$) μmol. guanine g^{-1} tissue. This is 17 times the purine content (5.2 ± 1.9 ($n = 3$) μmol. urate g^{-1} tissue) of the midgut gland of *Coenobita brevimanus* (Linton et al. 2005). It appears that purinotely and guanine metabolism has only evolved in *Birgus latro* and is not a synapomorphic character of the Coenobitidae.

2.2.3 Potential Steps in the Evolution of Purinotely as Displayed by the Terrestrial Crustacea

As *Birgus latro* is the only known terrestrial crustacean that utilises purinotely as its primary mode of nitrogenous excretion, we can propose the following steps in the evolution of purinotely:

1. Overproduction of purines, uric acid and guanine (beyond the requirements of nucleic acid synthesis and catabolism of nucleic acids to uric acid). There may be a change in the inhibition of the control steps of the de novo synthetic pathway.
2. Precipitation of uric acid as round spherules that do not puncture membranes within the midgut gland. Uncontrolled uric acid precipitation produces potentially damaging needles (McNabb and McNabb 1977). The problem may be solved by precipitation of uric acid around a nucleation site within intracellular vesicles. Proteins may provide a matrix for the deposition of urate crystals (Linton and Greenaway 1997a; Dillaman et al. 1999; Vega et al. 2007).
3. A mechanism for the periodic release or exocytosis of urate deposits from the midgut gland cells into the gland lumen. As the gut can handle solid materials, it represents an exaptation for the elimination of a solid pellet from the body.

Numerous terrestrial crustaceans (decapods and isopods) display the first character (◻ Table 2.1). These species also possess the ability to precipitate the urate as round spherules. For the majority of terrestrial crustaceans, urate is synthesised de novo within the midgut gland and deposited as spherules within spongy connective tissue cells (Wägele 1992; Linton and Greenaway 1997a). *Birgus latro* appears to be the only species, however, that is capable of coordinated secretion from this tissue into the gut. Nevertheless, the ability of other decapod species to concentrate purines in the midgut could be an exaptation for uricotely. Given this, and the paucity of work done on nitrogenous excretion in several families, particularly the Coenobitidae, terrestrial decapods offer a promising model in which to study the evolution of uricotely.

2.3   Storage and Detoxification of Nitrogenous Wastes

2.3.1 Ammonia Chemistry and Toxicity

A limitation of ammonotely is that ammonia is potentially toxic at quite low concentrations (>1–10 mM). Ammonia is a weak base which is in equilibrium with its conjugate acid (◻ Fig. 2.1). While at physiological pH the majority of ammonia exists as the

☐ **Table 2.1** Concentration of ammonia (mean ± sem) within the haemolymph of amphibious (☐) and terrestrial crustaceans (●)

Species	Haemolymph ammonia concentration	Reference
Amphibious Crustacea		
Austrothelphusa transversa ☐	≈500 µmol.L⁻¹	Linton and Greenaway (1995)
Cardisoma carnifex ☐	118 ± 14 µmol.L⁻¹, 2.67 ± 0.42 mmol.L⁻¹	Henry and Cameron (1981), Wood and Boutilier (1985), Wood et al. (1986)
Terrestrial Crustacea		
Gecarcoidea natalis ●	1.46 ± 0.12 mmol.L⁻¹	Greenaway (1991)
Gecarcoidea lalandii ●	238 ± 17 µmol.L⁻¹	Henry and Cameron (1981)
Geograpsus grayi ●	1.92 ± 0.68 mmol.L⁻¹	Greenaway (1991)
Ocypode quadrata ●	860 ± 390 µmol.L⁻¹	De Vries et al. (1994)
Birgus latro ●	174 ± 8 µmol.L⁻¹	Henry and Cameron (1981)
Coenobita brevimanus ●	196 ± 4 µmol.L⁻¹	Henry and Cameron (1981)
Coenobita perlatus ●	282 ± 11 µmol.L⁻¹	Henry and Cameron (1981)
Terrestrial Isopoda		
Porcellio scaber ●	1.28 ± 0.46 mmol.L⁻¹	Wright and O'Donnell (1993)

ionic form, a small but significant fraction comprises the uncharged base which has the potential to diffuse across membranes. For example, at the physiological pH of the haemolymph of terrestrial crustaceans (pH ≈ 7.6), 97.5 % of 'total ammonia' will comprise NH_4^+ and 2.5 % will comprise NH_3 (Varley and Greenaway 1994; De Vries et al. 1994). Ammonia exerts its toxicity through reacting with α-ketoglutarate to form glutamate. This drains intermediates from the citric acid cycle and hence disrupts energy metabolism. It also has the potential to dissipate the proton gradient across the inner mitochondrial membrane. As ammonium can substitute for potassium, it can be transported into cells by the Na^+/K^+ ATPase and thus could also disrupt electrochemical gradients, epithelial transport and the functioning of neurons (reviewed in ▶ Chap. 1 (Chew and Ip 2014)).

Terrestrial crustaceans maintain ammonia within the haemolymph at relatively high concentrations (0.1–2 mmol.L⁻¹) compared to aquatic species (100–200 µmol.L⁻¹) (☐ Table 2.1) (▶ Chap. 1). The mitochondria are the sites of detoxification in mammals (via the urea cycle) and birds and reptiles (via uric acid synthesis) (Campbell 1995; Chew and Ip 2014). In these animals, ammonia is initially detoxified by incorporating it into carbamoyl phosphate and glutamine, respectively. In Crustacea, the mitochondria may efficiently detoxify ammonia by incorporating it into non-toxic intermediates such as non-essential amino acids. The next section of the review describes the temporary storage of nitrogenous wastes within non-toxic molecules in amphibious and terrestrial Crustacea.

2.3.2 Amphibious T$_{1-3}$ Species: Storage and Mobilisation of Nitrogenous Wastes During Emersion and Re-immersion

Detoxification and temporary storage of nitrogenous wastes during periods where excretion is not possible have been studied in amphibious species such as *Carcinus maenas* (T$_1$) and *Cardisoma carnifex* (T$_3$) and in terrestrial isopods. In amphibious species, ammonia levels within the body are regulated. When emersed for 72 h, *Carcinus maenas* (T$_1$) showed no significant accumulation of ammonia within its tissues (Durand and Regnault 1998). The concentration of ammonia in the haemolymph increased from 81.2 ± 6.9 to 134.8 μmol.L^{-1} over the initial 12 h and then remained constant for the remainder of emersion. Similarly, during emersion of the amphibious *Cardisoma carnifex* (T$_3$) for 192 h, the ammonia concentration in the haemolymph doubled between 36 and 72 h then returned to starting levels (1.26 mmol.L^{-1}) for the remainder of the period (Wood et al. 1986). In the absence of water, waste nitrogen in these species is probably incorporated into non-essential amino acids by transamination. In *Carcinus maenas*, specific amino acids are synthesised and accumulated in the muscle (Durand et al. 1999). Glycine was the major accumulated amino acid, together with smaller amounts of alanine, glutamine and arginine. Total amino acid levels increased up to 24 h and then remained constant for the remainder of emersion, indicating suppression of amino acid metabolism. Arginine is considered to be an essential amino acid, and its accumulation may represent degradation of phosphoarginine which is a high energy intermediate involved in buffering the energy charge of the cell. Production of alanine may be explained by either the glucose alanine cycle or by the partial oxidation of other amino acids, as observed in fish (Chew and Ip 2014). Amino acids were not accumulated in the hepatopancreas or the haemolymph (Durand and Regnault 1998; Durand et al. 1999).

Upon re-immersion, *Carcinus maenas* showed a decrease in amino acid levels in muscle, while levels in the hepatopancreas and haemolymph increased (Durand et al. 1999). Ammonia levels within the muscle also increased from 3.18 ± 0.43 to 5.15 mmol.L^{-1} in first 6 h, while the haemolymph ammonia concentration initially decreased following re-immersion and then increased, reaching 317.9 ± 37.4 μmol.L^{-1} after 6 h and remained high after 24 h (Durand et al. 1999). Deamination of amino acids within the muscle may partially explain the decrease in amino acids and accompanying increase in ammonia within this tissue (Durand and Regnault 1998). Alternatively, amino acids could be transported to the hepatopancreas to undergo transamination and deamination which could explain the subsequent increase in haemolymph levels (Durand et al. 1999). Similarly, following re-immersion of *Cardisoma carnifex*, the onset of branchial ammonia excretion was accompanied by an increase in haemolymph ammonia levels from 1.26 to 7 mmol.L^{-1} (Wood et al. 1986). *Carcinus maenas* apparently suppresses protein catabolism during emersion as the amount of ammonia excreted upon re-immersion is less than that excreted by control immersed crabs over the same time period (Durand and Regnault 1998).

2.3.3 Terrestrial T$_4$ Species: Storage and Remobilisation of Nitrogenous Wastes Between Excretory Bouts

Similar to amphibious crabs, terrestrial isopods temporarily accumulate waste nitrogen as amino acids between excretory bouts (Wright et al. 1994, 1996; Wright and Peña-Peralta 2005). In *Porcellio scaber*, 'ammonia loading' in a high-ambient partial pressure of ammonia (pNH$_3$) did not result in a significant increase in haemolymph ammonia levels, indicating detoxification and storage (Wright et al. 1994, 1996). Over the same period, animals showed significant increases in tissue levels of the non-essential amino acids, glutamate, glutamine and glycine. These amino acids are accumulated in the hepatopancreas and the body wall, including the somatic muscle (Wright et al. 1996; Nakamura and Wright 2013). Given that terrestrial isopods excrete most ammonia during the day, they would be predicted to accumulate N-sequestering amino acids at night *in vivo,* and this is borne out by experimental data. In the terrestrial isopod, *Armadillidium vulgare,* the concentrations of glutamine in the body wall and hepatopancreas increased 4.2 and 7.1-fold (24.5–102.4 and 140.3–993.4 $\mu mol.g^{-1}$), respectively between dusk and dawn (Wright and Peña-Peralta 2005). The amphibious isopods, *Tylos punctatus* and *Alloniscus perconvexus,* similarly had significant increases in whole-animal glutamine levels at dawn relative to dusk, consistent with diurnal ammonia excretion (Nakamura and Wright 2013).

The biochemical mechanism for the formation of glutamate and glutamine due to ammonia loading in the isopod, *Porcellio scaber* should be considered as it illustrates the cost of temporarily storing nitrogen within glutamine. Within tissues, the deamination of glutamate to α-ketoglutarate is catalysed by glutamate dehydrogenase. According to Le Chatelier's principle, an increase in ammonia concentration during ammonia loading will increase the formation of glutamate (☐ Fig. 2.4). Glutamate could then serve as the

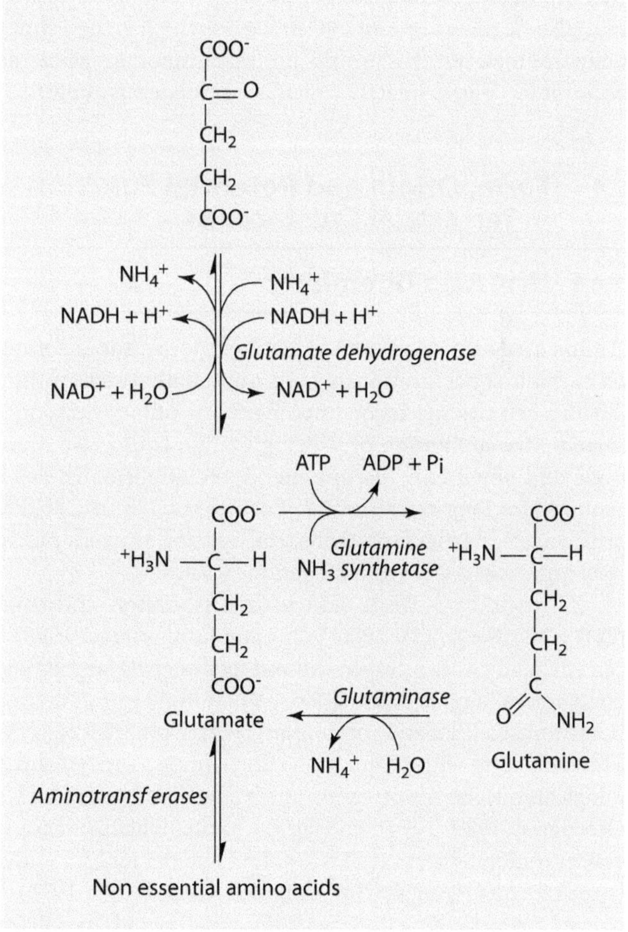

☐ **Fig. 2.4** Potential detoxification of ammonia through the synthesis and catabolism of glutamate and glutamine. The reaction catalysed by glutamate dehydrogenase is near equilibrium. Therefore, if ammonia is present, it will combine with α-ketoglutarate to produce glutamate. Glutamate may either be the substrate for aminotransferases (also in equilibrium) or used to synthesise glutamine. Glutamine is deaminated to form glutamate (Newsholme and Leech 1983; Moran et al. 2014)

substrate for transaminases and lead to the formation of other non-essential amino acids. Glutamate is a precursor of glutamine which is synthesised by glutamine synthetase and involves the hydrolysis of 1 molecule of ATP (☐ Fig. 2.4). Thus, the net energetic cost of the storage of two ammonia molecules via synthesis of glutamine is 1 ATP (☐ Fig. 2.4).

Glutaminase catalyses the hydrolysis of glutamine to glutamate and ammonia and is present in high activities within the body wall of terrestrial isopods (☐ Fig. 2.4) (Wieser 1972b; Nakamura and Wright 2013). Measured activities were significantly higher in four species of terrestrial isopods ($0.55–0.85 \ \mu mol.min^{-1}.g^{-1}$) than in the amphibious species *Alloniscus perconvexus, Tylos punctatus* and *Ligia occidentalis* ($0.25–0.3 \ \mu mol.min^{-1}.g^{-1}$) (Nakamura and Wright 2013) and probably reflect the greater temporal restrictions for volatilizing ammonia in terrestrial habitats. During bouts of ammonia gas excretion, glutamine and other N-sequestering amino acids are catabolised to release ammonia into the haemolymph. This causes the ammonia concentration within this compartment to rise from 1 $mmol.L^{-1}$ to as much as 50 $mmol.L^{-1}$ and creates a favourable gradient for its secretion into the pleon fluid and subsequent volatilization (Wieser 1972b; Wright and O'Donnell 1993; Wright et al. 1994, 1996; Wright and Peña-Peralta 2005). Potential ammonia production from glutamine catabolism in *Armadillidium vulgare* exceeds the measured excretion rates by a factor of 60 (Wright and Peña-Peralta 2005). This has been suggested to be necessary to create a sufficient partial pressure of NH_3 to drive volatilization of NH_3 from the pleon fluid (Wright and Peña-Peralta 2005). Some of the glutamine may also be incorporated into uric acid which is then stored in purine deposits (discussed below). However, this would be only minor as whole-animal urate levels would only account for approximately 1 % of accumulated glutamine.

2.4 Form, Origin and Potential Function of Urate Deposits in Terrestrial Crustaceans

2.4.1 Uric Acid Chemistry

Amino acids are only suitable for short-term storage of nitrogenous wastes. In the longer term, high concentrations would potentially cause an increase in osmolarity and hence perturb cell volume regulation. Synthesis of large amounts of amino acids such as glutamate by the amination of α-ketoglutarate could also drain intermediates from the TCA cycle and potentially disrupt energy metabolism. Therefore, another compound may be required for long-term storage. Purines such as uric acid may be a possible option. Many crustaceans, particularly terrestrial species, possess purine deposits, and these deposits may represent a store of nitrogenous wastes.

Uric acid is a weak acid with dissociation constants of $pK_1 = 5.26$ and $pK_2 = 10.3$ (☐ Fig. 2.5) (Seegmiller 1980; Wang and Königsberger 1998). With a N:C ratio of 4:5, uric acid can efficiently store nitrogen without sequestering large amounts of carbon, particularly when compared to amino acids (☐ Fig. 2.5). Monovalent urate can form salts with cations such as potassium, sodium and ammonium (Porter 1963; McNabb and McNabb 1977). Uric acid has a low solubility, with a saturation concentration at pH 7 and 25 °C of 9.75 $mmol.L^{-1}$. At physiological concentrations, cations such as sodium ($\approx 0.3 \ mol.L^{-1}$ for terrestrial gecarcinid crabs; Greenaway 1989, 1994) will salt out urate, which further reduces its solubility (saturation concentration of urate at 25 °C, pH = 7 and 0.3 $mol.L^{-1}$ sodium = 12.3 $\mu mol.L^{-1}$; calculations based on equations given by Wang and Königsberger 1998). Urate concentrations can exceed true solubility because they can form stable colloids in solution (Porter 1963). Storage of waste

⬛ Fig. 2.5 Molecular structure and acid base equilibria for the disassociation constants of uric acid at 25 °C (Seegmiller 1980; Wang and Königsberger 1998)

nitrogen as precipitated urate salts would ameliorate osmotic problems. However, a potential problem could arise if urate precipitated out as needle-like crystals which could potentially puncture membranes (Seegmiller 1980). Decapods overcome this problem by precipitating urate as protein-complexed spherules (Linton and Greenaway 1997a).

2.4.2 Presence and Anatomy of the Urate Deposits

Urate deposits are frequently observed within the tissues of aquatic, amphibious and terrestrial decapods, amphipods and isopods (⬛ Table 2.2). When maintained in the laboratory, the viscera of terrestrial gecarcinids such as *Gecarcoidea natalis*, *Cardisoma guanhumi* and *Gecarcinus lateralis* are quite frequently white with large deposits (Gifford 1968; Wolcott and Wolcott 1984; Wolcott and Wolcott 1987; Linton and Greenaway 2000). White urate deposits are also observed in animals collected in the field, but are rarely extensive (⬛ Fig. 2.6a). It was initially thought that the urate existed free within the haemolymph and was excreted with the moult (Gifford 1968). However, this is not the case, as the cast exuviae of moulted crabs contains insignificant amounts of urate (Linton and Greenaway 2000). In *Gecarcoidea natalis* and potentially in other gecarcinid and terrestrial decapods, the urate exists as dense intracellular spherules within the spongy connective tissue which is distributed throughout the body (⬛ Fig. 2.6b, c) (Linton and Greenaway 1997a). The urate spherules are contained within membrane-bound vesicles (⬛ Fig. 2.6d, e). Electron dense structures are commonly observed within smaller vesicles and may serve as nucleation sites for the precipitation of urate (Linton and Greenaway 1997a). Similar membrane-bound urate concretions are also present in the reserve inclusion cells (R cells) of the midgut gland of *Birgus latro* (⬛ Fig. 2.3c, d) (Dillaman et al. 1999). Urate deposits are probably not a derived feature of terrestrial decapods given they are also present in aquatic species. Intracellular urate deposits have been reported within the spongy connective tissue of the lobster, *Homarus americanus*, and the blue crab, *Callinectes sapidus* (Johnson 1980; Battison 2013). However, the deposits in *Homarus americanus* were considered by Battison (2013) to be a pathology, perhaps a disruption in nucleic acid catabolism and excretion.

Terrestrial and aquatic isopods also possess uric acid deposits within their body wall. In the terrestrial isopod, *Oniscus asellus*, 93 % of the total body urate is contained within the body wall at a concentration of 6.7 μmol g^{-1} wet wt (Hartenstein 1968). Small levels of urate (0.48 μmol g^{-1}) were also reported in the supralittoral amphipod *Orchestia* sp. and similar amounts (0.42–0.59 μmol g^{-1}) in two littoral *Marinogammarus* spp. (Dresel and Moyle 1950). The histology of these deposits is unknown. However, in the freshwater isopod, *Asellus aquaticus*, urate is stored in specialised Zenker's cells associated with the fat body flanking the gut and comprises intracellular membrane-bound concretions (Wägele 1992).

◘ Table 2.2 Species of aquatic (○), amphibious (□) and terrestrial crustaceans (●) with internal urate deposits

Species	Reference
	Decapoda
Anomura	
Birgus latro ●	Henry and Cameron (1981), Greenaway and Morris (1989), Linton et al. (2005)
Coenobita brevimanus ●	Henry and Cameron (1981), Linton et al. (2005)
Coenobita perlatus ●	Henry and Cameron (1981)
Brachyura	
Callinectes sapidus ○	Johnson (1980)
Austrothelphusa transversa □	Linton and Greenaway (1995)
Gecarcoidea natalis ●	Greenaway (1991), Linton and Greenaway (1997a)
Gecarcoidea lalandii ●	Henry and Cameron (1981)
Gecarcinus lateralis ●	Wolcott and Wolcott (1984)
Cardisoma guanhumi □	Horne (1968), Gifford (1968), Wolcott and Wolcott (1987)
Cardisoma carnifex □	Henry and Cameron (1981)
Geograpsus grayi ●	Greenaway (1991)
Astacidea	
Homarus americanus ○	Battison (2013)
Dendrobranchiata	
Marsupenaeus japonicas ○	Cheng et al. (2004, 2005)
	Isopoda
Oniscus asellus ●	Hartenstein (1968), Dresel and Moyle (1950)
Porcellio laevis ●	Dresel and Moyle (1950)
Armadillidium vulgare ●	Dresel and Moyle (1950)
Asellus aquaticus ○	Wägele (1992)
	Amphipoda
Niphargus schellenbergi ○	Graf and Michaut (1975)

Purine stores appear to be a general characteristic of isopods and have been reported in aquatic, amphibious and terrestrial species (Hartenstein 1968; Wright et al. 1997; Howe and Wright, 2009, unpublished data). In a selection of eight terrestrial isopod species, which were field collected in June and July (North American Summer), HPLC analysis of whole-animal purines revealed uric acid, together with smaller amounts of xanthine, in most species (Howe and Wright, 2009, unpublished; ◘ Table 2.3). Guanine and hypo-xanthine were not detected. Xanthine is likely present as an intermediate in urate syn-thesis (► Sect. 4.3.3). Whole-animal urate levels (μmol.g^{-1}) are markedly lower than in

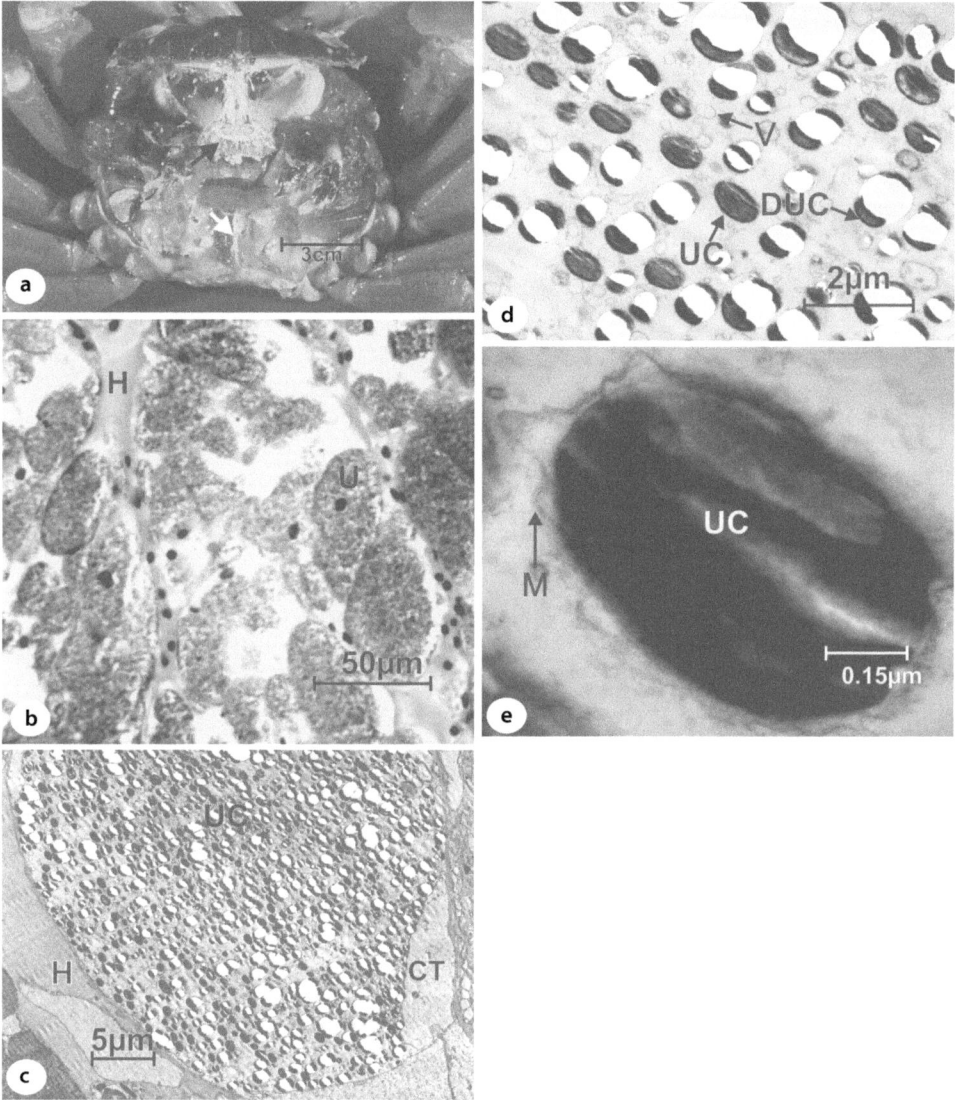

◻ **Fig. 2.6** Anatomy of urate deposits within the viscera and tissues of *Gecarcoidea natalis*. (**a**) *Gecarcoidea natalis* with the carapace removed. White deposits consisting of uric acid are present in the spongy connective tissue surrounding muscle and ossicles of the proventriculus (*black arrow*) and in the subepidermal connective tissue (*white arrow*). (**b**) Spongy connective tissue from *Gecarcoidea natalis*. Tissue contains numerous urate cells filled with granular urate inclusions. Tissue is well vascularised with haemolymph spaces (*H*) surrounding the cells. Tissue is stained for urate using methenamine silver staining and counterstained with haematoxylin and eosin (Bancroft 1975). (**c–e**) Electron micrographs of a mature urate cell from *Gecarcoidea natalis*. (**c**) Urate cell with urate crystals (*UC*) scattered throughout the cytoplasm. Adjacent to the cell is a haemolymph lacuna and a spongy connective tissue cell (*CT*) devoid of urate concretions. (**d**) High-power TEM of the cytoplasm of the urate cell with mature ovoid urate crystals (*UC*) and smaller round vesicles (*V*). Some of the urate crystals were damaged during processing (*DUC*). (**e**) Membrane-bound vacuole containing an electron dense urate crystal which appears to have cracked during tissue processing (Figure modified and reproduced with permission from Linton and Greenaway (1997a))

Table 2.3 Concentrations (µmol.g⁻¹) of purines in aquatic (○), amphibious (☐) and terrestrial (●) isopods (Howe and Wright, 2009, unpublished data) and decapods

Species	Uric acid	Xanthine	Guanine	References
Isopoda				
Ligiidae				
Ligia occidentalis (whole body) (µmol.g⁻¹ wet wt) ☐	0.55 ± 0.1 (26)	0.39 ± 0.1 (26)	ND (26)	
Ligidium lapetum (whole body) (µmol.g⁻¹ wet wt) ●	0.42 ± 0.08 (18)	0.092 ± 0.02 (18)	ND (18)	
Tylidae				
Tylos punctatus (whole body) (µmol.g⁻¹ wet wt) ☐	1.1 ± 0.2 (6)	0.13 ± 0.04 (6)	ND (6)	
Helleria brevicornis (whole body) (µmol.g⁻¹ wet wt) ●	111.6 ± 25.2 (4)			
Alloniscidae				
Alloniscus perconvexus (whole body) (µmol.g⁻¹ wet wt) ☐	1.7 ± 0.2 (22)	0.08 ± 0.01 (22)	ND (22)	
Porcellionidae				
Porcellio dilatatus (whole body) (µmol.g⁻¹ wet wt) ●	0.24 ± 0.05 (14)	0.024 ± 0.005 (14)	ND (14)	
Porcellionides floria (whole body) (µmol.g⁻¹ wet wt) ●	3.1 ± 0.8 (9)	0.021 ± 0.01 (9)	ND (9)	
Armadillidiidae				
Armadillidium vulgare (whole body) (µmol.g⁻¹ wet wt) ●	1.5 ± 0.1 (143)	0.11 ± 0.01 (143)	ND (143)	
Armadillidae				
Venezillo arizonicus (µmol.g⁻¹ wet wt) ●	6.95 ± 0.48 (6)			
Decapoda				
Brachyura				
Gecarcoidea natalis (whole body) (µmol.g⁻¹ dry wt) ●	27.5 ± 4.9 (10)–115.6 (10)			Linton and Greenaway (1997b, 2000)
Gecarcinus lateralis (whole body) (µmol.g⁻¹ dry wt) ●	78.5 ± 20.8 (9)–198.1 ± 20.8 (10)			Wolcott and Wolcott (1984)

				Reference
Cardisoma guanhumi (whole body) (µmol.g⁻¹ dry wt) ☐	17.9±6.5 (11)–367.6±25.0 (11)			Wolcott and Wolcott (1987)
Cardisoma guanhumi (whole body) (µmol.g⁻¹ wet wt) ☐	136.4±69.5 (3)			Gifford (1968)
Austrothelphusa transversa (whole body) (µmol.g⁻¹ wet wt) ☐	19.1±9.3 (3)	0.033±0.012 (3)	1.1±0.6 (3)	Linton and Greenaway (1995)
Anomura				
Birgus latro ●				
Midgut gland (µmol.g⁻¹ tissue wet wt)	65.7±12.0 (5)		26.1±3.9 (5)	Linton et al. (2005)
Muscle (µmol.g⁻¹ tissue wet wt)	35.0±9.1 (5)		15.1±3.4 (5)	Linton et al. (2005)
Spongy connective tissue (µmol.g⁻¹ tissue wet wt)	169.6±24.5 (5)		23.3±6.7 (5)	Linton et al. (2005)
Coenobita brevimanus ●				
Midgut gland (µmol.g⁻¹ tissue wet wt)	5.2±1.9 (3)		ND (3)	Linton et al. (2005)
Muscle (µmol.g⁻¹ tissue wet wt)	24.6±5.9 (3)		ND (3)	Linton et al. (2005)
Spongy connective tissue (µmol.g⁻¹ tissue wet wt)	266.1±62.5 (3)		ND (3)	Linton et al. (2005)
Dendrobranchiata				
Marsupenaeus japonicus ○				
Midgut gland (µmol.g⁻¹ tissue wet wt) (control animals)	1.5±0.3 (6)–3.1±0.1 (8)			Cheng et al. (2004, 2005)
Epidermis (µmol.g⁻¹ tissue wet wt) (control animals)	0.7 (8)–0.8±0.1 (6)			Cheng et al. (2004, 2005)

Data are expressed as mean ± sem (n). Urate content expressed as either whole-body content or tissue content and as either µmol per gram dry weight or µmol per gram wet weight

ND not detected

gecarcinid crabs, with the notable exception of the tylid *Helleria brevicornis* (see below). As with the decapods, there is a general trend of increasing purine storage in more terrestrial species. The urate forms discrete plaques in the epidermis, often visible in the integument as yellow spots; microdissection and assay of these plaques in *Armadillidium vulgare* (Howe and Wright, 2009, unpublished) confirm their urate composition. Urate plaques are primarily localised in the dorsal integument, and this tissue contained 93.4 % of whole-animal urate in *Oniscus asellus* (Hartenstein 1968). A small but measureable fraction of urate is present in aqueous solution; haemolymph urate in *Armadillidium vulgare* = 2.3 ± 0.9 µmol L^{-1} ($n = 5$).

Helleria brevicornis inhabits Mediterranean oak woodland and is the only terrestrial species in the primarily marine-littoral family Tylidae. Unlike other isopod species, *Helleria brevicornis* accumulates substantial whole-animal urate stores, localised in the dorsal body wall (◻ Table 2.3). The nitrogen storage in these deposits (µmol.g^{-1}) is two orders of magnitude larger than in other isopods that have been studied and comparable to that of the gecarinid crabs (◻ Table 2.3). Faecal urate excretion in *Helleria brevicornis* is very modest, comprising 0.157 µmol.g^{-1}.d^{-1} ($n = 3$ animals) or only about 3 % of total ammonia excretion over the same period (4.2 ± 0.96 µmol.g^{-1}.d^{-1}; $n = 35$ collection trials with seven animals), so like other isopods, this species remains primarily ammonotelic. Its substantial urate deposits may function in osmoregulation, sequestering haemolymph cations during dehydration as shown in the American cockroach *Periplaneta americana* (Hyatt and Marshall 1985a, b).

2.4.3 Origin and Synthesis of the Urate

De Novo Synthesis of Urate from Excess Dietary Nitrogen

It is clear, at least in gecarcinid crabs, that urate within the purine deposits is synthesised de novo from excess nitrogen assimilated from the diet. Gecarcinid crabs (*Gecarcoidea natalis, Gecarcinus lateralis* and *Cardisoma guanhumi*) fed a high-nitrogenous diet of leaves supplemented with either soya beans or casein assimilated more nitrogen and had a higher total body urate content than that of animals fed low-nitrogenous diet of leaf litter (Wolcott and Wolcott 1984; Wolcott and Wolcott 1987; Linton and Greenaway 1997b). Thus, uric acid levels are correlated with the nitrogen consumed, and this suggests that excess dietary nitrogen is incorporated into urate (Wolcott and Wolcott 1984, 1987; Linton and Greenaway 1997b). *Birgus latro* is also capable of purine de novo synthesis given its main nitrogenous excretory products are uric acid and guanine (Greenaway and Morris 1989; Linton et al. 2005). De novo synthesis of urate from dietary amino acid nitrogen has also been demonstrated directly from the incorporation of heavy nitrogen (N^{15}) from dietary glycine (Linton and Greenaway 1997b). In addition to *Birgus latro* and gecarcinid crabs, de novo synthesis of purines occurs in *Artemia salina* (Liras et al. 1992), and this suggests that the capacity for purine synthesis may be present in Crustacea generally, overturning previous dogma that crustaceans lacked this ability (Claybrook 1983).

In gecarcinids and *Birgus latro*, uric acid is apparently synthesised primarily in the midgut gland or hepatopancreas. This tissue possesses high activities of xanthine oxidoreductase, the last enzyme involved in urate synthesis (◻ Fig. 2.7; ◻ Table 2.4). Urate is then probably transported via the haemolymph to spongy connective tissue where it is stored

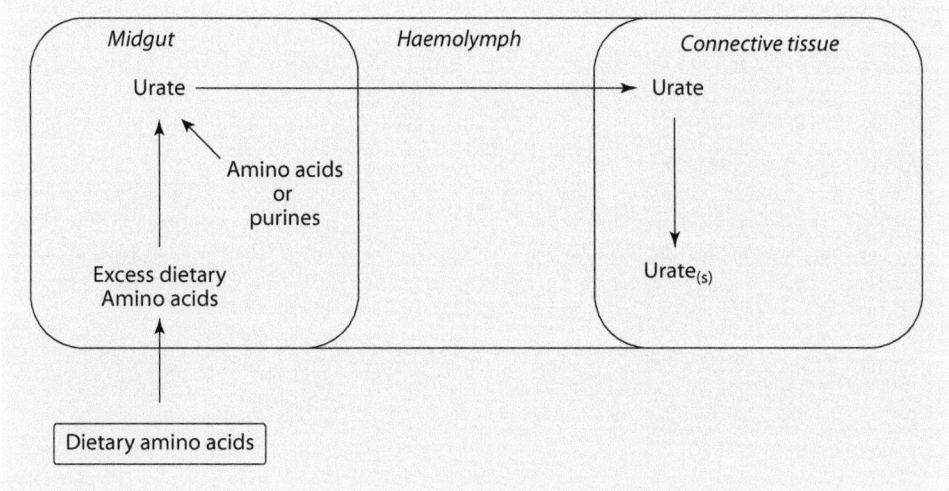

■ **Fig. 2.7** Synthesis and storage of urate within the tissues of *Gecarcoidea natalis*. Urate is synthesised *de novo* from excess dietary amino acids within the midgut gland. It is then transported to the spongy connective tissue, via the haemolymph, where it is precipitated as rounded intracellular crystals. Urate may also be synthesised from endogenous amino acids or purines

intracellularly (■ Fig. 2.7) (Linton and Greenaway 1998). The form in which urate is transported is unknown; it is unlikely to be as substantial amounts of a salt given the low solubility of sodium urate. Spongy connective tissue of *Gecarcoidea natalis* also possesses some xanthine oxidoreductase activity (■ Table 2.4), although its role in urate synthesis is unclear given its lower mass and lower xanthine oxidoreductase activities compared to the midgut gland (Linton and Greenaway 1998).

Urate deposits in isopods are likely to be synthesised de novo given that the enzymes involved in urate synthesis appear to be present in Crustacea generally (Liras et al. 1992; Linton and Greenaway 1997b). Xanthine oxidoreductase activity was not detected in *Oniscus asellus* (Hartenstein 1968), but given the significant activities measured in other crustaceans, this requires re-examination. Like the decapods, uric acid synthesis may occur in the midgut gland ('hepatopancreas') with storage (and possible catabolism) occurring in somatic epidermis and/or associated connective tissues ('body wall'). *Armadillidium vulgare* sampled from the same field population showed significant seasonal variation in urate levels (Howe and Wright, 2009 unpublished data), with a peak in September (North American Autumn) ($3.6 \pm 0.5\ \mu mol.g^{-1}$) and lowest levels observed in June (North American Summer) ($0.44 \pm 0.05\ \mu mol.g^{-1}$), indicating long-term synthesis and breakdown in accordance with some physiological role.

Form of Xanthine Oxidoreductase in Crustaceans

Xanthine oxidoreductase is the last enzyme in the metabolic pathway for the synthesis of uric acid. It catalyses the oxidation of hypoxanthine to xanthine and, subsequently, the oxidation of xanthine to uric acid (■ Fig. 2.8). Xanthine oxidoreductase can exist in two forms, either as an oxidase or a dehydrogenase. In the reaction catalysed by xanthine oxidase, oxygen accepts the electrons produced to form hydrogen peroxide (■ Fig. 2.8). Catalase, within peroxisomes, converts the hydrogen peroxide to water. In contrast, in the reaction catalysed by the dehydrogenase, NAD^+ is the electron acceptor which is

▢ Table 2.4 Xanthine oxidoreductase activities within the tissues of Crustacea

Species	Tissue	XDH	XO	References
Terrestrial species				
Gecarcoidea natalis	Midgut gland	58.87±4.76 (10)	ND	Linton and Greenaway (1998)
Gecarcoidea natalis	Branchiostegite (spongy connective tissue)	22.10±4.79 (10)	ND	Linton and Greenaway (1998)
Cardisoma guanhumi	Midgut gland	20.6±7.7 (7)	2.9±1.3 (2)	Lallier and Walsh (1991)
Gecarcinus lateralis	Midgut gland	1.7±1.3 (8)	2.2±1.1 (8)	Lallier and Walsh (1991)
Birgus latro	Midgut gland (reducing agent dithioerythritol + serine protease inhibitor (phenylmethylsulfonyl fluoride)	77.7±7.6 (5)	ND	Dillaman et al. (1999)
Birgus latro	Midgut gland (no reducing agent or protease inhibitor)	40.5±4.57 (3)	30.41±4.51 (3)	Dillaman et al. (1999)
Birgus latro	Midgut gland (no reducing agent or protease inhibitor)	ND	65.2±15.2– 81.5±32.4	Greenaway and Morris (1989)
Aquatic species				
Menippe mercenaria	Midgut gland	41.1±12.7 (11)	11.3±5.2 (11)	Lallier and Walsh (1991)
Callinectes sapidus	Midgut gland	17.5±7.2 (7)	3.4±0.7 (11)	Lallier and Walsh (1991)

Xanthine oxidoreductase activities are expressed as either xanthine dehydrogenase (XDH) or xanthine oxidase (XO). Xanthine dehydrogenase utilises NAD^+ as the electron acceptor, while xanthine oxidase utilises oxygen as the electron acceptor. Activities expressed as nmol urate produced.g^{-1} tissue wet weight.min^{-1}. Data expressed as mean ± sem (n)
ND not detected

reduced to NADH (▢ Fig. 2.8). NADH can then be oxidised in the electron transport chain (Kooij 1994). Xanthine oxidoreductase predominantly exists as a dehydrogenase, but can be converted into the oxidase form by limited proteolytic cleavage, catalysed by serine proteases such as trypsin and chymotrypsin and/or via oxidation (Della Corte and Stirpe 1968, 1972; Kooij 1994). Proteolysis can be inhibited with the serine proteinase inhibitor phenylmethylsulfonyl fluoride (PMSF), while oxidation can be prevented with a reducing agent such as dithioerythritol (DTT) or β-mercaptoethanol (Della Corte and Stirpe 1972). In Crustacea, xanthine oxidoreductase probably exists as the dehydrogenase. Xanthine dehydrogenase activities have been measured in both *Gecarcoidea natalis* and *Birgus latro* when the assay conditions have included both PMSF and dithioerythritol (▢ Table 2.4) (Linton and Greenaway 1998; Dillaman et al. 1999).

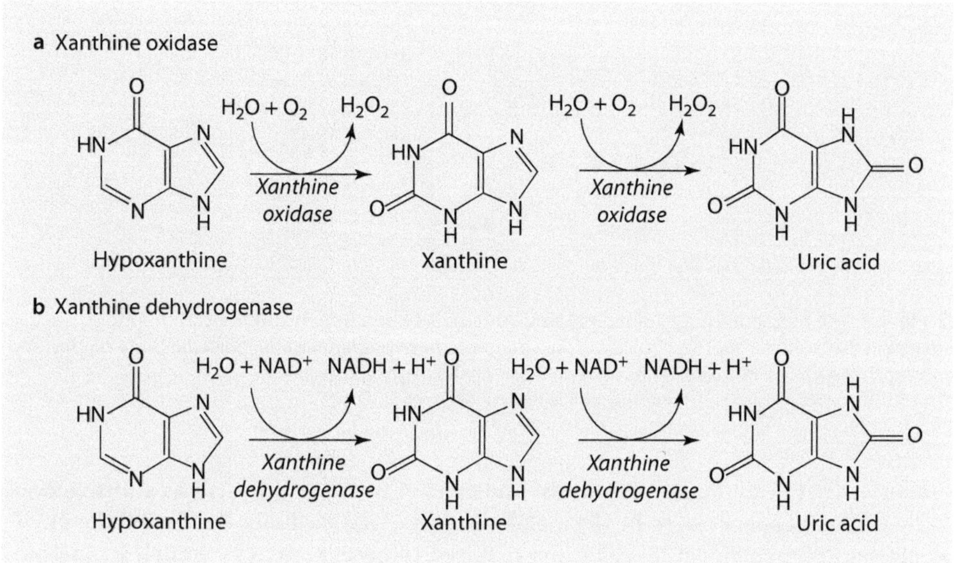

a Xanthine oxidase

b Xanthine dehydrogenase

☐ **Fig. 2.8** Oxidation reactions catalysed by the two forms of xanthine oxidoreductase, xanthine oxidase (**a**) and xanthine dehydrogenase (**b**). Xanthine oxidase uses oxygen as the electron acceptor, while xanthine dehydrogenase uses NAD^+ as the electron acceptor in the oxidation/reduction reaction

If these agents are omitted, the enzyme is converted to the oxidase form (Dillaman et al. 1999). Xanthine dehydrogenase is also the dominant form of xanthine oxidoreductase within the midgut gland of *Gecarcinus lateralis* and *Cardisoma guanhumi* (☐ Table 2.4) (Lallier and Walsh 1991). Xanthine oxidoreductase is a cytosolic enzyme, and it is likely that it is present in the cytoplasm of the midgut gland cells (Ichikawa et al. 1992; Kooij 1994; Dillaman et al. 1999).

Metabolic Cost of Synthesising Urate de novo

Xanthine oxidoreductase as a dehydrogenase has dramatic implications for the metabolic cost of urate synthesis. It means that synthesis of 1 molecule of uric acid requires 4 ATP, and this is much cheaper than previously thought. Of the four nitrogen atoms in the purine ring, two originate from glutamine, one from glycine and one from aspartate (☐ Fig. 2.9) (Berg et al. 2012). Of the five carbon atoms, two originate from glycine, two from N^{10} formyl tetrahydrofolate and one from bicarbonate (HCO_3^-) (☐ Fig. 2.9) (Berg et al. 2012). Glutamine and aspartate are catalytic compounds since only their amino groups ($-NH_3^+$) are incorporated into the purine ring (Berg et al. 2012). Their carbon skeletons can enter other metabolic pathways. Also since these amino acids are probably derived primarily from the diet, their cost of synthesis is not borne by the animal. Hence, they should not be included in the overall cost of purine synthesis.

Uric acid de novo synthesis begins with the synthesis of hypoxanthine via inosine monophosphate. As per the conventional purine de novo synthetic pathway, this requires 9 ATP (Berg et al. 2012). Hypoxanthine is then oxidised to xanthine and subsequently to urate. These reactions are catalysed by xanthine oxidoreductase. As the dehydrogenase, these reactions would produce two molecules of NADH. Assuming a phosphate:oxygen ratio of 2.5, these can be oxidised by the electron transport chain to produce 5 ATP (Berg et al. 2012). Hence, the net cost of urate synthesis is 9 ATP- 5 ATP = 4 ATP (Berg et al. 2012). The cost expressed per nitrogen molecule is 1 ATP,

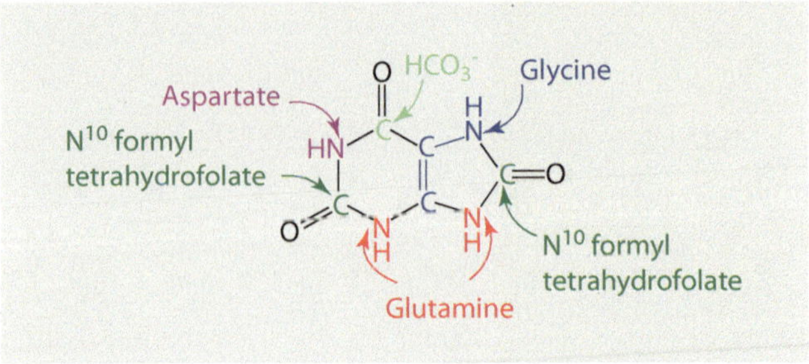

Fig. 2.9 The metabolic origin of atoms within uric acid. Four nitrogen and five carbons atoms are present in the purine ring. The nitrogen atoms originate from two glutamine molecules, one glycine and one aspartate. Of the five carbon atoms, two originate from glycine, two from N^{10} tetrahydrofolate and one from bicarbonate (Berg et al. 2012)

similar to that of glutamine synthesis. The cons of utilising uric acid as a nitrogenous waste product appear not to be the metabolic cost of its synthesis, but rather, the physiological and biochemical mechanisms required to precipitate it as complex spheroids and the metabolic costs of either storing it as a solid or to secreting it into the gut for excretion by the animal.

2.4.4 Hypothesised Function of the Urate Deposits: Storage Excretion or Temporary Nitrogen Store

Evidence Supporting Storage Excretion

The function of the urate deposits in Crustacea has not been clearly elucidated. They have generally been hypothesised to be involved in either storage excretion of nitrogenous wastes or as a nitrogen store. As a nitrogen store, urate could be degraded and utilised for synthesis of amino acids during situations of high nitrogen demand or at times of inadequate dietary nitrogen intake. There is substantial evidence against the urate being a nitrogen store. In *Gecarcoidea natalis* at least, storage excretion appears to be the most likely function given the presence of uricolytic enzymes within the tissues.

Uricase and urease are, respectively, the first and last enzymes involved in catabolising uric acid to ammonia (◘ Fig. 2.10). Uricase catalyses the oxidation of uric acid to form allantoin, while urease hydrolyses urea to ammonia (◘ Fig. 2.10). The branchiostegite, which is composed almost solely of spongy connective tissue, contains the highest uricase activities within *Gecarcoidea natalis* (◘ Tables 2.5) (Linton and Greenaway 1998). Uricase is also present in the midgut gland and gill epithelium, but activities in these tissues are lower and variable (◘ Table 2.4). Only 11 % of the animals assayed contained uricase activities within their midgut gland, while only 37.5 % of the animals assayed contained uricase activities within their gills (◘ Table 2.4) (Linton and Greenaway 1998). High urease activities were present in the gills of *Gecarcoidea natalis* in 100 % of the animals assayed (◘ Table 2.4). Urease activities were also present in midgut gland, but were lower and variable (present in only 75 % of the animals assayed) (◘ Table 2.4) (Linton and Greenaway 1998). The position of uricase suggests that uric acid catabolism in *Gecarcoidea natalis* probably begins in the spongy connective tissue (◘ Fig. 2.11). It is probably degraded to

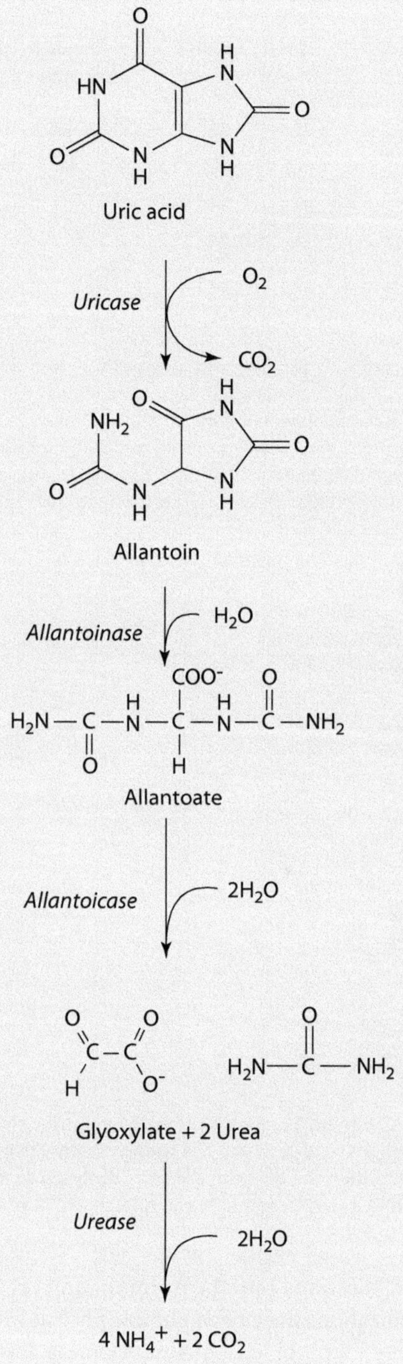

☐ **Fig. 2.10** Enzymatic steps and metabolites produced during the catabolism of urate by the uricolytic pathway

Table 2.5 Activities of enzymes involved in the catabolism (uricase and urease) of uric acid

	Midgut gland	Branchiostegite (spongy connective tissue)	Muscle	Gill
Uricase activity	3.66 ± 3.66^{b} (9)	$48.44 + 4.29^{a}$ (8)	ND (8)	13.51 ± 8.47^{b} (8)
% Animals with uricase	11.1	100		37.5
Urease activity	88.28 ± 78.35^{a} (8)	ND (8)	ND (8)	365.31 ± 37.21^{b} (8)
% Animals with urease	75			100

Reproduced from Linton and Greenaway (1998) with permission of Springer Science + Business media.
Enzyme activities (mean ± sem (n)) are expressed as the amount (nmoles) of substrate consumed. g^{-1} wet weight.min^{-1}. For each enzyme, the percentage of animals that activities were detected in is also given. Within a row, different superscript letters (a, b) indicate that the mean enzyme activities differed between tissues. ND enzyme activity not detected

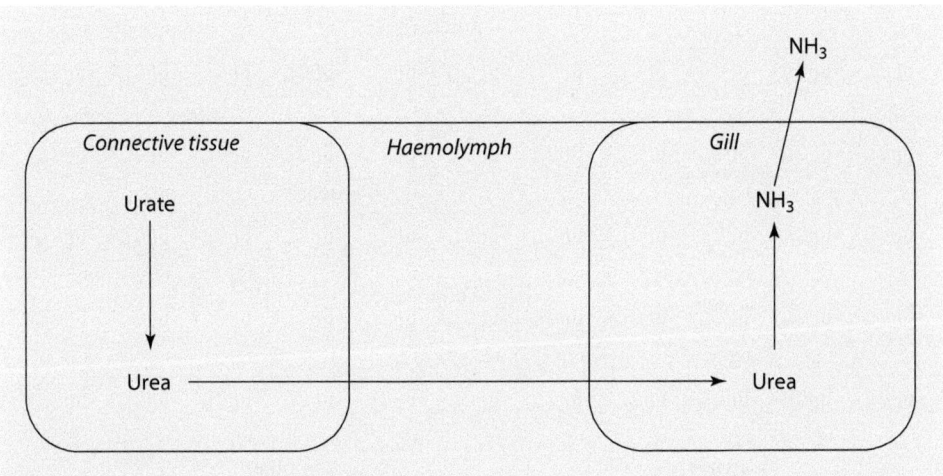

Fig. 2.11 Catabolism of urate and the excretion of the resultant nitrogen by *Gecarcoidea natalis*. Uric acid is catabolised to urea within the spongy connective tissue; it is then transported to the gills via the haemolymph; within the gills, the urea is hydrolysed to ammonia which is subsequently secreted into the urine during its reprocessing in the branchial chambers

urea which is secreted into the haemolymph (■ Fig. 2.11). Consistent with this is that high urea concentrations have been commonly but inconsistently observed in the haemolymph of numerous amphibious (*Austrothelphusa transversa* (0.25 ± 0.5–8.45 ± 4.24 mmol.L^{-1}) and terrestrial crustaceans (*Gecarcoidea natalis* (0.9 ± 1.3 mmol.L^{-1}), *Cardisoma carnifex* (up to 3.8 mmol.L^{-1}) and *Ocypode quadrata* (6.52 ± 1.02 mmol.L^{-1}) (Wood et al. 1986; Greenaway and Nakamura 1991; De Vries et al. 1994; Linton and Greenaway 1995; Taylor and Greenaway 1994). Urea could then be taken up by the gills and hydrolysed to ammonia given the high urease activities within this tissue (■ Fig. 2.11). The ammonia produced could then be excreted (■ Fig. 2.11) (Greenaway 1991).

Other species of aquatic and terrestrial crustaceans possess uricolytic enzymes and thus the potential to degrade urate. Low uricase activities are present in the midgut gland of the aquatic crabs, *Menippe mercenaria* and *Callinectes,* although not in the midgut gland of the terrestrial gecarcinids, *Cardisoma guanhumi* and *Gecarcinus lateralis* (Lallier and Walsh 1991). Like *Gecarcoidea natalis*, uricase activities in *Gecarcinus lateralis* may be present in the spongy connective tissue and intermittently expressed in the midgut gland. Terrestrial isopods such as *Oniscus asellus* also possess the ability to degrade urate to ammonia within the hepatopancreas, intestine and body wall, since these tissues possess both uricase and urease activities and the ability to catabolise urate-2-C^{14} (Hartenstein 1968).

Conditions which induce accumulation and degradation of the uric acid deposits have yet to be conclusively demonstrated in any terrestrial crustacean. Dehydration in gecarcinids may induce urate storage, as urine flow ceases, and therefore there is no route for ammonia excretion during the reprocessing of the urine within the branchial chambers. The Australian arid zone crab, *Austrothelphusa transversa* undergoes long-term aestivation for up to 3 years during drought (MacMillen and Greenaway 1978). During this time, the animal would not be able to excrete ammonia into water and thus may store nitrogenous wastes as urate deposits. Accumulation of uric acid deposits during aestivation and subsequent degradation upon emergence has been demonstrated in the apple-snail, *Pomaecea canaliculata* (Giraud-Billoud et al. 2011), and this may serve as a model system for examining storage excretion in aestivating decapods. It is also suggested that the urate may act as an antioxidant during the reperfusion of tissues upon arousal (Giraud-Billoud et al. 2011).

While storage excretion is plausible in some land crabs, the purine deposits of terrestrial isopods probably serve a different function. With their efficient ammonia volatilization and potential to couple this to water vapour absorption, isopods seem unlikely to face long-term constraints on nitrogen excretion. Their small size and relatively high water loss rates also tend to restrict them to humid microclimates which favour volatilization. It is therefore unclear if and when storage excretion would be advantageous. Seven-day ammonia-loading trials with *Armadillidium vulgare* did not result in significant increases in whole-animal urate (Howe and Wright, 2009, unpublished data), in contrast to the significant effect of ammonia loading on glutamine levels (Wright et al. 1994). A comparison of whole-animal urate levels in animals ranging from 10.3 to 105.1 mg in mass ($n = 31$) did not reveal any significant increase in mass-specific urate levels (μmol g^{-1}), contrary to what would be expected if animals utilise urate as a form of long-term storage excretion (Howe and Wright, 2009, unpublished data).

Possible Alternative Functions of Purine Stores

The low solubility of sodium and potassium urate (▶ Sect. 4.1) means that urate deposits could potentially function as cation stores. The American cockroach, *Periplaneta americana*, sequesters haemolymph Na^+ and K^+ as urate salts during dehydration (Hyatt and Marshall 1985a, b), and the large urate stores in *Helleria brevicornis* may function in a similar manner (▶ Sect. 4.2). The purine stores of other land isopods are too small to serve for osmoregulation (Wright et al. 1997). In freshwater isopods, urate cation stores could potentially be used to replenish haemolymph salts during periods of hypo-osmotic stress. *Asellus aquaticus* accumulates substantial uric acid stores (36–50 μmol g^{-1}) in the Zenker's organ (Dresel and Moyle 1950; Lockwood 1959; Wägele 1992) which probably serve such a function. The Zenker's organ cells contain 20–30 % of the total sodium in the animal, and the micromolar urate levels are sufficient to account for the sodium content assuming formation of the monovalent salt.

A further possible function of purine deposits is protection against oxidative damage. Uric acid is generally believed to serve as a key antioxidant at physiological concentrations in mammals (Becker 1993; Hooper 2000; Glantzounis et al. 2005) and may be an important antioxidant in crustaceans. The small urate stores in many species could serve as a mobilisable reserve to maintain stable aqueous concentrations. Several reactive oxygen species (ROS) are generated endogenously during cellular metabolism and include hydrogen peroxide (H_2O_2), superoxide ($\cdot O_2^-$) and hydroxyl ($\cdot OH$) radicals. Ubiquitous superoxide dismutases catalyse the breakdown of superoxide, while peroxide is reduced by urate under hydrophilic conditions, as well as by vitamin C (ascorbic acid), glutathione and carotenoids. A significant antioxidant function of urate in isopods and decapods would be consistent with the higher levels found in more metabolically active terrestrial species. Such an antioxidant system would aid in the preservation of tissues that would be required for the longevity of the gecarcinids ($\approx$30 years) and *Birgus latro* (50–100 years) (Wolcott 1988; Drew et al. 2013; Sato et al. 2013).

Evidence Against a Temporary Nitrogen Store

Urate deposits have also been hypothesised to be a temporary nitrogen store which is degraded and utilised during situations of high nitrogen demand such as oogenesis and moulting and with inadequate intake of dietary nitrogen. During oogenesis, there is substantial nitrogen investment into the ovaries and eggs. In *Gecarcoidea natalis*, mature ovaries and eggs constitute 9.3 ± 0.4 and $13.4 \pm 1.1\%$ of total body nitrogen, respectively (Linton and Greenaway 2000). In premoult, 30–60 % of the muscle within the chelae of gecarcinid crabs is degraded; this allows the chelae to be withdrawn through the basal part of the ischium during ecdysis (Skinner 1966; Mykles and Skinner 1990; Mykles 1992). As protein cannot be stored, it must be catabolised and the resulting nitrogen excreted. This would induce a nitrogen debt which must be repaid when the muscle is resynthesised during postmoult. A new cuticle is also synthesised during pre- and postmoult. Both oogenesis and moulting would incur a substantial nitrogen debt which must be compensated. Gecarcinid crabs are mainly herbivorous, feeding on mostly fallen green and brown leaf litter (Greenaway and Raghaven 1998). During oogenesis and moulting, the animal may not be able to meet its nitrogen requirements from this low-nitrogen diet and hence may require a nitrogen reserve. The urate could be degraded and the nitrogen released incorporated into amino acids. However, the animal would only be able to synthesise non-essential amino acids from urate nitrogen and could not meet total amino acid requirements from this store.

There is substantial evidence against such a function in *Gecarcoidea natalis*. In crabs fed on a nitrogen-free but energy-rich diet for an extended period (6 weeks), the body or non-urate nitrogen content of the crabs decreased while urate content increased (◘ Fig. 2.12) (Linton and Greenaway 2000). On this diet, the animal can meet its energy but not its nitrogen requirement. The decrease in non-urate nitrogen is thought to represent protein turnover and thus is an estimate of the animal's minimum nitrogen requirement (4.83 ± 1.68 mmol N. g^{-1} dry wt. d^{-1}). The increase in urate content is contrary to the hypothesis that it represents a nitrogen store (Linton and Greenaway 2000). The nitrogen within the newly synthesised urate probably originates from the catabolised body protein, and again this suggests that the urate deposits may represent storage of nitrogenous wastes (Linton and Greenaway 2000). In *Gecarcoidea natalis* undergoing early oogenesis and fed

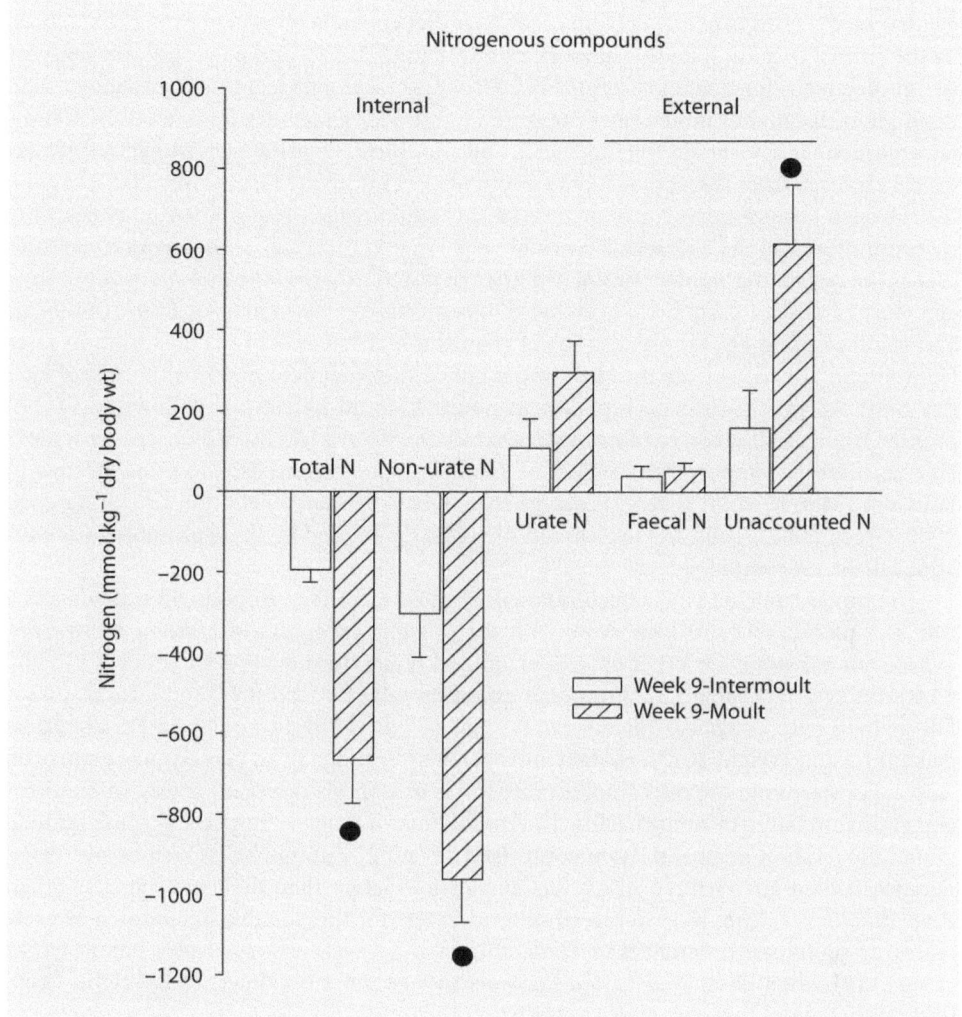

☐ **Fig. 2.12** Changes in the content of nitrogenous compounds (total nitrogen, non-urate nitrogen and urate nitrogen) of *Gecarcoidea natalis* maintained on a nitrogen-free but energy-rich diet for 9 weeks. Unaccounted nitrogen is presumably the nitrogen excreted by the animal. Week-9 intermoult: crabs that did not moult during the dietary period ($n = 7$). Week-9 moult: crabs that moulted during the dietary period. Closed circles indicate that the means for the week-9 moult differed significantly from the week-9 intermoult (1-way ANOVA $P < 0.05$) (Figure reproduced with permission from Linton and Greenaway (2000))

a leaf litter diet for 21 days, there was no change in the animal's body non-urate nitrogen and urate content (Linton and Greenaway 2000). Thus, the crabs were in nitrogen balance and were able to synthesis ovaries using nitrogen assimilated from the diet. There is no evidence that urate was degraded and the nitrogen utilised in oogenesis.

It appears that *Gecarcoidea natalis* does not require a nitrogen reserve as the animal is able to meet its nitrogen requirements for metabolism, growth and reproduction from a leaf litter diet. The intermoult minimum nitrogen requirement is 4.83 ± 1.68 mmol N.

g^{-1} dry wt. d^{-1}. This can be met from a leaf litter diet given the dry matter intakes required by the animal to assimilate this amount of nitrogen (5.2, 13.9 and 6.6 g kg^{-1} dry body wt d^{-1}, respectively, for intakes of *Erythrina*, *Ficus* and rainforest leaf litters) are lower than the highest dry matter intake rates measured (15.0–19.2 g kg^{-1} dry body wt d^{-1}). A similar argument is also made for oogenesis and moulting. Ovaries are synthesised slowly over 3 months, from the end of July to September (Linton and Greenaway 2000). Again, the nitrogen requirements for oogenesis (minimum nitrogen requirement plus rate of N incorporation into the ovaries = 8.6 mmol N. g^{-1} dry wt. d^{-1}) can be met from a leaf litter diet as the dry matter intake rates of leaf litter required to assimilate this amount of nitrogen (9.2, 24 and 11 g kg^{-1} dry body wt d^{-1}, respectively, for intakes of *Erythrina*, *Ficus* and rainforest leaf litters) are also lower than the highest measured consumption rates (Linton and Greenaway 2000). Moulting incurs a nitrogen debt of 658 ± 126 mmol kg^{-1} dry body wt which must be repaid postmoult. It could be recouped slowly over 1–3 months from small excesses of nitrogen assimilated from a leaf litter diet. This is achievable as there is temporal separation of moulting and oogenesis. Moulting occurs in January to March, while oogenesis occurs from July to September (Linton and Greenaway 2000; Green 2004). Thus, the increase in nitrogen demanded by these physiological situations does not overlap.

The modest purine stores accumulated by most terrestrial isopods do not suggest a role as a mobilisable nitrogen store, with the possible exception of *Helleria brevicornis*. Whole-animal stores for other species examined range from 0.24 to 3.1 μmol g^{-1} (0.96–12.4 μmol N g^{-1}; ◘ Table 2.3). Ammonia excretion rates for terrestrial and littoral species fall in the range of approximately 5–15 μmol g^{-1} d^{-1} (Wright and Peña-Peralta 2005; Nakamura and Wright 2013), so daily nitrogen turnover equals or exceeds urate nitrogen stores. Furthermore, the micromolar urate levels in isopods represent a very small nitrogen reserve relative to amino acids. In *Armadillidium vulgare*, respective whole-animal glutamine levels at dusk and dawn comprised 52 and 214 μmol N g^{-1} (Wright and Peña-Peralta 2005), a nitrogen repository 17- and 71-fold larger than their urate stores. When *Armadillidium vulgare* were deprived of food to test for the possible breakdown of urate stores, no significant differences in whole-animal urate levels were seen after 1 and 2 weeks ($n = 21$, 19) when compared to fed controls ($n = 19$, $n = 15$) (Howe and Wright, 2009, unpublished data).

2.5   Conclusion

Terrestrial crustaceans have largely retained the ancestral pattern of ammonia excretion. Amphibious species (T_{1-3}) excrete ammonia into water during periodic immersion (◘ Fig. 2.13). Terrestrial species (T_4) excrete ammonia into an excretory fluid; either in urine produced by the antennal gland (*Ocypode*) or 'P', urine which is passed into the branchial chamber for reprocessing (mainly salt reabsorption) (Gecarcinids and grapsid crabs) (◘ Fig. 2.13). Isopods (T_4) excrete ammonia into branchial fluid and may couple this to water vapour absorption. Both grapsids (T_4) and isopods (T_4) are able to volatilise ammonia. Only one terrestrial crustacean, the robber or coconut crab, *Birgus latro* (T_4), is purinotelic, like insects and arachnids (◘ Fig. 2.13); *Birgus latro* excretes a white faecal pellet which consists of a mixture of guanine and uric acid. Despite this difference, *Birgus latro* has a number of characters of purine metabolism which are shared by other terrestrial crustaceans. These include an overproduction of purine and storage within the

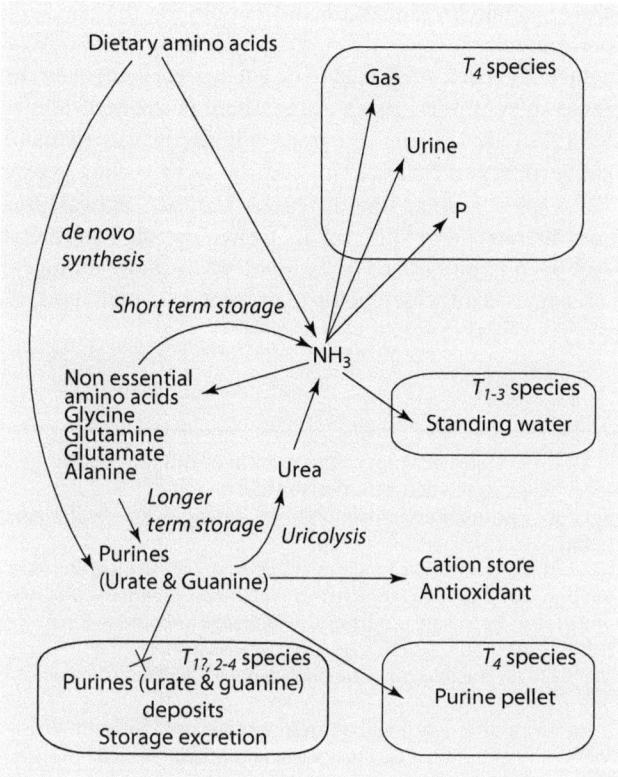

◘ Fig. 2.13 Overview of the nitrogenous waste metabolism of amphibious (T_{1-3} species) and terrestrial (T_4 species) crustaceans. Catabolism of excess dietary protein results in the production of ammonia. In the majority of amphibious and terrestrial crustaceans, the ammonia is then excreted. The ammonia may be excreted into standing water, as a gas, within the urine produced by the antennal gland or within the 'P' (fluid released after the urine is reprocessed by the gills within the branchial chambers). Purinotely has evolved in one species, the robber crab, *Birgus latro*. It excretes a white purine faecal pellet. Within the animal, nitrogenous wastes may be temporarily stored within non-essential amino acids (short term) or within purine deposits (long term). These nitrogenous compounds could potentially be catabolised to ammonia. Purine deposits may also represent storage secretion, cation storage during dehydration or an antioxidant to prevent oxidative damage. *Rounded rectangles* represent end products of nitrogenous waste metabolism

spongy connective tissue. Differences include guanine production and excretion of purine into the lumen of the midgut.

Crustaceans store nitrogenous wastes as non-toxic intermediates between excretory bouts. In the species so far studied, non-essential amino acids such as glycine, glutamine, glutamate and alanine appear to be the main compounds accumulated (◘ Fig. 2.13).

Urate deposits are present in numerous terrestrial decapod and isopod species. Urate in crabs is synthesised de novo from excess dietary nitrogen and stored intracellularly as round membrane-bound spherules within spongy connective tissue (◘ Fig. 2.13). Circumstantial evidence suggests that these deposits also represent storage of nitrogenous wastes (◘ Fig. 2.13). They may be temporary as crustaceans possess the ability to catabolise these deposits via the uricolytic pathway and excrete the resulting nitrogen

(◼ Fig. 2.13). However, catabolism of urate and conditions that cause it have yet to be demonstrated experimentally.

Urate deposits have also been suggested to be a temporary nitrogen store which can be degraded and utilised to synthesis amino acids. There is strong evidence against this as urate deposits are not degraded during situations of high nitrogen demand (oogenesis and moulting) or negative nitrogen balance. Indeed the herbivorous gecarcinid land crab, *Gecarcoidea natalis*, (a species with a low-nitrogenous intake) appears not to require such a store as it can meet its nitrogen requirements from a mainly leaf litter diet. In isopods, the urate stores have been hypothesised to be involved in cation storage during dehydration and perhaps an antioxidant which helps to prevent oxidative tissue damage.

References

Aylward GH, Findlay TJV (1971) SI chemical data. John Wiley and Sons, Gladesville

Bancroft JD (1975) Histochemical techniques. Butterworths, London

Battison AL (2013) Subcuticular urate accumulation in an American lobster *(Homarus americanus)*. Vet Pathol 50(3):451–456

Becker BF (1993) Towards the physiological function of uric acid. Free Radical Biol Med 14(6):615–631

Berg J, Tymoczko J, Stryer L, Gatto GJ (2012) Biochemistry, 7th edn. Freeman and Company, New York

Bishop J (1963) The Australian freshwater crabs of the family Potamonidae (Crustacea: Decapoda). Mar Freshw Res 14(2):218–238

Broly P, Deville P, Maillet S (2013) The origin of terrestrial isopods (Crustacea: Isopoda: Oniscidea). Evol Ecol 27:461–476

Campbell JW (1995) Excretory nitrogen metabolism in reptiles and birds. In: Walsh PJ, Wright P (eds) Nitrogen metabolism and excretion. CRC Press, Boca Raton, pp 147–178

Cheng SY, Lee WC, Chen JC (2004) Increase of uricogenesis in the kuruma shrimp *Marsupenaeus japonicus* reared under hyper-osmotic conditions. Comp Biochem Physiol B Biochem Mol Biol 138(3):245–253

Cheng SY, Lee WC, Chen JC (2005) An increase of uricogenesis in the kuruma shrimp *Marsupenaeus japonicus* under nitrite stress. J Exp Zool A Comp Exp Biol 303(4):308–318

Chew SF, Ip YK (2014) Excretory nitrogen metabolism and defence against ammonia toxicity in air-breathing fishes. J Fish Biol 84(3):603–638

Claybrook DL (1983) Nitrogen metabolism. In: Mantel LH (ed) The biology of crustacea, volume 5 internal anatomy and physiological regulation, vol 5. Academic, New York, pp 163–213

Dall W, Moriarty DJ (1983) Functional aspects of nutrition and digestion. In: Mantel LH (ed) The biology of crustacea, vol 5. Academic, New York, pp 215–261

Davie PJF (2002) Crustacea: Malacostraca: Eucarida (Part 2): Decapoda- Anomura, Brachyura. In: Wells A, Houston WWK (eds) Zoological catalogue of Australia, vol 19.3B. CSIRO Publishing, Melbourne, p xiv 641

De Vries MC, Wolcott DL (1993) Gaseous ammonia evolution is coupled to reprocessing of urine at the gills of ghost crabs. J Exp Zool 267(2):97–103

De Vries MC, Wolcott DL, Holliday CW (1994) High ammonia and low pH in the urine of the ghost crab, *Ocypode quadrata*. Biol Bull 186(3):342–348

Dela-Cruz J, Morris S (1997a) Respiratory, acid–base, and metabolic responses of the Christmas Island blue crab, *Cardisoma hirtipes* (Dana), during simulated environmental conditions. Physiol Zool 70(1):100–115

Dela-Cruz J, Morris S (1997b) Water and ion balance and nitrogenous excretion as limitations to terrestrial excursion in the Christmas Island Blue Crab, *Cardisoma hirtipes* (Dana). J Exp Zool 279(6):537–548

Della Corte E, Stirpe F (1968) The regulation of rat-liver xanthine oxidase: activation by proteolytic enzymes. FEBS Lett 2(2):83–84

Della Corte E, Stirpe F (1972) The regulation of rat liver xanthine oxidase. Involvement of thiol groups in the conversion of the enzyme activity from dehydrogenase (type D) into oxidase (type O) and purification of the enzyme. Bio J 126:739–745

Dillaman RM, Greenaway P, Linton SM (1999) Role of the midgut gland in purine excretion in the robber crab, *Birgus latro* (Anomura: Coenobitidae). J Morphol 241(3):227–235

Dresel EIB, Moyle V (1950) Nitrogenous excretion of amphipods and isopods. J Exp Zool 27(2):210–225

Drew MM, Smith MJ, Hansson BS (2013) Factors influencing growth of giant terrestrial robber crab *Birgus latro* (Anomura: Coenobitidae) on Christmas Island. Aquat Biol 19(2):129–141

Durand F, Regnault M (1998) Nitrogen metabolism of two portunid crabs, *Carcinus maenas* and *Necora puber*, during prolonged air exposure and subsequent recovery: a comparative study. J Exp Zool 201(17):2515–2528

Durand F, Chausson F, Regnault M (1999) Increases in tissue free amino acid levels in response to prolonged emersion in marine crabs: an ammonia-detoxifying process efficient in the intertidal *Carcinus maenas* but not in the subtidal *Necora puber*. J Exp Zool 202(16):2191–2202

Gifford CA (1968) Accumulation of uric acid in the land crab, *Cardisoma guanhumi*. Am Zool 8(3):521–528

Giraud-Billoud M, Abud MA, Cueto JA, Vega IA, Castro-Vazquez A (2011) Uric acid deposits and estivation in the invasive apple-snail, *Pomacea canaliculata*. Comp Biochem Physiol B Biochem Mol Biol 158(4):506–512

Glantzounis GK, Tsimoyiannis EC, Kappas AM, Galaris DA (2005) Uric acid and oxidative stress. Curr Pharm Des 11:4145–4151

Graf F, Michaut P (1975) Chronologie du développement et évolution du stockage de calcium et des cellules à urates chez *Niphargus schellenbergi* Karaman. Int J Speleol 7:247–272

Green PT (2004) Field observations of moulting and moult increment in the red land crab, *Gecarcoidea natalis* (Brachyura, Gecarcinidae), on Christmas Island (Indian Ocean). Crustaceana 77(1):125–128

Green JW, Harsch M, Barr L, Prosser CL (1959) The regulation of water and salt by fiddler crab, *Uca pugnax* and *Uca pugilator*. Biol Bull 116(1):76–87

Greenaway P (1984) Survival strategies in desert crabs. In: Cogger HG, Cameron E (eds) Arid Australia. Australian Museum, Sydney, pp 145–152

Greenaway P (1989) Sodium balance and adaptation to fresh water in the amphibious crab *Cardisoma hirtipes*. Physiol Zool 62(3):639–653

Greenaway P (1991) Nitrogenous excretion in aquatic and terrestrial crustaceans. Mem Queensland Mus 31:215–227

Greenaway P (1994) Salt and water balance in field populations of the terrestrial crab, *Gecarcoidea natalis*. J Crustacean Biol 14(3):438–453

Greenaway P, Morris S (1989) Adaptations to a terrestrial existence by the robber crab, *Birgus latro*. L. III. Nitrogenous excretion. J Exp Biol 143(1):333–346

Greenaway P, Nakamura T (1991) Nitrogenous excretion in two terrestrial crabs *Gecarcoidea natalis* and *Geograpsus grayi*. Physiol Zool 64:767–786

Greenaway P, Raghaven S (1998) Digestive strategies in two species of leaf-eating land crabs (Brachyura: Gecarcinidae) in a rain forest. Physiol Zool 71(1):36–44

Hartenstein R (1968) Nitrogen metabolism in the terrestrial isopod *Oniscus asellus*. Am Zool 8:507–519

Hartnoll RG (1988) Evolution, systematics, and geographical distribution. In: Burggren WW, McMahon BR (eds) Biology of the land crabs. Cambridge University Press, Cambridge, pp 6–54

Henry RP, Cameron JN (1981) A survey of blood and tissue nitrogen compounds in terrestrial decapods of Palau. Journal of Experimental Zoology 218 (1):83–88. doi:10.1002/jez.1402180110

Henry RP, Cameron JN (1982) The distribution and partial characterization of carbonic anhydrase in selected aquatic and terrestrial decapod crustaceans. J Exp Zool 221:309–321

Hooper DC (2000) Uric acid, a peroxynitrite scavenger, inhibits CNS inflammation, blood-CNS barrier permeability changes, and tissue damage in a mouse model of multiple sclerosis. FASEB J 14(5):691–698

Horne FR (1968) Nitrogen excretin in crustacea-I. The herbivorous land crab *Cardisoma guanhumi* latreille. Comp Biochem Physiol 26(2):687–695

Hyatt AD, Marshall AT (1985a) Water and ion balance in the tissues of the dehydrated cockroach, *Periplaneta americana*. J Insect Physiol 31(1):27–34

Hyatt AD, Marshall AT (1985b) X-ray microanalysis of cockroach fat body in relation to ion and water regulation. J Insect Physiol 31(6):495–508

Ichikawa M, Nishino T, Nishino T, Ichikawa A (1992) Subcellular localization of xanthine oxidase in rat hepatocytes: high-resolution immunoelectron microscopic study combined with biochemical analysis. J Histochem Cytochem 40(8):1097–1103

Johnson PT (1980) Histology of the blue crab, *Callinectes sapidus*. A model for the decapoda. Praeger, New York

Kirby PK, Harbaugh RD (1974) Diurnal patterns of ammonia release in marine and terrestrial isopods. Comp Biochem Physiol A Physiol 47(4):1313–1322

Kooij A (1994) A re-evaluation of the tissue distribution and physiology of xanthine oxidoreductase. Histochem J 26(12):889–915

Lallier FH, Walsh PJ (1991) Activities of uricase, xanthine oxidase, and xanthine dehydrogenase in the hepatopancreas of aquatic and terrestrial crabs. J Crustacean Biol 11(4):506–512

Linton SM, Greenaway P (1995) Nitrogenous excretion in the amphibious crab *Holthuisana transversa* under wet and dry conditions. J Crustacean Biol 15(4):633–644

Linton SM, Greenaway P (1997a) Intracellular purine deposits in the gecarcinid land crab, *Gecarcoidea natalis*. J Morphol 231:101–110

Linton SM, Greenaway P (1997b) Urate deposits in the Gecarcinid land crab, *Gecarcoidea natalis* are synthesised *de novo* from excess dietary nitrogen. J Exp Zool 200:2347–2354

Linton SM, Greenaway P (1998) Enzymes of urate synthesis and catabolism in the Gecarcinid land crab *Gecarcoidea natalis*. Exper Biol Online 3:5

Linton SM, Greenaway P (2000) The nitrogen requirements and dietary nitrogen utilization for the gecarcinid land crab, *Gecarcoidea natalis*. Physiol Biochem Zool 73(2):209–218

Linton S, Wilde JE, Greenaway P (2005) Excretory and storage purines in the anomuran land crab *Birgus latro*: guanine and uric acid. J Crustacean Biol 25(1):100–104

Liras A, Rotllán P, Llorente P (1992) De novo purine biosynthesis in the crustacean Artemia: influence of salinity and geographical origin. J Comp Physiol B 162(3):263–266

Little C (1983) The colonisation of land. Origins and adaptations of terrestrial animals. Cambridge University Press, Cambridge

Lockwood APM (1959) The extra-haemolymph sodium of *Asellus aquaticus* (L.) J Exp Biol 36:562–565

MacMillen RE, Greenaway P (1978) Adjustments of energy and water metabolism to drought in an Australian arid-zone crab. Physiol Zool 51(3):230–240

McNabb RA, McNabb FMA (1977) Avian urinary precipitates: their physical analysis, and their differential inclusion of cations (Ca, Mg) and anions (Cl). Comp Biochem Physiol 56A:621–625

Moran LA, Horton RA, Scrimgeour G, Perry M (2014) Principles of biochemistry (Pearson New International Edition). Pearson Essex, UK

Morris S, Van Aardt WJ (1998) Salt and water relations, and nitrogen excretion, in the amphibious African freshwater crab *Potamonautes warreni* in water and in air. J Exp Zool 201(6):883–893

Mykles DL (1992) Getting out of a tight squeeze: enzymatic regulation of claw muscle atrophy in molting. Am Zool 32(3):485–494

Mykles DL, Skinner DM (1990) Atrophy of crustacean somatic muscle and the proteinases that do the job. A review. J Crustacean Biol 10(4):577–594

Nakamura M, Wright J (2013) Discontinuous ammonia excretion and glutamine storage in littoral Oniscidea (Crustacea: Isopoda): testing tidal and circadian models. J Comp Physiol B 183(1):51–59

Newsholme EA, Leech AR (1983) Biochemistry for the medical sciences. Wiley, Chichester

Ng PKL, Davie PJF (2012) The blue crab of Christmas Island, *Discoplax celeste*, new species (Crustacea: Decapoda: Brachyura: Gecarcinidae). Raffles Bulletin Zool 60(1):89–100

O'Donnell MJ, Wright JC (1995) Nitrogen excretion in terrestrial crustaceans. In: Walsh PJ, Wright P (eds) Nitrogen metabolism and excretion. CRC Press, Boca Raton, pp 105–118

Porter P (1963) Physico-chemical factors involved in urate calculus formation. Res Vet Sci 4:580–591

Sato T, Yoseda K, Abe O, Shibuno T, Takada Y, Dan S, Hamasaki K (2013) Growth of the coconut crab *Birgus latro* estimated from mark-recapture using passive integrated transponder (PIT) tags. Aquat Biol 19(2):143–152

Seegmiller JE (1980) Diseases of purine and pyrimidine metabolism. In: Bondy PK, Rosenburg LE (eds) Metabolic control and disease. Sauders and Company, Philadelphia, pp 777–937

Skinner DM (1966) Breakdown and reformation of somatic muscle during the molt cycle of the land crab, *Gecarcinus lateralis*. J Exper Zool 163(2):115–123

Spicer JI, Moore PG, Taylor AC (1987) The physiological ecology of land invasion by the Talitridae (Crustacea: Amphipoda). P Royal Soc London B Biol Sci 232(1266):95–124

Taylor HH, Greenaway P (1994) The partitioning of water and solutes during evaporative water loss in three terrestrial crabs. Physiol Zool 67(3):539–565

Varley D, Greenaway P (1994) Nitrogenous excretion in the terrestrial carnivorous crab *Geograpsus grayi*: site and mechanism of excretion. J Exp Zool 190(1):179–193

Vega IA, Giraud-Billoud M, Koch E, Gamarra-Luques C, Castro-Vazquez A (2007) Uric acid accumulation within intracellular crystalloid corpuscles of the midgut gland in *Pomacea canaliculata* (Caenogastropoda, Ampullariidae). Veliger 48(4):276–283

Wägele JW (1992) Isopoda. In: Microscopic anatomy of invertebrates, volume 9 crustacea. Wiley-Liss, New York, pp 529–617

Wang Z, Königsberger E (1998) Solubility equilibria in the uric acid–sodium urate–water system12. Thermochimica Acta 310(1–2):237–242

Weihrauch D, Morris S, Towle DW (2004) Ammonia excretion in aquatic and terrestrial crabs. J Exp Zool 207:4491–4504

Wieser W (1972a) Oxygen consumption and ammonia excretion in *Ligia beaudiana* M.-E. Comp Biochem Physiol 43A:869–876

Wieser W (1972b) A glutaminase in the body wall of terrestrial isopods. Nature 239(5370):288–290

Wieser W, Schweizer G (1970) A re-examination of the excretion of nitrogen by terrestrial isopods. J Exp Zool 52(2):267–274

Wieser W, Schweizer G, Hartenstein R (1969) Patterns in the release of gaseous ammonia by terrestrial isopods. Oecologia 3(3–4):390–400

Wolcott TG (1988) Ecology. In: Burggren WW, McMahon BR (eds) Biology of the land crabs. Cambridge University Press, Cambridge

Wolcott DL (1991) Nitrogen excretion is enhanced during urine recycling in two species of terrestrial crab. J Exper Zool 259(2):181–187

Wolcott DL, Wolcott TG (1984) Food quality and cannibalism in the red land crab *Gecarcinus lateralis*. Physiol Zool 57(3):318–324

Wolcott DL, Wolcott TG (1987) Nitrogen limitation in the herbivorous land crab *Cardisoma guanhumi*. Physiol Zool 60(2):262–268

Wood CM, Boutilier RG (1985) Osmoregulation, ionic exchange, blood chemistry, and nitrogenous waste excretion in the land crab *Cardisoma* carnifex: a field and laboratory study. Biol Bull 169(1):267–290

Wood CM, Boutilier RG, Randall DJ (1986) The physiology of dehydration stress in the land crab, *Cardisoma carnifex*: respiration, ionoregulation, acid–base balance and nitrogenous waste excretion. J Exp Zool 126(1):271–296

Wright JC, Machin J (1990) Water vapour absorption in terrestrial isopods. J Exp Zool 154(1):13–30

Wright JC, O'Donnell MJ (1992) Osmolality and electrolyte composition of pleon fluid in *Porcellio scaber* (Crustacea, Isopoda, Oniscidea): implications for water vapour absorption. J Exp Zool 164(1):189–203

Wright JC, O'Donnell MJ (1993) Total ammonia concentrations and pH of haemolymph, pleon fluid and maxillary urine in *Porcellio* scaber Lattreille (Isopoda, Oniscidea): relationships to ambient humidity and water vapour uptake. J Exp Biol 176(1):233–246

Wright J, Peña-Peralta M (2005) Diel variation in ammonia excretion, glutamine levels, and hydration status in two species of terrestrial isopods. J Comp Physiol B 175(1):67–75

Wright JC, O'Donnell MJ, Reichert J (1994) Effects of ammonia loading on *Porcellio scaber*; glutamine and glutamate synthesis, ammonia excretion and toxicity. J Exp Zool 188:143–157

Wright JC, O'Donnell MJ and Sazgar, S (1997) Haemolymph osmoregulation and the fate of sodium and chloride during dehydration in terrestrial isopods. J Insect Physiol 43:795–807

Wright JC, Caveney S, O'Donnell MJ, Reichert J (1996) Increases in tissue amino acid levels in response to ammonia stress in the terrestrial isopod, *Porcellio scaber* Latr. J Exp Zool 274:265–274

Gill Ion Transport ATPases and Ammonia Excretion in Aquatic Crustaceans

Francisco A. Leone, Malson N. Lucena, Daniela P. Garçon, Marcelo R. Pinto, and John C. McNamara

© Springer International Publishing Switzerland 2017
D. Weihrauch, M. O'Donnell (eds.), *Acid-Base Balance and Nitrogen Excretion in Invertebrates*,
DOI 10.1007/978-3-319-39617-0_3

3.1   Summary

Crustaceans inhabit diverse biotopes, often subject to alterations that constitute a severe challenge to their homeostatic mechanisms. These challenges have driven the evolution of biochemical and physiological processes that have enabled their survival in such niches. Ion-transporting enzymes like the (Na+, K+)-ATPase and V(H+)-ATPase present in the gill epithelia underpin the ion regulatory abilities of these highly diversified organisms. The present chapter examines the structure and function of these two gill ATPases that also participate actively in ammonia excretion. We summarize current knowledge on their role in osmotic and ionic regulation and associated with ontogenetic changes. We analyze the effects of polyamines on (Na+, K+)-ATPase activity and phosphoenzyme formation, aiming to provide insights into the biochemical bases of physiological homeostasis in crustaceans. We examine future perspectives that should provide a better understanding of the role of gill ATPases in active ammonia excretion.

3.2   The (Na+, K+)-ATPase: Structure, Function, Mechanisms, and Regulation

Jens Skou discovered the (Na+, K+)-ATPase in 1954. Interested in studying the membrane ATPase from squid giant axons at Aarhus University, curiously, Skou had no access to squid. Instead, he isolated nerves from the pereiopods of the green shore crab *Carcinus maenas* that, like the giant axon, lack a myelin sheath. He homogenized the nerves and, after differential centrifugation, isolated a membrane fraction. His systematic kinetic measurements of the combined effects of sodium and potassium ions revealed that the enzyme had two binding sites, one at which sodium ions were required for activation and another where potassium ions activated the enzyme after sodium ion binding (Skou 1957).

The ATPases are membrane transporters termed "pumps" and perform the primary active transport of ions employing the hydrolysis of a phosphate group from the ATP molecule (Post 1999). They are classified into four main groups: the P-, F-, M-, and V(H+)-ATPases (Pedersen and Amzel 1993). The P-ATPases bind the γ phosphate group of ATP at a specific aspartate residue present in the invariant amino acid sequence DKTGT (I/L) (Bublitz et al. 2010; Palmgren and Nissen 2011; Chourasia and Sastry 2012). The P-ATPase superfamily can be further classified into five families (I–V) according to the alignment of a conserved sequence of 159 amino acids: group II is the most investigated and includes the (Ca^{2+})-, (Na+, K+)-, and (H+, K+)-ATPase pumps (Bublitz et al. 2010; Palmgren and Nissen 2011; Chourasia and Sastry 2012).

The (Na+, K+)-ATPase (EC 3.6.1.37) is a member of the $P_{II}c$ subfamily (Palmgren and Nissen 2011) and underpins many vital functions, such as generating membrane potential, secondary active transport, neural signaling, maintenance of thermogenesis, ionic balance, and pH regulation (Poulsen et al. 2010a; Tidow et al. 2010; Zheng et al. 2011; Aperia 2012; Clarke et al. 2013). The (Na+, K+)-ATPase or sodium-potassium pump uses the energy provided by hydrolysis of the γ phosphate from ATP to establish electrochemical gradients across the plasma membrane of all animal cells, simultaneously transporting Na+ from the intracellular *milieu* and K+ from the extracellular fluid against their respective gradients (Skou 1957; Bublitz et al. 2010; Palmgren and Nissen 2011).

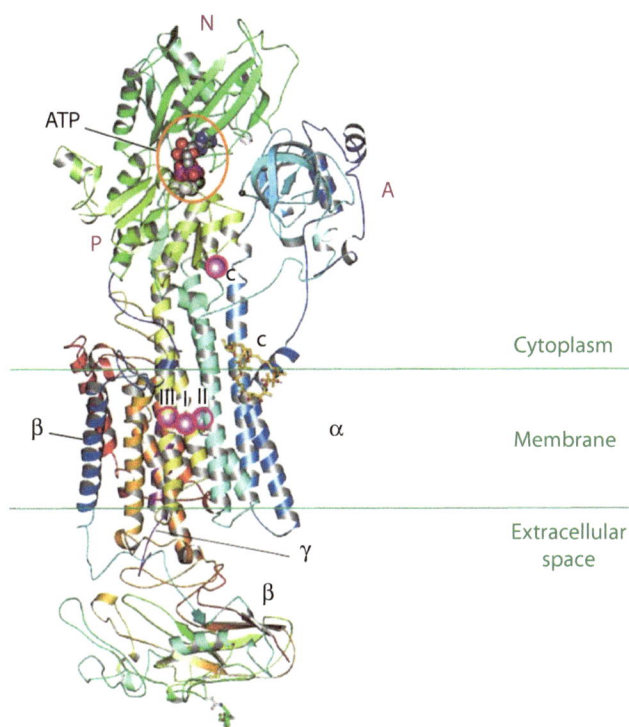

Fig. 3.1 Molecular architecture of the (Na+, K+)-ATPase αβγ complex. The molecular architecture of the (Na+, K+)-ATPase consists of an α-, a β-, and a γ-subunit (FXYD protein). The cytoplasmic catalytic α-subunit exhibits A (actuator), N (nucleotide binding), and P (phosphorylation) domains. α-Helices are represented by ribbons and β-strands by wires with arrows that follow the backbone of the α-carbon loops. The β-subunit is indicated in dark blue and the γ-subunit in purple. Red circles represent Na+-binding sites (I, II, III) in the transmembrane helices. The Na+-binding site in the cytoplasmic P domain (C) plays a regulatory role. The horizontal green lines delimit the cytoplasmic and extracellular membrane leaflets (Modified from Kanai et al. (2013))

The basic functional unit of the (Na+, K+)-ATPase (Fig. 3.1) consists of a single α-subunit responsible for catalytic activity, a glycosylated β-subunit, and in some cells, a peptide from the FXYD family, the γ-subunit (Kaplan 2002; Garty and Karlish 2006; Morth et al. 2007; Geering 2008; Poulsen et al. 2010a). The α-subunit, of about 110 kDa, contains the ATP-binding site, the phosphorylation site, and the amino acids essential for the binding of sodium and potassium ions (Geering 2000; Horisberger 2004; Capendeguy and Horisberger 2005).

In this subunit both the N- and C-terminals face the cytoplasm and are separated by ten transmembrane segments, designated M1 to M10, which give rise to intra- and extra-cellular loops (Jorgensen et al. 2003; Toustrup-Jensen and Vilsen 2005). The cytoplasmic loops are organized into three main domains: the P or phosphorylation domain, the A or actuator domain, and the N or nucleotide-binding domain (Horisberger 2004; Toustrup-Jensen et al. 2009; Bublitz et al. 2010; Palmgren and Nissen 2011). The A domain, formed by segments M2-M3, plays an important role in (Na+, K+)-ATPase affinity for ATP, acting as a regulator (Daly et al. 1997; Kaplan 2002). The M4-M5 loop contains the N and P domains that include amino acid residues important for ATP binding to the enzyme and the aspartate residue, which is phosphorylated during the catalytic cycle of the enzyme (Hebert et al. 2003; Belogus et al. 2009; Bublitz et al. 2010). Enzyme affinity for Na+ is modulated by interaction between the α-subunit C-terminus and the cytoplasmic loop formed by segments M8-M9 and M5 (Toustrup-Jensen et al. 2009; Yaragatupalli et al. 2009; Morth et al. 2011). Residues from the extracellular loop formed by segments M3-M4 greatly influence enzyme affinity for K+, suggesting a likely binding site for these ions here (Eguchi et al. 2005; Morth et al. 2011).

The β-subunit is a highly glycosylated, type II membrane protein constituted by a single transmembrane segment whose N-terminus is oriented to the cytoplasm (Cohen et al. 2005; Durr et al. 2009; Morth et al. 2007, 2011). It influences α-subunit affinity for Na^+ and K^+ and has a large extracellular domain; the presence of sugars imparts stability against proteases (Martin 2005; Purhonen et al. 2006; Morth et al. 2007). Besides stabilizing the enzyme in the E2 conformation, a constituent YXXYF domain is important for interaction between the α- and β-subunits (Hasler et al. 2001; Durr et al. 2009; Shinoda et al. 2009; Morth et al. 2011). In the absence of the β-subunit, the α-subunit is retained in the endoplasmic reticulum, suggesting a chaperone role for the β-subunit (Laughery et al. 2003; Toustrup-Jensen et al. 2009).

In vertebrates, the (Na^+, K^+)-ATPase may be associated with a proteolipid of about 60 amino acid residues and an M_r of ≈ 7 kDa, the γ-subunit (or FXYD2 peptide), belonging to the FXYD family (Therien and Blostein 2000; Crambert et al. 2004; Füzesi et al. 2005; Lubarski et al. 2007). The FXYD proteins are small, tissue-specific proteins exhibiting a single transmembrane segment located adjacent to segments M2-M6-M9 of the α-subunit, whose N-terminus faces the extracellular medium (Dempski et al. 2008; Geering 2008; Morth et al. 2011). The γ-subunit modulates the kinetic properties and the stability of the (Na^+, K^+)-ATPase (Geering 2008; Cirri et al. 2011; Mishra et al. 2011; Shindo et al. 2011; Yoneda et al. 2013) and controls both the activity and cation affinity of the enzyme in crustacean gill microsomes (Silva et al. 2012). The FXYD2 peptide may regulate (Na^+, K^+)-ATPase activity by modulating the effective membrane surface electrostatics near the ion-binding sites (Gong et al. 2015).

The products of different genes, four α-subunit isoforms and three β-subunit isoforms, have been identified in vertebrates (Lingrel et al. 2003; Blanco 2005; Pressley et al. 2005; Jimenez et al. 2010, 2011; Parekh et al. 2010; Schaefer et al. 2011; Lucas et al. 2012; Azarias et al. 2013; Lai et al. 2013; Mladinov et al. 2013; Radzyukevich et al. 2013; McDermott et al. 2015). These isoforms may be combined yielding isoenzymes with different kinetic properties (Segall et al. 2000; Mobasheri et al. 2000; Pressley et al. 2005; Mijatovic et al. 2007; Pierre et al. 2008; Tokhtaeva et al. 2012).

The best structurally characterized P-ATPase is the sarco/endoplasmic reticulum Ca^{2+}-ATPase (SERCA) (Toyoshima et al. 2000, 2004; Toyoshima and Mizutani 2004; Jensen et al. 2006; Toyoshima and Cornelius 2013). Owing to its more complex structure, the (Na^+, K^+)-ATPase is less well characterized (Pedersen et al. 2010; Toyoshima et al. 2011; Toyoshima and Cornelius 2013). The first crystal structure of the (Na^+, K^+)-ATPase from pig kidney outer medulla was obtained at 3.5 Å resolution and was based on the enzyme in the E2 conformation (Morth et al. 2007). High-resolution structural studies of shark rectal gland (Na^+, K^+)-ATPase have identified the amino acids essential for K^+-binding sites (Shinoda et al. 2009) and the binding sites for ouabain (Ogawa et al. 2009) using an αβγ complex in a state analogous to the E2-2 K^+ conformation. A high-affinity binding site for ouabain has been identified in studies using the enzyme in the ouabain-bound E2 conformation (Yatime et al. 2011). Structures obtained in the E1 conformation have enabled the identification of Na^+-binding sites (Kanai et al. 2013; Nyblom et al. 2013).

Two glycine residues, Gli_{93} and Gli_{94}, both located in transmembrane segment M1 are important for interaction of the α-subunit with Na^+ and K^+ (Einholm et al. 2005). Thr_{774}, Val_{920}, and Glu_{954}, located in transmembrane segments M5, M8, and M9, respectively, are important for Na^+ transport (Imagawa et al. 2005). Tyr_{1017}, Tyr_{768}, Arg_{935}, and Lys_{768} also influence interaction of Na^+ with the enzyme (Toustrup-Jensen et al. 2009). The third binding site for Na^+ is located in the C-terminal tail that preserves the KETYY motif, Tyr

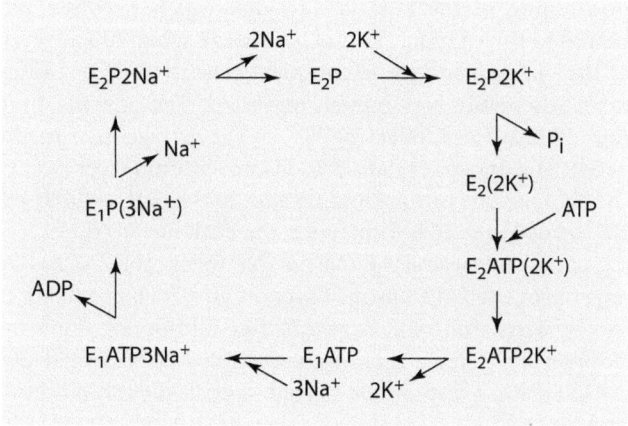

Fig. 3.2 Albers-Post scheme describing the reaction mechanism of the (Na$^+$, K$^+$)-ATPase. Occluded ions or membrane-occluded forms are shown in parentheses

residues being essential for occlusion and Na$^+$ affinity (Meier et al. 2010; Poulsen et al. 2010a; Vedonato and Gadsby 2010; Poulsen et al. 2012; Paulsen et al. 2013). The extracellular loop between segments M3-M4 includes two highly conserved residues, Glu$_{319}$ and Leu$_{318}$, apparently involved in cation selectivity, specifically for K$^+$ (Eguchi et al. 2005; Ratheal et al. 2010). Particularly important are deprotonated amino acid residues, critical for K$^+$ selectivity (Yu et al. 2011). A large Asp$_{369}$-containing cytoplasmic loop of about 430 amino acid residues between segments M4-M5, representing 30 % of the polypeptide chain, undergoes phosphorylation during the reaction cycle (Kaplan 2002; Jorgensen et al. 2003; Horisberger 2004). Asp$_{443}$ binds ATP, establishing ionic interactions with an Arg$_{544}$ residue (Middleton et al. 2011). The movements of segment M10 and its cytoplasmic extension (C-terminal tail) regulate ion selectivity and transport through the K$^+$ sites (Mahmmoud et al. 2014).

The most accepted model explaining the catalytic cycle of P-ATPases is the E1-E2 conformational model of Albers-Post (**Fig. 3.2**), based on conformational transitions between the two principal conformational states of the enzyme (Alberts 1967; Post et al. 1972; Kaplan 2002; Scheiner-Bobis 2002; Clarke 2009).

The cycle is a result of both affinity changes between E1 and E2 and the accessibility of ions to their binding sites in the transmembrane segments (Apell et al. 2011; Toyoshima et al. 2011; Gadsby et al. 2012). The E1 conformation shows high affinity for intracellular Na$^+$ and ATP, while the E2 conformation exhibits high affinity for extracellular K$^+$ (Jorgensen et al. 2003; Horisberger 2004; Khalid et al. 2010; Poulsen et al. 2010b; Monti et al. 2013). ATP binding leads to the convergence of the nucleotide binding and phosphorylation domains, transferring the enzyme from the "E1-open" to the "E1-closed" conformation, ready for phosphorylation (Petrushanko et al. 2014). According to this model, the reaction cycle initiates with the binding of an ATP molecule to form the E2(2K) conformation, with two K$^+$ occluded within the enzyme (Jorgensen et al. 1998, 2003; Kaplan 2002; Horisberger 2004). ATP binding to its site accelerates the conformational change of E2ATP(2K) to E1ATP2K, in which the two K$^+$ are deoccluded with the simultaneous reorientation of the cation-binding sites from the extracellular side into the cytoplasm. The release of the two K$^+$ into the cytoplasm, followed by their replacement by two Na$^+$, leads to the E1ATP2Na form. The binding of a third cytoplasmic Na$^+$ gives rise to the E1ATP3Na form, causing rearrangement of the α-subunit transmembrane segments (Kaplan 2002) that is propagated to the cytoplasmic domain, inducing appropriate positioning of the aspartate residue which becomes phosphorylated by ATP already bound to the enzyme (Rice et al. 2001; Kaplan 2002; Horisberger 2004). Phosphorylation of the enzyme and the consequent release of ADP causes a transition to the E1P(3Na) form where the Na^+ remain occluded. The enzyme undergoes rapid isomerization to the E2P2Na form, releasing Na^+ to the extracellular medium (Kaplan 2002; Horisberger 2004). The phosphorylation and isomerization steps occur coupled to the reorientation of the cytoplasmic cation-binding sites to the extracellular medium. The deocclusion of two Na^+ leads to the E2P form, releasing both ions to the extracellular fluid. Finally, the binding of two extracellular K^+ to the E2P form catalyzes its dephosphorylation, returning the enzyme to the E2(K) form (Myers et al. 2011), restarting the catalytic cycle. Site-directed mutagenesis studies have revealed an additional, nontransporting K^+ site located in the cytoplasmic domain (Sorensen et al. 2004; Schack et al. 2008).

Ouabain, a cardiotonic steroid, specifically inhibits both the (Na^+, K^+)-ATPase activity and Na^+ and K^+ transport (Kaplan 2002; Nesher et al. 2007; Cornelius and Mahmmoud 2009; Lingrel 2010; Liu and Xie 2010; Cornelius et al. 2011; Miles et al. 2013; Slingerland et al. 2013). Ouabain binds with high affinity to E2P, and its interaction with the enzyme involves α-subunit residues and to a lesser extent β-subunit residues (Ogawa et al. 2009; Yatime et al. 2011; Sánchez-Rodriguéz et al. 2015), the most relevant being Asp_{121} located in segment M1 (Ogawa et al. 2009; Sandtner et al. 2011; Cornelius et al. 2013; Laursen et al. 2015). The phosphorylated form of the enzyme, in the presence of the K^+ ion and the orientation of Val_{329}, shows a higher binding affinity for ouabain (Khalid et al. 2014). Ouabain interacts with the (Na^+, K^+)-ATPase from the extracellular membrane surface, and its binding site, inserted into the membrane, is located in a cavity formed by transmembrane segments M1, M2, M4, M5, and M6, very close to the K^+-binding site (Ogawa et al. 2009; Yatime et al. 2011; Laursen et al. 2015). The sugar and the hydroxyl moieties of the steroid ring are both essential for the potential action of ouabain (Cornelius et al. 2013). Recently, selectivity of (Na^+, K^+)-ATPase isoforms for digitalis-like compounds that contain no sugar groups has been shown (Weigand et al. 2014).

Owing to its similarity with the tetrahedral structure of inorganic phosphate, orthovanadate is a powerful inhibitor of vertebrate P-ATPases in the nanomolar range (Boxenbaum et al. 1998; Fedosova et al. 1998; Montes et al. 2012). Orthovanadate binds covalently to the aspartate residue, blocking the reaction cycle and maintaining the enzyme in a state similar to E2 (Fedosova et al. 1998; Montes et al. 2012). Interaction of orthovanadate with the enzyme results in an enzyme-orthovanadate complex in an open or closed conformation (Montes et al. 2012).

The (Na^+, K^+)-ATPase participates in a variety of secondary active transport processes and is subject to multiple, tissue-specific regulatory mechanisms (Therien and Blostein 2000). Such regulation involves a series of complex factors acting in the short and long term (Glynn 2002; Pressley et al. 2005; Mijatovic et al. 2007). Short-term regulation involves direct effects on the kinetic behavior of the enzyme and on enzyme translocation from the plasma membrane to intracellular storage locations and is dependent on intra- and extracellular ATP, Na^+, and K^+ concentrations (Therien and Blostein 2000; Feraille et al. 2003). Adjustment also can be triggered by enzyme phosphorylation/dephosphorylation or by increasing affinity for the Na^+ (Mijatovic et al. 2007). Long-term adjustment usually involves changes in gene transcription, mRNA stability and translation, isoform expression, and protein degradation (Seok et al. 1998; Therien and Blostein 2000; Colina et al. 2010; Karitskaya et al. 2010; Sottejeau et al. 2010). (Na^+, K^+)-ATPase activity also can be regulated by molecules such as aldosterone (Salyer et al. 2013), insulin (Oubaassine et al. 2012), protein kinase A (Lecuona et al. 2013), protein kinase C (Gallo et al. 2010; Wengert et al. 2013), and phosphatases (El-Beialy et al. 2010) that act both in the short and long term through complex signaling cascades. Endobains, inhibitors similar to ouabain, are also endogenous regulators of the (Na^+, K^+)-ATPase (Therien and Blostein 2000; Scheiner-Bobis 2002; Hansen 2003).

3.3 The Crustacean Gill (Na^+, K^+)-ATPase as a Biochemical Marker of Adjustment to Salinity

The animal kingdom embraces 30 phyla, all of marine origin. Of these, 16 have successfully occupied freshwater, and seven phyla have conquered the terrestrial environment (Lee and Bell 1999). The colonization of fresh water and low salinity habitats by marine species constitutes one of the most challenging evolutionary transitions that have taken place during the history of life on Earth (Lee et al. 2011). Low and/or variable salinities represent a major physicochemical barrier to the invasion of osmotically demanding habitats like estuaries (Henry et al. 2012). In counterpoint, the evolutionary and ecological advantages of exploiting the estuarine environment lie in its nutrient-rich habitats (Gross 1972). Any species that can tolerate its inherent osmotic challenges can exploit an estuarine niche that is rich in resources with little competition. Such positive selection pressures may have driven the appearance of biochemical and physiological mechanisms that enabled survival in low salinity (Henry et al. 2012).

Crustaceans arose in the primitive sea about 600 million years ago and today constitute an essentially marine group (Barnes 2000). During evolution, various crustacean groups have confronted the challenge of invading environments of salinity lower than seawater and from estuary or land to fresh water (Schubart and Diesel 1998). Decapod crustaceans invaded fresh water around 3.4 million years ago (Schubart et al. 1998) and occupy and exploit virtually all habitats available on Earth from the marine environment to deserts (McNamara and Faria 2012).

Crustaceans have evolved a diversity of osmoregulatory strategies that enable them to occupy many different environments and habitats (Prosser 1973). *Euryhaline* crustaceans tolerate and survive ample variations in the salinity of their external environment, while *stenohaline* crustaceans are restricted to limited variation in their ambient salinity (Péqueux 1995; Randall et al. 2000). Regarding osmotic regulation of their body fluids, crustaceans may both osmoconform and/or osmoregulate (Péqueux 1995; Lucu et al. 2000). Within certain limits, *osmoconformers* hold the osmotic concentration of their hemolymph or extracellular fluid similar to that of the environment. *Osmoregulators* actively maintain their hemolymph osmolalities between species-specific limits, either isosmotic to or below or above that of their external environment (Péqueux 1995; Lucu et al. 2000). Such osmoregulators may also osmoconform over the more extreme parts of their range of salinity tolerance.

In general, marine crustaceans are stenohaline osmoconformers, while osmoregulating species that have colonized intertidal zones, estuaries, and fresh water exhibit varying degrees of euryhalinity and efficient mechanisms of osmotic hyperregulation (Péqueux 1995; Onken and McNamara 2002; Lucu and Towle 2003). These diverse patterns of osmoregulatory capability may have resulted from adaptation to specific environmental conditions, genetic drift, and natural selection or may represent the retention of an ancestral ability (McNamara and Faria 2012).

Marine stenohaline crustaceans typically show a lower lethal salinity limit near 18‰ salinity (Mantel and Farmer 1983). Strong regulators can maintain elevated osmotic gradients in fresh water and can spend their entire adult life in this environment. The blue crab *Callinectes sapidus* (Cameron 1978) and the Chinese mitten crab *Eriocheir sinensis* (Wang et al. 2012b) are perhaps the best examples of strongly osmoregulating species. *Callinectes sapidus* is found in salinities from 40 to 0‰ salinity and can survive direct transfer from 35 to 0‰ salinity in the laboratory. This blue crab maintains a gradient greater than 600 mOsm/kg H_2O above ambient when acclimated to fresh water (Cameron 1978). Moderate osmoregulators can survive in low salinity but not in fresh water. One such moderate regulator is the green shore crab, *Carcinus maenas*, which is found in salinities as low as 10‰ and can survive laboratory acclimation to 8‰ salinity (Zanders 1980) in which it maintains its hemolymph osmolality approximately 350 mOsm/kg H_2O above ambient (Zanders 1980). A weak osmoregulator like *C. similis* has a lower salinity limit near 15‰ salinity (Henry et al. 2012), suffers 80% mortality when gradually acclimated to 5‰ salinity in the laboratory, and maintains a maximum hemolymph-medium osmotic gradient of 250 mOsm/kgH_2O (Engel 1977).

Two distinct processes overall osmotic homeostasis of the body fluids in crustaceans: *anisosmotic extracellular regulation* maintains the osmolality, composition, and volume of the hemolymph or extracellular fluid, while *isosmotic intracellular regulation* adjusts the volume and composition of the intracellular fluid or cytosol (Florkin and Schoffeniels 1969; Péqueux 1995; Lang and Waldegger 1997; Wehner et al. 2003; Henry et al. 2012). Anisosmotic extracellular regulation is affected by changes in the permeability of different body surfaces to ions and water and by the active absorption or secretion of salts (Mantel and Farmer 1983). Isosmotic intracellular regulation consists of adjustments in the titers of intracellular osmotic effectors such as ions, peptides, and free amino acids, affecting water movement down its osmotic gradient (Schmidt-Nielsen 1996).

In decapod crustaceans, the gut, the antennal glands, and the gills and other structures found in the branchial chambers are involved in ion regulatory and excretory functions (Holliday and Miller 1984; Péqueux 1995; Lima et al. 1997; Khodabandeh et al. 2005a, b; Freire et al. 2008). The antennal glands regulate body fluid volume, maintaining the concentrations of certain organic and inorganic solutes, especially divalent ions like magnesium and sulfate and also produce dilute urine through NaCl reabsorption in some freshwater species (Baldwin and Kirschner 1976; Mantel and Farmer 1983; Holliday and Miller 1984; Péqueux 1995; Freire et al. 2008). Nevertheless, rarely have the structure and function of the antennal glands been described in detail in different groups (Freire et al. 2008; Khodabandeh et al. 2005a, b; Tsai and Lin 2014; McNamara et al. 2015).

Crustacean gills provide a selective interface between the external environment and the internal *milieu*, constituting a multifunctional organ serving in gas exchange, osmolyte transport, nitrogenous waste excretion, and volume and acid-base regulation (Burnett et al. 1985; Gilles and Péqueux 1985; Henry and Wheatly 1992; Taylor and Taylor 1992; Péqueux 1995; Weihrauch et al. 2004a; Freire et al. 2008; Henry et al. 2012). The (Na^+, K^+)-ATPase is the main force driving the transepithelial movement of monovalent ions across the gill epithelia of most anisosmotic crustaceans (Lucu and Towle 2003). While numerous reviews are available for the vertebrate enzyme, they are few for the crustacean (Na^+, K^+)-ATPase (Towle 1984; Harris and Bayliss 1988; Lucu 1990; Towle 1990; Onken and Riestenpatt 1998; Lucu and Towle 2003; Leone et al. 2005; Freire et al. 2008; Henry et al. 2012).

In decapod crustaceans, the gills are located in the branchial chambers found on either side of the cephalothorax and often constitute appendices positioned between the pleural wall and the branchiostegites (Taylor and Taylor 1992). Their fine epithelia possess several different cell types, including those specialized in ion exchange, the ionocytes (Taylor and Taylor 1992; Péqueux 1995).

The capability of crustaceans for anisosmotic extracellular osmoregulation depends on the transbranchial transport of Na^+ and Cl^- between the external environment and the hemolymph (Péqueux 1995; Kirschner 2004; Freire et al. 2008; Henry et al. 2012). This regulatory function is energetically driven mainly by the (Na^+, K^+)- ATPase located in the basal membranes of the gill ionocytes that actively exchange Na^+ and K^+ between their cytosol and the hemolymph (Kirschner 2004; Freire et al. 2008; Sáez et al. 2009). Ultracytochemical and fluorescence studies have shown that the crustacean gill (Na^+, K^+)-ATPase is restricted to the basal membrane of the ionocytes (Towle and Kays 1986; McNamara and Torres 1999), and electrophysiological experiments using inhibitors have confirmed this localization (Lucu and Siebers 1987; Burnett and Towle 1990; Luquet et al. 2002; Riestenpatt et al. 1996). An apical driving force furnished by the $V(H^+)$-ATPase also plays an important role in osmoregulatory ion transport, particularly in freshwater crustaceans (Onken and Putzenlechner 1995; Morris 2001; Weihrauch et al. 2004b). In addition to these two ATPases, a variety of membrane transporters such as apical Na^+/H^+ and Cl^-/HCO_3^- exchangers and the $Na^+/K^+/2Cl^-$ symporter, apical and basal K^+ and Cl^- channels (Kirschner 2004; Freire et al. 2008; Sáez et al. 2009; Towle et al. 2011; Yang et al. 2011), and particularly, apical Na^+ channels (Zeiske et al. 1992) also contribute to ion movements and thus to osmotic regulation. Carbonic anhydrase that catalyzes the reversible hydration of CO_2 to H^+ and HCO_3^- used by the apical exchangers (Perry 1986) also plays a critical role in ion uptake (Henry 1988a,b; Skaggs and Henry 2002). The distribution of such carriers in the ionocyte membranes depends on the osmotic niche occupied by each species (McNamara and Faria 2012).

Early studies of (Na^+, K^+)-ATPase in the gills of euryhaline hyper-osmoregulating crabs found that enzyme activities, measured in homogenates or in purified membranes, increased significantly in response to salinity variation (Mantel and Olson 1976; Towle et al. 1976; Siebers et al. 1982). Posterior gills showed a higher specific (Na^+, K^+)-ATPase activity than anterior gills, suggesting their specialization for osmoregulatory ion transport (Neufeld et al. 1980; Siebers et al. 1982; Wanson et al. 1984; Holliday 1985; Winkler 1986; Welcomme and Devos 1988; Harris and Santos 1993). Supporting evidence came from ultrastructural studies of gill epithelia showing differentiated cells with greatly augmented apical and basal surface areas and elevated mitochondrial densities in the lamellae of the posterior but not anterior gills (Copeland and Fitzjarrell 1968; Goodman and Cavey 1990; Compère et al. 1989). These characteristics are now considered typical of an ion-transporting function in crustacean epithelia.

Compared to the vertebrate enzyme, the structure of the crustacean gill (Na^+, K^+)-ATPase is poorly known. Like other (Na^+, K^+)-ATPases, the crustacean enzyme possesses an α-subunit of 95–104 kDa M_r and a β-subunit of 38–40 kDa M_r (Peterson et al. 1978; Lucu and Flik 1999; Furriel et al. 2000; Towle et al. 2001; Masui et al. 2002; Gonçalves et al. 2006; Garçon et al. 2007; Lucena et al. 2012; Leone et al. 2015a). The native holoenzyme is a tetramer consisting of two α- and two β-subunits of 270–280 kDa M_r (Peterson and Hokin 1981).

The α-subunit is the catalytic and ouabain-binding component of the crustacean (Na^+, K^+)-ATPase (Skou and Esmann 1992). While the β-subunit is presumed to assist in the correct folding and membrane targeting of the α-subunit, its role in modulating enzyme activity has been little investigated (Lucu and Towle 2003; Leone et al. 2005). A third subunit, the γ- subunit, is associated with some (Na$^+$, K$^+$)-ATPases (Forbush et al. 1978) and apparently modulates Na$^+$ and K$^+$ binding to the enzyme (Arystarkhova et al. 1999). An FXYD2 peptide or γ-subunit is associated with the gill (Na$^+$, K$^+$)-ATPase from the blue crab *C. danae* and plays an important role in regulating enzyme activity (Silva et al. 2012).

The complete primary structure has been deduced for the α-subunits of only three crustacean species, *Artemia franciscana*, *C. sapidus*, and *Homarus americanus*, revealing elevated interspecific identities and some identity with the vertebrate α2 and α3 isoforms, including the avian enzyme (Baxter-Lowe et al. 1989; Macías et al. 1991; Pressley 1992; Towle et al. 2001; Parrie and Towle 2002). Two different genes (α1 and α2) code for α-subunit isoforms in *A. franciscana* (Baxter-Lowe et al. 1989; Macías et al. 1991) that exhibit only eight transmembrane segments, the native holoenzyme being a diprotomer (αβ)$_2$ (Peterson and Hokin 1981; Lucu and Towle 2003). The (Na$^+$, K$^+$)-ATPase α-subunit is highly conserved among crustacean species. The *C. sapidus* enzyme (AF327439) is 94 % identical to that of *H. americanus* (Towle et al. 2001), and close identities have been revealed for the partial (Na$^+$, K$^+$)-ATPase α-subunit sequence from *M. amazonicum* (GQ329698) with those for the estuarine crab *Neohelice granulata* (AF548369.1), the diadromous freshwater crab *Eriocheir sinensis* (AF301158.1), the intertidal crab *Pachygrapsus marmoratus* (DQ173925.2), the marine lobster *Homarus americanus* (AY140650.1), the true freshwater crab *Dilocarcinus pagei* (AF409119.1), and several species of *Uca* (JX854012-14, KC954714-19). The β-subunit of the *A. franciscana* (Na$^+$, K$^+$)-ATPase, the only β-subunit sequence known for crustaceans, is similar to that of vertebrates with respect to its single transmembrane domain and C-terminus (Bhattacharyya et al. 1990).

(Na$^+$, K$^+$)-ATPase activity in gill homogenates is optimal over a range of pH values from 7.0 to 7.7 (Neufeld et al. 1980; Corotto and Holliday 1996; Kosiol et al. 1988; D'Orazio and Holliday 1985; Holliday 1985), and a considerable number of studies have kinetically characterized the crustacean enzyme (Gache et al. 1977; Wanson et al. 1984; Holliday 1985; D'Orazio and Holliday 1985; Wilkie 1997; Harris and Bayliss 1988; Péqueux 1995; Corotto and Holliday 1996; Furriel et al. 2000; Masui et al. 2002; Wilder et al. 2000; Lucu and Towle 2003; Garçon et al. 2007; Mendonça et al. 2007; Lucena et al. 2012; Leone et al. 2012; Pinto et al. 2016; Gonçalves et al. 2006; Belli et al. 2009; Faleiros et al. 2010; França et al. 2013).

Since 1997, we have used the (Na$^+$, K$^+$)-ATPase in microsomal fractions from crustacean gills as a biochemical marker to examine adaptation of this group to environments of different salinities. We first characterized the enzyme from the gills of the freshwater shrimp *Macrobrachium olfersii* (Furriel et al. 2000), showing that the detergent-free preparation was a convenient material in which to examine the properties of the (Na$^+$, K$^+$)-ATPase in vitro since the native interactions between the enzyme and the membrane bilayer are apparently preserved.

Our kinetic characterization of the (Na$^+$, K$^+$)-ATPase from the posterior gills of fresh caught (33 ‰) euryhaline crab *Callinectes danae* revealed a high-affinity binding site for ATP, similar to the vertebrate enzyme (Masui et al. 2002). Later, we also demonstrated this site in the enzymes of the hermit crab *Clibanarius vittatus* (Gonçalves et al. 2006; Lucena et al. 2012) and freshwater shrimp *M. amazonicum* (Santos et al. 2007), suggesting that this high-affinity ATP-binding site is characteristic of the crustacean enzyme. We also showed that the contribution of the high- and low-affinity binding sites for ATP on the (Na$^+$, K$^+$)-ATPase from *C. danae*, relative to the maximum reaction rate, is the same for both sodium and potassium (❏ Fig. 3.3). However, there is a potassium concentration that triggers the appearance of high-affinity sites for ATP (Masui et al. 2008). Further, acclimation of *C. danae* to 15 ‰ salinity caused the high affinity ATP sites to disappear (Masui et al. 2009).

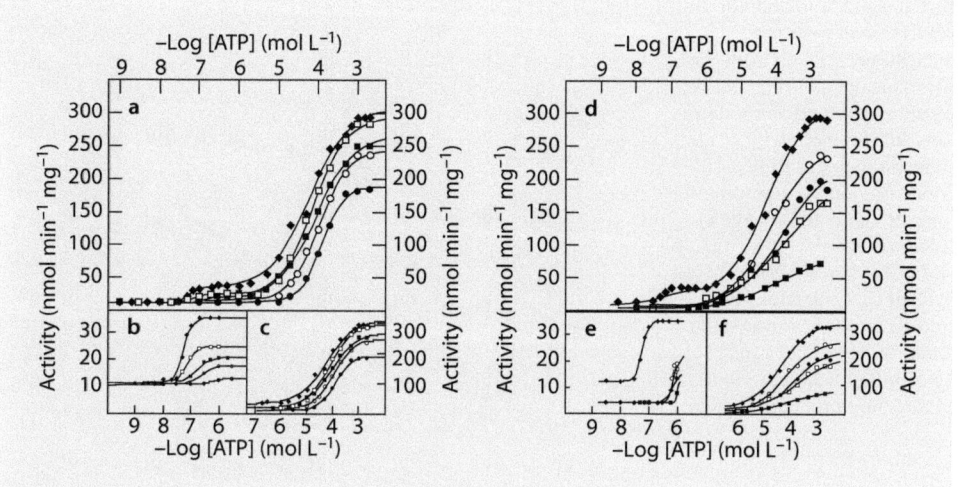

☐ **Fig. 3.3** Effect of ATP concentration on (Na⁺, K⁺)-ATPase activity in *C. danae* gill tissue at different Na⁺ and K⁺ concentrations. (**a**) ATP saturation curve at different NaCl concentrations. (**b**) Effect of sodium ions on ATP hydrolysis by high-affinity binding sites. (**c**) Effect of sodium ions on ATP hydrolysis by low-affinity binding sites. NaCl concentrations: (●) 20 mM, (○) 35 mM, (■) 50 mM, (☐) 70 mM, (◆) 100 mM. (**d**) ATP saturation curve at different KCl concentrations. (**e**) Effect of potassium ions on ATP hydrolysis by high-affinity binding sites. (**f**) Effect of potassium ions on ATP hydrolysis by low-affinity binding sites. KCl concentrations: (■) 1 mM, (☐) 2 mM, (●) 3 mM, (○) 5 mM, (◆) 10 mM (Reproduced from Masui et al. (2008) with permission)

We first described the synergistic stimulation of gill (Na⁺, K⁺)-ATPase activity by K⁺ (or NH₄⁺) and NH₄⁺ (or K⁺) in the blue crab *Callinectes danae* (☐ Fig. 3.4) (Masui et al. 2002, 2005a). Except for the freshwater crab *Dilocarcinus pagei* (Furriel et al. 2010), species-specific synergistic stimulation by NH₄⁺ plus K⁺ occurs in various crustaceans: *M. amazonicum* (Santos et al. 2007; Leone et al. 2014), *M. olfersii* (Furriel et al. 2004), *M. rosenbergii* (França et al. 2013), *Clibanarius vittatus* (Gonçalves et al. 2006; Lucena et al. 2012), *Callinectes ornatus* (Garçon et al. 2007, 2009), and *Xiphopenaeus kroyeri* (Leone et al. 2015a).

Although the (Na⁺, K⁺)-ATPase exhibits a high specificity for ATP, it also hydrolyses *p*-nitrophenylphosphate (Glynn 1985). Similar to the vertebrate enzyme, we first characterized K⁺-phosphatase activity in a gill microsomal fraction from the freshwater shrimp *M. olfersii* (Furriel et al. 2001). The use of *p*-nitrophenylphosphate as a substrate to kinetically characterize the K⁺-phosphatase activity of the crustacean gill enzyme has proved a convenient alternative for comparative studies of (Na⁺, K⁺)-ATPase activity not only in *M. olfersii* (Furriel et al. 2001; Mendonça et al. 2007) but also in the portunid crab *C. danae* (Masui et al. 2003; 2005b), in the diadromous palaemonid shrimp *M. amazonicum* (Belli et al. 2009) and the intertidal hermit crab *Clibanarius vittatus* (Lucena et al. 2012). The kinetic characterization of K⁺-phosphatase activity in juvenile and adult *M. amazonicum* helped to elucidate osmoregulatory ability in these two life cycle stages (Leone et al. 2013). Further, for 21‰ salinity-acclimated *C. ornatus*, *p*-nitrophenylphosphate hydrolysis is synergistically stimulated by K⁺ and NH₄⁺ (Garçon et al. 2013).

(Na⁺, K⁺)-ATPase-specific activity (Garçon et al. 2009) and relative expression of (Na⁺, K⁺)-ATPase α-subunit mRNA (Leone et al. 2015b) increase significantly in the posterior gills of *C. ornatus* transferred from seawater to a lower salinity (21‰). However,

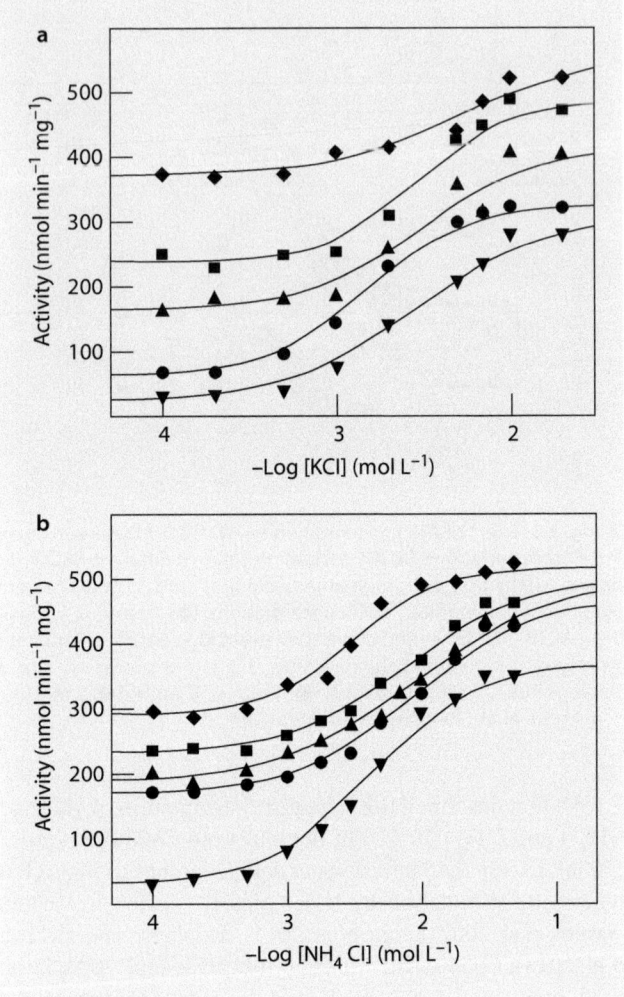

◘ Fig. 3.4 Stimulation of (Na$^+$, K$^+$)-ATPase activity in a microsomal fraction from the gill tissue of *Callinectes danae* by potassium and ammonium ions. (**a**) Stimulation by potassium ions at fixed concentrations of ammonium ions: (▼) none; (●) 1 mM; (▲) 5 mM; (■) 20 mM; (◆) 50 mM. (**b**) stimulation by ammonium ions at fixed concentrations of potassium ions: (▼) none; (●) 3 mM; (▲) 5 mM; (■) 7 mM; (◆) 10 mM (Reproduced from Masui et al. (2002) with permission)

gill (Na$^+$, K$^+$)-ATPase activity is unchanged in *C. danae* acclimated to 15 ‰ (Masui et al. 2009). Gill (Na$^+$, K$^+$)-ATPase and K$^+$-phosphatase activities decrease in the diadromous freshwater shrimps *M. olfersii* and *M. amazonicum* acclimated to 21 and 28 ‰ salinity (Lima et al. 1997; Mendonça et al. 2007; Belli et al. 2009; Faleiros et al. 2010), and gill (Na$^+$, K$^+$)-ATPase activity decreases by 20 % in the intertidal hermit crab *C. vittatus* acclimated to 45 ‰ (Lucena et al. 2012).

3.4   The V(H$^+$)-ATPase: Structure, Function, Mechanism, and Regulation

The regulation of pH in intracellular compartments and in the extracellular environment is critical for cellular homeostasis in general and in particular for ion transport across membranes, protein degradation, bone resorption, and sperm maturation (Nishi and Forgac 2002; Pietrement et al. 2006; Kane 2006; Wang and Hiesinger 2013). The preservation of

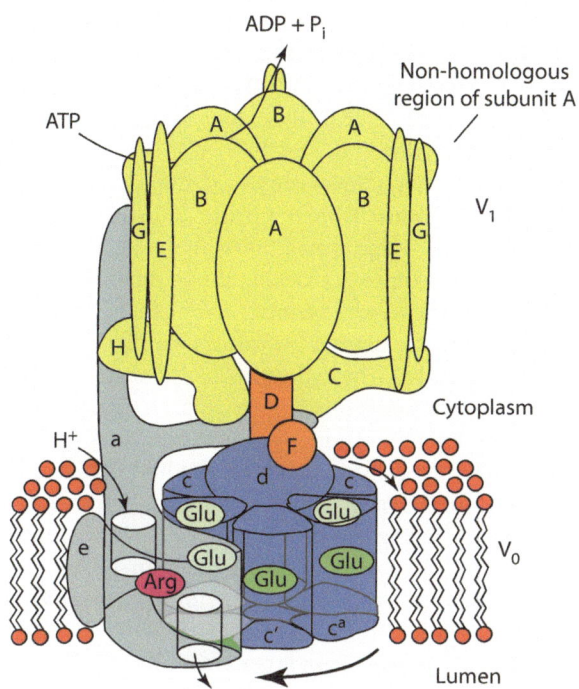

☑ **Fig. 3.5** Structure and mechanism of the V(H⁺)-ATPase. The figure shows the two domains, V_1 and V_0, of the V(H⁺)-ATPase. The peripheral V_1 domain consists of eight different subunits, A–H (*yellow and orange*) and is responsible for ATP hydrolysis. The integral V_0 domain is composed of six subunits a, c, c', c'', d, and e (*blue and gray*) and is involved in the translocation of protons across the membrane. ATP hydrolysis drives the rotation of a central rotor, which is composed of the D, F, d, and c, c', and c'' proteolipid subunits. Protons translocate through two hemi-channels and an arginine residue (*red*) of the a subunit. The V_1 and V_0 domains are connected by a central stalk, which is composed of subunits D, F, and d, and three peripheral stalks, which are composed of subunits C, E, G, H, and the N-terminal cytoplasmic domain of subunit a (Reproduced from Toei et al. (2010) with permission)

intracellular pH is effected by proton pumps dependent on ATP hydrolysis, the vacuolar or V(H⁺)-ATPases (Forgac 2007; Cipriano et al. 2008). V(H⁺)-ATPases are present in the plasma and endo-membranes of all eukaryotic cells, performing variable functions according to their specific locations (Nishi and Forgac 2002; Kane 2006; Forgac 2007; Jefferies et al. 2008; Toei et al. 2010). The enzyme is responsible for the acidification of lysosomes, endosomes, and vesicles and also generates the driving force for neurotransmitter uptake by secretory vesicles (Forgac 2007). The plasma membrane V(H⁺)-ATPase participates in many cellular processes (Wagner et al. 2004; Toyomura et al. 2003; Pietrement et al. 2006), and mutations in its subunits result in various human diseases (Smith et al. 2000; Frattini et al. 2000). V(H⁺)-ATPases are also associated with metastatic tumor cell formation (Sennoune et al. 2004; Gocheva et al. 2007; Kozik et al. 2013).

The V(H⁺)-ATPase is a large protein complex consisting of multiple subunits organized into two separate functional domains V_0 and V_1 (☑ Fig. 3.5). The integral V_0 domain is a 260-kDa integral membrane domain comprising six different subunits, in which the c subunit contains a reversibly protonated glutamate residue and participates in H^+ transport. ATP hydrolysis itself takes place in the V_1 structure, a 650-kDa peripheral cytoplasmic complex comprising eight different subunits (Forgac 2007; Cipriano et al. 2008; Jefferies et al. 2008; Toei et al. 2010; Scott et al. 2011; Benlekbir et al. 2012). The peripheral V_1 domain consists of three A subunits and three B subunits, arranged in alternating locations, forming a hexamer, exhibiting three catalytic sites for ATP hydrolysis and three ATP-modulation sites that bind but do not hydrolyze ATP. The remaining subunits form two shafts connecting the two domains: the rotational center shaft (D and F subunits) and a peripheral shaft (subunits C, E, G, and H) (Forgac 2007; Cipriano et al. 2008; Jefferies et al. 2008; Toei et al. 2010). Although all subunits are essential for enzyme activity, complexes lacking the H subunit are unstable and inactive, suggesting that this subunit affects the arrangement of the others (Hirata et al. 2003).

Proton translocation across the membrane occurs via the V_0 complex which consists of six subunits (a, c_{4-5}, c', c'', d, and e). The hydrophobic c, c', and c'' subunits are proteolipids and form transmembrane helices arranged in a ring structure within the membrane bilayer. Each possesses a glutamate residue that is reversibly protonated during proton transport. The a subunit has two distinct domains: the hydrophobic N-terminal domain facing the cytosol and a C-terminal portion consisting of 8–9 transmembrane helices that form two hemi-channels, one facing the cytosol, the other in the opposite direction. An arginine residue in the TM7 domain also participates in proton transport. Finally, a peripheral d subunit connects the c proteolipid ring to the central V_1 shaft (Forgac 2007; Cipriano et al. 2008; Jefferies et al. 2008; Toei et al. 2010).

The catalytic cycle of the V(H^+)-ATPase is generated by means of a rotational mechanism in which the A_3B_3 hexamer is stationary relative to the a subunit. The peripheral shafts formed by the C, E, G, and H subunits and the N-terminal domain of the a subunit provide communication between the two complexes. Each EG pair interacts with the C or H subunits and these with the a subunit. This set of subunit interactions ensures immobilization of V_1 relative to V_0 (Forgac 2007; Cipriano et al. 2008; Jefferies et al. 2008; Toei et al. 2010). According to this model, in a first step the proton moves from the cytosol through the hemi-channel of the a subunit and protonates the carboxyl group of the glutamate residue of the c subunit. ATP hydrolysis in the A subunit causes rotation of the c ring. After complete rotation of the c ring, the protonated glutamate residue approaches the arginine residue located in the a subunit hemi-channel. Finally, the arginine residue causes deprotonation of the glutamate residue, enabling the proton to reach the extracellular medium or organelle lumen (Cipriano et al. 2008; Toei et al. 2010).

Cells are capable of regulating V(H^+)-ATPase activity in both temporal and spatial modes (Forgac 2007; Toei et al. 2010). Their diversity of functions involves a complex set of regulatory mechanisms of enzyme activity, including reversible dissociation, and changes in coupling efficiency between ATP hydrolysis and proton transport (Cipriano et al. 2008; Jefferies et al. 2008). The most well-characterized regulatory mechanism of V(H^+)-ATPase activity is the reversible dissociation of the V_0 and V_1 domains found in yeast, insect, and mammalian cells (Forgac 2007; Cipriano et al. 2008; Jefferies et al. 2008; Toei et al. 2010; Oot and Wilkens 2012). This mechanism is triggered in response to decreased glucose levels, resulting in the uncoupling of ATP hydrolysis and proton transport (Forgac 2007; Cipriano et al. 2008; Jefferies et al. 2008; Toei et al. 2010). The C subunit plays a critical role in this reversible dissociation, since it separates completely from the V_1 and V_0 domains (Iwata et al. 2004; Jefferies et al. 2008). The nonhomologous domain of the a subunit also seems to be important for dissociation (Toei et al. 2010). A further regulatory mechanism is the reversible formation of disulfide bonds between conserved cysteine residues (Cys_{254} and Cys_{532}) present in the catalytic site of the A subunit, resulting in enzyme inactivation. The formation of such bridges prevents conformational changes in the enzyme (Forgac 2007; Jefferies et al. 2008; Toei et al. 2010). Various cell types regulate their $V(H^+)$-ATPase activity through the number of enzyme molecules present in the plasma membrane. This adjustment is effected by means of intracellular vesicles containing a large number of $V(H^+)$-ATPase molecules that fuse with the plasma membrane, increasing the number of enzyme molecules and consequently, proton flow (Forgac 2007; Cipriano et al. 2008; Jefferies et al. 2008; Toei et al. 2010). A final regulatory mechanism involves alterations in the coupling efficiency of the enzyme. Different *a* subunit isoforms differ in the efficiency with which protons are transported in response to ATP hydrolysis (Jefferies et al. 2008).

The investigation of specific inhibitors of $V(H^+)$-ATPase activity is important for developing drugs to treat various diseases related to $V(H^+)$-ATPase function (Toei et al. 2010). Two structurally related compounds that specifically inhibit $V(H^+)$-ATPase activity at nanomolar concentrations are bafilomycin A_1 and concanamycin (Dröse et al. 1993; Fernandes et al. 2006; Huss and Weczorek 2009; Osteresch et al. 2012). Bafilomycin A_1 is an antibiotic with a 16-atom lactone ring, isolated from *Streptomyces griseus*, while concanamycin isolated from *Streptomyces diastatochromogenes*, also an antibiotic, has a lactone ring of 18 carbon atoms (Dröse et al. 1993). Binding between enzyme and inhibitor takes place by interaction with the V_0 complex in which the *c* and *a* subunits are fundamental (Bowman et al. 2004; Huss and Weczorek 2009). This interaction blocks the transport of protons, since *c* ring rotation is prevented (Wang et al. 2005).

$V(H^+)$-ATPase isoforms in mammals exhibit different and independent functions. While they are expressed in a tissue-specific manner, one isoform is ubiquitous (Nishi et al. 2003; Paunescu et al. 2004; Forgac 2007; Jefferies et al. 2008; Toei et al. 2010; Murata et al. 2002; Smith et al. 2002; Wagner et al. 2004; da Silva et al. 2007; Sun-Wada and Wada 2010). In contrast, in yeast, with the exception of the *a* subunit, all other subunits are unique isoforms (Forgac 2007; Jefferies et al. 2008; Toei et al. 2010).

3.5 The Crustacean $V(H^+)$-ATPase and Its Role in Osmoregulation

The $V(H^+)$-ATPase has played a crucial role in the adaptation of crustaceans to fresh water (Onken and Putzenlechner 1995; Morris 2001; Weihrauch et al. 2004b; Tsai and Lin 2007; Faleiros et al. 2010; Firmino et al. 2011; Lee et al. 2011; McNamara and Faria 2012) and participates actively in osmoregulatory ion uptake across the gill epithelium of freshwater-tolerant crustaceans (Zare and Greenaway 1998; Towle and Weihrauch 2001; Weihrauch et al. 2001; Lee et al. 2011). The $V(H^+)$-ATPase is also involved in acid-base equilibria (Tresguerres et al. 2008) and ammonia excretion (Weihrauch et al. 2002, 2004b; Bianchini et al. 2008; Freire et al. 2008).

The localization and activity of the $V(H^+)$-ATPase are crucial for ion transport against the steep concentration gradients encountered by many crustacean species (Ehrenfeld and Klein 1997; Weihrauch et al. 2004b; Patrick et al. 2006; Tsai and Lin 2007). In freshwater shrimps, the $V(H^+)$-ATPase is located apically in the pillar cell flanges of the gill epithelium (Boudour-Boucheker et al. 2014; Pinto et al. 2016) and appears to complement the driving force of the (Na^+, K^+)-ATPase present in the invaginations of the intralamellar septal cells (McNamara and Torres 1999) in energizing the uptake of NaCl from fresh water (McNamara and Lima 1997; Faleiros et al. 2010; McNamara and Faria 2012). In *Macrobrachium olfersii* and *M. amazonicum*, the apical surface of the gill pillar cells is greatly augmented by a system of extensive evaginations associated with mitochondria in the subapical cytoplasmic (Freire and McNamara 1995; McNamara and Lima 1997; Belli et al. 2009; Faleiros et al. 2010; Boudour-Boucheker et al. 2014). These evaginations increase the area of apical membrane available for inclusion of the $V(H^+)$-ATPase and the Cl^-/HCO_3^- exchanger (Faleiros et al. 2010). In the inner branchiostegite epithelium, another site of ion transport, the $V(H^+)$-ATPase and (Na^+, K^+) ATPase are localized in the same cells, in electron-dense areas (Boudour-Boucheker et al. 2014). The $V(H^+)$-ATPase is also distributed intensely in the marginal channels of the gill lamellae in *M. amazonicum* (Lucena et al. 2015); this channel collects and distributes hemolymph through the capillaries and lacunae across each hemi-lamella and serves as a shunt under certain conditions (McNamara and Lima 1997).

In crabs, the $V(H^+)$-ATPase has been immunolocalized in the apical membranes of the gill ionocytes of *Eriocheir sinensis* (see Freire et al. 2008, Figure 11) and four marine species, including *Uca formosensis*, *Ocypode stimpsoni*, *Chasmagnathus convexus*, and *Helice formosensis*, all acclimated to low (5 ‰) salinity (Tsai and Lin 2007). Studies on the gills of *E. sinensis* (Gilles and Péqueux 1985) and *Carcinus maenas* (Compère et al. 1989) and the shrimp *Penaeus aztecus* (Foster and Howse 1978) show that cellular adaptation to dilute medium includes the development of an extensive system of apical evaginations associated with numerous mitochondria in the subapical cytoplasm of the lamellar epithelial ionocytes. In other crab species that do not survive in fresh water, such as *Scylla paramamosain*, *Macrophthalmus abbreviatus*, *Macrophthalmus banzai*, and *Uca lactea*, the $V(H^+)$-ATPase is located in the cytoplasm (Tsai and Lin 2007). In *C. maenas*, the $V(H^+)$-ATPase B subunit is also cytoplasmic and not localized in the apical membranes of the epithelial cells of anterior and posterior gills (Weihrauch et al. 2001), suggesting that the enzyme also can be associated with cytoplasmic vesicles. In this case, the $V(H^+)$-ATPase appears to acidify such vesicles by H^+ accumulation rather than generating an electrochemical proton gradient across the apical membrane as might be expected from its role in osmoregulatory NaCl transport (Larsen et al. 1992; Lin and Randall 1993; Zare and Greenaway 1998).

The crustacean $V(H^+)$-ATPase can mediate epithelial proton transport (Onken and Putzenlechner 1995; Weihrauch et al. 2001, 2002). However, while the enzyme appears to play an important osmoregulatory role in species like the freshwater-tolerant *E. sinensis*, in marine crabs like *C. maenas*, the $V(H^+)$-ATPase likely functions in the acidification of intracellular organelles and not in transbranchial NaCl uptake (Weihrauch et al. 2001). As seen for several apical antiporters, the $V(H^+)$-ATPase may contribute to Na^+ uptake in fresh water (Towle et al. 1997; Onken 1999). The outward pumping of H^+ by the $V(H^+)$-ATPase generates a large electrical potential across the apical ionocyte membrane, providing an inwardly directed electrical gradient for Na^+ entry via Na^+ channels in the apical membrane (Zare and Greenaway 1998; Morris 2001; Arata et al. 2002). These apical Na^+ channels have been demonstrated by Zeiske et al. (1992). Whole-body uptake of silver in *Daphnia magna* is dependent on Na^+ transport. In the larval gill epithelium, $V(H^+)$-ATPase activity coupled with a Na^+ channel provides the driving force for Na^+ uptake; Ag^+ competes with Na^+ for transport via this mechanism (Bianchini and Wood 2003). The absence of such a mechanism in adult *D. magna* results in less effective Ag^+ uptake by a putative Na^+ channel associated with an electrogenic $2Na^+/H^+$ exchanger (Shetlar and Towle 1989). Further, in *E. sinensis*, Cl^- transport through the apical Cl^-/HCO_3^- exchanger depends on the gradient produced by the $V(H^+)$-ATPase in the apical cell membranes

(Onken 1996). Although both the (Na^+, K^+)- and $V(H^+)$-ATPase are involved in Cl^- absorption in strong hyperregulators (Freire et al. 2008; Belli et al. 2009), the $V(H^+)$-ATPase appears to be essential for chloride uptake across the distal epithelium of the posterior gill lamellae in the freshwater crab *Dilocarcinus pagei* (Weihrauch et al. 2004b).

In addition to the electrophysiological methods used to investigate the role of the $V(H^+)$-ATPase in crustacean osmoregulation (Onken and McNamara 2002; Genovese et al. 2005), several kinetic studies have examined the crustacean gill $V(H^+)$-ATPase, which has been characterized kinetically in *M. amazonicum* (Faleiros et al. 2010), *D. pagei* (Firmino et al. 2011; Weihrauch et al. 2004b), *E. sinensis* (Onken and Putzenlechner 1995; Morris 2001), and *M. amazonicum* (Lucena et al. 2015). However, negligible $V(H^+)$-ATPase activity was encountered in *Callinectes danae* (Masui et al. 2002), *C. ornatus* (Garçon et al. 2009), and 21‰-acclimated *M. amazonicum* (Faleiros et al. 2010). Cultivated juvenile and adult *M. amazonicum* show similar gill $V(H^+)$-ATPase-specific activities (Lucena et al. 2015) that are 50% greater than in *E. sinensis* (Onken and Putzenlechner 1995; Morris 2001), *D. pagei* (Firmino et al. 2011), and wild *M. amazonicum* (Faleiros et al. 2010) but tenfold greater than in *U. formosensis* (Tsai and Lin 2007).

Long-term regulatory mechanisms of $V(H^+)$-ATPase activity operate effectively in *D. pagei*, since after 10-day exposure to 21‰ salinity, enzyme activity in posterior gills is threefold less than in fresh water (Firmino et al. 2011). This agrees with the decrease in $V(H^+)$-ATPase activity seen in postlarval estuarine shrimp *Litopenaeus vannamei* in response to increasing salinity (Pan et al. 2007) and the threefold increase after 7-day exposure of the fiddler crab *Uca formosensis* to 5‰ salinity (Tsai and Lin 2007). Such regulation in response to salinity change may involve the induction/inhibition of mRNA synthesis or retention, leading to alterations in the rate enzyme synthesis (Weihrauch et al. 2004b; Luquet et al. 2005; Tsai and Lin 2007; Faleiros et al. 2010). The $V(H^+)$-ATPase B subunit is highly conserved in *C. maenas, E. sinensis, C. sapidus, N. granulatus*, and *Cancer irroratus* (Weihrauch et al. 2004b). In *M. amazonicum* juveniles and adults, $V(H^+)$-ATPase B subunit mRNA expression is similar (Lucena et al. 2015); however, expression decreases fourfold in adults acclimated to 25‰ salinity (Faleiros et al. 2010). In *N. granulata* acclimated to 2‰ salinity, mRNA levels increase fourfold (Luquet et al. 2005), but there is no change in the relative abundance of $V(H^+)$-ATPase in *U. formosensis* (Tsai and Lin 2007) or in the shrimp *Halocaridina rubra* (Havird et al. 2014) in response to salinity changes. These findings suggest that in addition to alterations in mRNA expression, other mechanisms may regulate $V(H^+)$-ATPase activity.

Changes in pH are also important modulators of transepithelial solute transport, endocrine function, and cell growth and differentiation (Boron 1986). Maximum gill $V(H^+)$-ATPase activity is found at pH 7.5 in many crustaceans (Zare and Greenaway 1998; Onken and Putzenlechner 1995; Onken et al. 2000; Weihrauch et al. 2004b; Pan et al. 2007; Firmino et al. 2011; Lucena et al. 2015). Lower activities at lower pH values result from modification of subunit structure, revealing a hidden protonation site (Rastogi and Girvin 1999; Muller et al. 2002). In *L. vannamei* (Pan et al. 2007; Wang et al. 2012a), *N. granulatus* (Tresguerres et al. 2008), *Marsupenaeus japonicus* (Pan et al. 2010), and *Fenneropenaeus chinensis* (Pan et al. 2010), changes in extracellular pH modulate both $V(H^+)$-ATPase activity and expression.

The regulation of gill $V(H^+)$-ATPase activity during ontogenetic development has been little studied in crustaceans. During the early development of *M. japonicus* and *F. chinensis*, $V(H^+)$-ATPase activity increases from absent in the nauplii to stable levels in the zoeae (Pan et al. 2010). Kinetic studies have revealed that the $V(H^+)$-ATPase of adult *M.*

amazonicum has a threefold greater affinity for ATP than the juvenile enzyme, while the affinity for Mg^{2+} of the adult $V(H^+)$-ATPase is fivefold greater than in the juvenile (Lucena et al. 2015). Further, the inhibition constant for bafilomycin A_1 of adult *M. amazonicum* $V(H^+)$-ATPase is twofold less than in juveniles. These different affinities for ATP, Mg^{2+}, and bafilomycin A_1 for the gill $V(H^+)$-ATPase from different developmental stages of *M. amazonicum* suggest the expression of distinct isoenzymes that may play a role in ontogenetic changes in enzyme activity (Lucena et al. 2015). Although binding sites for ATP and Mg^{2+} are located in the A and B subunits in the V_1 complex (Kawasaki-Nishi et al. 2003; Nakanishi-Matsui et al. 2010; Toei et al. 2010), expression of different isoforms of other subunits can induce conformational changes that affect affinity for substrate and modulators.

The $V(H^+)$-ATPase also plays an important role in crustacean ammonia excretion. After transport or diffusion from the hemolymph into the gill ionocytes, the diffusion of cytoplasmic NH_3 into intracellular vesicles and its consequent protonation to NH_4^+ by $V(H^+)$-ATPase driven acidification allows excretion via an exocytic mechanism (Weihrauch et al. 2002, 2004a, 2009; Henry et al. 2012). The inhibition of $V(H^+)$-ATPase activity in brackish water-acclimated *C. maenas* reduces gill NH_4^+ excretion rate by 66 %, while inhibitors of the cytoskeletal microtubule network completely abolish active ammonia excretion (Weihrauch et al. 2002). Apparently, ammonia is trapped in its ionic form NH_4^+ in the acidified vesicles that are transported over the microtubule network to the target membrane. Gene expression of the $V(H^+)$-ATPase B subunit is upregulated in the marine crab *Metacarcinus magister* after short-term ammonia exposure (Martin et al. 2011). Consequently, increased $V(H^+)$-ATPase protein expression likely increases NH_3 protonation to NH_4^+, underpinning increased NH_4^+ excretion driven by the $V(H^+)$-ATPase. In *N. granulata*, $V(H^+)$-ATPase mRNA expression is a consequence of $V(H^+)$-ATPase participation in ammonium excretion across the gills and not of altered external salinity (Weihrauch et al. 2002, 2004a).

3.6   Gill ATPases and Ammonia Excretion in Crustaceans

The by-products generated during the metabolism of nitrogenous compounds like amino acids and nucleic acids are often toxic and cannot accumulate within the body to any great degree without serious consequences, including the death of an organism (Randall et al. 2000). Exposure of crustaceans to toxic quantities of ammonia disturbs many vital biological processes such as ionic regulation, cell permeability, and immune system functions (Young-Lai et al. 1991; Harris et al. 2001; Le Moullac and Haffner 2000). Elevated ammonia-N can rapidly cause severe damage to the gill structure of crustaceans, including necrosis, hyperplasia, epithelial damage, pillar cell disruption, and collapse of the gill lamellae (Rebelo et al. 2000; Romano and Zeng 2007). Long-term exposure to ammonia renders *Metacarcinus magister* completely unable to actively excrete ammonia via the gills (Martin et al. 2011).

The ammonia concentration in the aquatic environment is usually low owing to the action of nitrifying bacteria (Weihrauch et al. 1999). In unpolluted seawater, ammonia concentrations rarely exceed 5 μmol L^{-1} (Koroleff 1983) and are much lower than in the hemolymph of crustaceans, around 100 μmol L^{-1} (Rebelo et al. 1999). Benthic crustaceans, often buried in the sediment, may encounter ammonia concentrations up to 2–3 mmol L^{-1}, which can lead to diffusive NH_4^+ influx to the hemolymph (Rebelo et al.

1999; Weihrauch et al. 1999). However, during acute ammonia exposure, the crab *Neohelice granulata* maintains its hemolymph ammonia concentration lower than that of the external medium (Rebelo et al. 1999). Apparently, exposure to high ambient ammonia concentrations has led to the evolutionary selection of an efficient mechanism for active NH_4^+ excretion by the gill epithelia of crustaceans (Weihrauch et al. 2004a). Further, mRNA expression levels of the gill (Na^+, K^+)-ATPase, $V(H^+)$-ATPase, cation/H^+-exchanger and Rhesus-like ammonia transporter are downregulated when hemolymph ammonia levels increase to nearly 1 $mmoL^{-1}$ (Martin et al. 2011). Thus, effective detoxification or excretion of ammonia is critical to maintain cell function and body fluid ammonia levels within a tolerable range (Weihrauch et al. 2004a).

The excretion rate of ammonium has been used to evaluate the effect of various environmental factors on crustacean physiology (Jiang et al. 2000). A decrease in ambient osmolality causes a decrease in tissue amino acids and an increase in ammonium excretion (Lange 1972), while temperature and salinity significantly affect ammonium excretion in shrimp species like *Macrobrachium rosenbergii* (Stern et al. 1984), *Marsupenaeus japonicus* (Chen and Lai 1993), and *Litopenaeus vannamei* (Díaz et al. 2001). Ammonium excretion rates in *L. stylirostris* exposed to salinity fluctuations are linked to osmoregulation since these increase during hyperregulation and are reduced during hyporegulation (Díaz et al. 2004). In the crab *Portunus pelagicus* exposed to 40 mg L^{-1} NH_4^+, posterior gill (Na^+, K^+)-ATPase activity and ammonium excretion rates are higher at 15‰ salinity than at 45‰ salinity (Romano and Zeng 2010). Lowering ambient pH results in higher rates of ammonia and urea excretion (Al-Reasi et al. 2013).

Most aquatic crustaceans excrete ammonia (NH_3/NH_4^+) across their gill epithelia (Randall et al. 2000; Regnault 1987), mainly in its ionic form, NH_4^+. The inhibition of active transepithelial NH_4^+ flux by ouabain applied basally to the gill epithelia in various crab species strongly suggests the involvement of the (Na^+, K^+)-ATPase in this process (Lucu et al. 1989; Weihrauch et al. 1998, 1999). The ability to excrete ammonia against its gradient has important ecological consequences regarding the habitats available to these crustaceans (Weihrauch et al. 2004a).

Ammonia excretion across the gill epithelia is not completely understood (Weihrauch et al. 2004a). The most accepted model holds that excretion is linked to Na^+ transport and thus NH_4^+ passes from the hemolymph into the cytosol of the gill epithelial cells via the (Na^+, K^+)-ATPase in the basal membrane, substituting for K^+ (Weihrauch et al. 2004a). Such movement also may be effected by a basal ammonium transporter and/or Cs^+-sensitive K^+ channels (Weihrauch et al. 2002). Channel-like structures permeable to Na^+ and NH_4^+, sensitive to amiloride, and potentially involved in NH_4^+ extrusion, are found in the gill cuticle of *C. maenas* (Onken and Riestenpatt 2002; Weihrauch et al. 2002). Hydrated NH_4^+ and K^+ ions have the same ionic radius, 1.45 Å (Weiner and Hamm 2007) and, owing to their K^+-like behavior, ammonium ions can compete with K^+ for K^+-transporting proteins such as the (Na^+, K^+)-ATPase and K^+ channels (Choe et al. 2000). Further, Cl^- secretion by the $Na^+/K^+/2Cl^-$ cotransporter (NKCC) takes place in the presence of K^+ or NH_4^+, suggesting that in vertebrates at least the NKCC can also operate as a Na^+- NH_4^+- $2Cl^-$ cotransporter (Wall et al. 2001). In addition to the (Na^+, K^+)-ATPase, the $V(H^+)$-ATPase also plays an important role in gill NH_4^+ excretion. In the marine crab *M. magister* subjected to high external ammonia, expression of the gill $V(H^+)$-ATPase B subunit is upregulated (Martin et al. 2011). In *M. amazonicum*, increased total ammonia nitrogen increases $V(H^+)$-ATPase B subunit mRNA expression by 2.5-fold, while (Na^+, K^+)-ATPase α-subunit expression is unchanged (Pinto et al. 2016). These findings suggest the coupling

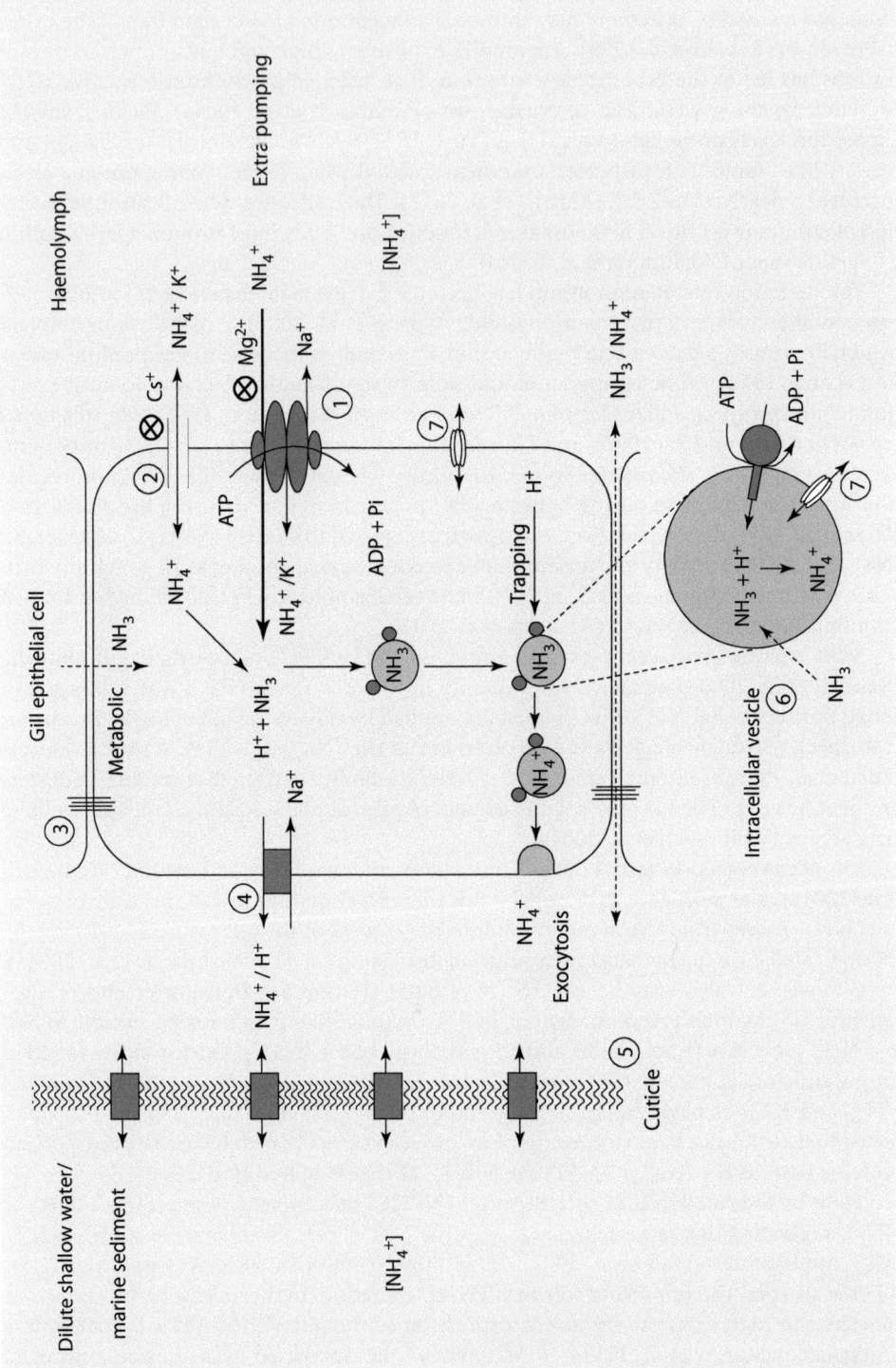

☐ **Fig. 3.6** Hypothetical model for active ammonia excretion across the gill epithelium of the euryhaline crab *Callinectes danae*. ① The main driving force for NH_4^+ transport into the cell is the (Na^+, K^+)-ATPase located in the basolateral membrane. An additional, magnesium-inhibitable, pumping force for NH_4^+ extrusion is represented by a second NH_4^+-binding site, which appears when the pump is fully saturated by K^+ ions. ② Cs^+-sensitive K^+ channels, located in the basal membrane, do not discriminate between K^+ and NH_4^+ ions. ③ Septate junction. ④ Putative $Na^+/NH_4^+(H^+)$ transporter, located in the apical membrane, particularly active during transcellular NH_4^+ extrusion. ⑤ Presumed, amiloride-sensitive, cation-permeable, channel-like structures in the cuticle allow the diffusion of NH_4^+ ions to the external medium. ⑥ Cytoplasmic NH_3 diffuses into intracellular vesicles containing a $V(H^+)$-ATPase proton pump, which acidifies their interior leading to NH_4^+ formation. NH_4^+ is extruded to the subcuticular space via exocytosis. ⑦ A rhesus-like protein, a supposed ammonia transporter of unknown localization, may support the vesicular acid-trapping mechanism (Reproduced from Masui et al. (2005a) with permission)

of NH_4^+ excretion by the gill epithelium to the activity of both ATPases and to the expression of V(H$^+$)-ATPase B subunit mRNA.

The gill (Na$^+$, K$^+$)-ATPase from the marine swimming crab *C. danae* is synergistically stimulated by NH_4^+ and K$^+$ (Masui et al. 2002, 2009). Catalytic activity is enhanced by up to 90 % when NH_4^+ ions are added simultaneously with K$^+$, compared to K$^+$ alone, suggesting that the two ions bind to different sites on the (Na$^+$, K$^+$)-ATPase. The gill (Na$^+$, K$^+$)-ATPase from the freshwater shrimp *M. olfersii* responds similarly (Furriel et al. 2004), suggesting in this case that at high NH_4^+ concentrations the enzyme exposes a new binding site for NH_4^+ that, after NH_4^+ binding, modulates pump activity independently of K$^+$. Synergistic stimulation of the gill (Na$^+$, K$^+$)-ATPase by K$^+$ plus NH_4^+ is seen also in *C. ornatus* (Garçon et al. 2009) although quite differently from *C. danae* and *M. olfersii*. When NH_4^+ substitutes for K$^+$, specific activity increases by 50 % compared to K$^+$ alone. At elevated NH_4^+ concentrations, the enzyme is fully active, regardless of K$^+$ concentration, and K$^+$ cannot displace NH_4^+ from its exclusive binding sites. Further, the binding of NH_4^+ to its specific sites induces an increase in enzyme apparent affinity for K$^+$, which may contribute to maintaining K$^+$ transport, assuring that exposure to elevated ammonia concentrations does not decrease intracellular K$^+$ levels (Garçon et al. 2009). The (Na$^+$, K$^+$)-ATPases from fresh caught hermit crab *Clibanarius vittatus* (Gonçalves et al. 2006) and *M. amazonicum* (Santos et al. 2007) respond similarly, suggesting that NH_4^+ binding to the enzyme can be species specific. The additional increase in (Na$^+$, K$^+$)-ATPase activity in *M. rosenbergii* (França et al. 2013) and 45 ‰-acclimated *C. vittatus* (Lucena et al. 2012) in the presence of K$^+$ plus NH_4^+ suggests distinct binding sites on the enzyme, both occupied simultaneously by these two ions, and that neither K$^+$ nor NH_4^+ can displace each other from their respective binding sites. While each ion stimulates enzyme activity in the presence of the other, $K_{0.5}$ remains unaffected. NH_4^+ binding seems to induce conformational changes that affect V_M with minor effects on $K_{0.5}$ (França et al. 2013; Lucena et al. 2012). Intriguingly, synergistic stimulation by NH_4^+ and K$^+$ is not seen in the freshwater crab *D. pagei* (Furriel et al. 2010) although each ion modulates affinity for the other. While distinct binding sites for NH_4^+ and K$^+$ are present on the *D. pagei* enzyme, both sites may be simultaneously occupied either by NH_4^+ or K$^+$, attaining similar maximum velocities. NH_4^+ binding to its site induces conformational changes that increase K$^+$ affinity but not activity; NH_4^+ can be dislocated from its binding site by increasing K$^+$ concentrations (Furriel et al. 2010). The absence of a synergistic effect induced by K$^+$ plus NH_4^+ in the pelagic shrimp *Xiphopenaeus kroyeri* suggests that both ions compete for the same site on the enzyme molecule (Leone et al. 2015a).

Given these observations, we have extended Weihrauch's model (Weihrauch et al. 2004a), proposing that the *C. danae* (Na$^+$, K$^+$)-ATPase is also responsible for directing NH_4^+ transport to the external environment (● Fig. 3.6) (Masui et al. 2005a).

The synergistically stimulated (extra-pumping) activity of the (Na$^+$, K$^+$)-ATPase is the main driving force for NH_4^+ transport into the cell, where it may replace K$^+$ or be transported via a second binding site, exposed when the enzyme is saturated with K$^+$. Basal K$^+$ channels, which do not discriminate between K$^+$ and NH_4^+, are also present in the gill epithelial cells, together with a Rhesus-like protein (Weihrauch et al. 2004a), a carrier that can mediate the transfer of NH_4^+ across cell membranes. Once in the cytoplasm, NH_4^+ can be discharged into the subcuticular space by exocytosis inside vesicles acidified by a V-(H$^+$)-ATPase or via a Na$^+$/NH_4^+(H$^+$) cotransporter present in the apical membrane. Finally, NH_4^+ diffuses across the cuticle through amiloride-sensitive cation-permeable structures.

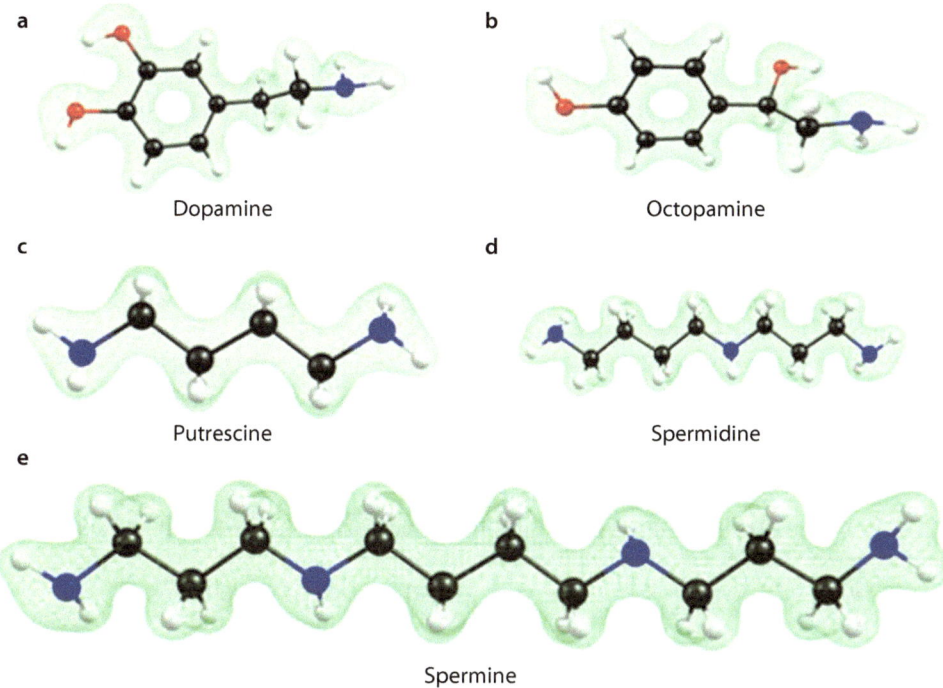

a   Dopamine

b   Octopamine

c   Putrescine

d   Spermidine

e   Spermine

◘ **Fig. 3.7** Schematic representation of the charge densities of various polyamines. (**a**) Dopamine; (**b**) octopamine; (**c**) putrescine; (**d**) spermine; (**e**) spermidine

3.7 A Role for Polyamines in Regulating Crustacean Gill (Na⁺, K⁺)-ATPase Activity

Polyamines are basic nitrogen compounds resulting from the decarboxylation of amino acids or by amination/transamination of aldehydes and ketones (Askar and Treptow 1986; Maijala and Eerola 1993). The polyamines putrescine, spermidine, and spermine (◘ Fig. 3.7) are present in almost all cells and play important roles in protein synthesis, cell division, and cell growth (Williams 1997).

They also alter membrane permeability and glucose and ion transport (Elgavish et al. 1984; Wild et al. 2007). Because they are fully protonated under physiological conditions, polyamines can interact with nucleic acids, especially RNA, ATP, specific proteins, and phospholipids (Watanabe et al. 1991; Igarashi and Kashiwagi 2010). Polyamines are synthesized by microorganisms, plants, and animals (Brink et al. 1990) and are directly or indirectly essential for normal cellular function and proliferation (for review, see Kalac 2009; Igarashi and Kashiwagi 2010; Pegg 2009). They are ubiquitous, and their localization within cells is not restricted to RNA- and DNA-rich structures. They are also present in the hemolymph and gills of crustaceans (Péqueux et al. 2002). Putrescine, spermidine, and spermine have been detected in the nanomolar range in gill tissue from both *E. sinensis* and *C. sapidus* (Lovett and Watts 1995; Péqueux et al. 2002).

Polyamines are closely associated with phospholipids (Toner et al. 1988) and with negatively charged plasma membrane proteins (Lin et al. 2006; Tassoni et al. 1996) but not with cytoplasmic proteins. However, spermine in particular interacts with certain ion channels (Williams 1997). Intracellular spermine is responsible for the intrinsic gating and rectification by strong inward rectifying K^+ channels by directly plugging the ion channel pore. These K^+ channels regulate resting membrane potential in both excitable and nonexcitable cells and the threshold excitability of neurons and muscle cells (Ficker et al. 1994; Lopatin et al. 1994). Intracellular spermine also causes inward rectification of some subtypes of Ca^{2+}-permeable glutamate receptors in the central nervous system, again by plugging the receptor channel pore; spermine may permeate the ion channel of such receptors.

Extracellular polyamines exert multiple effects on the N-methyl-d-aspartate (NMDA) subtype of glutamate receptor, including stimulation that increases receptor currents, and voltage-dependent block (Williams 1997). Activation of the NMDA receptor results in Ca^{2+} influx and the activation of nitric oxide synthase (NOS) leading to nitric oxide (NO) synthesis (Prast and Philippu 2001). Nitric oxide then stimulates guanylate cyclase (GC), increasing intracellular cyclic GMP (cGMP) titers (Rubin and Ferrendelli 1977), which activates cGMP-dependent protein kinase (PKG). PKG-mediated effects on neurotransmission include the inhibition of Ca^{2+} currents (Meriney et al. 1994) and an increase in firing rate and neurotransmitter release (Akamatsu et al. 1993). PKG also may be implicated in long-term potentiation (Zhuo et al. 1994) and in regulation of (Na^+, K^+)-ATPase activity (Munhoz et al. 2005). Polyamines may dually modulate the NMDA receptor (Williams 1997). MK-801, a noncompetitive NMDA receptor antagonist, and arcaine, an antagonist of the polyamine binding site on the NMDA receptor, prevent the inhibitory effect on (Na^+, K^+)-ATPase activity induced by spermidine (Carvalho et al. 2012). N-nitro-L-arginine methyl ester (L-NAME), 1 h-[1,2,4]-oxadiazole-[4,3-a]-quinoxalin-1-one (ODQ), and KT5823, inhibitors of NOS, GC, and PKG, respectively, prevent the inhibitory effect of spermidine on (Na^+, K^+)-ATPase activity (Carvalho et al. 2012). Spermidine decreases (Na^+, K^+)-ATPase activity hippocampus slices, but not in homogenates, suggesting that NMDA receptor/NO/cGMP/PGK mechanisms mediate the effect of spermidine on (Na^+, K^+)-ATPase activity (Carvalho et al. 2012).

The cellular concentration of polyamines is finely regulated by biosynthesis, degradation, uptake, and excretion (Igarashi and Kashiwagi 1999; Wallace et al. 2003). In both prokaryotes and eukaryotes, polyamine levels increase during the cellular response to proliferative stimuli (Marton and Pegg 1995). ATP and sodium gradient-dependent transport across the cell membrane, and de novo polyamine biosynthesis are two general mechanisms that regulate intracellular polyamine content (Koenig et al. 1983; Seiler and Dezeure 1990; Aziz et al. 1994; Kobayashi et al. 1999). Igarashi and Kashiwagi (2010) have reviewed polyamine metabolism in mammalian cells. Putrescine is synthesized from ornithine by ornithine decarboxylase or from S-adenosylmethionine by S-adenosylmethionine decarboxylase; both enzymes are rate limiting (Nishimura et al. 2002; Pendeville et al. 2001). Spermidine is synthesized from putrescine by spermidine synthase. The conversion of spermidine to spermine is catalyzed by spermine synthase, while both spermidine/spermine N1-acetyltransferase and acetylpolyamine oxidase catalyze the conversion of spermidine to spermine and that of spermidine to putrescine, respectively (Casero and Pegg 1993; Höltta 1977; Jänne et al. 2004). Regulation of cellular polyamine concentration is strongly influenced by a protein known as antizyme, which inhibits the activity of ornithine decarboxylase and also stimulates degradation (Coffino 2001). Antizyme also inhibits polyamine absorption and increases excretion (Suzuki et al. 1994); however, this mechanism is not yet fully elucidated (Igarashi and Kashiwagi 2010).

Polyamine deficiency causes decreases in DNA polymerase, thymidine kinase, DNA ligase I, and flap endonuclease activities, linked to strong inhibition of DNA synthesis (Johansson et al. 2008). The ability to stabilize various cellular and subcellular particles may constitute another physiological action of polyamines (Tabor and Tabor 1984; Jellink and Perry 1967). Osmotic stress apparently causes an increase in total cellular polyamine content with a marked increase in spermidine (Jantaro et al. 2003). Polyamines also inhibit (Na^+, K^+)-ATPase activity in various vertebrate tissues by binding to amino acid residues on the enzyme (Heinrich-Hirsch et al. 1977; Robinson et al. 1986; Kuntzweiler et al. 1995; Quarfoth et al. 1978; Tashima et al. 1978).

The regulation of crustacean gill (Na^+, K^+)-ATPase activity is likely controlled by neuroendocrine factors (Morris and Edwards 1995). Polyamines affect (Na^+, K^+)-ATPase activity (Silva et al. 2008; Garçon et al. 2011) and are involved in osmotic and ionic regulation by interacting directly with the (Na^+, K^+)-ATPase (Lovett and Watts 1995). In *C. sapidus*, gill polyamine levels respond differentially to salinity change, each polyamine responding differently (Lovett and Watts 1995; Watts et al. 1994). Putrescine concentration in all gills is significantly lower in crabs acclimated to low salinity (10‰) than in crabs acclimated to high salinity (35‰) (Lovett and Watts 1995), in contrast to increased putrescine concentrations in *Artemia franciscana* after hypo-osmotic exposure (12‰) (Watts et al. 1994). The induction in *A. franciscana* of ornithine decarboxylase, an enzyme that converts L-ornithine to putrescine, represents one of the first enzyme systems activated in euryhaline organisms during hypo-osmotic stress (Watts et al. 1996).

Polyamine titers, particularly putrescine, are inversely proportional to (Na^+, K^+)-ATPase activity in *Artemia* nauplii acclimated to different salinities (Lee 1992). In *C. sapidus* and *A. franciscana*, spermidine concentration does not differ significantly among acclimation salinities or gills (Lovett and Watts 1995; Watts et al. 1994). In contrast, spermine concentration in gills six and seven of crabs acclimated to low salinity is significantly greater than in crabs acclimated to high salinity, with no changes in gills three and four, regardless of acclimation (Lovett and Watts 1995). This may result from stimulation of cell differentiation in osmoregulatory gills six and seven (Lovett and Watts 1995), since the role of polyamines in cellular differentiation is well known (Heby 1989). Anterior and posterior gills from *E. sinensis* show differences in polyamine concentration (Péqueux et al. 2002). Under optimal Na^+ (100 mmol L^{-1}) and K^+ (10 mmol L^{-1}) concentrations, increasing concentrations of putrescine, spermine, and spermidine inhibit *C. danae* gill (Na^+, K^+)-ATPase by 20 % (Silva et al. 2008). At tenfold lower ion concentrations, spermine, and spermidine, but not putrescine, cause 40 % inhibition (Silva et al. 2008). Inhibition of *C. ornatus* gill (Na^+, K^+)-ATPase activity by spermidine and spermine is concentration dependent (Garçon et al. 2011) in contrast to *C. danae* (Silva et al. 2008). $K_{0.5}$ values for Na^+, K^+, and NH_4^+ in *C. ornatus* gill (Na^+, K^+)-ATPase activity are markedly affected by spermine and spermidine (Garçon et al. 2011) in contrast to the *C. danae* enzyme where spermidine affects the $K_{0.5}$ for Na^+ only (Silva et al. 2008). Putrescine has a negligible effect on *C. ornatus* gill (Na^+, K^+)-ATPase (Garçon et al. 2011) compared to the 20 % inhibition seen for *C. danae* (Silva et al. 2008). Putrescine inhibits the *C. sapidus* (Na^+, K^+)-ATPase by increasing the apparent $K_{0.5}$ for Na^+ (Lovett and Watts 1995). For 15‰- and 28‰-acclimated *C. danae*, the efficiency of spermidine inhibition was greater than spermine (Garçon unpublished data) as seen for the *C. ornatus* gill enzyme (Garçon et al. 2011); in 40‰-acclimated *C. danae*, spermine inhibition is greater than spermidine (Garçon unpublished data). The effect of spermine and spermidine on the Na^+ and K^+ sites on the *C. ornatus* enzyme (Garçon

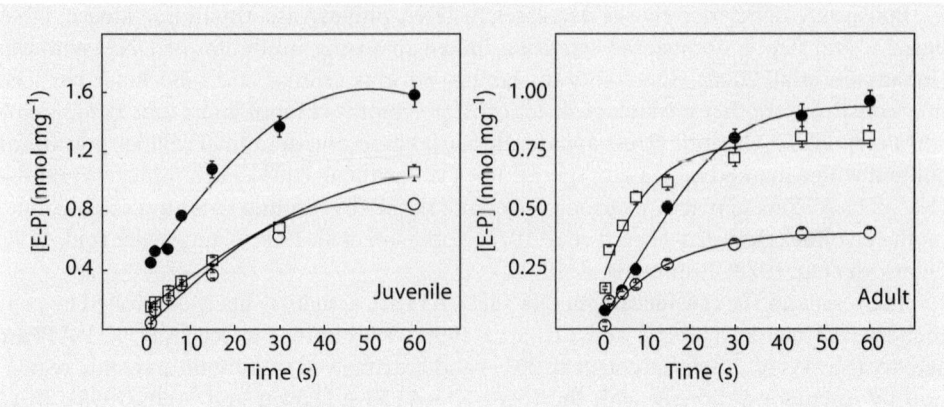

◘ Fig. 3.8 Effect of spermidine and putrescine on [γ-^{32}P]ATP phosphorylation of the gill (Na$^+$, K$^+$)-ATPase from juvenile and adult *M. amazonicum*. *Juvenile*: (●) control; (○) 10 mmol L^{-1} spermidine; (□) 25 mmol L^{-1} putrescine. *Adult*: (●) control; (○) 20 mmol L^{-1} spermidine; (□) 50 mmol L^{-1} putrescine (Reproduced from Lucena et al. (2016) with permission)

et al. 2011) differs considerably from that for the *C. danae* enzyme, where spermidine inhibits pumping activity by competing with Na$^+$ at the Na$^+$-binding site, also inhibiting enzyme dephosphorylation (Silva et al. 2008). Apparently, the effect of spermine and spermidine concentration on inhibition of *C. danae* gill (Na$^+$, K$^+$)-ATPase activity is a consequence of both size and charge since, at physiological pH, spermine has a greater charge density (Silva et al. 2008). Based on the availability of positive charges within its structure, the polyamine showing the greatest charge density should exhibit greater interaction with the cation-binding domain of the (Na$^+$, K$^+$)-ATPase. However, for the *C. ornatus* enzyme, the fact that spermidine was a stronger inhibitor than spermine suggests that polyamine inhibition of (Na$^+$, K$^+$)-ATPase activity is dependent on both the ionic charge radius of the molecule and the aliphatic chain length (Garçon et al. 2011). Both spermidine and putrescine, but not spermine, show important effects on the formation, stability, and dephosphorylation of the phosphoenzyme in juvenile and adult *M. amazonicum* and on K$_I$ values for spermidine inhibition of (Na$^+$, K$^+$)-ATPase activity (Lucena et al. 2016). The time course of phospho-intermediate (EP) formation (◘ Fig. 3.8) is greater in adult than juvenile *M. amazonicum* with spermidine but similar with putrescine (Lucena et al. 2016).

Maximal EP formation for (Na$^+$, K$^+$)-ATPase from juvenile shrimp decreased 46 % and 32 % in the presence of 10 mmol L^{-1} spermidine or 25 mmol L^{-1} putrescine, respectively (Lucena et al. 2016). In adults, maximal EP levels decreased 50 % and 8 % in the presence of 20 mmol L^{-1} spermidine or 50 mmol L^{-1} putrescine, respectively. With both spermidine and putrescine, dephosphorylation rates were higher for adults than juveniles and were always higher than controls (Lucena et al. 2016).

Concluding, polyamines appear to influence osmotic and ionic regulation in crustaceans during acclimation to different salinities. It is difficult to envisage a physiologically important role for spermine in modulating the (Na$^+$, K$^+$)-ATPase present in the ionocyte plasma membranes given that this polyamine is restricted to the cell nucleus (Tabor and Tabor 1984). Spermine may function as a salvage compound and may provide a reserve pool to be converted back to putrescine, as seen in vertebrates (Jänne et al. 1991). Thus, like the species-specific differences seen in ouabain affinity for the (Na^+, K^+)-ATPase (Pressley 1992), the kinetic modulation of crustacean Na,K-ATPase activity by polyamines also seems to be species specific (Lucena et al. 2016).

3.8 Crustacean Ontogeny and the (Na^+, K^+)-ATPase: The Freshwater Shrimp *Macrobrachium amazonicum* as a Model

The establishment of a species in a given niche depends on the ability of each of its developmental stages to adapt to their respective environments. The occupation of fresh water from more concentrated media requires the ability to maintain an elevated hemolymph osmolality and that a species be able to complete its life cycle independently of access to marine or brackish waters (Charmantier 1998). Many species may still be actively invading fresh water, since some of their ontogenetic stages still depend on brackish or marine waters to develop or reproduce (Tan and Choong 1981; Dugger and Dobkin 1975; Read 1982; Moreira et al. 1983; Péqueux 1995).

Crustacean embryos are enveloped by the embryonic yolk membrane and the corium (Stronberg 1972). During early embryogenesis, the eggs may be incubated on the female's pleiopods in sealed pouches and are osmotically protected (Charmantier 1998; Charmantier and Charmantier-Daures 2001). Osmoregulatory capacity then develops gradually and persists throughout the life cycle. However, in some species the eggs are exposed to the external environment, and the egg membranes osmotically protect the embryo. In this case, the embryos osmoconform during embryogenesis and acquire osmoregulatory capacity during the postembryonic stages (Charmantier 1998).

Ontogenetic differences in osmoregulatory capacity result from differential salt tolerances and involve morphological changes that determine the capability of each developmental stage to adjust physiologically and to occupy niches of different salinities (Charmantier 1998; Anger 2003; Boudour-Boucheker et al. 2013). Three basic osmoregulatory patterns are encountered during crustacean ontogenesis: (i) osmoregulatory ability is weak, varies little with development and is not altered by metamorphosis; (ii) adult osmoregulatory ability is established in the first postembryonic stage; and (iii) metamorphosis marks the appearance of adult osmoregulatory ability (Charmantier 1998).

The ontogeny of osmoregulation has been examined in many different crustaceans including *A. franciscana* (Conte et al. 1977; Peterson et al. 1978; Escalante et al. 1995), *Callianassa jamaicense* (Felder et al. 1986), *Penaeus japonicus* (Bouaricha et al. 1994), *Palaemonetes argentinus* (Charmantier and Anger 1999; Ituarte et al. 2005, 2008), *M. rosenbergii* (Wilder et al. 2001), *Crangon crangon* (Cieluch et al. 2005), *M. olfersii* and *M. amazonicum* (Augusto et al. 2007), *Rhithropanopeus harrisii* (Kalber and Costlow 1966), *Hemigrapsus sexdentatus* (Seneviratna and Taylor 2006), *H. crenulatus* (Taylor and Seneviratna 2005; Seneviratna and Taylor 2006) and *H. edwardsii* (Taylor and Seneviratna 2005), *Homarus gammarus* (Haond et al. 1999; Lignot and Charmantier 2001; Khodabandeh et al. 2006), and *Astacus leptodactylus* (Susanto and Charmantier 2000, 2001). The relationship between survival and salinity (0.16–44.3‰) in different developmental stages (zoeae, megalopae, juveniles and adults) has been examined in *Eriocheir sinensis* (Cieluch et al. 2007). The adult pattern of osmoregulation develops in *E. sinensis* through two molts: (i) from a moderately hyper-isoregulating zoea 1 to the moderately hyper-/hypo-regulating megalopa and (ii) from the megalopa to a strongly euryhaline, hyper-/hypo-regulating first juvenile crab stage.

One of the best-studied species is the palaemonid shrimp *Macrobrachium amazonicum* that exhibits at least two geographically isolated populations, one in the Amazon estuary in northern Brazil and the other from the seasonally inundated flood plains of the Pantanal in southwestern Brazil (Charmantier and Anger 2011). There are considerable differences in salinity tolerance and osmoregulatory ability between such populations. In the estuarine population, all ontogenetic stages are strong hyper-osmoregulators in brackish water (1–17‰ salinity) and hypo-osmoregulators at higher salinities. Except for the first postembryonic stage, hyper-osmoregulation in fresh water is absent in zoea I and weak in the other larvae and early juveniles. This complex pattern is consistent with a diadromous life cycle: hatching in rivers, rapid larval downstream transport to brackish estuarine waters, and active upstream migration of later juveniles and adults (Magalhães 1985; Moreira et al. 1986; Magalhães and Walker 1988; Odinetz-Collart 1991; Augusto et al. 2007; Anger 2013). In contrast, the Pantanal population spends its entire life cycle in fresh water (Anger and Hayd 2010; Hayd and Anger 2013). All life cycle stages survive in fresh water and hyper-osmoregulate in fresh and brackish waters (0.2–17‰ salinity). Mortality increases during experimental exposure to higher salinities, and no ontogenetic stage is able to survive in seawater. Hypo-osmoregulatory ability is completely absent in the Pantanal population. These differences between estuarine and fully limnic inland shrimps are suggestive of an initial species diversification attributable to continued genetic isolation of the Amazon and Pantanal populations, probably since the Late Miocene or Pliocene (Charmantier and Anger 2011; Vergamini et al. 2011; Anger 2013).

Further, gills are absent, and the (Na^+, K^+)-ATPase is localized along the inner epithelium of the branchiostegite in the newly hatched zoeae I of both populations. There are also notable differences in gill development and (Na^+, K^+)-ATPase expression between the intermediate (zoea V) and later larval stages (decapodid) in the two populations. (Na^+, K^+)-ATPase activity is found in the gills and branchiostegites of juveniles, with no differences between populations. (Na^+, K^+)-ATPase is found in the antennal glands of all developmental stages on hatching in both populations. The hypo-osmoregulatory ability of the early developmental stages of the Amazon population may be linked to ion transport across the inner branchiostegite epithelium, in which the (Na^+, K^+)-ATPase is absent or weak in the Pantanal population. In this population, fully functional gills expressing (Na^+, K^+)-ATPase appear to be essential for efficient hyper-osmoregulation in the late developmental stages (Boudour-Boucheker et al. 2013). In both fresh water and high salinity medium (25‰), (Na^+, K^+)-ATPase labeling is restricted to the gill intralamellar septal cells, while immunogold labeling reveals the (Na^+, K^+)-ATPase along the basal infoldings of the membrane as seen in *M. olfersii* (McNamara and Torres 1999). In the branchiostegite, (Na^+, K^+)-ATPase immunofluorescence is found in the cells of the internal epithelium, independently of salinity (Boudour-Boucheker et al. 2014).

The relationship between ontogeny and kinetic characteristics of the (Na^+, K^+)-ATPase has been systematically analyzed in microsomal homogenates of whole zoeae I and decapodid III (zoea IX) and whole-body and gill homogenates of juvenile and adult *M. amazonicum* reared from the estuarine population (Leone et al. 2012). The considerable differences in modulation by Na^+ (zoea I), K^+ and ATP (decapodid III) of (Na^+, K^+)-ATPase activity suggest an important role in the osmoregulatory capability of the larval

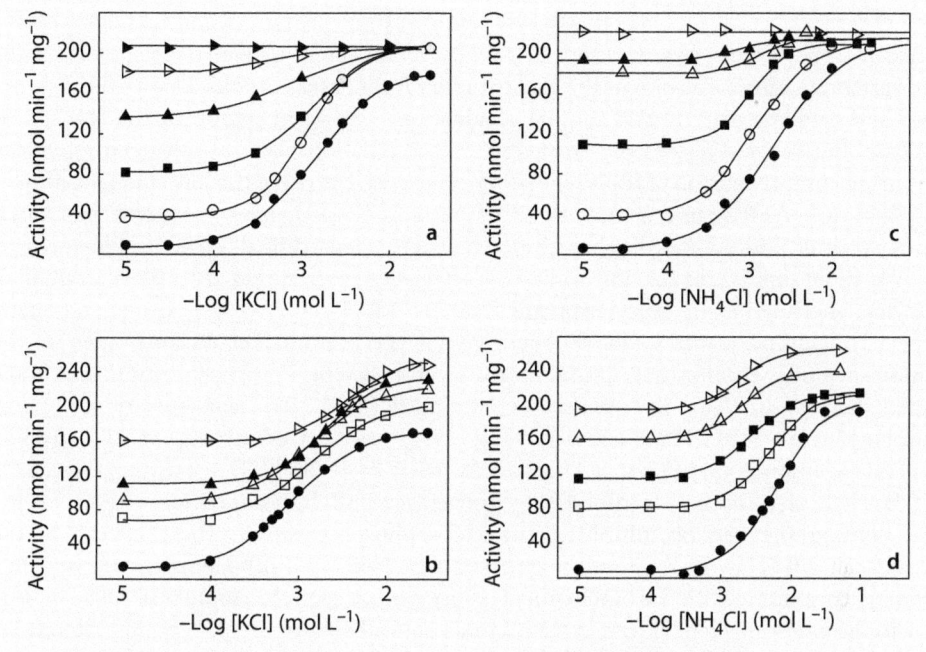

◘ **Fig. 3.9** Modulation by K^+ plus NH_4^+ of microsomal (Na^+, K^+)-ATPase activity in gill tissue from juvenile and adult *M. amazonicum*. (**a**) Juveniles. (**b**) adults. NH_4^+ concentration: (●) none, (○) 0.3 mmol L^{-1}, (■) 1 mmol L^{-1}, (□) 2 mmol L^{-1}, (△) 3 mmol L^{-1}, (▲) 5 mmol L^{-1}, (▷) 10 mmol L^{-1}, (►) 30 mmol L^{-1}. (**c**) Juveniles. (**d**) Adults. K^+ concentration: (●) none, (○) 0.4 mmol L^{-1}, (□) 0.5 mmol L^{-1}, (■) 2 mmol L^{-1}, (△) 5 mmol L^{-1}, (▲) 10 mmol L^{-1}, (▷) 20 mmol L^{-1} (Reproduced from Leone et al. (2014) **with** permission)

stages. The NH_4^+-stimulated, ouabain-insensitive ATPase activity seen in zoea I and decapodid III may reflect a stage-specific means of ammonia excretion since functional gills are absent in the early larval stages (Leone et al. 2012).

(Na$^+$, K$^+$)-ATPase activity in zoea I and decapodid III of estuarine *M. amazonicum* is synergistically stimulated by K^+ at fixed NH_4^+ concentrations (Leone et al. 2014). The ≈ 90 % synergistic stimulation of (Na$^+$, K$^+$)-ATPase activity by K^+ in the presence of NH_4^+ seen in zoea I may constitute part of an osmo-protective mechanism since zoea I is strongly euryhaline and can survive well at salinities ranging from 0 to 28 ‰ salinity (McNamara et al. 1983). The modulation of (Na$^+$, K$^+$)-ATPase activity is ontogenetic stage specific (◘ Fig. 3.9) and particularly distinct between juveniles and adults (Leone et al. 2014).

Although both gill enzymes exhibit two different sites for K^+ and NH_4^+ binding, in the juvenile enzyme these two sites are equivalent: binding by both ions results in slightly stimulated activity compared to that of a single ionic species. In the adult enzyme, the sites are not equivalent: when one ion occupies its specific binding site, (Na$^+$, K$^+$)-ATPase activity is stimulated synergistically by ≈ 50 % on binding of the complementary ion (Leone et al. 2014). These data suggest that the gill enzyme may be regulated by NH_4^+ during ontogenetic development in *M. amazonicum*.

3.9   Conclusions and Perspectives

Since its discovery in 1954, the (Na$^+$, K$^+$)-ATPase has been investigated intensively. While the vertebrate enzyme is one of the most studied P-ATPases, various aspects of its structure and catalytic activity still require elucidation. The participation of the (Na$^+$, K$^+$)-ATPase and other ATPases in crustacean osmoregulation also has attracted attention, requiring comprehension of their regulatory processes. However, the low concentration of the (Na$^+$, K$^+$)-ATPase in crustacean gills, representing less than 5 % total protein, constitutes a technical limitation that has hindered progress regarding its structural features.

We show that while sharing some characteristics in common with the mammalian enzyme, modulation of crustacean gill (Na$^+$, K$^+$)-ATPase activity is unique in many aspects, including features not yet reported for the mammalian enzyme, such as the appearance of high-affinity ATP-binding sites at a specific K$^+$ concentration; a proportional increase in (Na$^+$, K$^+$)-ATPase activity at elevated NH_4^+ concentrations; "extra-pumping" (Na$^+$, K$^+$)-ATPase activity at high NH_4^+ concentrations; and species specific, synergistic stimulation of (Na$^+$, K$^+$)-ATPase activity by K$^+$ and NH_4^+. Further, (Na$^+$, K$^+$)-ATPase hydrolytic activity is inhibited by polyamines both by competition with Na$^+$ at the Na$^+$-binding site and by inhibiting enzyme dephosphorylation. These effects on the crustacean gill (Na$^+$, K$^+$)-ATPase may be both species- and ontogenetic stage specific, possibly correlating with the biochemical adjustment of each developmental stage or species to the ammonia concentration found in its natural environment. Whether alterations in gill (Na$^+$, K$^+$)-ATPase activity reflect different regulatory mechanisms and/or transport-regulating factors needs to be investigated. Future studies also should focus on the expression of (Na$^+$, K$^+$)-ATPase isoforms during ontogenesis and their role in active ammonia excretion.

References

Akamatsu N, Inenaga K, Yamashita H (1993) Inhibitory effects of natriuretic peptides on vasopressin neurons mediated through cGMP and cGMP-dependent protein kinase in vitro. J Neuroendocrinol 5:517–522

Alberts RW (1967) Biochemical aspects of active transport. Annu Rev Biochem 6:727–756

Al-Reasi HA, Yusuf U, Smith DS, Wood CM (2013) The effect of dissolved organic matter (DOM) on sodium transport and nitrogenous waste excretion of the freshwater cladoceran (Daphnia magna) at circum-neutral and low pH. Comp Biochem Physiol 158C:207–215

Anger K (2003) Salinity as a key parameter in the larval biology of decapods crustaceans. Invertebr Reprod Dev 43:29–45

Anger K (2013) Neotropical macrobrachium (Caridea: Palaemonidae): on the biology, origin, and radiation of freshwater-invading shrimp. J Crustac Biol 33:151–183

Anger K, Hayd L (2010) Feeding and growth in early larval shrimp Macrobrachium amazonicum from the Pantanal, southwestern Brazil. Aquat Biol 9:251–261

Apell HJ, Benz G, Sauerbrunn D (2011) Proton diet for the sodium pump. Biochemistry 50:409–418

Aperia A (2012) 2011 Homer Smith Award: to serve and protect: classic and novel roles for Na$^+$, K$^+$-adenosine triphosphatase. J Am Soc Nephrol 23:1283–1290

Arata Y, Nishi T, Kawasaki-Nishi S, Shao E, Wilkens S, Forgac M (2002) Structure, subunit function and regulation of the coated vesicle and yeast vacuolar (H$^+$)-ATPases. Biochim Biophys Acta 1555:71–74

Arystarkhova E, Wetzel RK, Asinovski NK, Sweadner KJ (1999) The gamma subunit modulates Na$^+$ and K$^+$ affinity of the renal Na, K- ATPase. J Biol Chem 274:33183–33185

Askar A, Treptow H (1986) Biogene amine in Lebensmitteln. Vorkommen, Bedeutung und Bestimmung. Eugen Ulmer GmbH and Co, Stuttgart

Augusto A, Greene LJ, Laure HJ, McNamara JC (2007) The ontogeny of isosmotic intracellular regulation in the diadromous, freshwater palaemonid shrimps, *Macrobrachium amazonicum* and *M. olfersi* (Crustacea, Decapoda). J Crustac Biol 27:626–634

Azarias G, Kruusmagi M, Connor S, Akkuratov EE, Liu XL, Lyons D, Brismar H, Broberger C, Aperia A (2013) A specific and essential role for Na, K-ATPase α3 in neurons co-expressing α1 and α3. J Biol Chem 288:2734–2743

Aziz SM, Olson JW, Gillespie MN (1994) Multiple polyamine transport pathways in cultured pulmonary artery smooth muscle cells: regulation by hypoxia. Am J Respir Cell Mol Biol 10:160–166

Baldwin GF, Kirschner LB (1976) Sodium and chloride regulation in *Uca* adapted to 175 percent sea water. J Exp Zool 49:158–171

Barnes DB (2000) Zoologia dos invertebrados, 4th edn. Editora Roca, São Paulo

Baxter-Lowe LA, Guo JZ, Bergstrom EE, Hokin LE (1989) Molecular cloning of the Na, K-ATPase α-subunit in developing brine shrimp and sequence comparison with higher organisms. FEBS Lett 257:181–187

Belli NM, Faleiros RO, Firmino KC, Masui DC, Leone FA, McNamara JC, Furriel RP (2009) Na, K-ATPase activity and epithelial interfaces in gills of the freshwater shrimp *Macrobrachium amazonicum* (Decapoda, Palaemonidae). Comp Biochem Physiol 152A:431–439

Belogus T, Haviv H, Karlish SJD (2009) Neutralization of the charge on Asp(369) of Na⁺, K⁺-ATPase triggers E-1 < - > E-2 conformational changes. J Biol Chem 248:31038–31051

Benlekbir S, Bueler SA, Rubinstein JL (2012) Structure of the vacuolar-type ATPase from *Saccharomyces cerevisiae* at 11-A resolution. Nat Struct Mol Biol 19:1356–1362

Bhattacharyya KK, Bergstrom EE, Hokin LE (1990) Molecular cloning of the β-subunit of the Na, K-ATPase in the brine shrimp, *Artemia:* the cDNA-derived amino acid sequence shows low homology with the β-subunits of vertebrates except in the single transmembrane and the carboxy-terminal domains. FEBS Lett 269:233–238

Bianchini A, Wood CM (2003) Mechanism of acute silver toxicity in *Daphnia magna*. Environ Toxicol Chem 22:1361–1367

Bianchini A, Lauer MM, Nery LE, Colares EP, Monserrat JM, Dos Santos Filho EA (2008) Biochemical and physiological adaptations in the estuarine crab *Neohelice granulata* during salinity acclimation. Comp Biochem Physiol 151A:423–436

Blanco G (2005) Na, K-ATPase subunit heterogeneity as a mechanism for tissue-specific ion regulation. Semin Nephrol 25:292–303

Boron WF (1986) Intracellular pH regulation in epithelial cells. Annu Rev Physiol 48:377–388

Bouaricha N, Charmantier-Daures M, Thuet P, Trilles J-P, Charmantier G (1994) Ontogeny of osmoregulatory structures in the shrimp *Penaeus japonicus* (Crustacea, Decapoda). Biol Bull 186:29–40

Boudour-Boucheker N, Boulo V, Lorin-Nebel C, Elguero C, Grousset E (2013) Adaptation to freshwater in the palaemonid shrimp *Macrobrachium amazonicum*: comparative ontogeny of osmoregulatory organs. Cell Tissue Res 353:87–98

Boudour-Boucheker N, Boulo V, Charmantier-Daures M, Grousset E, Anger K, Charmantier G, Lorin-Nebel C (2014) Differential distribution of V-type H⁺-ATPase and Na⁺/K⁺-ATPase in the branchial chamber of the palaemonid shrimp *Macrobrachium amazonicum*. Cell Tissue Res 357:195–206

Bowman EJ, Graham LA, Stevens TH, Bowman BJ (2004) The bafilomycin/concanamycin binding site in subunit c of the V-ATPases from *Neurospora crassa* and *Saccharomyces cerevisiae*. J Biol Chem 279:33131–33138

Boxenbaum N, Daly SE, Javaid ZZ, Lane LK, Blostein R (1998) Changes in the steady-state conformational equilibrium resulting from cytoplasmic mutations of the (Na⁺, K⁺)-ATPase α-subunit. J Biol Chem 273:23086–23092

Brink B, Ten B, Damink C, Joosten HML, Huis In't Veld JHJ (1990) Occurrence and formation of biologically active amines in foods. Int J Food Microbiol 11:73–84

Bublitz M, Poulsen H, Morth JP, Nissen P (2010) In and out of the cation pumps: P-Type ATPase structure revisited. Curr Opin Struct Biol 20:431–439

Burnett LE, Towle DW (1990) Sodium ion uptake by perfused gills of the blue crab *Callinectes sapidus*: effects of ouabain and amiloride. J Exp Biol 149:293–305

Burnett LE, Dunn TN, Infantino JRL (1985) The function of carbonic anhydrase in crustacean gills. In: Gilles R, Gilles-Baillien M (eds) Transport processes, iono- and osmoregulation. Springer, Berlin, pp 159–168

Cameron JN (1978) NaCl balance in blue crabs, *Callinectes sapidus*, in freshwater. J Comp Physiol 123B:127–135

Capendeguy O, Horisberger JD (2005) The role of the third extracellular loop of the Na⁺, K⁺-ATPase α-subunit in a luminal gating mechanism. J Physiol 565:207–218

Carvalho FB, Mello CF, Marisco PC, Tonello R, Girardi BA, Ferreira J, Oliveira MS, Rubin MA (2012) Spermidine decreases Na⁺, K⁺-ATPase activity through NMDA receptor and protein kinase G activation in the hippocampus of rats. Eur J Pharmacol 684:79–86

Casero RA Jr, Pegg AE (1993) Spermidine/spermine*N*1-acetyltransferase—the turning point in polyamine metabolism. FASEB J 7:653–661

Charmantier G (1998) Ontogeny of osmoregulation in crustaceans: a review. Invertebr Reprod Dev 33:177–190

Charmantier G, Anger K (1999) Ontogeny of osmoregulation in the palaemonid shrimp *Palaemonetes argentinus* (Crustacea: Decapoda). Mar Ecol Prog Ser 181:125–129

Charmantier G, Anger K (2011) Ontogeny of osmoregulatory patterns in the South American shrimp *Macrobrachium amazonicum*: loss of hypo-regulation in a land-locked population indicates phylogenetic separation from estuarine ancestors. J Exp Mar Biol Ecol 396:89–98

Charmantier G, Charmantier-Daures M (2001) Ontogeny of osmoregulation in crustaceans: the embryonic phase. Am Zool 41:1078–1089

Chen JC, Lai SH (1993) Effects of temperature and salinity on oxygen consumption and ammonia-N excretion of juvenile *Penaeus japonicus* Bate. J Exp Mar Biol Ecol 165:161–170

Choe H, Sackin H, Palmer LG (2000) Permeation properties of inward-rectifier potassium channels and their molecular determinants. J Gen Physiol 115:391–404

Chourasia M, Sastry N (2012) The nucleotide, inhibitor, and cation binding sites of P-type II ATPases. Chem Biol Drug Des 79:617–627

Cieluch U, Charmantier G, Grousset E, Charmantier-Daures M, Anger K (2005) Osmoregulation, immunolocalization of Na⁺/K⁺- ATPase, and ultrastructure of branchial epithelia in the developing brown shrimp, *Crangon crangon* (Decapoda, Caridea). Physiol Biochem Zool 78:1017–1025

Cieluch U, Anger K, Charmantier-Daures M, Charmantier G (2007) Osmoregulation and immunolocalization of Na⁺/K⁺-ATPase during the ontogeny of the mitten crab *Eriocheir sinensis* (Decapoda, Grapsoidea). Mar Ecol Prog Ser 329:169–178

Cipriano DJ, Wang Y, Bond S, Hinton A, Jefferies KC, Qi J, Forgac M (2008) Structure and regulation of the vacuolar ATPases. Biochim Biophys Acta 1777:599–604

Cirri E, Katz A, Mishra NK, Belogus T, Lifshitz Y, Garty H, Karlish SJD, Apell HJ (2011) Phospholemman (FXYD1) raises the affinity of the human r1β1 isoform of Na, K-ATPase for Na ions. Biochemistry 50:3736–3748

Clarke RJ (2009) Mechanism of allosteric effects of ATP on the kinetics of P-type ATPases. Eur Biophys J 39:3–17

Clarke RJ, Catauro M, Rasmussen HH, Apell HP (2013) Quantitative calculation of the role of the Na⁺, K⁺-ATPase in thermogenesis. Biochim Biophys Acta 1827:1205–1212

Coffino P (2001) Antizyme, a mediator of ubiquitin-independent proteasomal degradation. Biochimie 83:319–323

Cohen E, Goldshleger R, Shainskaya A, Tal DM, Ebel C, Maire M, Karlish SJD (2005) Purification of Na⁺, K⁺-ATPase expressed in *Pichia pastoris* reveals an essential role of phospholipid-protein interactions. J Biol Chem 280:16610–16618

Colina C, Palavicini JP, Srikumar D, Holmgren M, Rosenthal JJC (2010) Regulation of Na⁺/K⁺ ATPase transport velocity by RNA. PLoS Biol 8:1–9

Compère P, Wanson S, Pequeux A, Gilles R, Goffinet G (1989) Ultrastructural changes in the gill epithelium of the green crab *Carcinus maenas* in relation to the external salinity. Tissue Cell 21:299–318

Conte FP, Droukas PC, Ewing RD (1977) Development of sodium regulation and de novo synthesis of Na⁺-K⁺-activated ATPase in larval brine shrimp *Artemia salina*. J Exp Zool 202:339–362

Copeland DE, Fitzjarrell AT (1968) The salt absorbing cells in the gills of the blue crab (*Callinectes sapidus*, Rathbun) with notes on modified mitochondria. Z Zellforsch 92:1–22

Cornelius F, Mahmmoud YA (2009) Interaction between cardiotonic steroids and Na, K-ATPase: effects of pH and ouabain induced changes in enzyme conformation. Biochemistry 48:10056–10065

Cornelius F, Yasser A, Mahmmoud YA, Toyoshima C (2011) Metal fluoride complexes of Na, K-ATPase: characterization of fluoride-stabilized phosphoenzyme analogues and their interaction with cardiotonic steroids. J Biol Chem 286:29882–29892

Cornelius F, Kanai R, Toyoshima C (2013) Moiety and steroid hydroxyls of cardiotonic steroids in binding to Na, K-ATPase. J Biol Chem 288:6602–6616

Corotto FS, Holliday CW (1996) Branchial Na, K-ATPase e osmoregulation in the purple shore crab *Hemigrapsus nudus* (Dana). Comp Biochem Physiol 113A:361–368

Crambert G, Li CM, Swee LK, Geering K (2004) FXYD7, mapping of functional sites involved in endoplasmic reticulum export, association with e regulation of (Na+, K+)-ATPase. J Biol Chem 279:30888–30895

D'Orazio SE, Holliday CW (1985) Gill Na, K-ATPase osmoregulation in these fiddler crabs, *Uca pugilator*. Physiol Zool 58:364–373

Da Silva N, Shum WW, El-Annan J, Paunescu TG, McKee M, Smith PJ, Brown D, Breton S (2007) Relocalization of the V-ATPase B2 subunit to the apical membrane of epididymal clear cells of mice deficient in the B1 subunit. Am J Physiol 293:C199–C210

Daly SE, Blostein R, Lane LK (1997) Functional consequences of a posttransfection mutation in the H2–H3 cytoplasmic loop of the alpha subunit of Na, K-ATPase. J Biol Chem 272:6341–6347

Dempski RE, Lustig J, Friedrich T, Bamberg E (2008) Structural arrangement and conformational dynamics of the γ subunit of the Na+/K+-ATPase. Biochemistry 47:257–266

Díaz F, Farfan C, Sierra E, Re AD (2001) Effects of temperature and salinity fluctuation on the ammonium excretion and osmoregulation of juveniles of *Penacus vannamei*, Boone. Mar Freshw Behav Physiol 34:93–104

Díaz F, Re AD, Sierra E, Díaz-Iglesias E (2004) Effects of temperate and salinity fluctuation on the oxygen consumption, ammonium excretion and osmoregulation of the blue shrimp *Litopenaeus stylirostris* (Stimpson). J Shellfish Res 23:903–910

Dröse S, Bindsei KU, Bowman EJ, Siebers A, Zeeck A, Altendor K (1993) Inhibitory effect of modified bafilomycins and concanamycins on P- and V-type adenosinetriphosphatases. Biochemistry 32:3902–3906

Dugger DM, Dobkin S (1975) A contribution to the larval development of *Macrobrachium olfersii* (Wiegaman, 1836) (Decapoda, Palaemonidae). Crustaceana 29:1–30

Durr KL, Tavraz NN, Dempski RE, Bamberg E, Friedrich T (2009) Functional significance of E2 state stabilization by specific α/β-Subunit interactions of Na, K- and H, K-ATPase. J Biol Chem 284:3842–3854

Eguchi H, Takeda K, Schwarz W, Shirahata A, Kawamura M (2005) Involvement in K+ access of Leu318 at the extracellular domain flanking M3 e M4 of the Na+, K+-ATPase α-subunit. Biochem Biophys Res Commun 330:611–614

Ehrenfeld J, Klein U (1997) The key role of the H+ V-ATPase in acid-base balance and Na+ transport processes in frog skin. J Exp Biol 200:247–256

Einholm AP, Toustrup-Jensen M, Eersen JP, Vilven B (2005) Mutation of Gly-94 in transmembrane segment M_1 of Na+, K+-ATPase interferes with Na+ e K+ binding in E_2P conformation. Proc Natl Acad Sci U S A 102:11254–11259

El-Beialy W, Galal N, Deyama Y, Yoshimura Y, Suzuki K, Tei K, Totsuka Y (2010) Regulation of human and pig renal Na+, K+-ATPase activity by tyrosine phosphorylation of their a1-subunits. J Membr Biol 233:119–126

Elgavish A, Wallace RW, Pillion DJ, Meezan E (1984) Polyamines stimulate D-glucose transport in isolated renal brush-border membrane vesicles. Biochim Biophys Acta 777:1–8

Engel DW (1977) Comparison of the osmoregulatory capabilities of two portunid crabs, *Callinectes sapidus* and *C. similis*. Mar Biol 41:275–279

Escalante R, García-Sáez A, Sastre L (1995) In situ hybridization analyses of Na, K ATPase α-subunit expression during early larval development of *Artemia franciscana*. J Histochem Cytochem 43:391–399

Faleiros RO, Goldman MHS, Furriel RPM, McNamara JC (2010) Differential adjustment in gill Na+/K+- and V-ATPase activities and transporter mRNA expression during osmorregulatory acclimation in the cinnamom shrimp *Macrobrachium amazonicum* (Decapoda, Palaemonidae). J Exp Biol 213:3894–3905

Fedosova NU, Cornelius F, Klodos I (1998) E2P phosphoforms of (Na+, K+)-ATPase. I: comparison of phosphointermediates formed from ATP e Pi by their reactivity toward hydroxylamine e vanadate. Biochemistry 37:13634–13642

Felder JM, Felder DL, Hand SC (1986) Ontogeny of osmoregulation in the estuarine ghost shrimp *Callianassa jamaicense* var. *louisianensis* Schmitt (Decapoda, Thalassinidea). J Exp Mar Biol Ecol 99:91–105

Feraille E, Mordasini D, Gonin S, Deschenes G, Vinciguerra M, Doucet A, Veewalle A, Summa V, Verrey F, Martin PY (2003) Mechanism of control of (Na+, K+)-ATPase in principal cells of the mammalian collecting duct. Ann N Y Acad Sci 986:570–578

Fernandes F, Loura LSM, Fedorov A, Dixon N, Kee TP, Pietro M, Hemminga MA (2006) Binding assays of inhibitors towards selected V-ATPase domains. Biochim Biophys Acta 1758:1777–1786

Ficker E, Taglialatela M, Wible BA, Henley CM, Brown AM (1994) Spermine and spermidine as gating molecules for inward rectifier K+ channels. Science 266:1068–1072

Firmino KCS, Faleiros RO, Masui DC, McNamara JC, Furriel RPM (2011) Short- and long-term, salinity-induced modulation of V-ATPase activity in the posterior gills of the true freshwater crab, *Dilocarcinus pagei* (Brachyura, Trichodactylidae). Comp Biochem Physiol 160:24–31

Florkin M, Schoffeniels E (1969) Molecular approaches to ecology. Academic, New York

Forbush B III, Kaplan JH, Hoffman JF (1978) Characterization of a new photoaffinity derivative of ouabain: labeling of the large polypeptide and of a proteolipid component of the Na, K-ATPase. Biochemistry 17:3667–3676

Forgac M (2007) Vacuolar ATPases: rotary proton pumps in physiology and pathophysiology. Mol Cell Biol 8:917–929

Foster CA, Howse HD (1978) Morphological study on gills of brown shrimp, *Penaeus aztecus*. Tissue Cell 10:77–92

França JL, Pinto MR, Lucena MN, Garçon DP, Valenti WC, McNamara JC (2013) Subcellular localization and kinetic characterization of a gill (Na^+, K^+)-ATPase from the giant freshwater prawn *Macrobrachium rosenbergii*. J Membr Biol 246:529–543

Frattini A, Orchard PJ, Sobacchi C, Giliani S, Abinun M, Mattsson JP, Keeling DJ, Andersson AK, Wallbrandt P, Zecca L, Notarangelo LD, Vezzoni P, Villa A (2000) Defects in TCIRG1 subunit of the vacuolar proton pump are responsible for a subset of human autosomal recessive osteopetrosis. Nat Genet 25:343–346

Freire CA, McNamara JC (1995) Fine structure of the gills of the freshwater shrimp *Macrobrachium olfersii* (Decapoda): effect of acclimation to high salinity medium and evidence for involvement of the lamellar septum in ion uptake. J Crustac Biol 15:103–116

Freire CA, Onken H, McNamara JC (2008) A structure–function analysis of ion transport in crustacean gills and excretory organs. Comp Biochem Physiol 151A:272–304

Furriel RPM, McNamara JC, Leone FA (2000) Characterization of (Na^+, K^+)-ATPase in gill microsomes of the freshwater shrimp *Macrobrachium olfersii*. Comp Biochem Physiol 126B:303–315

Furriel RPM, McNamara JC, Leone FA (2001) Nitrophenylphosphate as a tool to characterize gill (Na^+, K^+)-ATPase activity in hyperregulating Crustacea. Comp Biochem Physiol 130A:665–676

Furriel RP, Masui DC, McNamara JC, Leone FA (2004) Modulation of gill Na^+, K^+-ATPase activity by ammonium ions: putative coupling of nitrogen excretion and ion uptake in the fresh water shrimp *Macrobrachium olfersii*. J Exp Zool 301:63–74

Furriel RPM, Firmino KCS, Masui DC, Faleiros RO, Torres AH, McNamara JC (2010) Structural e biochemical correlates of (Na^+, K^+)-ATPase driven ion uptake across the true freshwater crab, *Dilocarcinus pagei* Brachyura, Trichodactylidae. J Exp Zool 313A:508–523

Füzesi M, Gottschalk KE, Lindzen M, Shainskaya A, Küster B, Garty H, Karlish SDJ (2005) Covalent cross-links between the γ subunit (FXYD2) and α and β-Subunits of Na, K-ATPase. J Biol Chem 280:18291–18301

Gache C, Rossi B, Lazdunski M (1977) Mechanistic analysis of the (Na^+, K^+)-ATPase using new pseudosubstrates. Biochemistry 16:2957–2965

Gadsby DC, Bezanilla F, Rakowski RF, De Weer P, Holmgren M (2012) The dynamic relationships between the three events that release individual Na^+ ions from the Na^+/K^+-ATPase. Nat Commun 3:699

Gallo LC, Davel APC, Xavier FE, Rossoni LV (2010) Time-dependent increases in ouabain-sensitive Na^+, K^+-ATPase activity in aortas from diabetic rats: the role of prostanoids and protein kinase C. Life Sci 87:302–308

Garçon DP, Masui DC, Mantelatto FLM, McNamara JC, Furriel RPM, Leone FA (2007) K^+ and NH_4^+ modulate gill (Na^+, K^+)-ATPase activity in the blue crab, *Callinectes ornatus*: fine tuning of ammonia excretion. Comp Biochem Physiol 147A:145–155

Garçon DP, Masui DC, Mantelatto FLM, McNamara JC, Furriel RPM, Leone FA (2009) Hemolymph ionic regulation e adjustments in gill (Na^+, K^+)-ATPase activity during salinity acclimation in the swimming crab *Callinectes ornatus* (Decapoda, Brachyura). Comp Biochem Physiol 154A:44–55

Garçon DP, Lucena MN, França JL, McNamara JC, Fontes CFL, Leone FA (2011) Na^+, K^+-ATPase activity in the posterior gills of the blue crab, *Callinectes ornatus* (Decapoda, Brachyura): modulation of ATP hydrolysis by the biogenic amines spermidine and spermine. J Membr Biol 244:9–20

Garçon DP, Lucena MN, Pinto MR, Fontes CFL, McNamara JC, Leone FA (2013) Synergistic stimulation by potassium and ammonium of K^+-phosphatase activity in gill microsomes from the crab *Callinectes ornatus* acclimated to low salinity: novel property of a primordial pump. Arch Biochem Biophys 530:55–63

Garty H, Karlish SJD (2006) Role of FXYD proteins in ion transport. Annu Rev Physiol 68:431–459

Geering K (2000) Topogenic motifs in P-type ATPases. J Membr Biol 174:181–190

Geering K (2008) Functional roles of Na, K-ATPase subunits. Curr Opin Nephrol Hypertens 17:526–532

Genovese G, Ortiz N, Urcola MR, Luquet CM (2005) Possible role of carbonic anhydrase, V-H$^{(+)}$-ATPase, and Cl(-)/HCO3- exchanger in electrogenic ion transport across the gills of the euryhaline crab *Chasmagnathus granulatus*. Comp Biochem Physiol 142A:362–369

Gilles R, Péqueux AJR (1985) Ion transport in crustacean gills: physiological and ultrastructural approaches. In: Gilles R, Gilles-Baillien M (eds) Transport processes, iono- and osmoregulation. Springer, Heidelberg, pp 136–158

Glynn IM (1985) The (Na$^+$K$^+$)-transporting adenosine triphosphatase. In: Martonosi AN (ed) The enzymes of biological membranes, vol 3. Plenum Press, New York, pp 35–114

Glynn IM (2002) A hundred years of sodium pumping. Annu Rev Physiol 64:1–18

Gocheva V, Joyce J, Johanna A (2007) Cysteine cathepsins and the cutting edge of cancer invasion. Cell Cycle 6:60–64

Gonçalves RR, Masui DC, McNamara JC, Mantellato FLM, Garcon DP, Furriel RPM, Leone FA (2006) A kinetic study of the gill (Na$^+$, K$^+$)-ATPase, and its role in ammonia excretion in the intertidal hermit crab, *Clibanarius vittatus*. Comp Biochem Physiol 145A:346–356

Gong XM, Ding Y, Yao Y, Marassi FM (2015) Structure of the Na, K-ATPase regulatory protein FXYD2b in micelles: implications for membrane-water interfacial arginines. Biochim Biophys Acta 1848:299–306

Goodman SH, Cavey MJ (1990) Organization of a phyllobranchiate gill from the green shore crab *Carcinus maenas* (Crustacea, Decapoda). Cell Tissue Res 260:495–505

Gross MG (1972) Oceanography: a view of the earth. Prentice Hall, Englewood Cliffs

Hansen O (2003) No evidence for a role in signal-transduction of Na$^{(+)}$/K$^{(+)}$-ATPase interaction with putative endogenous ouabain. Eur J Biochem 270:1916–1919

Haond C, Bonnal L, Sandeaux R, Charmantier G, Trilles JP (1999) Ontogeny of intracellular isosmotic regulation in the European lobster *Homarus gammarus* (L.). Physiol Biochem Zool 72:534–544

Harris RR, Bayliss D (1988) Gill (Na$^+$, K$^+$)-ATPases in decapod crustaceans: distribution and characteristics in relation to Na$^+$ regulation. Comp Biochem Physiol 90A:303–308

Harris RR, Santos MCF (1993) Sodium uptake and transport (Na$^+$K$^+$)ATPase changes following Na$^+$ depletion and low salinity acclimation in the mangrove crab *Ucides cordatus* (L.). Comp Biochem Physiol 105A:35–42

Harris RR, Coley S, Collins S, McCabe R (2001) Ammonia uptake and its effects on ionoregulation in the freshwater crayfish, *Pacifastacus leniusculus* (Dana). J Comp Physiol 171B:681–693

Hasler U, Crambert G, Horisberger JD, Geering K (2001) Structural and functional features of the transmembrane domain of the Na, K-ATPase b subunit revealed by tryptophan scanning. J Biol Chem 276:16356–16364

Havird JC, Santos SR, Henry RP (2014) Osmoregulation in the Hawaiian anchialine shrimp *Halocaridina rubra* (Crustacea: Atyidae): expression of ion transporters, mitochondria-rich cell proliferation and hemolymph osmolality during salinity transfers. J Exp Biol 217:2309–2320

Hayd L, Anger K (2013) Reproductive and morphometric traits of *Macrobrachium amazonicum* (Decapoda: Palaemonidae) from the Pantanal, Brazil, suggests initial speciation. Rev Biol Trop (Int J Trop Biol) 61:39–57

Hebert H, Purhonen P, Thomsen K, Vorum H, Maunsbach AB (2003) Renal Na, K-ATPase structure from cryo-electron microscopy of two dimensional crystals. Ann N Y Acad Sci 986:9–16

Heby O (1989) Polyamines and cell differentiation. In: Bachrach U, Heimer YM (eds) The physiology of polyamines, vol 1. CRC Press, Boca Raton, pp 83–94

Heinrich-Hirsch B, Ahlers J, Peter HW (1977) Inhibition of Na$^+$, K$^+$-ATPase from chick brain by polyamines. Enzyme 22:235–241

Henry RP (1988a) Subcellular distribution of carbonic anhydrase activity in the gill of the blue crab, *Callinectes sapidus*. J Exp Zool 245:1–8

Henry RP (1988b) Multiple functions of gill carbonic anhydrase. J Exp Zool 248:19–24

Henry RP, Wheatly MG (1992) Interaction of respiration ion regulation, and acid-base balance in the everyday life of aquatic crustaceans. Am Zool 32:407–416

Henry RP, Lucu C, Onken H, Weihrauch D (2012) Multiple functions of the crustacean gill: osmotic/ionic regulation, acid-base balance, ammonia excretion, and bioaccumulation of toxic metals. Front Physiol 3:431

Hirata T, Iwamoto-Kihara A, Sun-Wada GH, Okajima T, Wada Y, Futai M (2003) Subunit rotation of vacuolar-type proton pumping ATPase: relative rotation of the G and C subunits. J Biol Chem 278:23714–23719

Holliday CW (1985) Salinity-induced changes in gill Na, K-ATPase activity in the mud fiddler crab *Uca pugnax*. J Exp Zool 233:199–208

Holliday CW, Miller DS (1984) PAH excretion in 2 species of cancroids crab *Cancer-irroratus* and *Cancer-borealis*. Am J Physiol 246:R364–R368

Höltta E (1977) Oxidation of spermidine and spermine in rat liver: purification and properties of poly-amine oxidase. Biochemistry 16:91–100

Horisberger JD (2004) Recent insights into the structure e mechanism of the sodium pump. Physiology 19:377–388

Huss M, Weczorek M (2009) Inhibitors of V ATPases: old and new players. J Exp Biol 212:341–346

Igarashi K, Kashiwagi K (1999) Polyamine transport in bacteria and yeast. Biochem J 344:33–42

Igarashi K, Kashiwagi K (2010) Modulation of cellular function by polyamines. Int J Biochem Cell Biol 42:39–51

Imagawa T, Yamamoto T, Kaya S, Sakaguchi K, Taniguchi K (2005) Thr-774 (transmembrane segment M5), Val-920 (M8), and Glu-954 (M9) are involved in Na$^+$transport, and Gln-923 (M8) is essential for Na, K-ATPase activity. J Biol Chem 280:18736–18744

Ituarte RB, Spivak ED, Anger K (2005) Effects of salinity on embryonic development of *Palaemonetes argentinus* (Crustacea: Decapoda: Palaemonidae) cultured in vitro. Invertebr Reprod Dev 47:213–223

Ituarte RB, Mañanes AAL, Spivak ED, Anger K (2008) Activity of Na$^+$, K$^+$-ATPase in a freshwater shrimp, *Palaemonetes argentinus* (Caridea, Palaemonidae): ontogenetic and salinity-induced changes. Aquat Biol 3:283–290

Iwata M, Imamura H, Stambouli E, Ikeda C, Tamakoshi M, Nagata K, Makyio H, Hankamer B, Barber J, Yoshida M, Yokoyama K, Iwata S (2004) Crystal structure of a central stalk subunit C and reversible association/dissociation of vacuole-type ATPase. Proc Natl Acad Sci U S A 101:59–64

Jänne J, Alhonen L, Leinonen P (1991) Polyamines – from molecular biology to clinical applications. Ann Med 23:241–259

Jänne J, Alhonen L, Pietilä M, Keinänen TA (2004) Genetic approaches to the cellular functions of poly-amines in mammals. Eur J Biochem 271:877–894

Jantaro S, Maenpaa P, Mulo P, Incharoensakdi A (2003) Content e biosynthesis of polyamines in salt and osmotically stressed cells of *Synechocystis sp* PCC 6803. FEMS Microbiol Lett 228:129–135

Jefferies KC, Cipriano DJ, Forgac M (2008) Function, structure and regulation of the vacuolar (H$^+$)-ATPases. Arch Biochem Biophys 476:33–42

Jellink PH, Perry G (1967) Effect of polyamines on the metabolism of [16-14C]estradiol by rat-liver micro-somes. Biochim Biophys Acta 137:367–374

Jensen AML, Sorensen TLM, Olesen C, Moller JV, Nissen P (2006) Modulatory and catalytic modes of ATP binding by the calcium pump. EMBO J 25:2305–2314

Jiang DH, Lawrence AL, Neill WH, Gong H (2000) Effects of temperature and salinity on nitrogen excretion by *Litopenaeus vannamei* juveniles. J Exp Mar Biol Ecol 253:193–209

Jimenez T, Sanchez G, Wertheimer E, Blanco G (2010) Activity of the Na, K-ATPase α4 isoform is important for membrane potential, intracellular Ca^{2+}, and pH to maintain motility in rat spermatozoa. Reproduction 139:835–845

Jimenez T, Sanchez G, McDermott JP, Nguyen AN, Kumar TR, Blanco G (2011) Increased expression of the Na, K-ATPase alpha4 isoform enhances sperm motility in transgenic mice. Biol Reprod 84:153–161

Johansson VM, Miniotis MF, Hegardt C, Jönsson G, Staaf J, Berntsson PS, Oredsson SM, Alm K (2008) Effect of polyamine deficiency on proteins involved in Okazaki fragment maturation. Cell Biol Int 32:1467–1477

Jorgensen PL, Nielsen JM, Rasmussen JH, Pedersen PA (1998) Structure-function relationships of E-1-E-2 transitions e cation binding in Na, K-pump protein. Biochim Biophys Acta 1365:65–70

Jorgensen PL, Hakansson KO, Karlish SJD (2003) Structure and mechanism of functional sites and their interactions. Annu Rev Physiol 65:817–849

Kalac P (2009) Recent advances in the research on biological roles of dietary polyamines in man. J Appl Biomed 7:65–74

Kalber FA, Costlow JD (1966) The ontogeny of osmoregulation and its neurosecretory control in the deca-pod crustacean *Rhithropanopeus harrisii*. Am Zool 6:221–229

Kanai R, Ogawa H, Vilsen B, Cornelius F, Toyoshima C (2013) Crystal structure of a Na$^+$-bound Na$^+$, K$^+$-ATPase preceding the E1 state. Nature 502:201–207

Kane PM (2006) The where, when and how of organelle acidification by the yeast vacuolar H$^+$-ATPase. Microbiol Mol Biol Rev 70:177–191

Kaplan JH (2002) Biochemistry of Na, K-ATPase. Annu Rev Biochem 71:511–535

Karitskaya I, Aksenov N, Vassilieva I, Zenin V, Marakhova I (2010) Long-term regulation of Na, K-ATPase pump during T-cell proliferation. Eur J Phys 460:777–789

Kawasaki-Nishi S, Nishi T, Forgac M (2003) Proton translocation driven by ATP hydrolysis in V-ATPases. FEBS Lett 545:76–85

Khalid M, Fouassier G, Apell HJ, Cornelius F, Clarke RJ (2010) Interaction of ATP with the phosphosphoenzyme of the Na^+, K^+-ATPase. Biochemistry 49:1248–1258

Khalid M, Suliman R, Ahmed R, Salin H, Clarke RJ (2014) The high and low affinity binding sites of digitalis glycosides to Na, K-ATPase. Arab J Sci Eng 39:75–85

Khodabandeh S, Kutnik M, Aujoulat F, Charmantier G, Charmantier-Daures M (2005a) Ontogeny of the antennal glands in the crayfish *Astacus leptodactylus* (Crustacea, Decapoda): immunolocalization of Na^+, K^+-ATPase. Cell Tissue Res 319:167–174

Khodabandeh S, Charmantier G, Charmantier-Daures M (2005b) Ultrastructural studies and Na^+, K^+-ATPase immunolocalization in the antennal urinary glands of the lobster *Homarus gammarus* (Crustacea, Decapoda). J Histochem Cytochem 53:1203–1214

Khodabandeh S, Charmantier G, Charmantier-Daures M (2006) Immunolocalization of Na^+, K^+-ATPase in osmoregulatory organs during the embryonic and post embryonic development of the lobster *Homarus gammarus*. J Crustac Biol 26:515–523

Kirschner LB (2004) The mechanism of sodium chloride uptake in hyperregulating aquatic animals. J Exp Biol 207:1439–1452

Kobayashi M, Fujisaki H, Sugawara M, Iseki K, Miyazaki K (1999) The presence of an Na^+/spermine antiporter in the rat renal brush-border membrane. J Pharm Pharmacol 51:279–284

Koenig H, Goldstone A, Lu CY (1983) Polyamines regulate calcium fluxes in a rapid plasma membrane response. Nature 305:530–534

Koroleff F (1983) Determination of ammonia. In: Grasshoff K, Ehrhart M, Kremling K (eds) Methods of seawater analysis. Verlag Chemie, Weinheim, pp 150–151

Kosiol B, Bigalke T, Graszynski K (1988) Purification e characterization of gill Na^+, K^+-ATPase in the freshwater crayfish *Orconectes limosus* Rafinesque. Comp Biochem Physiol 89B:171–177

Kozik P, Hodson NA, Sahlender DA, Simecek N, Soromani C, Wu J, Collinson LM, Robinson MS (2013) A human genome-wide screen for regulators of clathrin-coated vesicle formation reveals an unexpected role for the V-ATPase. Nat Cell Biol 15:50–60

Kuntzweiler TA, Wallick ET, Johnson CL, Lingrel JB (1995) Glutamic-acid-327 in the sheep alpha-1 isoform of Na^+, K^+-ATPase stabilizes a K^+-induced conformational change. J Biol Chem 270:2993–3000

Lai FF, Madan N, Ye QQ, Duan QM, Li ZC, Wang SM, Si SY, Xie ZJ (2013) Identification of a mutant α1 Na/K-ATPase that pumps but is defective in signal transduction. J Biol Chem 288:13295–13304

Lang F, Waldegger S (1997) Regulating cell volume. Am Sci 85:456–463

Lange R (1972) Some recent work on osmotic, ionic and volume regulation in marine animals. Oceanograph mar. J Cell Biol 10:97–136

Larsen EH, Willumsen NJ, Christoffersen BC (1992) Role of proton pump of mitochondria-rich cells for active transport of chloride ions in toad skin epithelium. J Physiol 450:203–216

Laughery MD, Todd ML, Kaplan JH (2003) Mutational analysis of alpha-beta subunit interactions in the delivery of Na, K-ATPase heterodimers to the plasma membrane. J Biol Chem 278:34794–34803

Laursen M, Gregersen GL, Yatime L, Nissen P, Fedesova NU (2015) Structures and characterization of digoxin- and bufalin-bound Na^+, K^+-ATPase compared with the ouabain-bound complex. Proc Natl Acad Sci U S A 112:1755–1760

Le Moullac G, Haffner P (2000) Environmental factors affecting immune responses in Crustacea. Aquaculture 191:121–131

Lecuona E, Sun H, Chen J, Trejo HE, Baker MA, Sznajder JI (2013) Protein kinase A-Ia regulates Na, K-ATPase endocytosisin alveolar epithelial cells exposed to high CO_2 concentrations. Am J Respir Cell Mol Biol 48:626–634

Lee KJ (1992) Developmental regulation of Na,K-ATPase activity in the brine shrimp *Artemia*: response to salinity and the potential regulatory role of polyamines. M.S. thesis, University of Alabama at Birmingham, pp 66

Lee CE, Bell MA (1999) Causes and consequences of recent freshwater invasions by saltwater animals. Trends Ecol Evol 14:284–288

Lee CE, Kiergaard M, Gelembiuk GW, Eads BD, Posavi M (2011) Pumping ions: rapid parallel evolution of ionic regulation following habitat invasions. Evolution 65:2229–2244

Leone FA, Furriel RPM, McNamara JC, Mantelatto FLM, Masui DC, Rezende LA, Gonçalves RR, Garçon DP (2005) (Na^+, K^+)-ATPase from crustacean gill microsomes: a molecular marker to evaluate adaptation to biotopes of different salinity. Trends Comp Biochem Physiol 11:1–15

Leone FA, Masui DC, Bezerra TMS, Garçon DP, Valenti WC, Augusto AS, McNamara JC (2012) Kinetic analysis of gill (Na⁺, K⁺)-ATPase activity in selected ontogenetic stages of the Amazon river shrimp, *Macrobrachium amazonicum* (Decapoda, Palaemonidae): interactions at ATP- and cation-binding sites. J Membr Biol 245:201–215

Leone FA, Lucena MN, Garçon DP, Pinto MR, McNamara JC (2013) Gill (Na⁺, K⁺)-ATPase in the diadromous palaemonid shrimp *Macrobrachium amazonicum*: kinetic characterization of K⁺-phosphatase activity in juveniles and adults. Trends Comp Biochem Physiol 17:13–28

Leone FA, Bezerra TMS, Garçon DP, Lucena MN, Pinto MR, Fontes CFL, McNamara JC (2014) Modulation by K⁺ plus NH₄⁺ of microsomal (Na⁺, K⁺)-ATPase activity in selected ontogenetic stages of the diadromous river shrimp *Macrobrachium amazonicum* (Decapoda, Palaemonidae). PLoS One 9(2):e89626

Leone FA, Lucena MN, Rezende LA, Garçon DP, Pinto MR, Mantelatto FLM, McNamara JC (2015a) Kinetic characterization and immunolocalization of (Na⁺, K⁺)-ATPase activity from gills of the marine seabob shrimp *Xiphopenaeus kroyeri* (Decapoda, Penaeidae). J Membr Biol 248:257–272

Leone FA, Garçon DP, Lucena MN, Faleiros RO, Azevedo SV, Pinto MR, McNamara JC (2015b) Gill specific (Na⁺, K⁺)-ATPase activity and α-subunit mRNA expression during low-salinity acclimation of the ornate crab *Callinectes ornatus* (Decapoda, Brachyura). Comp Biochem Physiol 186B:59–67

Lignot JH, Charmantier G (2001) Immunolocalization of NA⁺, K⁺-ATPase in the branchial cavity during the early development of the European Lobster *Homarus gammarus* (Crustacea, Decapoda). J Histochem Cytochem 49:1013–1023

Lima AG, McNamara JC, Terra WR (1997) Regulation of hemolymph osmolytes and gill Na⁺/K⁺-ATPase activities during acclimation to saline media in the freshwater shrimp *Macrobrachium olfersii* (Wiegmann, 1836) (Decapoda, Palaemonidae). J Exp Mar Biol Ecol 215:81–91

Lin H, Randall DJ (1993) H⁺-ATPase activity in crude homogenates of fish gill tissue: inhibitor sensitivity and environmental and hormonal regulation. J Exp Biol 180:163–174

Lin X, Fenn E, Veenstra RD (2006) An amino-terminal lysine residue of rat connexin 40 that is required for spermine block. J Physiol 570:251–269

Lingrel JB (2010) The physiological significance of the cardiotonic steroid/ouabain-binding site of the Na, K-ATPase. Annu Rev Physiol 72:395–412

Lingrel J, Moseley A, Dostanic I, Cougnon M, He SW, James P, Woo A, O'connor K, Neumann J (2003) Functional roles of the α isoforms of the (Na⁺, K⁺)-ATPase. Ann N Y Acad Sci 986:354–359

Liu J, Xie ZJ (2010) The sodium pump and cardiotonic steroids-induced signal transduction protein kinases and calcium-signaling microdomain in regulation of transporter trafficking. Biochim Biophys Acta 1802:1237–1245

Lopatin AN, Makhina EN, Nichols CG (1994) Potassium channel block by cytoplasmic polyamines as the mechanism of intrinsic rectification. Nature 372:366–369

Lovett DL, Watts SA (1995) Changes in polyamine levels in response to acclimation salinity in gills of the blue-crab *Callinectes-sapidus* rathbun. Comp Biochem Physiol 110B:115–119

Lubarski I, Karlish SJD, Garty H (2007) Structural e functional interactions between FXYD5 and the Na⁺-K⁺-ATPase. Am J Physiol 293:F1818–F1826

Lucas TFG, Amaral LS, Porto CS, Quintas LEM (2012) NaC/KC-ATPase a1 isoform mediates ouabain-induced expression of cyclin D1 and proliferation of rat Sertoli cells. Reproduction 144:737–745

Lucena MN, Garçon DP, Mantelatto FLM, Pinto MR, McNamara JC, Leone FA (2012) Hemolymph ion regulation and kinetic characteristics of the gill (Na⁺, K⁺)-ATPase in the hermit crab *Clibanarius vittatus* (Decapoda, Anomura) acclimated to high salinity. Comp Biochem Physiol 161B:380–391

Lucena MN, Pinto MR, Garcon DP, McNamara JC, Leone FA (2015) A kinetic characterization of the gill V(H(⁺))-ATPase in juvenile and adult *Macrobrachium amazonicum*, a diadromous palaemonid shrimp. Comp Biochem Physiol 181B:15–25

Lucena MN, Pinto MR, Garçon DP, Fontes CFL, McNamara JC, Leone FA (2016) (Na⁺, K⁺)-ATPase of juvenile and adult *Macrobrachium amazonicum*: effect of polyamines on phophoenzyme-linked partial reactions. Submitted

Lucu C (1990) Ionic regulatory mechanisms in crustacean gill epithelia. Comp Biochem Physiol 97A:297–306

Lucu C, Flik G (1999) (Na⁺, K⁺)-ATPase and Na⁺, Ca²⁺ exchange activities in gills of hyperregulating *Carcinus maenas*. Am J Physiol 276:R490–R499

Lucu C, Siebers D (1987) Linkage of Cl⁻ fluxes with ouabain sensitive Na/K exchange through *Carcinus gill* epithelia. Comp Biochem Physiol 87A:807–811

Lucu C, Towle DW (2003) (Na⁺, K⁺)-ATPase in gills of aquatic crustacea. Comp Biochem Physiol 135A:195–214

Lucu C, Devescovi M, Siebers D (1989) Do amiloride and ouabain affect ammonia fluxes in perfused *Carcinus* gill epithelia? J Exp Zool 249:1–5

Lucu C, Devescovi M, Skaramuca B, Kozul V (2000) Gill NaK-ATPase in the spiny lobster *Palinurus elephas* and other marine osmoconformers. Adaptiveness of enzymes from osmoconformity to hyperregulation. J Exp Mar Biol Ecol 246:163–178

Luquet CM, Postel U, Halperin J, Urcola MR, Marques R, Siebers D (2002) Transepithelial potential differences and Na^+ flux in isolated perfused gills of the crab *Chasmagnathus granulatus* (Grapsidae) acclimated to hyper- and hypo-salinity. J Exp Biol 205:71–77

Luquet CM, Weihrauch D, Senek M, Towle DW (2005) Induction of branchial ion transporter mRNA expression during acclimation to salinity change in the euryhaline crab *Neohelice granulatus*. J Exp Biol 208:3627–3636

Macías MT, Palmero I, Sastre L (1991) Cloning of a cDNA encoding an *Artemia franciscana* Na/K ATPase α-subunit. Gene 105:197–204

Magalhães C (1985) The larval development of palaemonids from amazon region reared in the laboratory. 1. *Macrobrachium amazonicum* (Heller, 1862) (Crustacea, Decapoda). Amazoniana, series Limnologia et oecologia regionalis systemae fluminis amazonas 9:247–274

Magalhães C, Walker I (1988) Larval development and ecological distribution of central Amazonian palaemonid shrimps (Decapoda, Caridea). Crustaceana 55:279–292

Mahmmoud YA, Shattock M, Cornelius F, Pavlovic D (2014) Inhibition of K^+ transport through Na^+, K^+-ATPase by Capsazepine: role of membrane span 10 of the alpha subunit in the moduklation of ion gating. PLoS One 9:e96909

Maijala RL, Eerola SH (1993) Contaminant lactic acid bacteria of dry sausages produce histamine and tyramine. Meat Sci 35:387–395

Mantel LH, Farmer LL (1983) Osmotic and ionic regulation. In: Mantel LH (ed) The biology of the Crustacea, vol 5, Internal anatomy and physiological regulation. Academic, New York, pp 53–161

Mantel LH, Olson JR (1976) Studies on the Na^+K^+-activated ATPase of crab gills. Am Zool 16:223

Martin DW (2005) Structure-function. Relationships in the Na^+, K^+-pump. Semin Nephrol 198:282–291

Martin M, Fehsenfeld S, Sourial MM, Weihrauch D (2011) Effects of high environmental ammonia on branchial ammonia excretion rates and tissue Rh-protein mRNA expression levels in sea water acclimated Dungeness crab *Metacarcinus magister*. Comp Biochem Physiol 160A:267–277

Marton LJ, Pegg AE (1995) Polyamines as targets for therapeutic intervention. Annu Rev Pharmacol Toxicol 35:55–91

Masui DC, Furriel RPM, McNamara JC, Mantelatto FLM, Leone FA (2002) Modulation by ammonium ions of gill microsomal (Na^+, K^+)-ATPase in the swimming crab *Callinectes danae*: a possible mechanism for regulation of ammonia excretion. Comp Biochem Physiol 132C:471–482

Masui DC, Furriel RPM, Mantelatto FLM, McNamara JC, Leone FA (2003) Gill (Na^+, K^+)-ATPase from the blue crab *Callinectes danae*: modulation of K^+-phosphatase activity by potassium and ammonium ions. Comp Biochem Physiol 134B:631–640

Masui DC, Furriel RPM, Mantelatto FLM, McNamara JC, Leone FA (2005a) K^+-phosphatase activity of gill (Na^+, K^+)-ATPase from the blue crab, *Callinectes danae*: low-salinity acclimation and expression of the alpha-subunit. J Exp Zool 303A:294–307

Masui DC, Furriel RPM, Silva ECC, Mantelatto FLM, McNamara JC, Barrabin H, Scofano HM, Fontes CFL, Leone FA (2005b) Gill microsomal (Na^+, K^+)-ATPase from the blue crab *Callinectes danae*: interactions at cationic sites. Int J Biochem Cell Biol 37:2521–2535

Masui DC, Silva ECC, Mantelatto FLM, McNamara JC, Barrabin H, Scofano HM, Fontes CFL, Furriel RPM, Leone FA (2008) The crustacean gill (Na^+, K^+)-ATPase: allosteric modulation of high- and low-affinity ATP-binding sites by sodium and potassium. Arch Biochem Biophys 479:139–144

Masui DC, Mantelatto FLM, McNamara JC, Furriel RPM, Leone FA (2009) (Na^+, K^+)-ATPase activity in gill microsomes from the blue crab, *Callinectes danae*, acclimated to low salinity: novel perspectives on ammonia excretion. Comp Biochem Physiol 153A:141–148

McDermott J, Sánchez G, Nangia AK, Blanco G (2015) Role the human Na, K-ATPase alpha 4 in sperm function, derived from studies in transgenic mice. Mol Reprod Dev 82:167–181

McNamara JC, Faria SC (2012) Evolution of osmoregulatory patterns and gill ion transport mechanisms in the decapod Crustacea: a review. J Comp Physiol 182:997–1014

McNamara JC, Lima AG (1997) The route of ion and water movements across the gill epithelium of the freshwater shrimp *Macrobrachium olfersii* (Decapoda, Palaemonidae): evidence from ultrastructural changes induced by acclimation to saline media. Biol Bull 192:321–331

McNamara JC, Torres AH (1999) Ultracytochemical location of Na$^+$/K$^+$-ATPase activity and effect of high salinity acclimation in gill and renal epithelia of the freshwater shrimp *Macrobrachium olfersii* (Crustacea, Decapoda). J Exp Zool 284:617–628

McNamara JC, Moreira GS, Moreira PS (1983) The effect of salinity on respiratory metabolism, survival and moulting in the first zoea of *Macrobrachium amazonicum* (Heller) (Crustacea, Palaemonidae). Hydrobiologia 101:239–242

McNamara JC, Freire CA, Torres AH, Coelho SC (2015) The conquest of fresh water by the palaemonid shrimps: an evolutionary history scripted in the osmoregulatory epithelia of the gills and antennal glands. Biol J Linn Soc 114:673–688

Meier S, Tavraz NN, Durr KL, Friedrich T (2010) Hyperpolarization-activated inward leakage currents caused by deletion or mutation of carboxy-terminal tyrosines of the Na$^+$/K$^+$-ATPase α subunit. J Gen Physiol 135:115–134

Mendonça NN, Masui DC, McNamara JC, Leone FA, Furriel RPM (2007) Long- term exposure of the freshwater shrimp *Macrobrachium olfersii* to elevated salinity: effects on gill (Na$^+$, K$^+$)-ATPase α-subunit expression e K$^+$-phosphatase activity. Comp Biochem Physiol 146A: 534–543

Meriney SD, Gray DB, Pilar GR (1994) Somatostatin-induced inhibition of neuronal Ca^{2+} current modulated by cGMP-dependent protein kinase. Nature 369:336–339

Middleton DA, Hughes E, Esmann M (2011) The conformation of ATP within the Na, K-ATPase nucleotide site: a statistically constrained analysis of REDOR solid-state NMR sata. Angew Chem Int Ed 50:7041–7044

Mijatovic T, vanQuaquebeke E, Delest B, Debeir O, Darro F, Kiss R (2007) Cardiotonic steroids on the road to anticancer therapy. Biochim Biophys Acta 1776:32–57

Miles AJ, Fedosova NU, Hoffmann SV, Wallace BA, Esmann M (2013) Stabilisation of Na, K-ATPase structure by the cardiotonic steroid ouabain. Biochem Biophys Res Commun 435:300–305

Mishra NK, Peleg Y, Cirri E, Belogus T, Lifshitz Y, Voelker DR, Apell HJ, Garty H, Karlish SJD (2011) FXYD proteins stabilize Na, K-ATPase amplification of specific phosphatidylserine-protein interactions. J Biol Chem 286:9699–9712

Mladinov D, Liu Y, Mattson DL, Liang M (2013) Micro RNAs contribute to the maintenance of cell-type-specific physiological characteristics: miR-192 targets Na$^+$/K$^+$-ATPase. Nucleic Acids Res 41: 1273–1283

Mobasheri A, Avila J, Cozar-Castellano I, Brownleader MD, Trevan M, Francis MJO, Lamb JF, Martin-Vasallo P (2000) (Na$^+$, K$^+$)-ATPase isozyme diversity; comparative biochemistry e physiological implications of novel functional interactions. Biosci Rep 20:51–91

Montes MR, Monti JLE, Rossi RC (2012) E2 → E1 transition and Rb + release induced by Na$^+$ in the Na$^+$/K$^+$-ATPase. Vanadate as a tool to investigate the interaction between Rb$^+$ and E2. Biochim Biophys Acta 1818:2087–2093

Monti JLE, Montes MR, Rossi RC (2013) Alternative cycling modes of the Na$^+$/K$^+$-ATPase in the presence of either Na$^+$ or Rb$^+$. Biochim Biophys Acta 1828:1374–1383

Moreira GS, McNamara JC, Shumway SE, Moreira PS (1983) Osmoregulation and respiratory metabolism in Brazilian *Macrobrachium* (Decapoda, Palaemonidae). Comp Biochem Physiol 74A:57–62

Moreira GS, McNamara JC, Moreira PS (1986) The effect of salinity on the upper thermal limits of survival and metamorphosis during larval development in *Macrobrachium amazonicum* (Heller) (Decapoda, palaemonidae). Crustaceana 50:231–238

Morris S (2001) Neuroendocrine regulation of osmoregulation e the evolution of air-breathing in decapods crustaceans. J Exp Biol 204:979–989

Morris S, Edwards T (1995) Control of osmoregulation via regulation of Na$^+$/K$^+$-ATPase activity in the amphibious purple shore crab, *Leptograpsus variegatus*. Comp Biochem Physiol 112C: 129–136

Morth JP, Pedersen BP, Toustrup-Jensen MS, Sorensen TLM, Petersen J, Eersen JP, Vilsen B, Nissen P (2007) Crystal structure of the sodium–potassium pump. Nature 450:1043–1050

Morth JP, Pedersen BP, Buch-Pedersen MJ, Andersen JP, Vilsen B, Palmgren MG, Nissen P (2011) A structural overview of the plasma membrane Na$^+$, K$^+$-ATPase and H$^+$-ATPase ion pumps. Nat Rev Mol Cell Biol 60:60–70

Muller O, Bayer MJ, Peters C, Andersen JS, Mann M, Mayer A (2002) The Vtc proteins in vacuole fusion: coupling NSF activity to V(0) trans-complex formation. EMBO J 21:259–269

Munhoz CD, Kawamoto EM, de Sa Lima L, Lepsch LB, Glezer I, Marcourakis T, Scavone C (2005) Glutamate modulates sodium–potassium-ATPase through cyclic GMP and cyclic GMP-dependent protein kinase in rat striatum. Cell Biochem Funct 23:115–123

Murata Y, Sun-Wada GH, Yoshimizu T, Yamamoto A, Wada Y, Futai M (2002) Differential localization of the vacuolar H+ pump with G subunit isoforms (G1 and G2) in mouse neurons. J Biol Chem 277:36296–36303

Myers SL, Cornelius F, Apell HJ, Clarke RJ (2011) Kinetics of K+ occlusion by the phosphoenzyme of the Na+, K+-ATPase. Biophys J 100:70–79

Nakanishi-Matsui M, Sekiya M, Nakamoto RK, Futai M (2010) The mechanism of rotating proton pumping ATPases. Biochim Biophys Acta 1797:1343–1352

Nesher M, Shpolansky U, Rosen H, Lichtstein D (2007) The digitalis-like steroid hormones: new mechanisms of action and biological significance. Life Sci 80:2093–2107

Neufeld GJ, Holliday CW, Pritchard JB (1980) Salinity adaptation of gill Na, K-ATPase in the blue crab, *Callinectes sapidus*. J Exp Zool 211:215–224

Nishi T, Forgac M (2002) The vacuolar (H+)-ATPases—nature's most versatile proton pumps. Nat Rev Mol Cell Biol 3:94–103

Nishi T, Kawasaki-Nishi S, Forgac M (2003) Expression and function of the mouse V-ATPase d subunit isoforms. J Biol Chem 278:46396–46402

Nishimura K, Nakatsu F, Kashiwagi K, Ohno H, Saito T, Igarashi K (2002) Essential role of S-adenosylmethionine decarboxylase in mouse embryonic development. Genes Cells 7:41–47

Nyblom M, Poulsen H, Gourdon P, Reinhard I, Andersson M (2013) Crystal structure of Na+, K+-ATPase in the Na+-bound state. Science 342:123–127

Odinetz-Collart O (1991) Stratégie de reproduction de *Macrobrachium amazonicum* en Amazonie Centrale (Decapoda, Caridea, Palaemonidae). Crustaceana 61:253–270

Ogawa H, Shinoda T, Cornelius F, Toyoshima C (2009) Crystal structure of the sodium-potassium pump (Na+, K+-ATPase) with bound potassium e ouabain. Proc Natl Acad Sci U S A 166:13742–13747

Onken K (1996) Active and electrogenic absorption of Na+ and Cl- across posterior gills of *Eriocher sinensis*: influence of short-term osmotic variations. J Exp Biol 199:901–910

Onken H (1999) Active NaCl absorption across split gill lamellae of posterior gills of Chinese crabs (*Eriocher sinensis*) adapted to different salinities. Comp Biochem Physiol 123A:377–384

Onken H, McNamara JC (2002) Hyperosmoregulation in the red freshwater crab *Dilocarcinus pagei* (Brachyura, Trichodactylidae): structural and functional asymmetries of the posterior gills. J Exp Biol 205:167–175

Onken H, Putzenlechner M (1995) A V-ATPase drives active, electrogenic and Na+-independent Cl-absorption across the gills of *Eriocheir sinensis*. J Exp Biol 198:767–774

Onken H, Riestenpatt S (1998) NaCl absorption across split gill lamellae of hyperregulating crabs: transport mechanisms and their regulation. Comp Biochem Physiol 119A:883–893

Onken H, Riestenpatt S (2002) Ion transport across posterior gills of hyperosmoregulating shore crabs (*Carcinus maenas*): amiloride blocks the cuticular Na+ conductance and induces current-noise. J Exp Biol 205:523–531

Onken H, Schobel A, Kraft J, Putzenlechner M (2000) Active NaCl absorption across split lamellae of posterior gills of the chinese crab *Eriocheir sinensis*: stimulation by eyestalk extract. J Exp Biol 203:1373–1381

Oot RA, Wilkens S (2012) Subunit interactions at the V1-Vo interface in yeast vacuolar ATPase. J Biol Chem 287:13396–13406

Osteresch C, Bender T, Grond S, Zezschwitz P, Kunze B, Jansen R, Huss M, Wieczorek H (2012) The binding site of the V-ATPase inhibitor Apicularen is in the vicinity of those for Bafilomycin and Archazolid. J Biol Chem 287:31866–31876

Oubaassine R, Weckring M, Kessler L, Breidert M, Roegel JC, Eftekhari P (2012) Insulin interacts directly with Na+/K+ATPase and protects from digoxin toxicity. Toxicology 299:1–9

Palmgren MG, Nissen P (2011) P-Type ATPases. Annu Rev Biophys 40:243–266

Pan LQ, Zhang LJ, Liu HY (2007) Effects of salinity and pH on ion-transport enzyme activities, survival and growth of *Litopenaeus vannamei* postlarvae. Aquaculture 273:711–720

Pan L, Li L, Zhang L (2010) Variations of ion-transport enzyme activities during early development of the shrimps *Fenneropenaeus chinensis* and *Marsupenaeus japonicus*. J Ocean Univ Chin 9:76–80

Parekh A, Campbell AJM, Djouhri L, Fang X, Mcmullan S, Berry C, Acosta C, Lawson SN (2010) Immunostaining for the α3 isoform of the Na+/K+-ATPase is selective for functionally identified muscle spindle afferents in vivo. J Physiol 588:4131–4143

Parrie L, Towle DW (2002) Induction of Na++K+-ATPase alpha subunit mRNA in branchial tissues of the American lobster *Homarus americanus*. J Integr Comp Biol 42:1291

Patrick ML, Aimanova K, Sanders HR, Gill SS (2006) P-type Na^+/K^+-ATPase and V-type H^+-ATPase expression patterns in the osmoregulatory organs of larval and adult mosquito *Aedes aegypti*. J Exp Biol 209:4638–4651

Paulsen PA, Jurkowski W, Apostolov R, Lindahl E, Nissen P, Poulsen H (2013) The C-terminal cavity of the Na, K-ATPase analyzed by docking and electrophysiology. Mol Membr Biol 30:195–205

Paunescu TG, da Silva N, Marshansky V, McKee M, Breton S, Brown D (2004) Expression of the 56-kDa B2 subunit isoform of the vacuolar $H\beta$-ATPase in proton secreting cells of the kidney and epididymis. Am J Physiol 287:C149–C162

Pedersen PL, Amzel LM (1993) ATP synthases. Structure, reaction center, mechanism, e regulation of one nature's most unique machines. J Biol Chem 268:9937–9940

Pedersen BP, Morth JP, Nissen P (2010) Structure determination using poorly diffracting membrane-protein crystals: the H^+-ATPase and Na^+, K^+-ATPase case history. Acta Crystallogr 66D:309–313

Pegg AE (2009) Mammalian polyamine metabolism and function. IUBMB Life 61:880–894

Pendeville H, Carpino N, Marine JC, Takahashi Y, Muller M, Martial JA, Cleveland JL (2001) The ornithine decarboxylase gene is essential for cell survival during early murine development. Mol Cell Biol 21:6549–6558

Péqueux A (1995) Osmotic regulation in crustaceans. J Crustac Biol 15:1–60

Péqueux A, Labras P, Cann-Moisan C, Coroff J, Sebert P (2002) Polyamines, indolamines, and catecholamines in gills and haemolymph of the euryhaline crab. *Eriocheir sinensis* effects of high pressure and salinity. Crustaceana 75:567–578

Perry SF (1986) Carbon dioxide excretion in fishes. Can J Zool 64:565–572

Peterson GL, Hokin LE (1981) Molecular weight and stoichiometry of the sodium- and potassium- activated adenosine triphosphatase subunits. J Biol Chem 256:3751–3761

Peterson G, Ewing R, Conte F (1978) Membrane differentiation and de novo synthesis of the Na^+K^+-activated adenosine triphosphatase during development of *Artemia salina* nauplii. Dev Biol 67:90–98

Petrushanko JY, Mitkevich VA, Anashkira AA, Klimanova EA, Dergousova EA, Lopina OD, Makarov AA (2014) Critical role of γ-phosphate in structural transition of Na, K-ATPase upon ATP binding. Sci Rep 4:5165

Pierre SV, Sottejeau Y, Gourbeau J, Sanchez G, Shidyak KA, Blanco VG (2008) Isoform-specificity of Na, K-ATPase-mediated ouabain signaling. Am J Physiol 294:F859–F866

Pietrement C, Sun-Wada GH, Silva ND, McKee M, Marshansky V, Brown D, Futai M, Breton S (2006) Distinct expression patterns of different subunit isoforms of the V-ATPase in the rat epididymis. Biol Reprod 74:185–194

Pinto MR, Lucena MN, Faleiros RO, Almeida EA, McNamara JC, Leone FA (2016) Effects of ammonia stress in the Amazon river shrimp *Macrobrachium amazonicum* (Decapoda, Palaemonidae). Aquatic Toxicol 170:13–23.

Post RL (1999) Active transport and pumps. Curr Top Membr 48:397–417

Post RL, Hegyvary C, Kume S (1972) Activation by adenosine triphosphate in the phosphorylation kinetics of sodium and potassium ion transport adenosine triphosphatase. J Biol Chem 247:6530–6540

Poulsen H, Morth P, Egebjerg J, Nissen P (2010a) Phosphorylation of the Na^+, K^+-ATPase and the H^+, K^+-ATPase. FEBS Lett 584:2589–2595

Poulsen H, Khandelia H, Morth P, Bublitz M, Mouritsen OG, Egebjerg J, Nissen P (2010b) Neurological disease mutations compromise a C-terminal ion pathway in the Na1/K1-ATPase. Nature 467:99–102

Poulsen H, Nissen P, Mouritsen OG, Khandelia H (2012) Protein kinase A (PKA) phosphorylation of Na/K--ATPase opens intracellular C-terminal water pathway leading to third Na-binding site in molecular dynamics simulations. J Biol Chem 287:15959–15965

Prast H, Philippu A (2001) Nitric oxide as modulator of neuronal function. Prog Neurobiol 64:51–68

Pressley TA (1992) Ionic regulation of Na^+-ATPase, K^+-ATPase expression. Semin Nephrol 12:67–71

Pressley TA, Duran MJ, Pierre SV (2005) Regions conferring isoform-specific function in the catalytic subunit of the Na, K-pump. Front Biosci 10:2018–2026

Prosser CL (1973) Comparative animal physiology. WB Saunders, Philadelphia

Purhonen P, Thomsen K, Maunsbach AB, Hebert H (2006) Association of renal Na, K-ATPase alpha-subunit with the beta- and gamma-subunits based on cryoelectron microscopy. J Membr Biol 214:139–146

Quarfoth G, Ahmed K, Foster D (1978) Effects of polyamines on partial reactions of membrane (Na^+K^+)-ATPase. Biochim Biophys Acta 526:580–590

Radzyukevich TL, Neumann JC, Rindler TN, Oshiro N, Goldhamer DJ, Lingrel JB, Heiny JA (2013) Tissue-specific role of the Na, K-ATPase $\alpha2$ isozyme in skeletal muscle. J Biol Chem 288:1226–1237

Randall D, Burggren W, French K (2000) Animal physiology – mechanisms and adaptations, 4th edn. WH Freeman, New York

Rastogi VK, Girvin ME (1999) Structural changes linked to proton translocation by subunit c of the ATP synthase. Nature 402:263–268

Ratheal IAM, Virgin GK, Yu H, Roux B, Gatto C, Artiga P (2010) Selectivity of externally facing ion-binding sites in the Na/K pump to alkali metals and organic cations. Proc Natl Acad Sci U S A 107: 18718–18723

Read GHL (1982) An ecophysiological study of the effects of changes in salinity and temperature on the distribution of *Macrobrachium petersi* (Hilgendorf) in the Keiskamma river and estuary. PhD thesis, Rhodes University

Rebelo MF, Santos EA, Monserrat JM (1999) Ammonia exposure of *Chasmagnathus granulata* (Crustacea-Decapoda) Dana, 1851. Accumulation in haemolymph and effects on osmoregulation. Comp Biochem Physiol 122A:429–435

Rebelo MF, Rodriguez EM, Santos EA, Ansaldo M (2000) Histopathological changes in gills of the estuarine crab *Chasmagnathus granulata* (Crustacea, Decapoda) following after exposure to ammonia. Comp Biochem Physiol 125C:157–164

Regnault M (1987) Nitrogen excretion in marine and freshwater crustacea. Biol Rev 62:1–24

Rice WJ, Young HS, Martin DW, Sachs JR, Stokes DL (2001) Structure of (Na+, K+)-ATPase at 11-angstrom resolution: comparison with Ca2+ -ATPase in E1 e E2 states. Biophys J 80:2187–2197

Riestenpatt S, Onken H, Siebers D (1996) Active absorption of Na+ and Cl− across the gill epithelium of the shore crab *Carcinus maenas*: voltage-clamp and ion-flux studies. J Exp Biol 199:1545–1554

Robinson JD, Leach CA, Robinson LJ (1986) Cation sites, spermine, e the reaction sequence of the (NA+K+)-dependent ATPase. Biochim Biophys Acta 856:536–544

Romano N, Zeng C (2007) Ontogenetic changes in tolerance to acute ammonia exposure and associated histological gill alterations during early juvenile development of the blue swimmer crab, *Portunus pelagicus*. Aquaculture 266:246–254

Romano N, Zeng C (2010) Survival, osmoregulation and ammonia-N excretion of blue swimmer crab *Portunus pelagicus* juveniles exposed to different ammonia-N and salinity combinations. Comp Biochem Physiol C Toxicol Pharmacol 151C:222–228

Rubin EH, Ferrendelli JA (1977) Distribution and regulation of cyclic nucleotide levels in cerebellum, in vivo. J Neurochem 29:43–51

Sáez AG, Lozano E, Zaldívar-Riverón A (2009) Evolutionary history of Na, K-ATPases and their osmoregulatory role. Genetica 136:479–490

Salyer SA, Parks J, Barati MT, Lederer ED, Clark BJ, Klein JD, Khundmiri SJ (2013) Aldosterone regulates Na+, K+ ATPase activity in human renal proximal tubule cells through mineralocorticoid receptor. Biochim Biophys Acta 1833:2143–2152

Sánchez-Rodriguéz JE, Khalili-Araghi F, Miranda P, Roux B, Holmgren M, Bezanilla F (2015) A structural rearrangement of the Na, K-ATPase traps ouabain within the external ion permeation pathway. J Mol Biol 427:1335–1344

Sandtner W, Egwolf B, Khalili-Araghi F, Sánchez-Rodrígues JE, Roux B, Bezanilla F, Holmgren M (2011) Ouabain binding site in a functioning Na/K ATPase. J Biol Chem 286:38177–38183

Santos LCF, Belli NM, Augusto A, Masui DC, Leone FA, McNamara JC, Furriel RPM (2007) Gill (Na+, K+)-ATPase in diadromous, freshwater palaemonid shrimps: species-specific kinetic characteristics and α-subunit expression. Comp Biochem Physiol 148A:178–188

Schack VR, Morth JP, Toustrup-Jensen MS, Anthonisen AN, Nissen P, Eersen JP, Vilsen B (2008) Identification e function of a cytoplasmic K+ site of the Na+, K+ -ATPase. J Biol Chem 283:27982–27990

Schaefer TL, Lingrel JB, Moseley AE, Vorhees CV, Williams MT (2011) Targeted mutations in the Na, K-ATPase alpha 2 isoform confer ouabain resistance and result in abnormal behavior in mice. Synapse 65:520–531

Scheiner-Bobis G (2002) The sodium pump – its molecular properties and mechanics of ion transport. Eur J Biochem 269:2424–2433

Schmidt-Nielsen K (1996) Fisiologia animal: Adaptação e Meio Ambiente, vol 1. Editora Santos, São Paulo

Schubart CD, Diesel R (1998) Osmoregulatory capacities and penetration into terrestrial habitats: a comparative study of Jamaican crabs of the genus *Armases* Abele, 199 (Brachyura: Grapsidae: Sesarminae). Bul Mar Sci 6:743–752

Schubart CD, Diesel R, Hedges SB (1998) Rapid evolution to terrestrial life in jamaican crabs. Nature 393:363–365

Scott C, Bissig G, Gruenberg J (2011) Duelling functions of the V-ATPase. EMBO J 30:4113–4115

Segall L, Daly SE, Boxenbaum N, Lane LK, Blostein R (2000) Distinct catalytic properties of the α 1, α 2 e α 3 isoforms of the rat (Na+, K+)-ATPase. Biophys J 78:78A

Seiler N, Dezeure F (1990) Polyamine transport in mammalian cells. Int J Biochem 22:211–218

Seneviratna D, Taylor HH (2006) Ontogeny of osmoregulation in embryos of intertidal crabs (Hemigrapsus sexdentatus and H. crenulatus, Grapsidae, Brachyura): putative involvement of the embryonic dorsal organ. J Exp Biol 209:1487–1501

Sennoune SR, Bakunts K, Martinez GM (2004) Vacuolar H+-ATPase in human breast cancer cells with distinct metastatic potential: distribution and functional activity. Am J Physiol 286:C1433–C1452

Seok JH, Kim JB, Hong JH, Sung JY, Hur GM, Lim K, Lee JH (1998) Regulation of Na, K-ATPase activity in renal basolateral membrane of 1-clip-1-kidney hypertensive rate. Biochem Mol Biol Int 46: 667–672

Shetlar RE, Towle DW (1989) Electrogenic sodium-proton exchange in membrane vesicles from crab (Carcinus maenas) gill. Am J Physiol 257:R924–R931

Shindo Y, Morishita K, Kotake E, Miura H, Carninci P, Kawai J, Hayashizaki Y, Hino A, Kanda T, Kusakabe Y (2011) FXYD6, a Na, K-ATPase regulator, is expressed in type II taste cells. Biosci Biotechnol Biochem 75:1061–1066

Shinoda T, Ogawa H, Cornelius F, Toyoshima C (2009) Crystal structure of the sodium-potassium pump at 2.4 Å resolution. Nature 459:446–450

Siebers D, Leweck K, Markus H, Winkler A (1982) Sodium regulation in the shore crab Carcinus maenas as related to ambient salinity. Mar Biol 69:37–43

Silva ECC, Masui DC, Furriel RPM, Mantelatto FLM, McNamara JC, Barrabin H, Leone FA, Scofano HM, Fontes CFL (2008) Regulation by the exogenous polyamine spermidine of Na, K-ATPase activity from the gills of the euryhaline swimming crab Callinectes danae (Brachyura, Portunidae). Comp Biochem Physiol 149B:622–629

Silva ECC, Masui DC, Furriel RP, McNamara JC, Barrabin H, Scofano HM, Perales J, Teixeira-Ferreira A, Leone FA, Fontes CFL (2012) Identification of a crab gill FXYD2 protein and regulation of crab microsomal Na, K-ATPase activity by mammalian FXYD2 peptide. Biochim Biophys Acta 1818:2588–2597

Skaggs HS, Henry RP (2002) Inhibition of carbonic anhydrase in the gills of two euryhaline crabs, Callinectes sapidus and Carcinus maenas by heavy metals. Comp Biochem Physiol 133C:605–612

Skou JC (1957) The influence of some cations on an adenosine triphosphatase from peripheral nerves. Biochim Biophys Acta 23:394–401

Skou JC, Esmann M (1992) The Na, K-ATPase. J Bioenerg Biomembr 24:249–261

Slingerland M, Cerella C, Guchelaar HJ, Diederich M, Gelderblom H (2013) Cardiac glycosides in cancer therapy: from preclinical investigations towards clinical trials. Invest New Drugs 31:1087–1094

Smith AN, Skaug J, Choate KA, Nayir A, Bakkaloglu A, Ozen S, Hulton SA, Sanjad SA, Al-Sabban EA, Lifton RP, Scherer SW, Karet FE (2000) Mutations in ATP6N1B, encoding a new kidney vacuolar proton pump 116-kD subunit, cause recessive distal renal tubular acidosis with preserved hearing. Nat Genet 26:71–75

Smith AN, Borthwick KJ, Karet FE (2002) Molecular cloning and characterization of novel tissue-specific isoforms of the human vacuolar V-ATPase C, G and d subunits, and their evaluation in autosomal recessive distal renal tubular acidosis. Gene 297:169–177

Sorensen TML, Moller JV, Nissen P (2004) Phosphoryl transfer e calcium ion occlusion in the calcium pump. Science 304:1672–1675

Sottejeau Y, Belliard A, Duran MJ, Pressley TA, Pierre SV (2010) Critical role of the isoform-specific region in R1-Na, K-ATPase trafficking and protein kinase C-dependent regulation. Biochemistry 49:3602–3610

Stern S, Borut A, Cohen D (1984) The effect of salinity and ion composition on oxygen consumption and nitrogen excretion of Macrobrachium rosenbergii (De Man). Comp Biochem Physiol 79A:271–274

Stronberg JO (1972) Cyathura polita (Crustacea, Isopoda), some embryological notes. Bull Mar Sci Gulf Carribb 22:463–483

Sun-Wada GH, Wada Y (2010) Vacuolar-type proton pump ATPases: roles of subunit isoforms in physiology and pathology. Histol Histopathol 25:1611–1620

Susanto GN, Charmantier G (2000) Ontogeny of osmoregulation in the crayfish Astacus leptodactylus. Physiol Biochem Zool 73:169–176

Susanto GN, Charmantier G (2001) Crayfish freshwater adaptation starts in eggs: ontogeny of osmoregulation in embryos of Astacus leptodactylus. J Exp Zool 289:433–440

Suzuki T, He Y, Kashiwagi K, Murakami Y, Hayashi S, Igarashi K (1994) Antizyme protects against abnormal accumulation and toxicity of polyamines in ornithine decarboxylase-overproducing cells. Proc Natl Acad Sci U S A 91:8930–8934

Tabor CW, Tabor H (1984) Polyamines. Annu Rev Biochem 53:749–790

Tan CH, Choong KY (1981) Effect of hyperosmotic stress on hemolymph protein, muscle ninhydrin positive substances and free amino acids in *Macrobrachium roserbergii* (de Man). Comp Biochem Physiol 70A:485–489

Tashima Y, Hasegawa M, Mizunuma H (1978) Activation of (Na$^+$K$^+$)-adenosine triphosphatase by spermine. Biochem Biophys Res Commun 82:13–18

Tassoni A, Antognoni F, Bagni N (1996) Polyamines binding to plasma membrane vesicles isolated from zucchini hypocotyls. Plant Physiol 110:817–824

Taylor HH, Seneviratna D (2005) Ontogeny of salinity tolerance and hyper osmoregulation by embryos of the intertidal crabs *Hemigrapsus edwardsii* and *Hemigrapsus crenulatus* (Decapoda, Grapsidae): survival of acute hyposaline exposure. Comp Biochem Physiol 140A:495–505

Taylor HH, Taylor EW (1992) Decapod Crustacea. In: Harrison FW, Humas AG (eds) Microscopic anatomy of invertebrates, vol 10. Wiley-Liss, New York, pp 203–293

Therien AG, Blostein R (2000) Mechanisms of sodium pump regulation. Am J Physiol 279:C541–C566

Tidow H, Aperia A, Nissen P (2010) How are ion pumps and agrin signaling integrated? Trends Biochem Sci 35:653–659

Toei M, Saum R, Forgac M (2010) Regulation and isoform function of the V-ATPases. Biochemistry 49:4715–4723

Tokhtaeva E, Clifford RJ, Kaplan JH, Sachs G, Vagin O (2012) Subunit isoform selectivity in assembly of Na, K-ATPase α-β heterodimers. J Biol Chem 287:26115–26125

Toner M, Vaio G, McLaughlin A, McLaughlin S (1988) Adsorption of cations to phosphatidylinositol 4,5-bisphosphate. Biochemistry 27:7435–7443

Toustrup-Jensen MS, Vilsen B (2005) Interaction between the catalytic site and the A-M3 linker stabilizes E2/E2P conformational states of Na$^+$, K$^+$-ATPase. J Biol Chem 280:10210–20218

Toustrup-Jensen MS, Holm R, Einholm AP, Schack VR, Morth JP, Nissen P, Eersen JP, Vilsen B (2009) The C terminus of Na$^+$, K$^+$-ATPase controls Na$^+$ affinity on both sides of the membrane through Arg935. J Biol Chem 284:18715–18725

Towle DW (1984) Regulatory functions of Na$^+$+K$^+$-ATPase in marine and estuarine animals. In: Péqueux A, Gilles R, Bolis L (eds) Osmoregulation in estuarine and marine animals. Springer, Berlin, pp 157–170

Towle DW (1990) Sodium transport systems in gills. In: Kinne RKH (ed) Comparative aspects of sodium cotransport systems. Karger Publishing, Basel, pp 241–263

Towle DW, Kays WT (1986) Basolateral localization of Na$^+$, K$^+$-ATPase in gill epithelium of two osmoregulating crabs, *Callinectes sapidus* and *Carcinus maenas*. J Exp Zool 239:311–318

Towle DW, Weihrauch D (2001) Osmoregulation by gills of euryhaline crabs: molecular analysis of transporters. Am Zool 41:770–780

Towle DW, Palmer GE, Harris JL (1976) Role of gill Na$^+$-+K$^+$-dependent ATPase in acclimation of blue crabs (*Callinectes sapidus*) to low salinity. J Exp Zool 196:315–321

Towle DW, Rushton ME, Heidysch D, Magnani JJ, Rose MJ, Amstutz A, Jordan MK, Shearer DW, Wu WS (1997) Sodium-proton antiporter in the euryhaline crab *Carcinus maenas*: molecular cloning, expression and tissue distribution. J Exp Biol 200:1003–1014

Towle DW, Paulsen RS, Weihrauch D, Kordylewski M, Salvador C, Lignot JH, Spanings-Pierrot C (2001) Na$^+$K$^+$-ATPase in gills of the blue crab *Callinectes sapidus*: cDNA sequencing and salinity related expression of α-subunit mRNA and protein. J Exp Biol 204:4005–4012

Towle DW, Henry RP, Terwilliger NB (2011) Microarray-detected changes in gene expression in gills of green crabs (*Carcinus maenas*) upon dilution of environmental salinity. Comp Biochem Physiol 6D:115–125

Toyomura T, Murata Y, Yamamoto A (2003) From lysosomes to the plasma membrane – localization of vacuolar type H$^+$-ATPase with the a3 isoform during osteoclast differentiation. J Biol Chem 248:22023–22030

Toyoshima C, Cornelius F (2013) New crystal structures of PII-type ATPases: excitement continues. Struct Biol 23:507–514

Toyoshima C, Mizutani T (2004) Crystal structure of the calcium pump with a bound ATP analogue. Nature 430:529–535

Toyoshima C, Nakasako M, Nomura H, Ogawa H (2000) Crystal structure of the calcium pump of sarcoplasmic reticulum at 2.6A resolution. Nature 405:647–655

Toyoshima C, Nomura H, Tsuda T (2004) Lumenal gating mechanism revealed in calcium pump crystal structures with phosphate analogues. Nature 432:361–368

Toyoshima C, Ryuta K, Flemming C (2011) First crystal structures of Na^+, K^+-ATPase: new light on the oldest ion pump. Structure 19:1732–1738

Tresguerres M, Parks SK, Sabatini SF, Goss GG, Luquet CM (2008) Regulation of ion transport by pH and [HCO_3^-] in isolated gills of the crab Neohelice (Chasmagnathus) granulata. Am J Physiol 294:R1033–R1043

Tsai JR, Lin HC (2007) V-type H1-ATPase e Na^+, K^+-ATPase in the gills of 13 euryhaline crabs during salinity acclimation. J Exp Biol 210:620–627

Tsai JR, Lin HC (2014) Functional anatomy and ion regulatory mechanisms of the antennal gland in a semi-terrestrial crab, Ocypode stimpsoni. Biol Open 3:409–417

Vedonato N, Gadsby DC (2010) The two C-terminal tyrosines stabilize occluded Na/K pump conformations containing Na or K ions. J Gen Physiol 136:163–182

Vergamini FG, Pileggi LG, Mantelatto FL (2011) Genetic variability of the Amazon River prawn Macrobrachium amazonicum (Decapoda, Caridea, Palaemonidae). Contrib Zool 80:67–83

Wagner CA, Finberg KE, Breton S, Marshansky V, Brown D, Geibel JP (2004) Renal vacuolar H^+-ATPase. Physiol Rev 84:1263–1314

Wall SM, Fischer MP, Mehta P, Hassell KA, Park SJ (2001) Contribution of the Na^+-K^+-$2Cl^-$ cotransporter NKCC1 to Cl^- secretion in rat OMCD. Am J Physiol 280:F913–F921

Wallace HM, Fraser AV, Hughes A (2003) A perspective of polyamine metabolism. Biochem J 376:1–14

Wang D, Hiesinger PR (2013) The vesicular ATPase: a missing link between acidification and exocytosis. J Cell Biol 203:171–173

Wang Y, Inoue T, Forgac M (2005) Subunit a of the yeast V-ATPase participates in binding of Bafilomycin. J Biol Chem 280:40481–40488

Wang L, Wang WN, Liu Y, Cai DX, Li JZ, Wang AL (2012a) Two types of ATPases from the Pacific white shrimp, Litopenaeus vannamei in response to environmental stress. Mol Biol Rep 39:6427–6438

Wang R, Zhuang P, Feng G, Zhang L, Huang X, Jia X (2012b) Osmotic and ionic regulation and Na^+/K^+-ATPase, carbonic anhydrase activities in mature Chinese mitten crab, Eriocheir sinensis H. Milne Edwards, 1853 (decapoda, brachyura) exposed to different salinities. Crustaceana 85:1431–1447

Wanson SA, Péqueux A, Roer RD (1984) Na^+ regulation and Na^+, K^+-ATPase activity in the euryhaline fiddler crab Uca minax (La Conte). Comp Biochem Physiol 79A:673–678

Watanabe S, Kusama-Eguchi K, Kobayashi H, Igarashi K (1991) Estimation of polyamine binding to macromolecules and ATP in bovine lymphocytes and rat liver. J Biol Chem 266:20803–20809

Watts SA, Lee KJ, Cline GB (1994) Elevated ornithine decarboxylase activity e polyamine levels during early development in the brine shrimp Artemia-franciscana. J Exp Zool 270:426–431

Watts SA, Yeh EW, Henry RP (1996) Hypoosmotic stimulation of ornithine decarboxylase activity in the brine shrimp Artemia franciscana. J Exp Zool 274:15–22

Wehner F, Olsen H, Tinel H, Kinnesaffran E, Kinne RKH (2003) Cell volume regulation: osmolytes, osmolyte transport, and signal transduction. Rev Physiol Biochem Pharmacol 148:1–80

Weigand KM, Messchaert M, Swarts HGP, Russel FGM, Koenderink JB (2014) Alternating hemiplegia of childhood mutations have a differential effect on Na^+, K^+-ATPase activity and ouabain binding. Biochim Biophys Acta 1842:1010–1016

Weihrauch D, Becker W, Postel U, Riestenpatt S, Siebers D (1998) Active excretion of ammonia across the gills of the shore crab Carcinus maenas and its relation to osmoregulatory ion uptake. J Comp Physiol 168B:364–376

Weihrauch D, Becker W, Postel U, Luck-Kopp S, Siebers D (1999) Potential of active excretion of ammonia in three different haline species of crabs. J Comp Physiol 169B:25–37

Weihrauch D, Ziegler A, Siebers D, Towle DW (2001) Molecular characterization of V-type $H^{(+)}$-ATPase (B-subunit) in gills of euryhaline crabs and its physiological role in osmoregulatory ion uptake. J Exp Biol 204:25–37

Weihrauch D, Ziegler A, Siebers D, Towle DW (2002) Active ammonia excretion across the gills of the green shore crab Carcinus maenas: participation of $Na^{(+)}/K^{(+)}$-ATPase, V-type $H^{(+)}$-ATPase and functional microtubules. J Exp Biol 205:2765–2775

Weihrauch D, McNamara JC, Towle DW, Onken H (2004a) Ion-motive ATPases and active, transbranchial NaCl uptake in the red freshwater crab, *Dilocarcinus pagei* (Decapoda, Trichodactylidae). J Exp Biol 207:4623–4631

Weihrauch D, Morris S, Towle DW (2004b) Ammonia excretion in aquatic terrestrial crabs. J Exp Biol 207:4491–4504

Weihrauch D, Wilkie MP, Walsh PJ (2009) Ammonia and urea transporters in gills of fish and aquatic crustaceans. J Exp Biol 212:1716–1730

Weiner ID, Hamm LL (2007) Molecular mechanisms of renal ammonia transport. Annu Rev Physiol 69:317–340

Welcomme L, Devos P (1988) Cytochrome *c* oxidase and Na$^+$-K$^+$ ATPase activities in the anterior and posterior gills of the shore crab *Carcinus maenas* L. after adaptation to various salinities. Comp Biochem Physiol 89B:339–341

Wengert M, Ribeiro MC, Abreu TP, Coutinho-Silva R, Leão-Ferreira LR, Pinheiro AAS, Caruso-Neves C (2013) Protein kinase C-mediated ATP stimulation of Na$^+$-ATPase activity in LLC-PK1cells involves a P2Y2 and/or P2Y4 receptor. Arch Biochem Biophys 535:136–142

Wild GE, Searles LE, Koski KG, Drozdowski LA, Begum-Hasan J, Thomson ABR (2007) Oral polyamine administration modifies the ontogeny of hexose transporter gene expression in the postnatal rat intestine. Am J Physiol 293:G453–G460

Wilder MN, Huong DTT, Atmomarsono M, Tran TTH, Phu TQ, Yang WJ (2000) Characterization of Na/K--ATPase in *Macrobrachium rosenbergii* and the effects of changing salinity on enzymatic activity. Comp Biochem Physiol 125A:377–388

Wilder MN, Huong DTT, Okuno A, Atmomarsono M, Yang WJ (2001) Ouabain sensitive Na/K-ATPase activity increases during embryogenesis in the giant freshwater prawn *Macrobrachium rosenbergii*. Fish Sci 67:182–184

Wilkie MP (1997) Mechanisms of ammonia excretion across fish gills. Comp Biochem Physiol 118A:39–50

Williams K (1997) Interactions of polyamines with ion channels. Biochem J 325:289–297

Winkler A (1986) The role of the transbranchial potential difference in hyperosmotic regulation of the shore crab *Carcinus maenas*. Helgoländer Meeresun 40:161–175

Yang WK, Kang CK, Chen TY, Chang WB, Lee TH (2011) Salinity-dependent expression of the branchial Na$^+$/K$^+$/2Clcotransporter and Na$^+$/K$^+$-ATPase in the sailfin molly correlates with hypoosmoregulatory endurance. J Comp Physiol 181B:953–964

Yaragatupalli S, Olivera JF, Gatto C, Artigas P (2009) Altered Na$^+$ transport after an intracellular α-subunit deletion reveals strict external sequential release of Na$^+$ from the Na/K pump. Proc Natl Acad Sci 106:15507–15512

Yatime L, Laursen M, Morth JP, Esmann M, Nissen P, Fedesova NU (2011) Structural insights into the high affinity binding of cardiotonic steroids to the Na$^+$, K$^+$-ATPase. J Struct Biol 174:296–306

Yoneda JS, Rigos CF, Ciancaglini P (2013) Addition of subunit c, K$^+$ ions, and lipid restores the thermal stability of solubilized Na, K-ATPase. Arch Biochem Biophys 530:93–100

Young-Lai WW, Charmantier-Daures M, Charmantier G (1991) Effect of ammonia on survival and osmoregulation in different life stages of the lobster *Homarus americanus*. Mar Biol 110:293–300

Yu H, Ratheal IM, Artiga P, Rouxi B (2011) Protonation of key acidic residues is critical for the K-selectivity of the Na/K pump. Nat Struct Mol Biol 18:1159–1164

Zanders IP (1980) Regulation of blood ions in *Carcinus maenas* (L.). Comp Biochem Physiol 65A:97–108

Zare S, Greenaway P (1998) The effect of moulting and sodium depletion on sodium transport and the activities of Na$^+$K$^+$-ATPase, and V-ATPase in the freshwater crayfish *Cherax destructor* (Crustacea: Parastacidae). Comp Biochem Physiol 119A:739–745

Zeiske W, Onken H, Schwarz H-J, Graszynski K (1992) Invertebrate epithelial Na$^+$ channels: amiloride-induced current-noise in crab gill. Biochim Biophys Acta 1105:245–252

Zheng J, Koh X, Hua F, Li G, Larrick JW, Bian JS (2011) Cardio protection induced by Na1/K1-ATPase activation involves extracellular signal-regulated kinase 1/2 and phosphoinositide 3-kinase/Akt pathway. Cardiovasc Res 89:51–59

Zhuo M, Hu Y, Schultz C, Kandel ER, Hawkins RD (1994) Role of guanylyl cyclase and cGMP-dependent protein kinase in long-term potentiation. Nature 368:635–639

Nitrogen Excretion and Metabolism in Insects

M.J. O'Donnell and Andrew Donini

© Springer International Publishing Switzerland 2017
D. Weihrauch, M. O'Donnell (eds.), *Acid-Base Balance and Nitrogen Excretion in Invertebrates*,
DOI 10.1007/978-3-319-39617-0_4

4.1 Introduction

Metabolism of proteins and nucleic acids results in the production of nitrogen-containing compounds. Deamination of amino acids, for example, yields α-keto acids that can be oxidized to carbon dioxide and water, but also ammonia. Ammonia may also be derived from the metabolism of urea or uric acid. Some of the resulting ammonia can be recycled through transamination reactions, by which glutamate is formed from α-ketoglutarate and glutamine is formed from glutamate. Any excess ammonia, however, is toxic; NH_4^+ may interfere with neuronal activity because it can pass through K^+ channels, whereas NH_3 may perturb the pH of cells and organelles, including mitochondria. Although some insects, such as the larvae of the sheep blowfly, tolerate high levels (10–20 mM) of ammonia in their hemolymph (Marshall and Wood 1990), in general, the excretory system functions to prevent accumulation of toxic levels of ammonia. A few species, discussed below, excrete much of their nitrogenous waste as ammonia, but insects generally are viewed as uricotelic, in keeping with the Baldwin–Needham hypothesis that terrestrial animals excrete nitrogenous excretion products with low solubility, such as urea, uric acid, or allantoin, as a means of conserving water (Baldwin and Needham 1934).

4.2 Uric Acid

4.2.1 Transport of Uric Acid by the Malpighian Tubules of Insects

Although uric acid is synthesized from either amino acid nitrogen or nucleic acid nitrogen, the former source is more important because insects generally ingest more protein than nucleic acids. The amino acids Gln, Glu, Asp, and Gly donate the four N-groups to the purine ring during uric acid biosynthesis in the fat body (❏ Fig. 4.1) (Barrett and Friend 1970; Bursell 1967). Uric acid, in spite of the metabolic cost of its synthesis (8 mols ATP/mol uric acid), is an ideal compound for elimination of excess nitrogen because it is only slightly soluble at physiological pH. The low solubility both reduces the toxicity of uric acid and also allows it to be excreted without the loss of much water. Blood-feeding insects must eliminate large amounts of nitrogenous waste produced as the protein-rich meal is metabolized. The tsetse fly, *Glossina morsitans*, for example, eliminates surplus nitrogen by excreting ~ 50 % of the dry weight of the ingested blood in the form of nitrogen-containing compounds (mainly uric acid, but also arginine and histidine) (Bursell 1965). In another blood feeder, *Rhodnius*

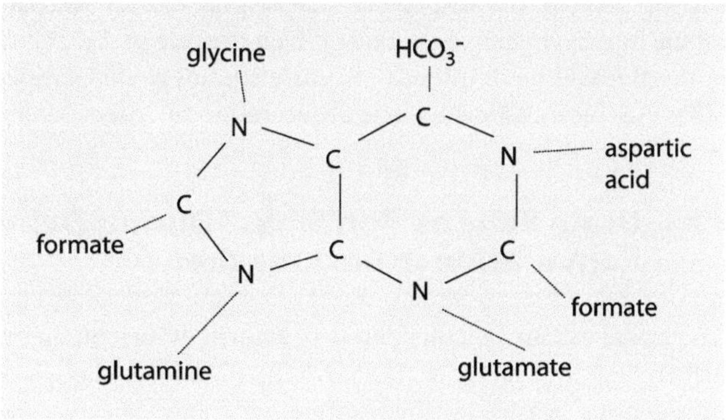

❏ **Fig. 4.1** Sources (*dashed lines*) of N and C atoms in the purine ring of uric acid

prolixus, urate is excreted at a rate equivalent to more than 15 % of the unfed body weight per day (O'Donnell et al. 1983). Urate transport has been described for the isolated lower Malpighian tubules of *Rhodnius*, as well as tubules of locusts, butterfly larvae, and mantids, suggesting that the mechanism is widespread (O'Donnell et al. 1983). Calculations of the electrochemical gradients for urate in the lower tubule cells versus the hemolymph and the lumen versus the cells indicate that urate secretion is likely to be an active process across both the basolateral and apical membranes (O'Donnell et al. 1983). Although most studies have dealt with nymphs or adult insects, uric acid deposition in the tubule lumen has been demonstrated in the embryo in *Drosophila*, prior to emergence of the first instar larva (Ainsworth et al. 2000). In *Calliphora*, urate oxidase activity in the Malpighian tubules converts uric acid to a related nitrogenous compound, allantoin, which is then excreted.

Most cockroaches excrete ammonia rather than uric acid, as described below, but a few species in subfamilies Blatellinae and Plecopterinae, as well as the wood cockroach *Cryptocercus*, excrete urates in the form of spherules that are found in the lumen of the Malpighian tubules (Lembke and Cochran 1988). Birefringent materials occur in vacuoles within the tubule cells of these species, suggesting a vesicular transport mechanism for urates.

4.2.2 Contrasts with Mechanisms of Urate Transport in Vertebrate Kidney

The mechanism of urate transport in insects is distinct from that seen in vertebrates; transport of urate by the Malpighian tubules is ouabain-insensitive and results in the precipitation of free uric acid instead of urate salts (Na^+, K^+). In the mammalian kidney, a sodium-dependent phosphate transporter (NPT1) acts as a membrane potential-driven urate exporter across the apical membrane of the kidney (Miyaji et al. 2013). Uptake of urate into the kidney cell may involve organic anion transporters OAT1 and OAT3, which exchange cellular dicarboxylates that are recycled through a Na^+-dependent transporter (Eraly et al. 2008). The transporters responsible for urate secretion by the Malpighian tubules are unknown, but members of the ABCG family are possible candidates. The ABCG family proteins are known as "half-type" ABC transporters comprising a single nucleotide-binding domain in the amino terminus and a single transmembrane domain in the carboxyl terminus. It has been proposed that a heterodimer of two ABCG members, Bm-ok and Bmwh3, are responsible for urate transport by the epidermal cells of the silkworm *Bombyx mori* (Wang et al. 2013). Bm-ok encodes a protein belonging to the ABCG family of ABC transporters, which includes Brown, Scarlet, and White. In *Drosophila*, White encodes an ABCG transporter that is responsible for secretion of cGMP into the Malpighian tubule lumen (Evans et al. 2008). Given the role of ABCG transporters in urate transport in *B. mori*, it will be of interest in future studies to determine the possible role of White and other ABCG transporters in urate secretion by insect Malpighian tubules.

4.2.3 Urate Salts as "Ion Sinks" During Dehydration/Rehydration in Cockroaches

Cockroaches can be considered as internally uricotelic but externally ammonotelic and uricotelic (Cochran et al. 1985; Mullins 2015). Although ammonium is the final form of nitrogenous waste in the excreta in many species, such as *Periplaneta americana* (see below), urate salts play a central role in ionoregulatory and N-balance physiology in these insects. Both K^+ and Na^+ are sequestered in the fat bodies during dehydration and then released during rehydration (Hyatt and Marshall 1977, 1985a). Measurements using X-ray microanalysis confirmed that urate crystals within the urocytes sequester both K^+ and Na^+ in water-deprived *P. americana* (Hyatt and Marshall 1985b). Storage of Na^+ and K^+ thus provides the insect with the ions needed to maintain hemolymph osmolality and ion balance when hemolymph volume increases after rehydration with fresh water.

In some species, insertion of the spermatophore may be accompanied by urates stored in specialized male accessory glands, forming a genital plug (Graves 1969; Roth and Dateo 1965). After the spermatophore is emptied, the female may consume the discarded spermatophore and associated urates, thus contributing to nitrogen balance.

4.2.4 Storage Excretion of Uric Acid in Larval Lepidoptera (Fat Body and Epidermis)

The excretory system is nonfunctional during larval–pupal metamorphosis of holometabolous insects, and uric acid is sequestered during this stage of the life cycle. Lepidopterans accumulate uric acid in the fat body during metamorphosis (Buckner 1982), whereas fly pupae store it in the fat body, intestine, and the Malpighian tubules in response to the cessation of excretion (Schwantes 1990). The switch from excretion to storage during the transition from the feeding to the wandering stage in larval tobacco hornworms (*Manduca sexta*) is hormonally controlled and requires an increase in the titer of 20-hydroxyecdysone and a decrease in the titer of juvenile hormone I (Buckner 1982). Fat body urate levels increase and hemolymph urate concentration declines at this time. Staining with reduced silver has demonstrated that urate is stored as discrete membrane-bound vacuoles in the fat body of *M. sexta* pupae. Electron micrographs reveal that the vacuoles contain tightly coiled fibers of crystalline uric acid or urate salts and that each fiber is enveloped with protein (Buckner et al. 1990). Uric acid is also stored in wings or larval cuticle in Lepidoptera, presumably to provide pigmentation (Tojo and Yushima 1972).

4.2.5 Uric Acid as a Protectant Against Oxidative Stress

Blood-feeding arthropods utilize several mechanisms to minimize the impact of multiple reactive oxygen species produced from reactions involving both the iron and the heme group that are released during digestion of hemoglobin (Graça-Souza et al. 2006). One of these mechanisms involves production of antioxidants of low molecular mass, such as urate. Although the Malpighian tubules of the blood-feeding hemipteran *Rhodnius prolixus* secrete urate at high levels after the blood meal, rates of synthesis and excretion are balanced so that high concentrations of urate (up to 5 mM) are retained in the hemolymph. Urate accounts for almost all of the scavenging of free radicals in the hemolymph (Souza et al. 1997). The synthesis of urate by the fat body increases in response to hemin injection but also when the insects are exposed to hyperoxia, confirming that urate release into the hemolymph is an antioxidant response. The signaling pathway leading to stimulation of urate synthesis by hemin in *R. prolixus* involves activation of protein kinase C (Graça-Souza et al. 1999).

4.3   Nitrogen Recycling

Insects feeding on wood, phloem, plant sap, and other nitrogen-deficient diets make use of microbial symbionts to recycle nitrogen, using ammonia as a substrate for the synthesis of essential amino acids (EAAs) which are then available to the host (Macdonald et al. 2012). In addition, the symbionts may also be capable of nitrogen fixation. At least 60 species of wood-feeding insects from three orders (Blattodea, Coleoptera, and Hymenoptera) harbor symbiotic prokaryotes which fix N_2 (Ulyshen 2015). Screening for the *nifH* gene, which encodes a major component of nitrogenase, suggests that multiple types of N-fixing bacteria and Archaea are present in the guts of wood-eating insects (Ulyshen 2015). Symbiont contributions to the N economies of their hosts thus represent a common solution to the problem of surviving on a diet of wood. The insect host benefits from the metabolic capabilities of the bacteria to fix N and obtains N-containing organic compounds from the endobacteria. The endosymbionts, in return, obtain a stable and protected habitat and a reliable supply of the required nutrients for their reproduction (Orona-Tamayo and Heil 2015). These symbioses not only contribute to insect biomass, but may also make a significant contribution to N-cycling in particular ecosystems (Orona-Tamayo and Heil 2015). As an alternative to bacterial endosymbionts, some insects which feed on wood augment their nitrogen balance by ingesting cord-forming fungi which translocate soil nitrogen into the decomposing wood.

Cockroaches accumulate uric acid in the fat body, especially on protein-rich diets. The deposits of uric acid are then depleted when the insects are transferred to a low-protein diet (Cochran et al. 1985). Increased gene expression for urate oxidase and glutamine synthetase in cockroaches deprived of dietary nitrogen is consistent with a role for uric acid as a reservoir of nitrogen that is mobilized when dietary nitrogen intake is limited (Patiño-Navarrete et al. 2014). Much of the diet of primitive cockroaches was low in nitrogen and, as a consequence, mechanisms to store excess nitrogen in the fat bodies were probably an adaptive response to exploit occasional nitrogen-rich food sources (Mullins 2015).

Cockroaches benefit from the contributions of two distinct systems of microbial symbionts to their N economy: mycetocytes in the fat body and microbes in the hindgut. The symbionts provide the insects with the capacity to recycle stored urates (Mullins 2015). Three cell types are present in the fat body: whereas the trophocytes function as centers of intermediary metabolism and storage, the mycetocytes contain symbiotic bacteroids implicated in synthesis of essential amino acids, and the urocytes store urates as distinct crystalline spherules. The bacterial symbiont *Blattabacterium* in the mycetocytes produces vitamins and metabolizes sulfur and sulfur-containing amino acids for the host. The bacterium, in turn, relies upon the host's tissues for production of the amino acids Gln, Asp, Pro, and Gly (Patiño-Navarrete et al. 2014). These amino acids are produced through a chimeric metabolic pathway in which enzymes are supplied by both the host and the endosymbiont. Uric acid is degraded to allantoin, allantoic acid and then urea using the host's enzymes in the fat body. The urea is then broken down by urease in the endosymbiont, and the resulting ammonia is synthesized into glutamine using glutamine synthetase supplied by the host.

In the brown planthopper *Nilaparvata lugens* (Hongoh and Ishikawa 1997), endosymbionts enable the host to use uric acid as a nitrogen source during starvation periods, as for cockroaches, but with the difference that the uricolytic activities are supplied by the symbionts. Homopterans such as *N. lugens* feed on plant saps which are characterized by low levels of N and unbalanced amino acid compositions. They utilize uric acid not just as a nitrogenous waste, but as an N-storage product when they ingest more nitrogen than they need. Uricolysis in the yeast-like fungal endosymbionts can then provide the host with the ammonia needed for amino acid synthesis. In the shield bug *Parastrachia japonensis*, uric acid is excreted by nymphs and reproductive adults but is retained by diapausing adults which survive without feeding for 10 months to 2 years. Uric acid is recycled as an amino acid source with the aid of *Erwinia*-like bacteria found only in the cecum of the midgut, and the uricase functions as a key enzyme during recycling (Kashima et al. 2006).

Endosymbiotic bacteria are also implicated in synthesis of some of the essential amino acids (EAAs) in aphids (Douglas 2006, 2015; Sasaki and Ishikawa 1995); symbionts essentially hydrolyze glutamine and asparagine into glutamic acid and aspartic acid, respectively (Leroy et al. 2011). Bacteria of the genus *Buchnera*, for example, contribute up to 90 % of the essential amino acids required by the pea aphid (Douglas 2006). It is worth noting the ratio of essential amino acids: nonessential amino acids in plant phloem sap are 1:4–1:20, considerably lower than the ratio of 1:1 in animal protein (Douglas 2006). The high levels of free proline and alanine in whiteflies are consistent with known roles for proline as a fuel, via oxidation to alanine, for flight (Gäde 1992). Glutamine metabolism appears to play a pivotal role as aphids adjust to changes in plant nitrogen status. Decreases in the pool of glutamine may allow the maintenance of the proline and alanine pools used for flight muscle metabolism (Crafts-Brandner 2002).

Recent studies have also revealed nitrogen recycling and host–symbiont sharing of biosynthetic pathways for synthesis of EAAs in the pea aphid *Acyrthosiphon pisum* (Macdonald et al. 2012). The bacterial symbiont *Buchnera aphidicola* is restricted to the cytoplasm of the bacteriocyte "host cell" and it supplies the host cell cytoplasm with carbon skeletons derived from the host's nonessential amino acids. Aphid transaminases then combine nitrogen derived from ammonia produced by the host cell metabolism with the carbon skeletons from the symbiont to synthesize essential amino acids. As in mosquitoes, glutamine synthetase/glutamine oxoglutarate aminotransferase (also known as glutamate synthase) (GS/GOGAT) plays an essential role in EAA synthesis.

Gut microbes also make important contributions to metabolic nitrogen utilization in the Asian long-horned beetle *Anoplophora glabripennis* (Motschulsky) (Coleoptera: Cerambycidae: Lamiini). The endosymbionts are important both in nitrogenous waste product (urea) recycling and in nitrogen fixation (Ayayee et al. 2014).

4.4   Ammonia Excretion in Terrestrial Insects

4.4.1 Locusts

Ammonium, in the excreta of the desert locust *Schistocerca gregaria,* is present in the form of a precipitate, so its elimination is compatible with conservation of water. It appears that most NH_4^+ is precipitated with organic anions because the concentration of ammonium in the excreta exceeds that of urate threefold (Harrison and Phillips 1992). The advantage of ammonium urate excretion is twofold; ammonium urate is less soluble than uric acid, and it also provides for the elimination of an additional N relative to uric acid. Whereas urate and organic ions are secreted by the Malpighian tubules, most of the NH_4^+ is transported across the ileum and into the hindgut lumen through amiloride-sensitive

Na^+/NH_4^+ exchange. Although the Malpighian tubules secrete fluids containing as much as 5 mM ammonia, the amount excreted by the tubules is less than 10 % of total NH_4^+ excreted. The ileum is also a major site of acid excretion through H^+ trapped by NH_3 as NH_4^+ (Lechleitner et al. 1989).

4.4.2 Cockroaches

Although 90 % of nitrogenous waste in the fecal pellets of some species of cockroaches consists of ammonia, much of the excreted ammonia appears to be derived from the actions of hindgut microflora on urate salts or other N compounds. Ammonium ions are the major cations contained in the feces and are excreted in increasing amounts as the dietary nitrogen levels increase. As a consequence of the relative toxicity of ammonia, water excretion and hence water uptake are increased on N-rich diets. In *P. americana*, for example, a threefold increase in dietary nitrogen is correlated with a 92 % increase in water intake. Ammonia may be secreted by the Malpighian tubules and released into the anterior end of the hindgut, where some of it may be recycled into metabolic nitrogen by gut microbial systems before excretion in the feces. Although the predominance of ammonia as a nitrogenous waste in cockroaches such as *P. americana* has been known since 1972, the means by which ammonia is transported into the hindgut, or is produced therein by microflora, is still unclear. Relative to our understanding of mycetocytes in nitrogen metabolism in the fat body, much less is known of the involvement of gut microbial systems in ammonia metabolism and subsequent absorption of potentially useful materials in the hindgut. Some of the bacteria in the hindgut are involved in nitrogen fixation, and their abundance increases in cockroaches on nitrogen poor diets, suggesting that the gut microbiota complement the activities of *Blattabacterium* in the fat body (Pérez-Cobas et al. 2015). Gut microbes may also be involved in synthesis of acetate and butyrate which may be absorbed across the hindgut wall (Mullins 2015). However, as the latter author noted: "The involvement of gut microbial systems in ammonia metabolism, coupled with subsequent absorption of potentially useful materials achieved by processes within the hindgut, is a much needed area of investigation".

4.4.3 Flesh Fly Larvae

It has been known since the early years of the last century that ammonia constitutes the majority of the nitrogenous waste of the blowfly *Calliphora vomitoria* (Weinland 1906). Subsequent work revealed that larvae reared on sterile media also excrete ammonia resulting from their own metabolism, but that ammonifying bacteria normally present on rotting meat are an additional source of ammonia in the larval environment (Hobson 1932). Larvae of the flesh fly *Sarcophaga bullata* also ingest ammonia as they burrow into and feed upon rotting flesh. Ammonia is absorbed across from the midgut and excreted across the hindgut epithelium. The larvae are also exposed to excess ammonia resulting from protein metabolism. Using an isolated hindgut preparation, Prusch showed that ammonia can be secreted against a lumen–alkaline pH gradient which would oppose NH_3 diffusion (Prusch 1972). NH_4^+ secretion is independent of that for K^+, which is also actively secreted into the lumen. Ammonium secretion is reduced ~ 60 % when either K^+ or Na^+ is removed, consistent with a link between alkali cation transport and NH_4^+ secretion.

4.4.4 *Drosophila*

Ammonia may accumulate to high levels in the medium ingested by larvae of the fruit fly *Drosophila melanogaster*, and the larvae must therefore have means to either detoxify or excrete ammonia. Active secretion of ammonium into the Malpighian tubule lumen is sufficient to maintain concentrations of ~1 mM ammonium in the hemolymph of larvae reared on diets containing 100 mM ammonium chloride (Browne and O'Donnell 2013). The rates of NH_4^+ transport by the Malpighian tubules are sufficient to clear the hemolymph content of NH_4^+ in ~3.5 h suggesting that the tubules play an important role in clearance of ammonia from the hemolymph, although other epithelia such as the hindgut may also contribute.

4.4.5 Mosquitoes

More than 70% of the amino acids derived from the digestion of blood meal by adult female mosquitoes (*Aedes aegypti*) is catabolized to provide energy and only 10% is used for egg protein production (Isoe and Scaraffia 2013; Zhou et al. 2004). Potentially toxic loads of ammonia resulting from high rates of amino acid catabolism are avoided through two mechanisms: 1) sequestration of ammonia through synthesis of glutamine and proline and 2) high rates of ammonia excretion in the feces (Goldstrohm et al. 2003; Scaraffia et al. 2005). Ammonia released by subsequent proline catabolism can be excreted, whereas the carbon skeleton is used for synthesis of other compounds or for energy production. Measurements of mRNA expression and the activity of several enzymes implicated in ammonia metabolism suggest that the fat body is the main tissue involved in ammonia detoxification and that both glutamine synthetase and glutamate synthase are important in the process of ammonia storage. Experimental inhibition of glutamine synthetase activity reduces hemolymph glutamine concentration and causes a corresponding increase in proline concentration. Inhibition of glutamate synthase, which contributes to the production of glutamate for proline synthesis, results in an increase in glutamine concentration and a corresponding decrease in proline concentration after feeding. Mosquitoes also synthesize urea through argininolysis and uricolysis. Detailed studies of the metabolic regulation of urea synthesis indicate links to the fixation, assimilation, and excretion of ammonia. For example, glutamine acts as a precursor in uric acid synthesis; measurements using the stable isotope ^{15}N indicate that N from the amide group of two glutamine molecules produces one molecule of uric acid labeled at two positions (Isoe and Scaraffia 2013; Scaraffia et al. 2008). Uric acid is either excreted or degraded by uricolysis to produce allantoin, allantoic acid, and then urea. Proline is used to sequester ammonia and is also linked to arginine metabolism; arginase cleaves arginine into urea and ornithine, which is then used to synthesize proline and several other amino acids (Isoe and Scaraffia 2013).

4.4.6 Lepidoptera

Analyses of ingestion and excretion of nitrogen by larvae of the lepidopteran *Mamestra brassicae*, an agricultural pest, have shown that high levels of nitrogen in host plants (e.g., after fertilizer application) enhances N metabolism in the larvae and leads to increased excretion of NH_4^+ (Kagata and Ohgushi 2011). The increase in NH_4^+-N in the frass originates from N metabolism of the insect, rather than from ammonia in the diet. Ammonia is a common component of the excreta of many terrestrial insects, comprising from ~9% to ~27% of total N in the frass. Insect herbivores can thus alter nitrogen dynamics in soil by transforming organic nitrogen into inorganic nitrogen and vice versa. For example, soil quality is enhanced by ammonium release during frass decomposition. Ammonium can then be taken up directly and utilized by plants without the lag associated with the normal breakdown and release of nitrogen from leaf litter (Kagata and Ohgushi 2011).

4.5   Uptake of N-Rich Compounds by Insects

4.5.1 *Manduca*

Two studies have described the uptake of ammonia by the midgut of *Manduca sexta* (Blaesse et al. 2010; Weihrauch 2006). It is proposed that ammonia uptake across the midgut and into the hemolymph provides an additional nitrogen source for rapidly growing larvae as they consume a diet low in protein. A model of active ammonia uptake across the columnar cell epithelium in the median midgut proposes that ammonia enters the columnar cell through one or both of two pathways (■ Fig. 4.2). Firstly, NH_4^+ may enter via an apical H^+/cation exchanger, driven by the steep pH gradient across this membrane (Dow 1992). *Manduca sexta* Na^+/H^+ exchanger 7, 9 (MsNHE7, 9) is a candidate transporter for this pathway because it exhibits an amiloride binding motif, consistent with observed amiloride sensitivity of ammonia uptake, and it may be localized to the apical membrane. The second pathway involves diffusion of gaseous ammonia along a gradient of P_{NH3} across the plasma membrane, possibly via one of the two identified (Rhesus) Rh proteins in the midgut tissue (GenBank accession nos. AY954627 and DQ864985). Dissociation of cytosolic NH_4^+ to H^+ and NH_3 would then allow diffusion of NH_3 into vesicles acidified by a V-type H^+-ATPase. Alternatively, a NHE capable of H^+/NH_4^+ exchange could sequester NH_4^+ in intracellular vesicles. NHE8, which has been identified as a subapical/vesicular transporter (Piermarini et al. 2009), is a potential candidate for such a role. Rhesus proteins (RhMS) are also richly expressed in the Malpighian tubules and hindgut, suggesting that these tissues contribute to elimination of any excess ammonia (■ Fig. 4.3).

4.5.2 "Mud-Puddling"

Multiple species of butterflies, moths, and homopterans and at least one orthopteran engage in a behavior known as "mud-puddling" in which they ingest fluids from sources such as moist ground or puddles, sea water, tears, excreta of other animals, or rotting carcasses. Although generally viewed as means of supplementing water intake or lowering body temperature, mud-puddling is also a means to augment the intake of sodium or nitrogen in species whose diet is deficient in these elements. The yellow-spined bamboo locust, *Ceracris kiangsu*, has been observed to visit puddles of human urine on hot summer days. Although the main N compound in human urine is urea, decomposition in high ambient temperatures produces ammonium bicarbonate. Behavioral studies revealed that urea is a repellent to *C. kiangsu*, whereas ammonium bicarbonate is an attractant (Shen et al. 2009). It appears that urine puddles or urine-soaked ground served as a source of both sodium and nitrogen in this species. Thus, although ammonia is generally considered

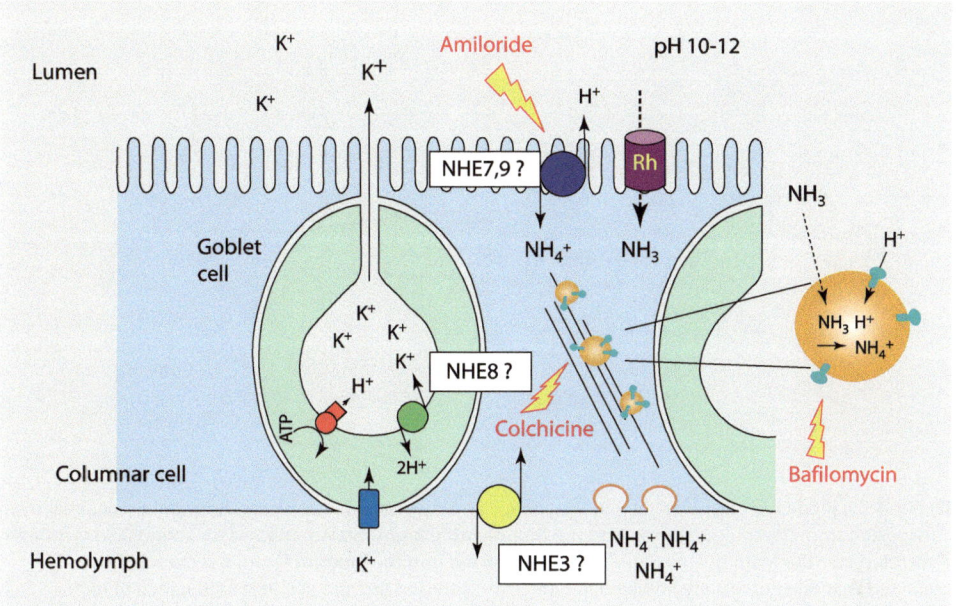

☐ **Fig. 4.2** Hypothetical model of active ammonia uptake across the columnar cell epithelium in the median midgut of the tobacco hornworm *Manduca sexta*. On the apical side, ammonia is transported as NH_4^+ via an apical amiloride-sensitive H^+/cation exchanger (possibly NHE7/9) and/or diffuses as gaseous ammonia through an Rh-like protein along an inwardly directed P_{NH3} across the plasma membrane. In the cytoplasm NH_4^+ dissociates into H^+ and NH_3 and gaseous ammonia diffuses into vesicles acidified by a V-type H^+-ATPase. The NH_4^+-loaded vesicles are transported along the microtubules network to the basal membrane, where they fuse with the membrane, releasing NH_4^+ into the hemolymph space. A participation of the basal NHE3 in the ammonia uptake process is unclear. Goblet cells, which are responsible for K^+ excretion, create a high luminal pH and are thereby indirectly involved in ammonia transport. NHE8 was found to be expressed apically/subapically in the goblet, but not in the columnar cells (personal communication, D. Weihrauch). Ammonia may also diffuse paracellularly (not shown). Pharmacological inhibitors of particular components in the ammonia transport mechanism are given with flash symbols pointing at sites of inhibition (Modified from Weihrauch et al. 2012)

to be a highly toxic waste product for most animals, species of beetle, moth, and mosquito recycle body ammonia as a resource for synthesis of amino acids, which in turn are used for protein biosynthesis (Honda et al. 2012). Both sexes of the swallowtail butterfly *Papilio polytes* utilize ammonia taken in at puddling sites such as surface water, dung, and carrion. Experiments following the fate of ^{15}N taken in during puddling shows that females incorporate the nitrogen into eggs, whereas males use it for protein replenishment in spermatozoa, seminal fluids, and thoracic muscle (Honda et al. 2012).

4.6 Ammonia Excretion in Aquatic Insects

In general, aquatic animals excrete nitrogenous waste primarily in the form of ammonia (NH_3/NH_4^+) (Weihrauch et al. 2012; Wright and Wood 2009; Larsen et al. 2014). This is because ammonia is highly soluble and thus aquatic animals take advantage of the abundance of water in their habitat to excrete nitrogenous waste in the form that requires the

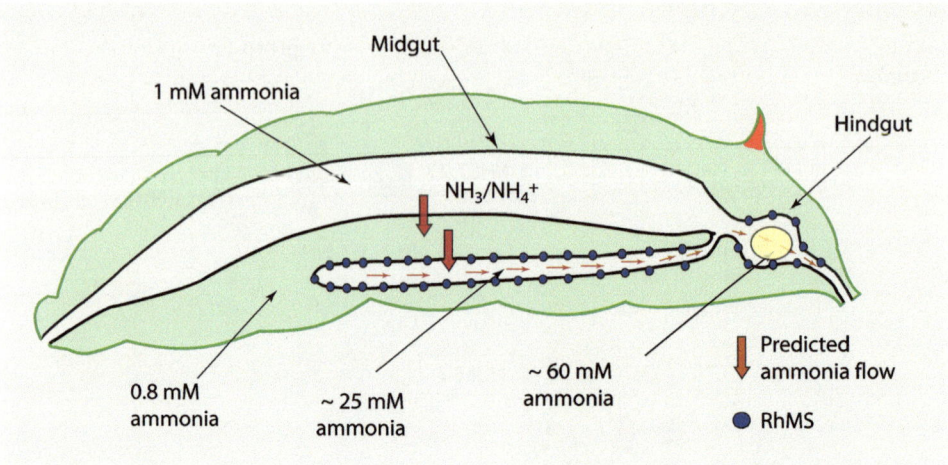

■ **Fig. 4.3** Predicted ammonia flow in the tobacco hornworm *Manduca sexta*. Ammonia is actively transported from the midgut lumen into the hemolymph. Ammonia not utilized for amino acid synthesis is secreted into the Malpighian tubules and transported into the hindgut for final secretion. It is predicted that Rh-proteins are involved in ammonia secretion process and in maintenance of high ammonia concentrations in the hindgut lumen. Concentrations of ammonia in midgut lumen, hemolymph, Malpighian tubules, and hindgut are shown (Modified from (Weihrauch et al. 2012))

least expenditure of metabolic energy. Insects appear no different; however, this is based on measurements of excretory products from a few select aquatic insects in orders Coleoptera, Odonata, Hemiptera, Trichoptera, and Diptera (Delaunay 1930; Staddon 1955, Donini and O'Donnell 2005). In species where specific life stages are aquatic (e.g., larvae) and others terrestrial (e.g., adult), the larvae excrete primarily ammonia, while the terrestrial stages excrete nitrogen waste in other forms such as uric acid (Staddon 1955). An exception is the adult, terrestrial blood-fed female mosquito, *Aedes aegypti*, where the molar ratios of uric acid, urea, and ammonia in the excreta are equal (Briegel 1986). In experiments in which adult females were fed NH_4Cl-rich fluids (80 mM), ammonia was the predominant (>80 %) N waste in the excreta (Scaraffia et al. 2005).

4.6.1   Anal Papillae of Larval Mosquitoes

Excretion in insects is accomplished through the combined actions of the Malpighian tubules and hindgut. In freshwater inhabiting mosquito larvae, the combined actions of these organs produces dilute urine thereby eliminating excess water while conserving ions. The larvae of many mosquito species also possess external organs called anal papillae. These are finger-like projections formed from the eversion of hindgut tissue and they surround the anus. The anal papillae are typically composed of a single cell layer thick syncytial epithelium that is covered by a thin cuticle that faces the aquatic environment (Edwards and Harrison 1983; Sohal and Copeland 1966). In freshwater species like *Aedes aegypti*, the apical membrane of the epithelial syncytium contains very regular and extensive infoldings that give rise to microvilli that project deep inside the cytoplasm, while the basolateral (lumen facing) membrane is also infolded but forms a dense network of channels that are continuous with the papilla hemocoel and extend into the epithelium. The cytoplasm contains numerous mitochondria which are densely packed near the microvilli of the apical membrane. This ultrastructure is consistent with ion transporting epithelia and functions to increase the surface area of the membranes available for solute exchange. The lumen of the anal papillae contains hemolymph and is continuous with the hemocoel of the body.

The ultrastructure of freshwater mosquito anal papillae matches their physiological function. The anal papillae play an important role in osmotic and ionic regulation by actively transporting ions into the hemolymph from the dilute freshwater environment (Stobbart 1971; Donini and O'Donnell 2005), and this function is regulated in response to salinity changes in the water in which the larvae reside (Donini et al. 2007). In addition to substantial measureable influx of Na^+, Cl^-, and K^+, relatively large effluxes of NH_4^+ and H^+ were detected at the anal papillae (Donini and O'Donnell 2005). The NH_4^+ efflux amounted to an estimated 360 nmol cm^{-2} h^{-1} which compares with 580 nmol cm^{-2} h^{-1} detected at the isolated locust hindgut (Thomson et al. 1988) and 220 nmol cm^{-2} h^{-1} for the cockroach (Mullins 1974). Therefore, the anal papillae of mosquito larvae play an important role in nitrogen waste excretion and are functionally similar to the gills of fish and crustaceans (Henry et al. 2012; Evans et al. 2005).

There are two families of ammonia transporters, the Rhesus glycoproteins (RhGP) which are found in all animals and implicated in ammonia excretion, and the methylammonium/ammonium transporters (MEP/Amt) found in unicellular prokaryotes, eukaryotic cells, plants, and invertebrate animals. The mosquito *Aedes aegypti* has two RhGPs (AeRh50-1, AeRh50-2) and at least one Amt (AeAmt1) (Weihrauch et al. 2012; Chasiotis et al. 2016). The AeAmt1 protein is expressed in the anal papillae and localized to the lumen-side membrane along with Na^+/K^+-ATPase. Although the transcripts of the RhGPs have been detected in the anal papillae (Weihrauch et al. 2012), the proteins have not been localized. Knockdown of AeAmt1 protein expression reduces ammonia excretion at the anal papillae (Chasiotis et al. 2016). It is suggested that NH_4^+ is driven into the cytoplasm through the AeAmt1 protein by the negative electrical gradient established by Na^+/K^+-ATPase (❏ Fig. 4.4). The NH_4^+ may then exit the apical side through the NHE3 or, if dissociated into NH_3 and H^+, through one of the RhGPs (Chasiotis et al. 2016). This is facilitated by the H^+ transporting functions of an apical V-type H^+-ATPase (VA) which pumps H^+ out of the cytoplasm creating an acidified boundary layer. This ensures that NH_3, exiting the anal papillae, is converted into NH_4^+ thereby maintaining a favorable partial pressure gradient for NH_3 to exit (Chasiotis et al. 2016). Both scenarios are supported by pharmacological inhibition of NHE3 and VA which results in reduced ammonia excretion by the anal papillae. Furthermore, pharmacological inhibition of carbonic anhydrase (CA) also reduces ammonia excretion suggesting that CA provides H^+ for the VA.

4.6.2 Ammonia Excretion by Larval Alderflies, Dragonflies, Stoneflies and Backswimmers

Although ammonia excretion in some aquatic insects has been measured, there are only a few studies that have gone beyond this to begin examining the physiology of ammonia excretion. The alderfly nymph, *Sialis lutaria*, excretes ~86 % of nitrogen waste as ammonia (Staddon 1955). The fluid in the hindgut contains relatively high amounts of ammonia (~136 mg N/100 mL), whereas fluid in the foregut had no measureable ammonia.

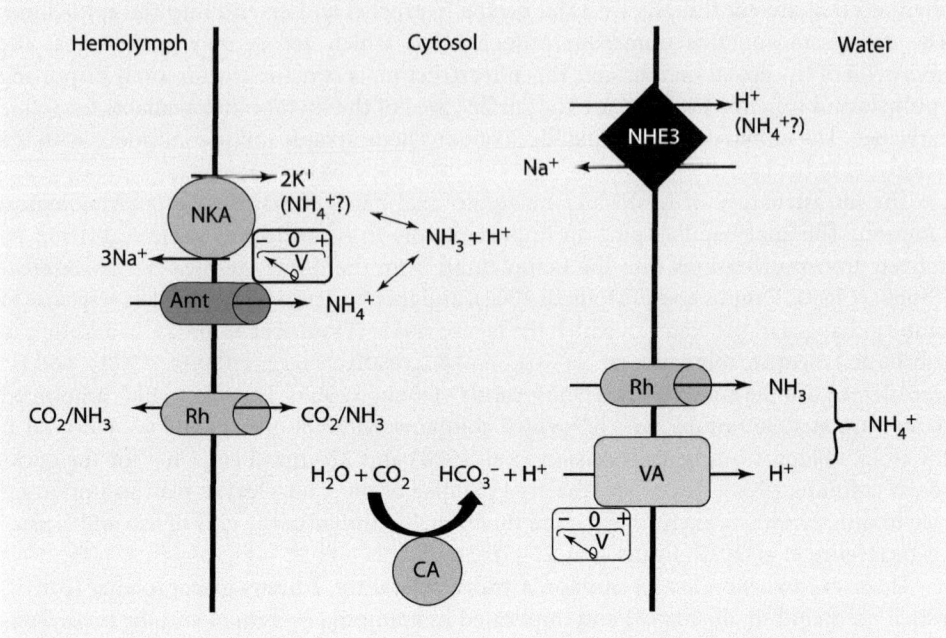

◻ Fig. 4.4 Proposed mechanism of ammonia excretion by anal papillae of larval *Aedes aegypti*. Na^+/K^+-ATPase (*NKA*) provides a cytosol negative voltage potential which serves to drive NH_4^+ from the hemolymph to the cytosol through AeAmt1 (*Amt*). NH_4^+ could also cross from hemolymph to cytosol directly through NKA substituting for K^+. CO_2 and NH_4 may enter the cytosol through one of the two RhGP-like proteins, AeRh50-1 and/or AeRh50-2 (Rh), if one or both are expressed in the basal membrane (not yet determined). NH_4^+ in the cytosol exits from the apical side to the water through the NHE3 in exchange for a cation (e.g., Na^+). NH_3 in the cytosol exits the apical side via one of the two RhGP-like proteins (*Rh*) (localization not yet determined) with the aid of an ammonia trapping mechanism, whereby apical V-type H^+ ATPase (*VA*) acidifies the papilla boundary layer. NHE3 may also participate in this mechanism by moving H^+ into the water. This would sustain the outward directed NH_3 gradient by converting NH_3 to NH_4^+. A cytoplasmic carbonic anhydrase (*CA*) can supply H^+ to the VA which is also likely participating in generating the cytosol negative potential that drives NH_4^+ into the cytosol on the basal side (Proposed model from Chasiotis et al. 2016)

Although fluid from the Malpighian tubules was not measured, it was suggested that the tubules were the source of ammonia excretion where the fluid would then pass into the hindgut for periodic expulsion (Staddon 1955). No ammonia excretion occurs across the body surface of the nymph. Furthermore, if excretion is prevented, there is no accumulation of ammonia in either the hemolymph or tissues, suggesting that *S. lutaria* nymphs can store ammonia in another form (e.g., glutamine) if excretion is not feasible. This is significant in that alderfly nymphs leave the water for some time prior to pupation.

Ammonia accounts for ~87% of the nitrogenous waste excreted by the dragonfly nymph *Aeshna cyanea* and uric acid makes up 8% (Staddon 1959). Exposure to aluminum or low pH results in a large reduction in ammonia excretion in another dragonfly nymph *Somatochlora cingulata*, resulting in the accumulation of ammonia and glutamate levels in tissues (Correa et al. 1985b). In contrast, exposure to trichloroacetic acid at environmentally relevant levels doubles the ammonia excretion rates (Correa et al. 1985a). The underlying physiological mechanisms behind these observations remain unstudied. The stonefly larva *Klaptopteryx kuscheli* is a leaf-litter shredder and shown to regulate internal carbon/nitrogen (C:N) at a level of ~5.5 regardless of the C:N ratio of the food (Balseiro and Albariño 2006). It was shown that as the C:N ratio decreases, the larvae excrete more ammonia.

In a study examining the regulation of water balance of a backswimmer *Notonecta glauca*, a carnivore, the relationship of water output and ammonia output exhibited a linear trend (Staddon 1963). This might be expected since ammonia is highly soluble and aquatic insects exploit the abundance of water in their habitat to excrete ammonia. The ammonia levels in the rectal fluid were ~75 mmol l^{-1} and contributed to ~75 % of total nitrogen in the fluid. In contrast, a similar study in another species of backswimmer *Corixa dentipes*, it was noted that the rectal fluid contained very variable and in some cases relatively low amounts of ammonia ranging from 21 to 85 mmol l^{-1} with an average of ~50 mmol l^{-1} (Staddon 1964). Little difference in ammonia levels were observed whether the fluid was forced out or allowed to be expelled naturally by the insect leading to the suggestion that the rectum plays no role in modification of the rectal fluid. Although it might be expected that water output, is at least in part, dependent on ammonia output, this was not the case in *Corixa dentipes*. Furthermore, the rate of water output far exceeded the volume required for the observed rate of ammonia output. The independence of water and ammonia output could be due to the fact that this insect is omnivorous and ingests mainly a liquid diet resulting in a copious amount of water ingestion which needs to be expelled regardless of ammonia levels; however, the results of the study on water intake of the insect were inconclusive (Staddon 1964). If this were the case, it could explain the relatively low concentrations of ammonia in rectal fluid of this species.

4.7 Excretion of Free Amino Acids

Silverleaf whiteflies (*Bemisia tabaci* Gennadius) and other phloem-feeding insects such as aphids excrete amino acids, especially nonessential amino acids such as glutamine, asparagine, glutamate, and aspartate (Douglas 2006). Whiteflies reared on well-fertilized cotton plants excrete an amount of amino nitrogen that exceeds by nearly twofold the total amino acid pool in the body. However, when flies are fed plants with low nitrogen levels, pools of individual amino acids in the body are adjusted and amino nitrogen is no longer excreted at significant rates (Crafts-Brandner 2002). Total nitrogen content of whiteflies reared on plants with reduced nitrogen content does not decline significantly, relative to whiteflies reared on plants with higher levels of nitrogen. However, the free amino acid content of whiteflies feeding on low nitrogen plants declines ~90 % relative to controls. It is important to note that the amino acid pools in tissues of whiteflies feeding on high nitrogen cotton plants are not closely related to the amino acid composition of the phloem sap. Whereas the predominant amino acids in whiteflies are glutamine (26 % of the total), alanine (19 %), proline (13 %), and glutamate (10 %), the predominant amino acids in phloem sap from high nitrogen cotton plants are aspartate, glutamate, and arginine, with relatively large amounts of glutamine and asparagine, and the excreted honeydew contains mostly asparagine (46 % of total amino acid content) and glutamine (12 %). These differences indicate that dietary amino nitrogen is rapidly assimilated into metabolites or protein.

References

Ainsworth C, Wan S, Skaer H (2000) Coordinating cell fate and morphogenesis in *Drosophila* renal tubules. Philos Trans R Soc Lond B Biol Sci 355(1399):931–937

Ayayee P, Rosa C, Ferry JG, Felton G, Saunders M, Hoover K (2014) Gut microbes contribute to nitrogen provisioning in a wood-feeding cerambycid. Environ Entomol 43(4):903–912

Baldwin E, Needham J (1934) Problems of nitrogen catabolism in invertebrates: The snail (*Helix pomatia*). Biochem J 28(4):1372–1392

Balseiro E, Albariño R (2006) C–N mismatch in the leaf litter–shredder relationship of an Andean Patagonian stream detritivore. J N Am Benthol Soc 25(3):607–615

Barrett F, Friend W (1970) Uric acid synthesis in *Rhodnius prolixus*. J Insect Physiol 16(1):121–129

Blaesse AK, Broehan G, Meyer H, Merzendorfer H, Weihrauch D (2010) Ammonia uptake in *Manduca sexta* midgut is mediated by an amiloride sensitive cation/proton exchanger: Transport studies and mRNA expression analysis of NHE7, 9, NHE8, and V-ATPase (subunit D). Comp Biochem Physiol A Mol Integr Physiol 157(4):364–376

Briegel H (1986) Protein catabolism and nitrogen partitioning during oogenesis in the mosquito *Aedes aegypti*. J Insect Physiol 32(5):455–462

Browne A, O'Donnell MJ (2013) Ammonium secretion by Malpighian tubules of *Drosophila melanogaster*: application of a novel ammonium-selective microelectrode. J Exp Biol 216(20):3818–3827

Buckner JS (1982) Hormonal control of uric acid storage in the fat body during last-larval instar of *Manduca sexta*. J Insect Physiol 28:987–993

Buckner JS, Newman SM (1990) Uric acid storage in the epidermal cells of *Manduca sexta*: localization and movement during the larval-pupal transformation. J Insect Physiol 36:219–229

Bursell E (1965) Nitrogenous waste products of tsetse fly, *Glossina morsitans*. J Insect Physiol 11:993–1001

Bursell E (1967) The excretion of nitrogen in insects. Adv Insect Physiol 4:33–67

Chasiotis H, Ionescu A, Misyura L, Bui P, Fazio K, Wang J, Patrick M, Weihrauch D, Donini A (2016) An animal homolog of plant Mep/Amt transporters promotes ammonia excretion by the anal papillae of the disease vector mosquito, *Aedes aegypti*. J Exp Biol 219:1346–1355

Cochran DG et al (1985) Nitrogenous excretion. In: Kerkut GA, Gilbert LI (eds) Comprehensive insect physiology. Pergamon Press, Oxford, pp 465–506

Correa M, Calabrese EJ, Coler RA (1985a) Effects of trichloroacetic acid, a new contaminant found from chlorinating water with organic material, on dragonfly nymphs. Bull Environ Contam Toxicol 34(1):271–274

Correa M, Coler RA, Yin C-M (1985b) Changes in oxygen consumption and nitrogen metabolism in the dragonfly *Somatochlora cingulata* exposed to aluminum in acid waters. Hydrobiologia 121(2):151–156

Crafts-Brandner S (2002) Plant nitrogen status rapidly alters amino acid metabolism and excretion in *Bemisia tabaci*. J Insect Physiol 48(1):33–41

Delaunay H (1931) L'excretion azotée des invertébrés. Biol Rev 6:265–301

Donini A, Gaidhu MP, Strasberg DR, O'Donnell MJ (2007) Changing salinity induces alterations in hemolymph ion concentrations and Na$^+$ and Cl$^-$ transport kinetics of the anal papillae in the larval mosquito, *Aedes aegypti*. J Exp Biol 210(Pt 6):983–992. doi:10.1242/jeb.02732

Donini A, O'Donnell MJ (2005) Analysis of Na$^+$, Cl$^-$, K$^+$, H$^+$ and NH$_4^+$ concentration gradients adjacent to the surface of anal papillae of the mosquito *Aedes aegypti*: application of self-referencing ion-selective microelectrodes. J Exp Biol 208(Pt 4):603–610. doi:10.1242/jeb.01422

Douglas A (2006) Phloem-sap feeding by animals: problems and solutions. J Exp Bot 57(4):747–754

Douglas AE (2015) Multiorganismal insects: diversity and function of resident microorganisms. Annu Rev Entomol 60:17–34

Dow JA (1992) pH gradients in lepidopteran midgut. J Exp Biol 172(Pt 1):355–375

Edwards H, Harrison J (1983) An osmoregulatory syncytium and associated cells in a freshwater mosquito. Tissue Cell 15(2):271–280

Eraly SA, Vallon V, Rieg T, Gangoiti JA, Wikoff WR, Siuzdak G, Barshop BA, Nigam SK (2008) Multiple organic anion transporters contribute to net renal excretion of uric acid. Physiol Genomics 33(2):180–192

Evans DH, Piermarini PM, Choe KP (2005) The multifunctional fish gill: dominant site of gas exchange, osmoregulation, acid-base regulation, and excretion of nitrogenous waste. Physiol Rev 85 (1):97–177. doi:85/1/97 [pii] 10.1152/physrev.00050.2003

Evans JM, Day JP, Cabrero P, Dow JA, Davies SA (2008) A new role for a classical gene: white transports cyclic GMP. J Exp Biol 211(Pt 6):890–899. doi:10.1242/jeb.014837

Gäde G (1992) The hormonal integration of insect flight metabolism. Zool Jahrb Abt für Allg Zool Physiol Tiere 96(2):211–225

Goldstrohm DA, Pennington JE, Wells MA (2003) The role of hemolymph proline as a nitrogen sink during blood meal digestion by the mosquito *Aedes aegypti*. J Insect Physiol 49(2):115–121 doi:S0022191002002676 [pii]

Graça-Souza AV, Maya-Monteiro C, Paiva-Silva GO, Braz GR, Paes MC, Sorgine MH, Oliveira MF, Oliveira PL (2006) Adaptations against heme toxicity in blood-feeding arthropods. Insect Biochem Mol Biol 36(4):322–335

Graça-Souza AV, Silva-Neto MA, Oliveira PL (1999) Urate synthesis in the blood-sucking insect *Rhodnius prolixus*: Stimulation by hemin is mediated by protein kinase C. J Biol Chem 274(14):9673–9676

Graves PN (1969) Spermatophores of the Blattaria. Ann Entomol Soc Am 62(3):595–602

Harrison JF, Phillips JE (1992) Recovery from acute haemolymph acidosis in unfed locusts: II. Role of ammonium and titratable acid excretion. J Exp Biol 165(1):97–110

Henry RP, Lucu C, Onken H, Weihrauch D (2012) Multiple functions of the crustacean gill: osmotic/ionic regulation, acid-base balance, ammonia excretion, and bioaccumulation of toxic metals. Front Physiol 3:431. doi:10.3389/fphys.2012.00431

Hobson R (1932) Studies on the nutrition of blow-fly larvae II. Role of the intestinal flora in digestion. J Exp Biol 9(2):128–138

Honda K, Takase H, Ōmura H, Honda H (2012) Procurement of exogenous ammonia by the swallowtail butterfly, *Papilio polytes*, for protein biosynthesis and sperm production. Naturwissenschaften 99(9): 695–703

Hongoh Y, Ishikawa H (1997) Uric acid as a nitrogen resource for the brown planthopper, *Nilaparvata lugens*: studies with synthetic diets and aposymbiotic insects. Zoolog Sci 14(4):581–586

Hyatt A, Marshall A (1977) Sequestration of haemolymph sodium and potassium by fat body in the water-stressed cockroach *Periplaneta americana*. J Insect Physiol 23(11):1437–1441

Hyatt A, Marshall A (1985a) Water and ion balance in the tissues of the dehydrated cockroach, *Periplaneta americana*. J Insect Physiol 31(1):27–34

Hyatt A, Marshall A (1985b) X-ray microanalysis of cockroach fat body in relation to ion and water regulation. J Insect Physiol 31(6):495–508

Isoe J, Scaraffia PY (2013) Urea synthesis and excretion in *Aedes aegypti* mosquitoes are regulated by a unique cross-talk mechanism. PLoS One 8(6):e65393

Kagata H, Ohgushi T (2011) Ingestion and excretion of nitrogen by larvae of a cabbage armyworm: the effects of fertilizer application. Agric Forest Entomol 13(2):143–148

Kashima T, Nakamura T, Tojo S (2006) Uric acid recycling in the shield bug, *Parastrachia japonensis* (Hemiptera: Parastrachiidae), during diapause. J Insect Physiol 52(8):816–825. doi:10.1016/j.jinsphys.2006.05.003

Larsen EH, Deaton LE, Onken H, O'Donnell M, Grosell M, Dantzler WH, Weihrauch D (2014) Osmoregulation and excretion. Compr Physiol 4(2):405–573. doi:10.1002/cphy.c130004

Lechleitner R, Audsley N, Phillips J (1989) Composition of fluid transported by locust ileum: influence of natural stimulants and luminal ion ratios. Can J Zool 67(11):2662–2668

Lembke HF, Cochran DG (1988) Uric acid in the Malpighian tubules of some blattellid cockroaches. Comp Biochem Physiol A Physiol 91(3):587–597

Leroy PD, Wathelet B, Sabri A, Francis F, Verheggen FJ, Capella Q, Thonart P, Haubruge E (2011) Aphid-host plant interactions: does aphid honeydew exactly reflect the host plant amino acid composition? Arthropod-Plant Interact 5(3):193–199

Macdonald SJ, Lin GG, Russell CW, Thomas GH, Douglas AE (2012) The central role of the host cell in symbiotic nitrogen metabolism. Proc R Soc Lond B Biol Sci 279(1740):2965–2973, rspb20120414

Marshall A, Wood R (1990) Ionic and osmotic regulation by larvae of the sheep blowfly, *Lucilia cuprina*. J Insect Physiol 36(9):635–639

Miyaji T, Kawasaki T, Togawa N, Omote H, Moriyama Y (2013) Type 1 sodium-dependent phosphate transporter acts as a membrane potential-driven urate exporter. Curr Mol Pharmacol 6(2):88–94

Mullins DE (1974) Nitrogen metabolism in the American cockroach: an examination of whole body ammonium and other cations excreted in relation to water requirements. J Exp Biol 61(3):541–556

Mullins DE (2015) Physiology of environmental adaptations and resource acquisition in cockroaches. Annu Rev Entomol 60:473–492

O'Donnell M, Maddrell S, Gardiner B (1983) Transport of uric acid by the Malpighian tubules of *Rhodnius prolixus* and other insects. J Exp Biol 103(1):169–184

Orona-Tamayo D, Heil M (2015) N fixation in insects: its potential contribution to N cycling in ecosystems and insect biomass. In: de Bruijn FJ (ed), Biological Nitrogen Fixation. John Wiley & Sons, Inc, Hoboken, NJ, USA. pp. 1141–1148. doi: 10.1002/9781119053095.ch112

Patiño-Navarrete R, Piulachs M-D, Belles X, Moya A, Latorre A, Peretó J (2014) The cockroach *Blattella germanica* obtains nitrogen from uric acid through a metabolic pathway shared with its bacterial endosymbiont. Biol Lett 10(7):20140407

Pérez-Cobas AE, Maiques E, Angelova A, Carrasco P, Moya A, Latorre A (2015) Diet shapes the gut microbiota of the omnivorous cockroach *Blattella germanica*. FEMS Microbiol Ecol 91(4):fiv022

Piermarini PM, Weihrauch D, Meyer H, Huss M, Beyenbach KW (2009) NHE8 is an intracellular cation/H$^+$ exchanger in renal tubules of the yellow fever mosquito *Aedes aegypti*. Am J Physiol Renal Physiol 296(4):F730–F750. doi:10.1152/ajprenal.90564.2008

Prusch RD (1972) Secretion of NH$_4$Cl by the hindgut of *Sarcophaga bullata* larva. Comp Biochem Physiol A Physiol 41(1):215–223. doi:10.1016/0300-9629(72)90049-7

Roth LM, Dateo GP (1965) Uric acid storage and excretion by accessory sex glands of male cockroaches. J Insect Physiol 11(7):1023–1029

Sasaki T, Ishikawa H (1995) Production of essential amino acids from glutamate by mycetocyte symbionts of the pea aphid, *Acyrthosiphon pisum*. J Insect Physiol 41(1):41–46

Scaraffia PY, Isoe J, Murillo A, Wells MA (2005) Ammonia metabolism in *Aedes aegypti*. Insect Biochem Mol Biol 35(5):491–503. doi:10.1016/j.ibmb.2005.01.012

Scaraffia PY, Tan G, Isoe J, Wysocki VH, Wells MA, Miesfeld RL (2008) Discovery of an alternate metabolic pathway for urea synthesis in adult *Aedes aegypti* mosquitoes. Proc Natl Acad Sci U S A 105(2):518–523

Schwantes U (1990) Uric acid during pupal and adult development of *Musca domestica* L. (Diptera). Zool Jb Physiol 94:1–18

Shen K, Wang H-J, Shao L, Xiao K, Shu J-P, Xu T-S, Li G-Q (2009) Mud-puddling in the yellow-spined bamboo locust, *Ceracris kiangsu* (Oedipodidae: Orthoptera): does it detect and prefer salts or nitrogenous compounds from human urine? J Insect Physiol 55(1):78–84

Sohal RS, Copeland E (1966) Ultrastructural variations in the anal papillae of *Aedes aegypti* (L.) at different environment salinities. J Insect Physiol 12(4):429–434

Souza AVG, Petretski JH, Demasi M, Bechara E, Oliveira PL (1997) Urate protects a blood-sucking insect against hemin-induced oxidative stress. Free Rad Biol Med 22(1):209–214

Staddon B (1963) Water balance in the aquatic bugs *Notonecta glauca* L. and *Notonecta marmorea* Fabr. (Hemiptera; Heteroptera). J Exp Biol 40(3):563–571

Staddon B (1964) Water balance in *Corixa dentipes* (Thoms.)(Hemiptera, Heteroptera). J Exper Biol 41(3):609–619

Staddon BW (1955) The excretion and storage of ammonia by the aquatic larva of *Sialis lutaria* (Neuroptera). J Exp Biol 32:84–94

Staddon BW (1959) Nitrogen excretion in the nymphs of *Aeshna cyanea* (Mull) (Odonata, Anisoptera). J Exp Biol 36:566–574

Stobbart RH (1971) Evidence for Na$^+$-H$^+$ and Cl$^-$-HCO$_3^-$ exchanges during independent sodium and chloride uptake by the larva of the mosquito *Aedes aegypti* (L.). J Exp Biol 54(1):19–27

Thomson RB, Thomson JM, Phillips JE (1988) NH$_4^+$ transport in acid-secreting insect epithelium. Am J Physiol 254(2 Pt 2):R348–R356

Tojo S, Yushima T (1972) Uric acid and its metabolites in butterfly wings. J Insect Physiol 18(3):403409–407422

Ulyshen MD (2015) Insect-mediated nitrogen dynamics in decomposing wood. Ecol Entomol. doi:10.1111/een.12176: 40.S1:97–112

Wang L, Kiuchi T, Fujii T, Daimon T, Li M, Banno Y, Kikuta S, Kikawada T, Katsuma S, Shimada T (2013) Mutation of a novel ABC transporter gene is responsible for the failure to incorporate uric acid in the epidermis of ok mutants of the silkworm, *Bombyx mori*. Insect Biochem Mol Biol 43(7):562–571

Weihrauch D (2006) Active ammonia absorption in the midgut of the Tobacco hornworm *Manduca sexta* L.: Transport studies and mRNA expression analysis of a Rhesus-like ammonia transporter. Insect Biochem Mol Biol 36(10):808–821

Weihrauch D, Donini A, O'Donnell MJ (2012) Ammonia transport by terrestrial and aquatic insects. J Insect Physiol 58(4):473–487. doi:10.1016/j.jinsphys.2011.11.005

Weinland E (1906) Ueber die Ausscheidung von Ammoniak durch die Laryen von Calliphoraund ueber eine Beziehung dieser Tatsache zu dem Entwicklungsstadium dieser Tiere. Ztschr Biol 47:232–250

Wright PA, Wood CM (2009) A new paradigm for ammonia excretion in aquatic animals: role of Rhesus (Rh) glycoproteins. J Exp Biol 212:2303–2312

Zhou G, Flowers M, Friedrich K, Horton J, Pennington J, Wells MA (2004) Metabolic fate of [14 C]-labeled meal protein amino acids in *Aedes aegypti* mosquitoes. J Insect Physiol 50(4):337–349

Nitrogen Excretion in Nematodes, Platyhelminthes, and Annelids

Alex R. Quijada-Rodriguez, Aida Adlimoghaddam, and Dirk Weihrauch

© Springer International Publishing Switzerland 2017
D. Weihrauch, M. O'Donnell (eds.), *Acid-Base Balance and Nitrogen Excretion in Invertebrates*,
DOI 10.1007/978-3-319-39617-0_5

5.1   Introduction

Maintenance of body fluid ion homeostasis is of extreme importance when inhabiting harsh environments, where organisms may experience either very ion-poor environments (i.e., freshwater and water film of soil particles) or constant fluctuations in environmental salinity as seen in estuaries and intertidal zones. Organisms inhabiting these environments typically actively osmoregulate through well-ventilated and vascularized tissues, which have a large surface area. Tissues important for osmoregulation such as the gills, the skin, and the digestive system have also been suggested to play a critical role in various other physiological and biochemical processes such as nitrogen balance, gas exchange, and acid-base regulation (Anderson et al. 2015; Cruz et al. 2013; Hwang 2009; Hwang et al. 2011; Larsen et al. 2014; Rubino et al. 2014; Shih et al. 2008; Weihrauch et al. 2009; Wilson et al. 2013; Wright and Wood 2009). The multifunctional role of these tissues implicates that many of these ionoregulatory processes may be directly linked, since some functions require the same transporter, channel, or ion pump, for instance, the Na^+/K^+-ATPase (Larsen et al. 2014). This means that environmental challenges affecting one process (e.g., NaCl uptake) will likely directly impact the ability of the tissue to regulate other physiological processes (e.g., acid-base regulation, nitrogen excretion). In this chapter we will summarize how the challenges posed by freshwater environments and the water film of soil particles influence nitrogen excretion strategies in the phyla Nematoda, Platyhelminthes, and Annelida.

5.1.1   Challenges of Inhabiting Freshwater Environments

Freshwater environments are characteristically known for having very low concentrations of dissolved ions. Due to the low ion concentration in freshwater, organisms inhabiting these environments face the problem of a large osmotic gradient between the body fluids and the environment, which drives a water influx by diffusion. As such, freshwater organisms secrete large amounts of hypoosmotic urine to maintain water balance while minimizing ion losses (Riegel 1968; Zerbst-Boroffka et al. 1997). Although freshwater vertebrates and invertebrates have kidneys or analogous structures for ion retention and very tight epithelia to reduce the paracellular loss of ions, the strong outwardly directed ion gradients faced by freshwater organisms still results in a passive loss of ions from the body fluids (Larsen et al. 2014). Consequently, freshwater organisms must maintain ion homeostasis through active NaCl uptake from the environment and consumption of food. In freshwater fish, uptake of ions through food has been shown to be essential for compensation of diffusive ion loss, as ion gradients across the intestine facilitate ion uptake relative to ion uptake from the environment (reviewed in Wood and Bucking 2011).

Investigations regarding osmoregulation in freshwater organisms have mainly been focused on fish with some studies also looking at crustaceans and amphibians (Bianchini and Wood 2008; Kirschner 2004; Kumai and Perry 2012; Marshall 2002; Onken and Mcnamara 2002; Onken and Putzenlechner 1995; Onken et al. 2000; Parks et al. 2008; Weihrauch et al. 2004a). Based on studies of these organisms, general hypothetical models for sodium and chloride uptake have been proposed (◘ Fig. 5.1). In terms of sodium transport, the Na^+/K^+-ATPase, which is present in most animal cells, has been established as one of the driving forces for Na^+ uptake in freshwater organisms. This transporter, which is discussed more thoroughly in ▶ Chap. 3, is localized on the basolateral membrane, where it

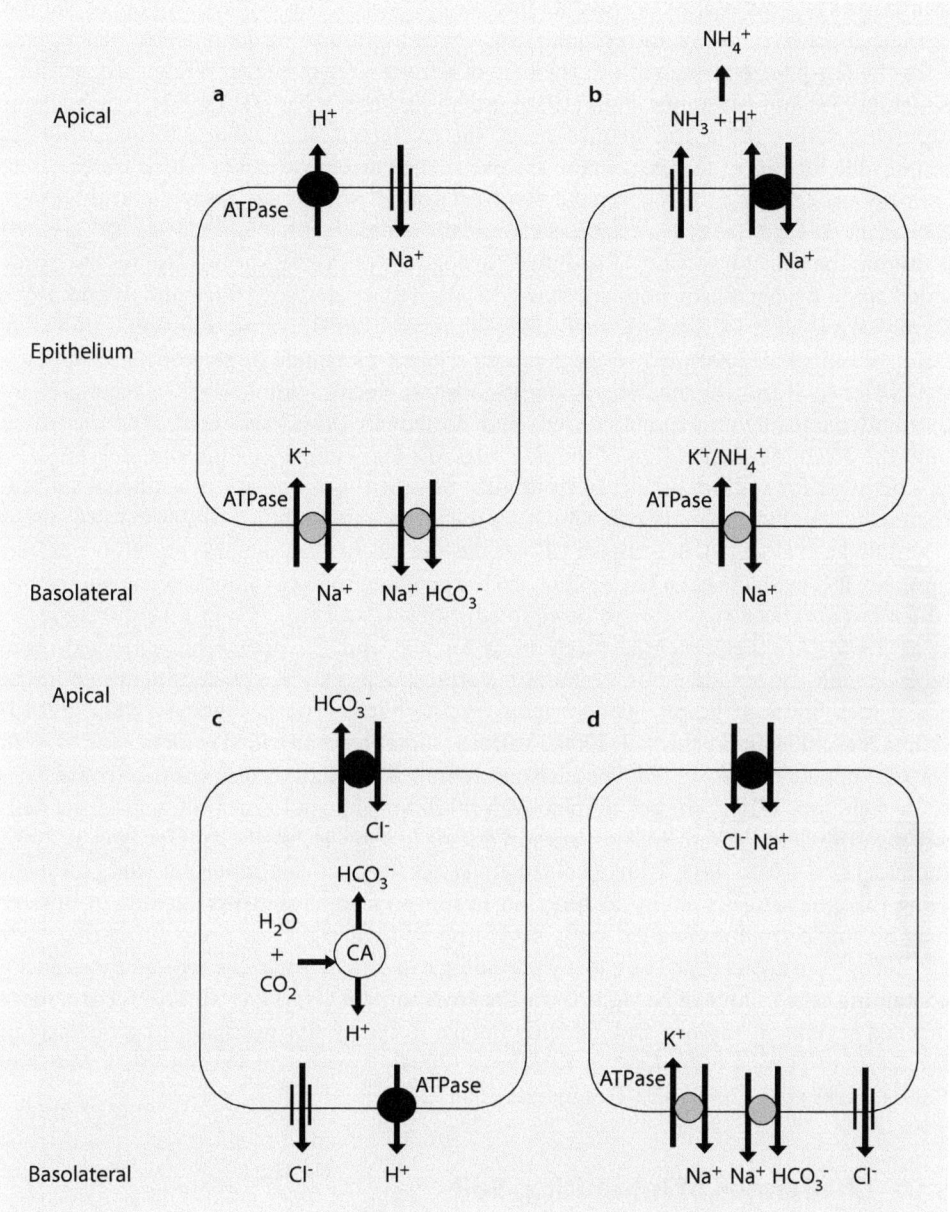

☐ **Fig. 5.1** Hypothetical models for Na⁺ and Cl⁻ uptake in freshwater fish. (**a**) V-ATPase and Na⁺/
K⁺-ATPase driven Na⁺ uptake model via Na⁺ channels. (**b**) Rh-NHE metabolon model for sodium uptake.
Here the transport of NH_3 by Rh proteins creates an apical alkalization which drives a proton secretion
and sodium uptake via NHEs. (**c**) Cl⁻ uptake model by Cl⁻/HCO_3^- exchanger coupled with basolateral
H⁺-ATPase. (**d**) Na⁺ and Cl⁻ uptake model by NCC coupled with Na⁺/K⁺-ATPase, Cl⁻ channel, and Na⁺/
HCO_3^- cotransporter. Models are adapted after Hwang et al. (2011), Kumai and Perry (2012), Parks et al.
(2008), and Tresguerres et al. (2006)

transports 3 Na$^+$ out of the cell and 2 K$^+$ into the cytoplasm. This active transport of Na$^+$ out of the cell generates a low intracellular Na$^+$ concentration to facilitate apical Na$^+$ uptake from the ion-poor environment in the case of a freshwater scenario. While early work by Krogh (1938) and Kerstetter et al. (1970) established that a Na$^+$/H$^+$ or NH$_4^+$ exchange is occurring across the apical membrane of the osmoregulating tissues, the transporters responsible for this exchange remain unsolved. The current debate of which transporters are responsible for apical Na$^+$ uptake revolves around whether an electroneutral Na$^+$/H$^+$ exchanger (◘ Fig. 5.1a) or Na$^+$ channel electrically coupled with a H$^+$-ATPase (◘ Fig. 5.1b) is driving the apical transport of sodium. Throughout the years, various studies have provided some evidences for both mechanisms (Boisen et al. 2003; Bury and Wood 1999; Dymowska et al. 2014; Edwards et al. 1999; Hirata et al. 2003; Horng et al. 2007; Shih et al. 2012; Wilson et al. 2000). However, the main arguments against these models lie in that a freshwater environment generates a condition where electroneutral Na$^+$/H$^+$ exchangers are thermodynamically unfavorable (Avella and Bornancin 1989; Parks et al. 2008). Further, until the recent demonstration of the potential role for acid-sensing ion channels (ASICs) as a pathway for sodium uptake in trout gills, there had been no apical epithelial sodium channel (ENaC) identified in fish, which are the more widely studied freshwater organisms (Dymowska et al. 2014). In terms of Cl$^-$ uptake, the mechanism remains relatively unclear; however, it is evident that a Cl$^-$/HCO$_3^-$ exchange is occurring in at least some organisms, which has also been suggested to be coupled with a H$^+$-ATPase (◘ Fig. 5.1c) (Tresguerres et al. 2006). Another potential mechanism for Na$^+$ and Cl$^-$ uptake has emerged more recently with the identification of Na$^+$/Cl$^-$ cotransporters (NCCs) being identified in the apical membrane of tilapia (*Oreochromis mossambicus*) type II ionocytes (◘ Fig. 5.1d) (Hiroi et al. 2008; Inokuchi et al. 2008). Various studies have provided evidence for NCC in Na$^+$/Cl$^-$ uptake not just in tilapia but also in zebrafish through scanning ion-selective electrode technique (SIET) studies coupled with inhibitors, knockdown studies, ionocyte density measurements, and mRNA expression during low ion acclimations (Horng et al. 2009; Inokuchi et al. 2008, 2009; Wang et al. 2009). However, the most important study for justifying this mechanism's ability to function in ion-poor environments such as freshwater may be an electrophysiological study by Horng and colleagues demonstrating that Na$^+$/Cl$^-$ fluxes from adjacent cells are likely providing a microenvironment, which allows NCC-containing cells to uptake Na$^+$/Cl$^-$ from the environment (Horng et al. 2009). For a more detailed review on sodium and chloride uptake in freshwater organisms, particularly in fish, refer to Hwang et al. (2011), Kirschner (2004), Kumai and Perry (2012), Marshall (2002), Parks et al. (2008) and Tresguerres et al. (2006).

5.1.2 Challenges of Inhabiting Soil

Under moist soil conditions, soil-inhabiting organisms such as nematodes and annelids face similar challenges to that occurring in freshwater environments. In soil, the concentration of ions relative to the body fluids of organisms inhabiting these environments is relatively low. For example, a study by Tavakkoli et al. (2011) measured soil Na$^+$ and Cl$^-$ concentrations of 1.9 mmol l^{-1}, which is approximately 40- and 25-fold lower than the coelomic fluid Na$^+$ and Cl$^-$ concentration of the earthworm (*Lumbricus terrestris*) (Diets and Alvarado 1970). Therefore, like in freshwater organisms, soil-dwelling nematodes and annelids would face strong outwardly directed ion gradients driving a loss of ions from the body fluids. In order to compensate for these passive losses, soil dwellers must actively uptake ions from the water film of soil particles and/or consume food. Further challenges to ion homeostasis of soil dwellers are experienced following rainfall; as the soil becomes flooded, a dilution of environmental ion concentration occurs generating even greater outwardly directed ion gradients. This dilution of environmental ion concentration would increase osmotic pressure driving water into the organisms similar to what is seen in freshwater organisms. One can assume that soil-dwelling nematodes and annelids likely counter this water uptake through excretion of very dilute urine to conserve ions and excrete water. However, unlike freshwater organisms, soil-dwelling organisms do not always have a high abundance of water present and may, in fact, encounter periods of desiccation. During desiccation, the water content of the soil decreases and therefore ion concentrations increase, which results in osmotic pressure driving water out of the organisms, thereby dehydrating the organisms (Roots 1955). In terms of ion balance, these organisms would still have the need to actively osmoregulate in order to maintain homeostasis as the decrease in body water would result in elevated body fluid ion concentration.

Unlike typically seen in freshwater environments, soil can have relatively high amounts of ammonia ranging from micromolar to millimolar concentrations (Nesdoly and Van Rees 1998). This high presence of ammonia typically serves as a major source of nitrogen for plant cells but can also be highly toxic to plants in abundance (Britto and Kronzucker 2002). For soil-inhabiting invertebrates such as nematodes and annelids, this high environmental ammonia provides a challenge as ammonia is highly toxic to animals (▶ see Sect. 5.1.3). In order to survive these high-ammonia environments, soil-dwelling invertebrates must either be very ammonia tolerant or have developed extremely efficient mechanisms to prevent influxes while excreting or detoxifying ammonia.

5.1.3 Nitrogenous Waste Excretion in Aquatic Invertebrates

While some exceptions may exist, the primary nitrogenous waste product of aquatic invertebrates is ammonia (Larsen et al. 2014; Wright 1995). For the purpose of this chapter, NH_3 refers to nonionic ammonia, NH_4^+ or ammonium to the ionic ammonia, and ammonia to the sum of both. In aquatic invertebrates, ammonia is formed through amino acid catabolism, where most amino acids are first transaminated into glutamate and subsequently deaminated into α-ketoglutarate and ammonia (Wright 1995). While primarily formed through transdeamination, other pathways for ammonia synthesis are seen in aquatic invertebrates and described for crustaceans in ▶ Chap. 1. In addition to ammonia, some aquatic invertebrates also excrete urea, although to a lesser extent (Adlimoghaddam et al. 2015; Hoeger et al. 1987; Martin et al. 2011; Quijada-Rodriguez et al. 2015). Urea production in aquatic invertebrates typically occurs through uricolysis or argininolysis (Wright 1995), while dietary urea may also comprise a portion of urea excreted.

Metabolic ammonia production within the body fluids can be highly toxic; therefore an effective excretion or detoxification of ammonia into molecules such as urea and uric acid is critical to prevent physiological distress. At least in mammals, the mode of ammonia toxicity to the central nervous system has been thoroughly studied (reviewed in Braissant et al. 2013). In aquatic invertebrates, the majority of studies on ammonia toxicity have been focused on crustaceans (▶ see Chap. 1). However, with invertebrates being such a large and diverse group, a paucity of data on the toxic effects of ammonia in the remaining invertebrate phyla exists. There are some general effects of ammonia that likely apply across all animals. For example, the uncoupling of proton gradients across the inner mitochondrial membrane can disrupt oxidative phosphorylation (O'Donnell 1997). Further, ammonia is capable of altering intracellular and/or intraorganelle pH, thereby disrupting the optimal pH required for proper function of proteins and organelles. In order to minimize the risk of accumulation of toxic ammonia, animals have specialized tissues capable of excreting nitrogenous wastes from the body (Cruz et al. 2013; Glover et al. 2013; Larsen et al. 2014; Quijada-Rodriguez et al. 2015; Rubino et al. 2014; Shih et al. 2008; Weihrauch et al. 2009; Wilson et al. 2013; Wright and Wood 2009).

As previously mentioned, tissues responsible for osmoregulation usually also play a role in other processes such as excretion of nitrogenous wastes and acid-base balance. In freshwater organism, the link between sodium uptake and ammonia excretion have been mainly focused on teleost fish (Kumai and Perry 2011; Shih et al. 2008, 2012; Tsui et al. 2009; Wright and Wood 2009; Wu et al. 2010; Zimmer et al. 2010). In freshwater organisms four key transporters are generally suggested to link sodium uptake and ammonia excretion; these include the basolateral Na^+/K^+-ATPase, apical V-type H^+-ATPase, Rhesus glycoproteins, and apical Na^+/H^+ exchanger. The functional role of these transporters with the exception of the Rhesus glycoprotein and Na^+/H^+ exchanger in ammonia excretion and sodium uptake will be covered in ▶ Chap. 3. For a review of the role of Rhesus proteins and Na^+/H^+ exchangers in linking ammonia excretion and sodium uptake, see Wright and Wood (2009), Weihrauch and colleagues (2009), and Kumai and Perry (2012).

5.2   Nitrogen Excretion in Annelids, Planarians, and Nematodes

5.2.1   Nitrogenous Waste Products of Planarians, Nematodes, and Annelids

Members of the phyla Annelida, Nematoda, and Platyhelminthes inhabit a variety of ecological niches including freshwater environments, marine systems, and terrestrial habitats, in addition to parasitizing both vertebrate and invertebrate hosts. Their success in these wide-ranging habitats is likely in part dependent on their adaptive ability to regulate ion homeostasis and eliminate toxic waste products (Adlimoghaddam et al. 2014, 2015; Diets and Alvarado 1970; Quijada-Rodriguez et al. 2015; Rothstein 1963; Weber et al. 1995; Weihrauch et al. 2012; Zerbst-Boroffka et al. 1997). Aquatic invertebrates primarily excrete their nitrogenous waste as ammonia (◘ Table 5.1, for crustaceans ▶ see Chap. 1). Thus far, aquatic annelids are no exception with both carnivorous and sanguivorous leeches exhibiting greater rates of ammonia excretion than urea (Quijada-Rodriguez et al. 2015; Tschoerner and Zebe 1989). In fact in an unfed state, the carnivorous ribbon leech (*Nephelopsis obscura*) excretes approximately 92% of its measured nitrogenous waste as ammonia. After feeding, nitrogenous excretion rates rapidly increased in both the carnivorous leech (*N. obscura*) and sanguivorous leech (*Hirudo medicinalis*), likely as a result to elevated protein catabolism and elimination of consumed nitrogenous wastes, i.e., ammonia and urea (Quijada-Rodriguez et al. 2015; Tschoerner and Zebe 1989). In *H. medicinalis*, feeding resulted in an elevated urea excretion rates from nearly undetectable levels to about 14 μmol individual^{-1} day^{-1}. While *H. medicinalis* experienced elevated urea excretion, in *N. obscura* urea excretion rates remain unaltered after feeding (A.R. Quijada-Rodriguez personal communication). It is noteworthy that in the aforementioned studies, excretion of other less commonly excreted (in aquatic organisms) nitrogenous products such as uric acid, amino acids, guanine, allantoin, etc., have not been measured and may contribute to a portion of nitrogenous waste excretion. In terms of members of the platyhelminthes, the excretion capabilities of nitrogenous wastes remain rela-

◻ **Table 5.1** Ammonia and urea excretion rates in a number of aquatic invertebrates excluding crustaceans

Species	Salinity	Ammonia (μmol/g/h)	Urea (μmol/g/h)	Source
Nematoda				
Caenorhabditis elegans	~350 mOsm	1.25	0.125	Adlimoghaddam et al. (2015)
Annelida				
Arenicola marina	SW	1.6	NM	Reitze and Schottler (1989)
Arenicola marina	BW	5	NM	Reitze and Schottler (1989)
Hirudo medicinalis	FW (fed)	1.25	0.58 μmol/ individual/h	Tschoerner and Zebe (1989)
Nephelopsis obscura	FW	0.166	0.014	Quijada-Rodriguez et al. (2015)
Platyhelminthes				
Schmidtea mediterranea	FW	0.70	NM	Weihrauch et al. (2012)
Echinodermata				
Tripneustes gratilla	SW	1.45	NM	Dy and Yap (2000)
Protoreaster nodosus	SW	0.492	NM	Dy and Yap (2000)
Ophiorachna incrassata	SW	0.361	NM	Dy and Yap (2000)
Eupentacta quinquesemita	BW	1.16	0.07	Sabourin and Stickle (1981)
Eupentacta quinquesemita	SW	2.12	0.03	Sabourin and Stickle (1981)
Strongylocentrotus droebachiensis	SW	1.75	0.05	Sabourin and Stickle (1981)
Strongylocentrotus droebachiensis	BW	1.95	0.09	Sabourin and Stickle (1981)
Mollusca				
Illex illecebrosus	SW	1.43	0.23	Hoeger et al. (1987)
Loligo forbesii	SW	10.9	0.42	Boucher-Rodoni and Mangold (1989)

(continued)

◘ **Table 5.1** (continued)

Species	Salinity	Ammonia (µmol/g/h)	Urea (µmol/g/h)	Source
Octopus rubescens	SW	0.25	0.052	Hoeger et al. (1987)
Acmaea scutum	SW	1.28	ND	Duerr (1968)
Acmaea digitalis	SW	0.112	ND	Duerr (1968)
Calliostoma ligatum	SW	2.53	ND	Duerr (1968)
Littorina sitkana	SW	1.78	ND	Duerr (1968)
Fusitriton oregonensis	SW	0.625	ND	Duerr (1968)
Thais lima	SW	0.119	ND	Duerr (1968)
Thais lamellosa	SW	0.208	ND	Duerr (1968)

► See Chap. 1 for crustaceans
SW is seawater, *BW* is brackish water, *FW* is freshwater, *NM* means not measured, and *ND* means measured but nothing significantly detected

tively unknown with the exception of relatively high levels of ammonia excreted by the planarian (*Schmidtea mediterranea*) (Weihrauch et al. 2012).

Similar to aquatic leeches and planarians, soil nematodes and annelids such as *Caenorhabditis elegans* and *Lumbricus terrestris* are predominantly ammonotelic (Adlimoghaddam et al. 2015; Cohen and Lewis 1949; Tillinghast et al. 1969). However, unlike leeches and planarians, soil-dwelling invertebrates have been shown to switch from ammonotelism to ureotelism based on feeding status and water availability. For example, in *L. terrestris* a switch to ureotelism has been reported when starved and water supply is limited (Tillinghast et al. 1969). Presumably this change to ureotelism may act as a counter measure to conserve water while maintaining an excretion of nitrogenous wastes. However, unlike in *L. terrestris*, all the enzymes of the urea cycle are not present in *C. elegans* (► see www.wormbase.org), so a similar change to ureotelism is less conceivable and remains to be investigated. During starvation, *C. elegans* experiences an approximately threefold decrease in ammonia excretion due to decreased protein metabolism but unchanged urea excretion (Adlimoghaddam et al. 2015). During starvation ammonia and urea excretion rates in *C. elegans* are about the same, while in *L. terrestris*, there is a marked change as rates of urea excretion increase (Adlimoghaddam et al. 2015; Tillinghast et al. 1969).

5.2.2 Tissues Potentially Involved in Excretion of Ammonia in Planarians, Nematodes, and Annelids

Epidermis/Hypodermis

In fish and crustaceans, oxygen exchange occurs through well-ventilated gills; however, in worm-type organisms like platyhelminthes, nematodes, and annelids, specialized gill-like structures are not always present. In these instances, these organisms tend to use their

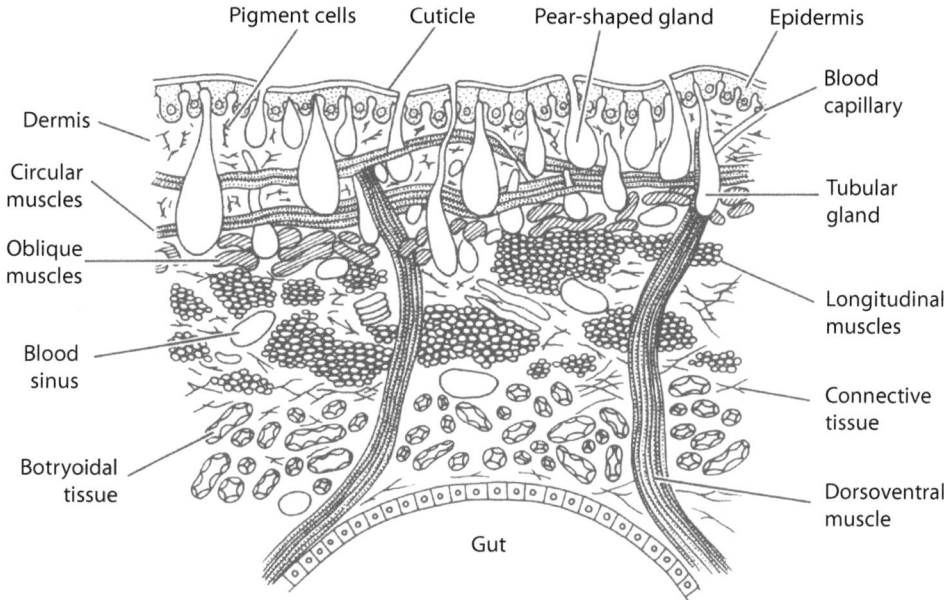

□ **Fig. 5.2** *Hirudo* body wall diagram demonstrating organization of the epidermis with epithelial cells at the surface and mucous glands in the subepidermal layer (This image is taken from Brusca and Brusca (2003) with permission)

outer dermal layer for enhanced oxygen exchange across the body wall. In annelids and planarians, the outer dermal layer is termed the epidermis while in nematodes it is known as the hypodermis, which is covered in a multilayered cuticle. The epidermis of planarians contains specialized cells called rhabdites, which are rod-shaped structures that secrete mucous onto the epithelial layer (Martin 1978). Similarly, annelids also contain a specialized cell known as gland cells that also secrete mucous. However, unlike in other annelids, leeches contain the gland cells within the subepidermal layer with openings leading to the epithelial surface (□ Fig. 5.2) (Ahmed and Rahemo 2013). The mucous-secreting cells of planarians and annelids could potentially play a critical role in ammonia excretion by creating a microenvironment for acid trapping of ammonia, a mechanism described in various freshwater organisms (Larsen et al. 2014). While annelids and planarians likely rely on secretion of mucous for the generation of an unstirred boundary layer, nematodes may utilize the subcuticular layer between the hypodermis and secreted cuticle or the multilayered cuticle itself (Peixoto and De Souza 1995) to form a microenvironment for ammonia trapping.

As previously mentioned, the epidermis/hypodermis of wormlike organisms likely plays a major role in ammonia excretion. SIET studies in *Caenorhabditis elegans* have revealed that the hypodermis of this nematode is involved in secretion of protons (□ see Fig. 5.3a) (Adlimoghaddam et al. 2015). Further, the hypodermis of *C. elegans* expresses both Rhr-1 and Rhr-2 proteins (□ Fig. 5.4) (Adlimoghaddam et al. 2016; Ji et al. 2006); therefore it is likely that the secretion of protons could drive an ammonia excretion by acid trapping. In leeches, elevated mRNA expression levels of the "primitive" Rhesus glycoprotein (Rh protein) in the skin relative to the rest of the body of *Nephelopsis obscura* implicate

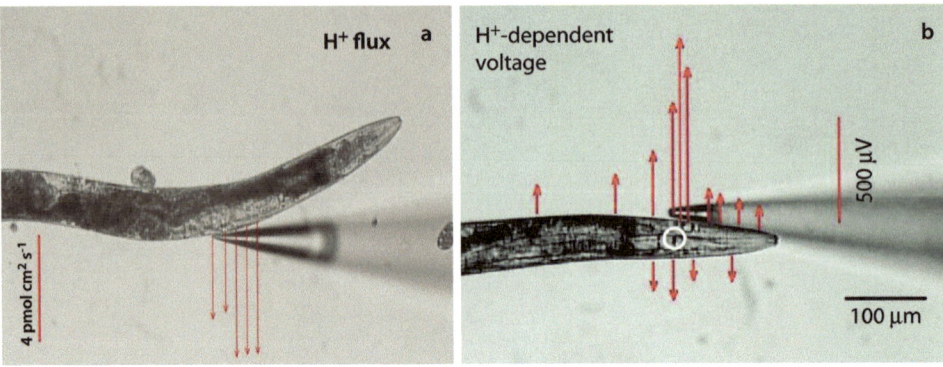

Fig. 5.3 Representative scans of area related to transport rates of H⁺ over the hypodermis (**a**) and voltage changes indicating H⁺ excretion by the excretory pore (**b**) of *Caenorhabditis elegans* recorded by SIET using H⁺-specific electrodes (This image is taken from Adlimoghaddam et al. (2016) and Adlimoghaddam et al. (2014) with permission)

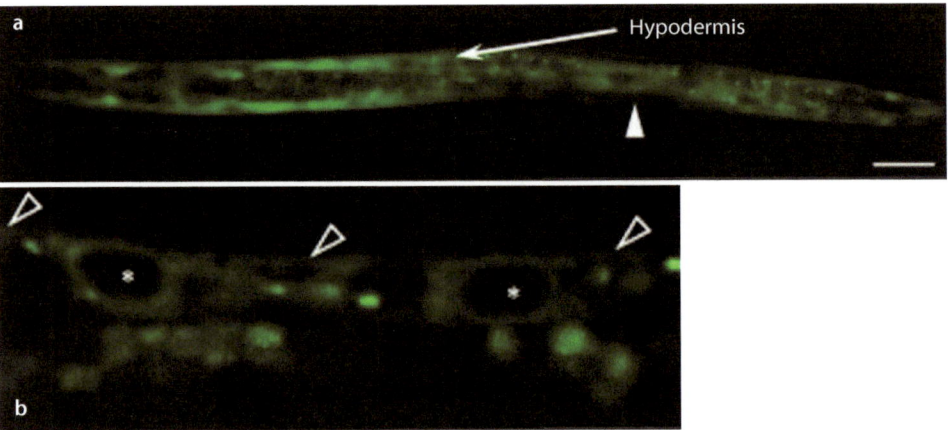

Fig. 5.4 *Caenorhabditis elegans* Rhr-2 tissue localization achieved through Rhr-2 promoter activated GFP expression. (**a**) GFP expression indicates Rhr-2 expression in the hypodermis and in the ventral nerve cord (*white arrow*). Scale bar = 50 μm. (**b**) Higher magnification image demonstrating single cells where * indicates cell nuclei. *Open arrows* show GFP expression concentrated at the apical membrane of the cell and cytoplasmic indicating localization of Rhr-2 in the hypodermis of *C. elegans* (This image is taken from Adlimoghaddam et al. (2016) with permission)

the skin as a site of ammonia excretion (Quijada-Rodriguez et al. 2015). Further, employing the isolated of the skin from *N. obscura* in Ussing chamber experiments provided direct evidence that the skin does indeed have a high capacity for ammonia transport, especially when compared to rates of ammonia transport by frog skin (Cruz et al. 2013; Quijada-Rodriguez et al. 2015). As in leeches, in situ hybridization in the planarian (*Schmidtea mediterranea*) suggests a high expression of an Rh protein in the epidermis (■ Fig. 5.5a, Weihrauch et al. 2012). Further, immunolocalization of the V-ATPase in the epidermis of *S. mediterranea* revealed a high abundance of the V-ATPase in the mucous-secreting rhabdites and in the epithelial cells indicating that the unstirred boundary layer is likely acidified (■ Fig. 5.5b).

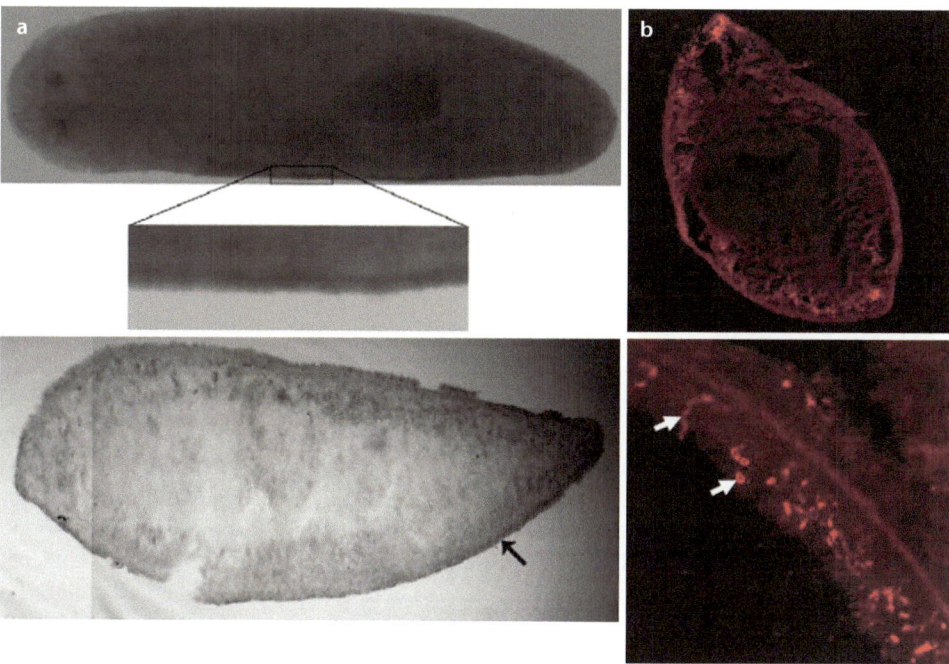

■ **Fig. 5.5** (**a**) Tissue expression of Rh-like protein in *Schmidtea mediterranea* by in situ hybridization. Application of the antisense riboprobe revealed staining over the entire whole animal mount. At 250× magnification, the strongest signal is seen at the edges on the epidermis. At 100× magnification the *dark arrow* indicates staining of what is likely the epidermis. (**b**) Immunolocalization of the V-ATPase B subunit in *S. mediterranea* whole body cross sections. Antibodies utilized were raised against the V-ATPase subunit B from *Manduca sexta* and demonstrated a distinct band of 56 kDa in western blot confirming functionality of the antibody. Fluorescence imaging demonstrates that the antiserum detected a signal in the epidermis with strong signals in rod-shaped structures, which are presumably the rhabdites (*white arrows*). Magnifications for images in **a** are from top to bottom 30×, 250×, and 100×. Magnifications for images in **b** are 100× for the top and 400× for the bottom image (These images are taken from Weihrauch et al. (2012) with permission)

Taken together, the abundance of Rhesus glycoproteins, evidence for potential ammonia trapping by V-ATPase, and direct measurement of ammonia transport at least in leech skin provide compelling evidence that the outer dermal layer of annelids, nematodes, and platyhelminthes play a role in the excretion of ammonia.

Intestine

In annelids, platyhelminthes, and nematodes, studies on the role of the intestine in nitrogen transport are very scarce. In many organisms, the intestine is typically separated into anatomically different sections leading to functional transport differences across the intestine. For example, in terms of nitrogen handling, the anterior, posterior, and mid-intestine of the rainbow trout (*Oncorhynchus mykiss*) exhibit different transport capacities and mechanisms (Rubino et al. 2014). The very small size of wormlike organisms and difficulty to obtain reliable preparations make a direct study of nitrogen transport across the intestine extremely difficult. However, few in vivo studies on the terrestrial earthworm (*Lumbricus terrestris*) have provided evidence for intestinal ammonia excretion at least in fed-state earthworms (Tillinghast 1967; Tillinghast et al. 2001). In this oligochaete, high ammonia concentrations can be detected in the intestine ($11-104$ mmol l^{-1}), with the highest intestinal ammonia concentration in the two posterior regions (Tillinghast 1967; Tillinghast et al. 2001). Early work by Tillinghast (1967) revealed that following defecation ammonia excretion rates increased while urea excretion rates were unaltered. Here it was suggested that in the fed earthworm, ammonia excretion must occur at least partly through the intestine. However, this study fails to account for the possibility of cutaneous transport of nitrogenous wastes, which as discussed above is a highly probable site for ammonia excretion.

Excretory System

In mammals, the kidney has been accepted to not only be essential for acid-base regulation and salt balance but also critical for ammonia handling (Weiner and Verlander 2014). While classical kidneys are not present in the annelids, platyhelminthes, and nematodes, analogous structures to the mammalian kidney can be found. In the annelids and platyhelminthes, these structures are called the metanephridia and protonephridia, respectively. On the other hand, in nematodes the entire excretory structure posses only 3 cells: the excretory cell, pore cell, and duct cell, which comprise the excretory system. The excretory cell has two channels that run down the length of the nematode and feeds through the duct cell to the pore cell which empties just anterior to the pharynx (Nelson et al. 1983).

The proto- and metanephridial system of platyhelminthes and annelids can be thought of as precursors to the mammalian nephron, which generally work through ultrafiltration (O'Donnell 1997; Zerbst-Boroffka et al. 1997). In annelids ultrafiltration occurs from the blood vessels into the coelom. Here the ciliated nephrostome opens into the coelom and funnels coelomic fluid into the metanephridia where the extracellular fluids are filtered resulting in the removal of solutes. Subsequently urine is formed, which is then altered by removal of water and solutes as it passes through the nephridia and eventually being excreted (O'Donnell 1997). The leech metanephridial system is slightly different in that coelomic vessels transfer coelomic fluid to the nephrostome, where funneling of the fluids occurs as would be seen in oligochaetes (◘ Fig. 5.6).

The basic structure of the metanephridia (◘ Fig. 5.6) consists of a nephridiopore that leads directly to the environment and a ciliated funnel called the nephrostome, which acts as a connection from the extracellular fluids to the metanephridia (Zerbst-Boroffka et al. 1997). In addition to these structures, the metanephridia have various lobes where urine is filtered and a urinary bladder that holds the final urine. For a more detailed look at the structure of the metanephridia, see Zerbst-Boroffka et al. (1997). Unlike the metanephridia, the protonephridia is a blind-ended tube that sits in the extracellular fluid and utilizes cilia to create a difference in pressure that drives fluid into the lumen of the protonephridia (O'Donnell 1997). Here, as the fluid passes through the protonephridia, solutes and water are removed from the filtrate and then secreted into the environment.

While no direct mechanistic studies investigating ammonia or urea transport in the nephridial systems of platyhelminthes or annelids have been performed, studies by Zerbst-Boroffka and colleagues have investigated the mechanism of chloride secretion in the sanguivorous leech (*Hirudo medicinalis*) (Zerbst-Boroffka et al. 1997). The hypothesized model of chloride secretion in this leech suggests the Na^+/K^+-ATPase, $Na^+/K^+/2Cl^-$, and anion/Cl^- exchanger are involved in basolateral transport of chloride. With NH_4^+ being able to substitute for K^+ in transporters due to similarity in hydrated ionic radius and charge, it is plausible that these transporters in the metanephridia could also mediate an

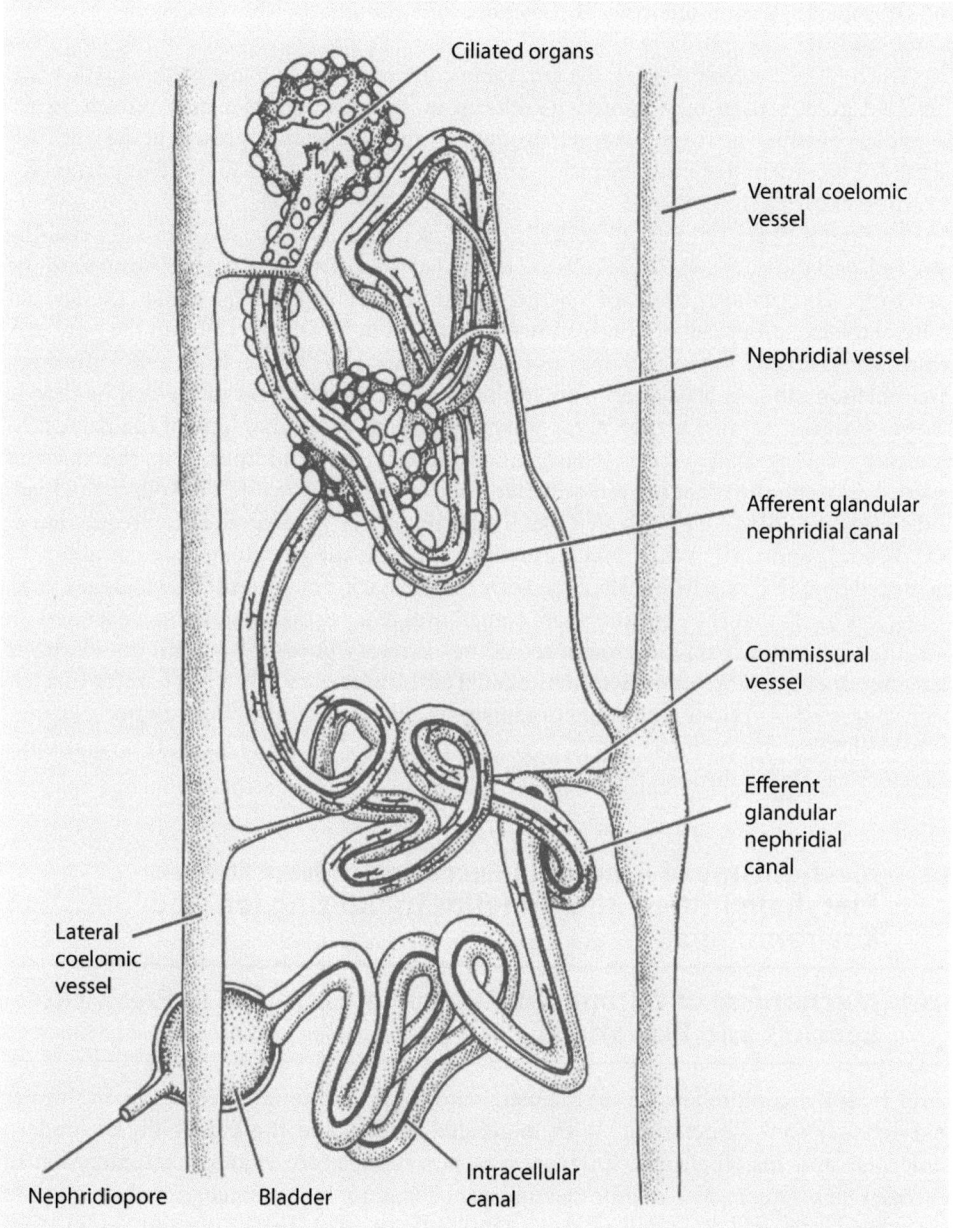

◻ Fig. 5.6 Structure of the leech metanephridia specifically from the genus *Erpobdella*. This figure shows the association between the coelomic channels and the metanephridia (This image is taken from Brusca and Brusca (2003) with permission)

ammonia transport (Knepper et al. 1989). The most prominent evidence for the involvement of the proto- and metanephridia in nitrogenous waste excretion are studies demonstrating direct measurement of both ammonia and urea in nephridial extracts from various platyhelminthes and annelids (Kulkarni et al. 1989; Lutz and Siddiq 1971; Webster and Wilson 1970). In the cestode (*Hymenolepis diminuta*), the urea content of the protonephridia was approximately tenfold greater than ammonia, while in the sanguivorous leech (*Poecilobdella viridis*), the ammonia content in the metanephridial extract was threefold greater than urea content (Kulkarni et al. 1989; Webster and Wilson 1970). Regardless of which nitrogenous waste product is more abundantly present in the nephridial fluid, it is evident that both the proto- and metanephridia are capable and important for excretion of nitrogenous wastes.

The excretory system in nematodes is composed of three cells: the excretory cell, the pore cell, and the duct cell. These cells are located at the anterior end of the worm with the excretory cell containing two canals, which run down the length of the nematode. The out route of the excretory cell has been shown to run through the duct cell into the pore cell, which leads directly to the excretory pore of the nematode (Nelson et al. 1983). Through laser ablation studies, it has been shown that the excretory system of the soil nematode (*Caenorhabditis elegans*) is critical for water balance, as ablation of any of the three cells results in swelling of the worms (Nelson and Riddle 1984). In addition to its role in water balance, through the scanning ion-selective electrode technique (SIET), Adlimoghaddam and colleagues (2014) demonstrated that the excretory cell is involved in ion regulation of Na^+, K^+, H^+, and Ca^{2+}. While NH_4^+ excretion through the excretory pore could not be measured by SIET due to interference with the K^+ background (Adlimoghaddam et al. 2014), it is likely that the excretory pore contributes to a portion of ammonia excreted. In fact, the high proton excretion measured at the excretory pore (◘ Fig. 5.3b) does indicate that ammonia trapping occurs within the canal of the excretory cell, a mechanism that has been suggested in various freshwater organisms (Larsen et al. 2014). Furthermore, expression of Rhr-1 in the excretory cell of *C. elegans* (McKay et al. 2003) suggests an ammonia excretory capability through the excretory pore.

5.3   Mechanisms of Ammonia Excretion in Nematodes, Platyhelminthes, and Annelids Inhabiting Ion-Poor Environments

5.3.1 Mechanism of Cutaneous Ammonia Excretion in Freshwater Leeches and Planarians

Until recently, comprehensive mechanistic studies of ammonia excretion in freshwater invertebrates were nonexistent. With increasing studies and the availability of modern molecular and histological techniques, it is becoming more evident that excretion of nitrogenous wastes in platyhelminthes and annelids is not merely occurring through their specialized intestinal and nephridial systems (Kulkarni et al. 1989; Lutz and Siddiq 1971; Tillinghast 1967; Tillinghast et al. 2001; Webster and Wilson 1970) but also through the epidermal tissues (Quijada-Rodriguez et al. 2015; Weihrauch et al. 2012). Thorough mechanistic studies of ammonia excretion in the platyhelminthes and annelids have been limited to two studies on the freshwater planarian (*Schmidtea mediterranea*) and freshwater carnivorous leech (*Nephelopsis obscura*) (Quijada-Rodriguez et al. 2015; Weihrauch et al. 2012). These two organisms provide ideal models for investigating ammonia excretion in freshwater organisms as the small size and easy maintenance of these organisms make them readily usable for various molecular and physiological techniques. Further, at least for leeches, the skin can be easily dissected and mounted in Ussing chambers for studies investigating ammonia flux capabilities directly across this tissue. This section will summarize the currently hypothesized mechanism of the cutaneous ammonia excretion in the freshwater planarian (*S. mediterranea*) and ribbon leech (*N. obscura*).

The Na^+/K^+-ATPase known to be the driving force for sodium uptake in freshwater organisms has also been shown to play a major role in cutaneous ammonia excretion for both *N. obscura* and *S. mediterranea* (Quijada-Rodriguez et al. 2015; Weihrauch et al. 2012). Here it is suggested to be localized to the basolateral membrane as shown for the ammonia-transporting tissues of various vertebrates and invertebrates (Braun et al. 2009; Nakada et al. 2007; Patrick et al. 2006; Towle et al. 2001; Tresguerres et al. 2008). Whole animal studies on *S. mediterranea* demonstrated that blocking this pump with ouabain, a specific inhibitor of the Na^+/K^+-ATPase, effectively reduces ammonia excretion by about 50 % (Weihrauch et al. 2012). Similarly, isolation of *N. obscura* skin in a modified Ussing chamber demonstrated that application of ouabain directly to the basolateral membrane reduces ammonia excretion by 39 % (Quijada-Rodriguez et al. 2015). In leeches and various other organisms, this pump has been shown to accept ammonium as a substrate in place of potassium and thereby likely acting as a direct transporter of ammonia across the basolateral membrane (Mallery 1983; Masui et al. 2002; Quijada-Rodriguez et al. 2015; Skou 1965; Wall and Koger 1994). Further, mRNA expression of the Na^+/K^+-ATPase α-subunit in *S. mediterranea* has been shown to increase following a high environmental ammonia exposure (1 mmol l^{-1} NH_4Cl, 48 h) (Weihrauch et al. 2012).

Having first been suggested in mammals to transport ammonia, the Rhesus glycoproteins are now recognized as of key importance in ammonia excretion (Marini et al. 2000). Early work by Weihrauch and colleagues revealed that this ammonia transporter of mammals was also expressed in the ionoregulatory tissue (e.g., gills) of aquatic crustaceans (Weihrauch et al. 2004b). Recently, X-ray crystallography suggested that the human RhCG promotes passage of gaseous ammonia and not ionic ammonia due to the hydrophobicity of the permeation pathway (Gruswitz et al. 2010). Since the initial discovery of Rh proteins as gaseous ammonia channels, subsequent studies have implied that these proteins also allow the transport of CO_2 (Endeward et al. 2007; Musa-aziz et al. 2009; Perry et al. 2010; Soupene et al. 2004). In the leech *N. obscura*, a single Rh protein has been identified (NoRhp) and studies employing yeast-based complementation assays demonstrated that this invertebrate Rh protein is indeed capable of transporting ammonia (Quijada-Rodriguez et al. 2015). While not yet characterized as an ammonia transporter, the Rh protein of *S. mediterranea* also likely transports ammonia as phylogenetic analysis of various invertebrate Rh proteins shows that they group within the Rhp1 cluster, where now three members of this cluster have been shown to transport ammonia (▶ see also Chap. 1) (Adlimoghaddam et al. 2015; Huang and Peng 2005; Pitts et al. 2014; Quijada-Rodriguez et al. 2015). Further, high environmental ammonia exposure to both *S. mediterranea* and *N. obscura* resulted in differential mRNA expression of the Rh protein, with an upregulation following a 2-day exposure and downregulation following a 7-day exposure in the planarian and leech, respectively (Adlimoghaddam et al. 2015; Quijada-Rodriguez et al. 2015). In vertebrates, both basolateral and apical localized Rh proteins have been confirmed (Larsen et al. 2014). Therefore, in the leech and planarian, Rh proteins may potentially provide both apical and basolateral routes for ammonia excretion; however, the subcellular localization of the ammonia transporter is not known to date.

Unlike the Na^+/K^+-ATPase and Rh proteins, the V-type H^+-ATPase (V-ATPase) is not capable of directly transporting ammonia but still plays a major role in apical ammonia excretion in freshwater organisms (Kumai and Perry 2011; Nawata et al. 2007; Shih et al. 2008; Tsui et al. 2009; Weihrauch et al. 2012). Diffusion of NH_3 across a membrane is heavily dependent on the partial pressure gradient of NH_3 (ΔP_{NH3}). Thus, the V-ATPase can promote ammonia excretion by creating a pH gradient across a membrane to generate a ΔP_{NH3} driving NH_3 across the membrane. Once transported across the membrane, NH_3 combines with free protons forming NH_4^+ which is incapable of freely diffusing across membranes effectively trapping ammonia. This mechanism of transporting ammonia known as "ammonia or acid trapping" and is likely facilitated by the NH_3 transport capability of Rh proteins (Larsen et al. 2014; Weihrauch et al. 2009; Wilson et al. 1994; Wright and Wood 2009). Ammonia trapping across the apical membrane has been extensively studied in freshwater fish and likely occurs in the majority of freshwater organisms (Cruz et al. 2013; Larsen et al. 2014; Shih et al. 2008; Weihrauch et al. 2009; Wilkie 2002; Wilson et al. 1994; Wright and Wood 2009). Blocking the V-ATPase in both planarians and leeches with concanamycin C effectively reduced cutaneous ammonia excretion (Quijada-Rodriguez et al. 2015; Weihrauch et al. 2012). Further, exposure to an acidic environment was shown to promote ammonia excretion suggesting an ammonia trapping mechanism similar to that seen in freshwater fish, where protons are actively pumped into the unstirred boundary layer creating an acidified microenvironment to promote an outwardly directed ΔP_{NH3} (Quijada-Rodriguez et al. 2015; Weihrauch et al. 2012). However, unlike in planarians and fish, acidification of the unstirred boundary layer in leeches was proposed to occur within the crypt of mucous-secreting cells (Quijada-Rodriguez et al. 2015). In this study, the presence of a buffer in the environment did not affect the ammonia excretion capability of *N. obscura* demonstrating that this organism was likely not manipulating its unstirred boundary layer at the epidermal surface but possibly elsewhere such as within the mucous-secreting cells. While in *S. mediterranea*, it has been shown that the V-ATPase is localized in both mucous-secreting rhabdites and apical membrane of epithelial cells (◘ Fig. 5.5b); it remains to be shown whether the crypts of mucous-secreting cells in the leech are truly acidified and whether the transport proteins necessary for ammonia excretion by mucous-secreting cells are present.

Similar to the V-ATPase, Na^+/H^+ exchangers (NHEs) can also promote ammonia excretion by generation of ΔP_{NH3} across a membrane. This group of exchangers is typically thought to be electroneutral transporters driven by the Na^+/K^+-ATPase. In freshwater organisms, the efficacy of these electroneutral exchangers has come into question due to thermodynamic constraints posed by the presence of low Na^+ concentration in freshwater (Avella and Bornancin 1989). One explanation for this thermodynamic enigma is the Rh-NHE functional metabolon (Wright and Wood 2009; Wu et al. 2010). Here it is proposed that the unstirred boundary layer on the apical surface of the membrane is alkalinized by excretion of NH_3 through Rh proteins that bind free protons, thereby generating a proton gradient to drive Na^+ uptake and H^+ excretion by NHEs. Based on the inhibitory effects of amiloride, an inhibitor of NHEs (Kleyman and Cragoe 1988), it has been suggested that NHEs may play a role in ammonia excretion in *S. mediterranea* (Weihrauch et al. 2012). However, unlike in the planarian and some fish, ammonia excretion in the leech (*N. obscura*) is unaffected by EIPA (0.1 mmol l^{-1}), an amiloride analog that inhibits NHEs (Kleyman and Cragoe 1988; Quijada-Rodriguez et al. 2015). While the permeability of the inhibitor during whole animal exposures may be in question, this same inhibitor (0.1 mmol l^{-1}) has been shown to decrease sodium uptake in *N. obscura*, so one can assume that the drug is likely inhibiting NHEs but at the same time that NHEs are rather not involved in ammonia excretion (A.R. Quijada-Rodriguez personal communication).

5.3.2 Mechanism of Ammonia Excretion in Soil Nematodes

Similarities to Freshwater Leeches and Planarians

As highlighted above, soil nematodes share many similarities to freshwater organisms in terms of the ionoregulatory challenges faced in their environment. Therefore it is conceivable that soil nematodes have adapted similar mechanisms to freshwater organism for ammonia excretion as well. As in freshwater leeches, the Na^+/K^+-ATPase in *C. elegans* is capable of accepting ammonium as a substrate, and the nematode likely utilizes this pump for basolateral transport of ammonia (Adlimoghaddam et al. 2015). Further, following high environmental ammonia exposure (2 day 1 $mmol\,l^{-1}\,NH_4Cl$), both mRNA expression of the α-subunit of this pump and Na^+/K^+-ATPase activity level increased (Adlimoghaddam et al. 2015).

In addition to the Na^+/K^+-ATPase, Rh proteins may serve as a route for basolateral and apical ammonia transport as *C. elegans* expresses two Rh proteins, Rhr-1 and Rhr-2, both of which are expressed in the hypodermis (◙ Fig. 5.4) (Ji et al. 2006; Adlimoghaddam unpublished). Following high environmental ammonia exposure, increased mRNA expression of Rhr-1 and Rhr-2 are seen in conjunction with an increased ammonia excretion suggesting a role for both of these transporters in ammonia excretion (Adlimoghaddam et al. 2015). Further, studies employing yeast-based functional complementation assays demonstrated that like the Rh protein of *N. obscura*, Rhr-1 is capable of transporting ammonia when expressed in yeast (Adlimoghaddam et al. 2015). While not yet functionally characterized, Rhr-2 also likely transports ammonia as it contains a set of conserved amino acid residues essential for ammonia conductance, while also having a high sequence similarity to the ammonia-transporting Rhr-1 (Zidi-Yahiaoui et al. 2009). In *C. elegans*, basolateral localization of the Rhr-1 protein is predicted due to its high expression level in the hypodermis relative to Rhr-2 and its broad expression across various tissues (Adlimoghaddam et al. 2015; Ji et al. 2006). These characteristics of Rhr-1 resemble that previously shown for the basolaterally localized Rhbg/RhBG of vertebrates (Cruz et al. 2013; Handlogten et al. 2005; Nawata et al. 2007).

As seen in the freshwater leeches and planarians, ammonia is believed to be transported across the apical membrane of the hypodermis *via* ammonia trapping. Through studies on wild-type *C. elegans*, a dependence on ammonia trapping across the hypodermis was hypothesized, as exposure to an acidic environment (buffered pH 5) promoted an enhanced ammonia excretion (Adlimoghaddam et al. 2015). In Rhr-2 knockout studies, Rhr-2 mutants were unable to promote an enhanced ammonia excretion following exposure to a low pH environment (buffered pH 5) as was seen in wild-type *C. elegans*, suggesting that the Rhr-2 protein is essential for ammonia trapping across the hypodermis and thus likely apically localized (Adlimoghaddam et al. 2016). The V-ATPase is also believed to be coupled with the Rhr-2 protein to promote ammonia trapping, as ammonia excretion is decreased by approximately 28 % following inhibition of this pump with concanamycin C, an inhibitor of the V-ATPase (Adlimoghaddam et al. 2015). Additionally, mRNA expression of the V-ATPase increases following exposure to 1 $mmol\,l^{-1}\,NH_4Cl$ after 2 days.

Another contributor to apical acidification for ammonia trapping could be the NHXs (Na^+/H^+ exchangers, cation proton antiporter 1 subfamily) and NHAs (Na^+/H^+-antiporter, cation proton antiporter 2 subfamily) of which *C. elegans* expresses 9 different NHXs (Nehrke and Melvin 2002) and 3 different NHAs (GB accession nos.: NP_509724, NP_509723, NP_507130). The hypodermis is known to express 3 NHXs, the NHXs 1, 3, and 4, of which NHX3 is localized in intracellular membranes and NHX4 to the basolateral membrane. Unlike the NHXs, tissue expression of the NHAs in *C. elegans* remains unknown. Further studies are necessary to determine the potential role of the NHXs and extremely understudied NHAs in the ammonia excretion mechanism of *C. elegans*.

Vesicular Transport of Ammonia

As an alternative mechanism to ammonia trapping across the apical membrane of the hypodermis, it is believed that vesicular transport of ammonia also occurs in *C. elegans* (Adlimoghaddam et al. 2015). Having first been suggested as mechanism of ammonia excretion in the gills of the green shore crab (*Carcinus maenas*) (▶ see Chap. 1, Weihrauch et al. 2002) and subsequently in the Chinese mitten crab (*Eriocheir sinensis*) (▶ Chap. 1) as well as in the midgut of the tobacco hornworm (*Manduca sexta*) (Weihrauch 2006), some ammonia could potentially also in *C. elegans* be excreted by exocytosis. Here the V-ATPase is proposed to generate a ΔP_{NH3} driving ammonia into acidified vesicles, which would act as an acidic microenvironment for ammonia trapping that would facilitate ammonia excretion independently of environmental conditions. In *C. elegans*, inhibition of the microtubule network by colchicine decreased ammonia excretion by 26 % implicating that at least a portion of the animals ammonia load is excreted *via* a vesicular transport of ammonia and subsequent exocytosis (Adlimoghaddam et al. 2015). In addition to vesicular acidification by the V-ATPase, the hypodermal expressed NHX3 which is localized to intracellular membranes (Nehrke and Melvin 2002) and may also facilitate ammonia trapping in vesicles. Ammonia trapping by the NHE orthologue NHX3 may either occur through diffusion or if co-localized with one of the Rh proteins may form a functional Rh-NHE metabolon as proposed by Wright and Wood for fish gills (Wright and Wood 2009).

Role of AMTs

Recent evidence has shown that ammonium transporters (AMTs) found in plants and bacteria to transport NH_4^+ and NH_3 (Gu et al. 2013; Khademi et al. 2004; Ludewig et al. 2002; Musa-aziz et al. 2009), respectively, are also found in the genomes and transcriptomes of invertebrates (Huang and Peng 2005). This discovery has opened the potential for another route of ammonia transport potentially independent of Rh proteins. Recently, expression of the mosquito (*Anopheles gambiae*) AMT in *Xenopus* oocytes demonstrated that the AMTs of invertebrates also transport ammonia and likely in the ionic form (Pitts et al. 2014). In *C. elegans*, 4 AMTs are expressed: AMT-1, AMT-2, AMT-3, and AMT-4; of these AMTs tissue expression is only known for AMT-3 which is found in the head nerves, tail nerves, ring nerves, and intestine (McKay et al. 2003). There remains the potential that the remaining 3 AMTs could be expressed in the hypodermis and promote either basolateral or apical transport of ammonia; however, tissue distribution and cellular localization of these AMTs remain to be performed.

5.4   Conclusion

Since the mid-1900s, some studies have investigated nitrogenous waste excretion in nematodes, platyhelminthes, and annelids. We now know that the epidermis, hypodermis, intestine, nephridial system, and nematode-specific excretory system are all potential sites for the excretion of nitrogenous waste products, which are predominantly either ammonia or urea. Based on the few comprehensive mechanistic studies in *Caenorhabditis elegans*, *Schmidtea mediterranea*, and *Nephelopsis obscura* preliminary models for the excretion of ammonia in invertebrates other than the decapod crustaceans and cephalopods are now being established (❑ Fig. 5.7). From these models, it is evident that the V-ATPase, Na⁺/K⁺-ATPase, and Rh proteins are crucial for ammonia excretion whether

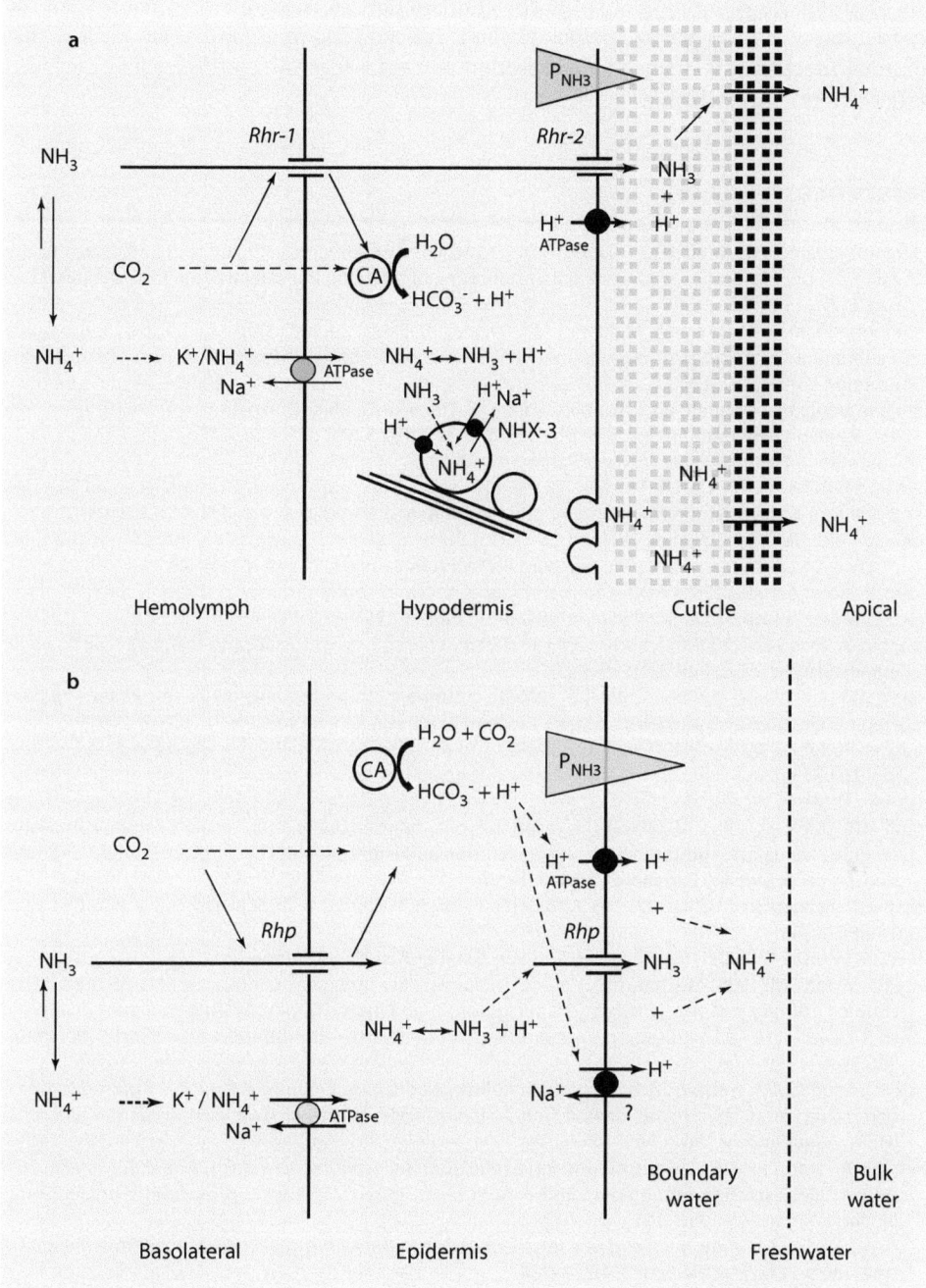

◼ Fig. 5.7 Hypothetical models for ammonia excretion in soil nematodes, freshwater planarians, and freshwater leeches. (**a**) Hypodermal ammonia excretion mechanism in the soil nematode (*Caenorhabditis elegans*). (**b**) Cutaneous ammonia excretion mechanism in the freshwater planarian (*Schmidtea mediterranea*) and freshwater leech (*Nephelopsis obscura*). The ? indicates potential involvement of NHEs in *S. mediterranea*. It is important to note that NHEs are suggested to not play a role in the *N. obscura* ammonia excretion model but present in the skin. Currently the role of each cell type, epithelial cell or mucous-secreting cell, in transepithelial ammonia excretion is unknown (The transport mechanisms are adapted from (Adlimoghaddam et al. 2015; Quijada-Rodriguez et al. 2015; Weihrauch et al. 2012) and described in the text above)

it is by ammonia trapping across the apical membrane or in acidified vesicles. With the broadening of studies across various phyla, it is becoming more and more evident that common mechanisms of ammonia excretion currently seen in vertebrates likely evolved early on in the invertebrates.

References

Adlimoghaddam A, Weihrauch D, O'Donnell MJ (2014) Localization of K^+, H^+, Na^+ and Ca^{2+} fluxes to the excretory pore in *Caenorhabditis elegans*: application of scanning ion-selective microelectrodes. J Exp Biol 217:4119–4122

Adlimoghaddam A, Boeckstaens M, Marini AM, Treberg JR, Brassinga AKC, Weihrauch D (2015) Ammonia excretion in *Caenorhabditis elegans*: mechanism and evidence of ammonia transport of the Rhesus protein CeRhr-1. J Exp Biol 218:675–683

Adlimoghaddam A, Donnell MJO, Kormish J, Banh S, Treberg JR, Merz D, Weihrauch D (2016) Ammonia excretion in *Caenorhabditis elegans*: physiological and molecular characterization of the rhr-2 knock-out mutant. Comp Biochem Physiol Part A 195:46–54

Ahmed ST, Rahemo ZIF (2013) Studies on the histology of the body wall of two species of leeches, *Erpobdella octoculata* and *Haemopis sanguisuga* (Annelida: Hirudinea). Adv J Biol Sci Res 1:1–7

Anderson WG, McCabe C, Brandt C, Wood CM (2015) Examining urea flux across the intestine of the spiny dogfish, *Squalus acanthias*. Comp Biochem Physiol A Mol Integr Physiol 181:71–78

Avella M, Bornancin M (1989) A new analysis of ammonia and sodium transport through the gills of the freshwater rainbow trout (*Salmo gairdneri*). J Exp Biol 142:155–175

Bianchini A, Wood CM (2008) Sodium uptake in different life stages of crustaceans: the water flea *Daphnia magna* Strauss. J Exp Biol 211:539–547

Boisen AM, Amstrup J, Novak I, Grosell M (2003) Sodium and chloride transport in soft water and hard water acclimated zebrafish (*Danio rerio*). Biochim Biophys Acta – Biomembr 1618:207–218

Boucher-Rodoni R, Mangold K (1989) Respiration and nitrogen excretion by the squid *Loligo forbesi*. Mar Biol 103:333–338

Braissant O, McLin VA, Cudalbu C (2013) Ammonia toxicity to the brain. J Inherit Metab Dis 36:595–612

Braun MH, Steele SL, Perry SF (2009) The responses of zebrafish (*Danio rerio*) to high external ammonia and urea transporter inhibition: nitrogen excretion and expression of rhesus glycoproteins and urea transporter proteins. J Exp Biol 212:3846–3856

Britto DT, Kronzucker HJ (2002) NH_4^+ toxicity in higher plants: a critical review. J Plant Physiol 159:567–584

Brusca RC, Brusca G (2003) Invertebrates, 2nd edn. Sinauer Associates, Sunderland, MA

Bury NR, Wood CM (1999) Mechanism of branchial apical silver uptake by rainbow trout is via the proton-coupled Na^+ channel. Am J Physiol – Regul Integr Comp Physiol 277:R1385–R1391

Cohen S, Lewis H (1949) The nitrogenous metabolism of the earthworm (*Lumbricus terrestris*). J Biol Chem 180:79–91

Cruz MJ, Sourial MM, Treberg JR, Fehsenfeld S, Adlimoghaddam A, Weihrauch D (2013) Cutaneous nitrogen excretion in the African clawed frog *Xenopus laevis*: effects of high environmental ammonia (HEA). Aquat Toxicol 136–137:1–12

Diets TH, Alvarado RH (1970) Osmotic and ionic regulation in *Lumbricus terrestris* L. Biol Bull 138:247–261

Duerr FG (1968) Excretion of ammonia and urea in seven species of marine prosobranch snails. Comp Biochem Physiol 26:1051–1059

Dy DT, Yap HT (2000) Ammonium and phosphate excretion in three common echinoderms from Philippine coral reefs. J Exp Mar Bio Ecol 251:227–238

Dymowska AK, Schultz AG, Blair SD, Chamot D, Goss GG (2014) Acid-sensing ion channels are involved in epithelial Na^+ uptake in the rainbow trout *Oncorhynchus mykiss*. Am J Physiol Cell Physiol 307:C255–C265

Edwards SL, Tse CM, Toop T (1999) Immunolocalisation of NHE3-like immunoreactivity in the gills of the rainbow trout (*Oncorhynchus mykiss*) and the blue-throated wrasse (*Pseudolabrus tetrious*). J Anat 195:465–469

Endeward V, Cartron JP, Ripoche P, Gros G (2007) RhAG protein of the Rhesus complex is a CO_2 channel in the human red cell membrane. FASEB J 22:64–73

Glover CN, Bucking C, Wood CM (2013) The skin of fish as a transport epithelium: a review. J Comp Physiol B 183:877–891

Gruswitz F, Chaudhary S, Ho JD, Schlessinger A, Pezeshki B, Ho C, Sali A, Westhoff CM, Stroud RM (2010) Function of human Rh based on structure of RhCG at 2.1 Å. Proc Natl Acad Sci U S A 107(21):9638–9643

Gu R, Duan F, An X, Zhang F, von Wirén N, Yuan L (2013) Characterization of AMT-mediated high-affinity ammonium uptake in roots of maize (Zea mays L.). Plant Cell Physiol 54:1515–1524

Handlogten ME, Hong S, Zhang L, Vander AW, Steinbaum ML, Campbell-thompson M, Weiner ID, Mary E, Allen W (2005) Expression of the ammonia transporter proteins Rh B glycoprotein and Rh C glycoprotein in the intestinal tract. Am J Physiol Gastrointest Liver Physiol 288(5):1036–1047

Hirata T, Kaneko T, Ono T, Nakazato T, Furukawa N, Hasegawa S, Wakabayashi S, Shigekawa M, Chang MH, Romero MF et al (2003) Mechanism of acid adaptation of a fish living in a pH 3.5 lake. Am J Physiol – Regul Integr Comp Physiol 284:R1199–R1212

Hiroi J, Yasumasu S, McCormick SD, Hwang PP, Kaneko T (2008) Evidence for an apical Na-Cl cotransporter involved in ion uptake in a teleost fish. J Exp Biol 211:2584–2599

Hoeger U, Mommsen TP, O'Dors R, Webber D (1987) Oxygen uptake and nitrogen excretion in two cephalopods, octopus and squid. Comp Biochem Physiol 87A:63–67

Horng JL, Lin LY, Huang CJ, Katoh F, Kaneko T, Hwang PP (2007) Knockdown of V-ATPase subunit A (atp6v1a) impairs acid secretion and ion balance in zebrafish (Danio rerio). Am J Physiol Regul Integr Comp Physiol 292:R2068–R2076

Horng JL, Hwang PP, Shih TH, Wen ZH, Lin CS, Lin LY (2009) Chloride transport in mitochondrion-rich cells of euryhaline tilapia (Oreochromis mossambicus) larvae. Am J Physiol Cell Physiol 297:C845–C854

Huang C, Peng J (2005) Evolutionary conservation and diversification of Rh family genes and proteins. Proc Natl Acad Sci 102:15512–15517

Hwang PP (2009) Ion uptake and acid secretion in zebrafish (Danio rerio). J Exp Biol 212:1745–1752

Hwang PP, Lee TH, Lin LY (2011) Ion regulation in fish gills: recent progress in the cellular and molecular mechanisms. Am J Physiol Regul Integr Comp Physiol 301:R28–R47

Inokuchi M, Hiroi J, Watanabe S, Lee KM, Kaneko T (2008) Gene expression and morphological localization of NHE3, NCC and NKCC1a in branchial mitochondria-rich cells of Mozambique tilapia (Oreochromis mossambicus) acclimated to a wide range of salinities. Comp Biochem Physiol A Mol Integr Physiol 151:151–158

Inokuchi M, Hiroi J, Watanabe S, Hwang PP, Kaneko T (2009) Morphological and functional classification of ion-absorbing mitochondria-rich cells in the gills of Mozambique tilapia. J Exp Biol 212:1003–1010

Ji Q, Hashmi S, Liu Z, Zhang J, Chen Y, Huang CH (2006) CeRh1 (rhr-1) is a dominant Rhesus gene essential for embryonic development and hypodermal function in Caenorhabditis elegans. Proc Natl Acad Sci U S A 103:5881–5886

Kerstetter TH, Kirschner LB, Rafuse DD (1970) On the mechanisms of sodium ion transport by the irrigated gills of rainbow trout (Salmo gairdneri). J Gen Physiol 56:342–359

Khademi S, O'Connell J III, Remis J, Robles-Colmenares Y, Miercke LJW, Stroud RM (2004) Mechanism of ammonia transport by Amt/MEP/Rh: structure of AmtB at 1.35 A. Science 305:1587–1594

Kirschner LB (2004) The mechanism of sodium chloride uptake in hyperregulating aquatic animals. J Exp Biol 207:1439–1452

Kleyman TR, Cragoe EJ (1988) Amiloride and its analogs as tools in the study of ion transport. J Membr Biol 105:1–21

Knepper MA, Packer R, Good DW (1989) Ammonium transport in the kidney. Physiol Rev 69:179–249

Krogh A (1938) The active absorption of ions in some freshwater animals. Zeitschrift Vergleichende Physiol 25:335–350

Kulkarni G, Kulkarni V, Rao AB (1989) Nephridial excretion of ammonia and urea in the freshwater leech, Poecilobdella viridis as a function of the temperature and photoperiod. Proc Indian Natn Acad B55:345–352

Kumai Y, Perry SF (2011) Ammonia excretion via Rhcg1 facilitates Na+ uptake in larval zebrafish, Danio rerio, in acidic water. Am J Physiol Regul Integr Comp Physiol 301:R1517–R1528

Kumai Y, Perry SF (2012) Mechanisms and regulation of Na(+) uptake by freshwater fish. Respir Physiol Neurobiol 184:249–256

Larsen EH, Deaton LE, Onken H, O'Donnell M, Grosell M, Dantzler WH, Weihrauch D (2014) Osmoregulation and excretion. Compr Physiol 4:405–573

Ludewig U, von Wirén N, Frommer WB (2002) Uniport of NH by the root hair plasma membrane ammonium transporter LeAMT1;1. J Biol Chem 277:13548–13555

Lutz P, Siddiq AH (1971) Nonprotein nitrogenous composition of the protonephridial fluid of the trematode *Fasciola gigantica*. Comp Biochem Physiol 40A:453–457

Mallery CH (1983) A carrier enzyme basis for ammonium excretion in teleost gill NH_4^+–stimulated Na^+ dependent ATPase activity in *Opsanus beta*. Comp Biochem Physiol 74:889–897

Marini A, Matassi G, Raynal V, André B, Cartron J, Chérif-zahar B (2000) The human Rhesus-associated RhAG protein and a kidney homologue promote ammonium transport in yeast. Nat Genet 26:341–344

Marshall WS (2002) $Na^{(+)}$, $Cl^{(-)}$, $Ca^{(?+)}$ and $Zn^{(?+)}$ transport by fish gills: retrospective review and prospective synthesis. J Exp Zool 293:264–283

Martin G (1978) Zoomorphologie a new function of rhabdites: mucus production for ciliary gliding. Zoomorphologie 91:235–248

Martin M, Fehsenfeld S, Sourial MM, Weihrauch D (2011) Effects of high environmental ammonia on branchial ammonia excretion rates and tissue Rh-protein mRNA expression levels in seawater acclimated Dungeness crab *Metacarcinus magister*. Comp Biochem Physiol A Mol Integr Physiol 160:267–277

Masui D, Furriel RP, McNamara J, Mantelatto FL, Leone F (2002) Modulation by ammonium ions of gill microsomal (Na^+, K^+)-ATPase in the swimming crab *Callinectes danae*: a possible mechanism for regulation of ammonia excretion. Comp Biochem Physiol Part C Toxicol Pharmacol 132:471–482

McKay SJ, Johnsen R, Khattra J, Asano J, Baillie DL, Chan S, Dube N, Fang L, Goszczynski B, Ha E et al (2003) Gene expression profiling of cells, tissues, and developmental stages of the nematode *C. elegans*. Cold Spring Harb Symp Quant Biol 68:159–169

Musa-aziz R, Chen L, Pelletier MF, Boron WF (2009) Relative CO_2/NH_3 selectivities of AQP1, AQP4, AQP5, AmtB, and RhAG. Proc Natl Acad Sci USA. 106:5406–5411

Nakada T, Westhoff CM, Kato A, Hirose S (2007) Ammonia secretion from fish gill depends on a set of Rh glycoproteins. FASEB J 21:1067–1074

Nawata CM, Hung CCY, Tsui TKN, Wilson JM, Wright PA, Wood CM (2007) Ammonia excretion in rainbow trout (*Oncorhynchus mykiss*): evidence for Rh glycoprotein and H^+–ATPase involvement. Physiol Genomics 31:463–474

Nehrke K, Melvin JE (2002) The NHX family of Na^+–H^+ exchangers in *Caenorhabditis elegans*. J Biol Chem 277:29036–29044

Nelson F, Riddle D (1984) Functional study of the *Caenorhabditis elegans* secretory-excretory system using laser microsurgery. J Exp Zool 231:45–56

Nelson FK, Albert PS, Riddle DL (1983) Fine structure of the *Caenorhabditis elegans* secretory—excretory system. J Ultrastruct Res 82:156–171

Nesdoly RG, Van Rees KCJ (1998) Redistribution of extractable nutrients following disc trenching on Luvisols and Brunisols in Saskatchewan. Can J Soil Sci 78(2):367–376

O'Donnell MJ (1997) Mechanisms of excretion and ion transport in invertebrates. In: Dantzler WH (ed) Comparative physiology. Oxford University Press, New York, pp 1207–1289

Onken H, Mcnamara JC (2002) Hyperosmoregulation in the red freshwater crab *Dilocarcinus pagei* (Brachyura, Trichodactylidae): structural and functional asymmetries of the posterior gills. J Exp Biol 205:167–175

Onken H, Putzenlechner MAX (1995) A V-ATPase drives active, electrogenic and Na^+–independent Cl^- absorption across the gills of *Eriocheir sinensis*. J Exp Biol 198:767–774

Onken H, Schöbel A, Kraft JAN, Putzenlechner MAX (2000) Active NaCl absorption across split lamellae of posterior gills of the Chinese crab *Eriocheir sinensis*: stimulation by eyestalk extract. J Exp Biol 203:1373–1381

Parks SK, Tresguerres M, Goss GG (2008) Theoretical considerations underlying $Na^{(+)}$ uptake mechanisms in freshwater fishes. Comp Biochem Physiol C Toxicol Pharmacol 148:411–418

Patrick ML, Aimanova K, Sanders HR, Gill SS (2006) P-type Na^+/K^+–ATPase and V-type H^+–ATPase expression patterns in the osmoregulatory organs of larval and adult mosquito *Aedes aegypti*. J Exp Biol 209:4638–4651

Peixoto CA, De Souza W (1995) Freeze-fracture and deep-etched view of the cuticle of *Caenorhabditis elegans*. Tissue Cell 27:561–568

Perry SF, Braun MH, Noland M, Dawdy J, Walsh PJ (2010) Do zebrafish Rh proteins act as dual ammonia-CO2 channels? J Exp Zool A Ecol Genet Physiol 313:618–621

Pitts RJ, Derryberry SL, Pulous FE, Zwiebel LJ (2014) Antennal-expressed ammonium transporters in the malaria vector mosquito *Anopheles gambiae*. PLoS One 9:e111858

Quijada-Rodriguez AR, Treberg JR, Weihrauch D (2015) Mechanism of ammonia excretion in the freshwater leech *Nephelopsis obscura*: characterization of a primitive Rh protein and effects of high environmental ammonia. Am J Physiol Regul Integr Comp Physiol 309(6):R692–R705. doi:10.1152/ajpregu.00482.2014

Reitze M, Schottler UDO (1989) The time dependence of adaption to reduced salinity in the lugworm *Arenicola marina* L. (Annelida: Polychaeta). Comp Biochem Physiol 93:549–559

Riegel BYJA (1968) Analysis of the distribution of sodium, potassium and osmotic pressure in the urine of crayfishes. J Exp Biol 48:587–596

Roots BI (1955) The water relations of earthworms II. Resistance to desiccation and immersion, and behaviour when submerged and when allowed choice of environment. J Exp Biol 33:29–44

Rothstein M (1963) Nematode biochemistry-III. Excretion products. Comp Biochem Physiol 9:51–59

Rubino JG, Zimmer AM, Wood CM (2014) Intestinal ammonia transport in freshwater and seawater acclimated rainbow trout (*Oncorhynchus mykiss*): evidence for a $Na^{(+)}$ coupled uptake mechanism. Comp Biochem Physiol A Mol Integr Physiol 183C:45–56

Sabourin TD, Stickle WB (1981) Effects of salinity on respiration and nitrogen excretion in two species of echinoderms. Mar Biol 65:91–99

Shih TH, Horng JL, Hwang PP, Lin LY (2008) Ammonia excretion by the skin of zebrafish (*Danio rerio*) larvae. Am J Physiol Cell Physiol 295:C1625–C1632

Shih TH, Horng JL, Liu ST, Hwang PP, Lin LY (2012) Rhcg1 and NHE3b are involved in ammonium-dependent sodium uptake by zebrafish larvae acclimated to low-sodium water. Am J Physiol Regul Integr Comp Physiol 302:R84–R93

Skou JC (1965) Enzymatic basis for active transport of Na^+ and K^+ across cell membrane. Physiol Rev 45:596–617

Soupene E, Inwood W, Kustu S (2004) Lack of the Rhesus protein Rh1 impairs growth of the green alga *Chlamydomonas reinhardtii* at high CO2. Proc Natl Acad Sci U S A 101:7787–7792

Tavakkoli E, Fatehi F, Coventry S, Rengasamy P, McDonald GK (2011) Additive effects of Na^+ and Cl^- ions on barley growth under salinity stress. J Exp Bot 62:2189–2203

Tillinghast EK (1967) Excretory pathways of ammonia and urea in the earthworm *Lumbricus terrestris* L. J Exp Zool 166:295–300

Tillinghast EK, McInnes D, Duffill R (1969) The effect of temperature and water availability on the output of ammonia and urea by the earthworm *Lumbricus terrestris* L. Comp Biochem Physiol 29:1087–1092

Tillinghast EK, O'Donnell R, Eves D, Calvert E, Taylor J (2001) Water-soluble luminal contents of the gut of the earthworm *Lumbricus terrestris* L. and their physiological significance. Comp Biochem Physiol Part A Mol Integr Physiol 129:345–353

Towle DW, Paulsen RS, Weihrauch D, Kordylewski M, Salvador C, Lignot J, Spanings-pierrot C, Island D, Cove S, College LF et al (2001) $Na^+ + K^+$ –ATPase in gills of the blue crab *Callinectes sapidus*: cDNA sequencing and salinity-related expression of α -subunit mRNA and protein. J Exp Biol 204:4005–4012

Tresguerres M, Katoh F, Orr E, Parks SK, Goss GG (2006) Chloride uptake and base secretion in freshwater fish: a transepithelial ion-transport metabolon? Physiol Biochem Zool 79:981–996

Tresguerres M, Parks SK, Sabatini SE, Goss GG, Luquet CM (2008) Regulation of ion transport by pH and [HCO3-] in isolated gills of the crab *Neohelice* (*Chasmagnathus*) *granulata*. Am J Physiol Regul Integr Comp Physiol 294:R1033–R1043

Tschoerner P, Zebe E (1989) Ammonia formation in the medicinal leech, *Hirudo medicinalis*-in vivo and in vitro investigations. Comp Biochem Physiol 94A:187–194

Tsui TKN, Hung CYC, Nawata CM, Wilson JM, Wright PA, Wood CM (2009) Ammonia transport in cultured gill epithelium of freshwater rainbow trout: the importance of Rhesus glycoproteins and the presence of an apical Na^+/NH_4^+ exchange complex. J Exp Biol 212:878–892

Wall SM, Koger LM (1994) NH_4^+ transport mediated by Na^+ –K^+ –ATPase in rat inner medullary collecting duct. Am J Physiol 267:F660–F670

Wang YF, Tseng YC, Yan JJ, Hiroi J, Hwang PP (2009) Role of SLC12A10.2, a Na-Cl cotransporter-like protein, in a Cl uptake mechanism in zebrafish (*Danio rerio*). Am J Physiol Regul Integr Comp Physiol 296:R1650–R1660

Weber WM, Blank U, Clauss W (1995) Regulation of electrogenic Na^+ transport across leech skin. Am J Physiol Regul Integr Comp Physiol 268:605–613

Webster LA, Wilson RA (1970) The chemical composition of protonephridial canal fluid from the cestode *Hymenolepis diminuta*. Comp Biochem Physiol 35:201–209

Weihrauch D (2006) Active ammonia absorption in the midgut of the tobacco hornworm *Manduca sexta* L.: transport studies and mRNA expression analysis of a Rhesus-like ammonia transporter. Insect Biochem Mol Biol 36:808–821

Weihrauch D, Ziegler A, Siebers D, Towle DW (2002) Active ammonia excretion across the gills of the green shore crab *Carcinus maenas*: participation of Na^+/K^+–ATPase, V-type H^+–ATPase and functional microtubules. J Exp Biol 205:2765–2775

Weihrauch D, McNamara JC, Towle DW, Onken H (2004a) Ion-motive ATPases and active, transbranchial NaCl uptake in the red freshwater crab, *Dilocarcinus pagei* (Decapoda, Trichodactylidae). J Exp Biol 207:4623–4631

Weihrauch D, Morris S, Towle DW (2004b) Ammonia excretion in aquatic and terrestrial crabs. J Exp Biol 207:4491–4504

Weihrauch D, Wilkie MP, Walsh PJ (2009) Ammonia and urea transporters in gills of fish and aquatic crustaceans. J Exp Biol 212:1716–1730

Weihrauch D, Chan AC, Meyer H, Döring C, Sourial M, O'Donnell MJ (2012) Ammonia excretion in the freshwater planarian *Schmidtea mediterranea*. J Exp Biol 215:3242–3253

Weiner ID, Verlander JW (2014) Ammonia transport in the kidney by Rhesus glycoproteins. Am J Physiol Renal Physiol 306:F1107–F1120

Wilkie MP (2002) Ammonia excretion and urea handling by fish gills: present understanding and future research challenges. J Exp Zool 293:284–301

Wilson RW, Wright PM, Munger S, Wood CM (1994) Ammonia excretion in freshwater rainbow trout (*Oncorhynchus mykiss*) and the importance of gill boundary layer acidification : Lack of evidence for Na^+/NH_4^+ exchange. J Exp Biol 191:37–58

Wilson JM, Laurent P, Tufts BL, Benos DJ, Donowitz M, Vogl AW, Randall DJ (2000) NaCl uptake by the branchial epithelium in freshwater teleost fish: an immunological approach to ion-transport protein localization. J Exp Biol 203:2279–2296

Wilson JM, Moreira-Silva J, Delgado ILS, Ebanks SC, Vijayan MM, Coimbra J, Grosell M (2013) Mechanisms of transepithelial ammonia excretion and luminal alkalinization in the gut of an intestinal air-breathing fish, *Misgurnus anguilliacaudatus*. J Exp Biol 216:623–632

Wood CM, Bucking C (2011) The role of feeding in salt and water balance. In: Grosell M, Farrell AP, Brauner CJ (eds) The multifunctional gut of fish. Academic, Amsterdam/Boston, pp 165–212

Wright PA (1995) Nitrogen excretion : three end products, many physiological roles. J Exp Biol 198:273–281

Wright PA, Wood CM (2009) A new paradigm for ammonia excretion in aquatic animals: role of Rhesus (Rh) glycoproteins. J Exp Biol 212:2303–2312

Wu SC, Horng JL, Liu ST, Hwang PP, Wen ZH, Lin CS, Lin LY (2010) Ammonium-dependent sodium uptake in mitochondrion-rich cells of medaka (*Oryzias latipes*) larvae. Am J Physiol Cell Physiol 298:C237–C250

Zerbst-Boroffka I, Bazin B, Wenning A (1997) Chloride secretion drives urine formation in leech nephridia. J Exp Biol 200:2217–2227

Zidi-Yahiaoui N, Callebaut I, Genetet S, Le Van Kim C, Cartron JP, Colin Y, Ripoche P, Mouro-Chanteloup I (2009) Functional analysis of human RhCG: comparison with *E. coli ammonium* transporter reveals similarities in the pore and differences in the vestibule. Am J Physiol Cell Physiol 297:C537–C547

Zimmer AM, Nawata CM, Wood CM (2010) Physiological and molecular analysis of the interactive effects of feeding and high environmental ammonia on branchial ammonia excretion and Na^+ uptake in freshwater rainbow trout. J Comp Physiol B 180:1191–1204

Acid–Base Regulation in Aquatic Decapod Crustaceans

Sandra Fehsenfeld and Dirk Weihrauch

© Springer International Publishing Switzerland 2017
D. Weihrauch, M. O'Donnell (eds.), *Acid-Base Balance and Nitrogen Excretion in Invertebrates*,
DOI 10.1007/978-3-319-39617-0_6

6.1   Summary

Aquatic decapod crustaceans live in a highly variable and constantly changing environment, permanently challenging their physiological homeostasis. One of the key processes considered ensuring physiological performance and function is the maintenance of acid–base balance. This chapter aims to provide a comprehensive overview of the challenges for aquatic decapod crustaceans' acid–base homeostasis, as well as the current knowledge regarding the respective mechanisms for acid–base regulation. Like many other marine organisms including fish and cephalopods, aquatic decapod crustaceans are capable of counteracting an acidosis or alkalosis of their bodily fluids mainly by modulating their hemolymph bicarbonate levels in order to buffer the pH. In addition, they adjust the excretion of acid and/or base equivalents, respectively. It is evident that ion transport mechanisms at the level of the gill epithelium contribute substantially to these acid–base regulatory processes, including the modulation of gene (mRNA) expression levels of distinct gill epithelial transcripts like carbonic anhydrase, Rhesus-like protein, Na^+/K^+-ATPase, V-(H^+)-ATPase and Cl^-/HCO_3^--exchanger. As a result of recently generated data mainly from gill perfusion experiments, a novel hypothetical working model for branchial acid–base regulation is put forward. It ties in general ion as well as ammonia regulatory mechanisms and accounts for the obvious linkage between these three processes.

6.2   The Importance of Acid–Base Homeostasis in Aquatic Decapod Crustaceans

Maintaining acid–base balance is fundamental for all living organisms, including decapod crustaceans (Henry and Wheatly 1992). Only slight disturbances in the concentration of acid–base equivalents resulting in shifts of pH in the intra- or extracellular fluids may impair properties of essential proteins and their regulation (i.e. enzymes, Somero 1986; respiratory proteins, Riggs 1988; Truchot 1975a) and ultimately lead to a disruption of basic physiological functions. Consequently, securing whole animal acid–base homeostasis not only includes the maintenance of a constant intra- and extracellular pH but also needs mechanisms in place for its re-adjustments after an acid–base disturbance. Factors to disrupt acid–base homeostasis in decapod crustaceans might include a variety of intrinsic as well extrinsic parameters like the internal acid load due to exercise (e.g. Booth et al. 1984; Rose et al. 1998), shifts in $CaCO_3$ handling during the moulting process (e.g. Mangum et al. 1985) and fluctuations of environmental parameters like salinity (e.g. Whiteley et al. 2001), temperature (e.g. McMahon et al. 1978; Whiteley and Taylor 1990), pO_2/pCO_2 (e.g. De Fur et al. 1983; Urbina et al. 2013) and ammonia (e.g. Cheng et al. 2013; Martin et al. 2011). The regulation of acid–base balance in aquatic decapod crustaceans therefore is a complex interaction of physiological and biochemical processes including respiratory gas exchange, ion regulation and overall adjustments of metabolism (Henry and Wheatly 1992).

Most studies on acid–base regulation in decapod crustaceans to date concentrate on whole animal extracellular acid–base status and its responses upon disturbances (▶ see Sect. 6.3). Few studies investigated acid–base regulation on a cellular level, and therefore, little is known about direct trans-branchial transport of acid–base equivalents in decapod crustaceans (▶ see Sect. 6.4). While most available acid–base related data has been collected on brachyuran hyper-regulating crabs (e.g. the green crab *Carcinus maenas*, the

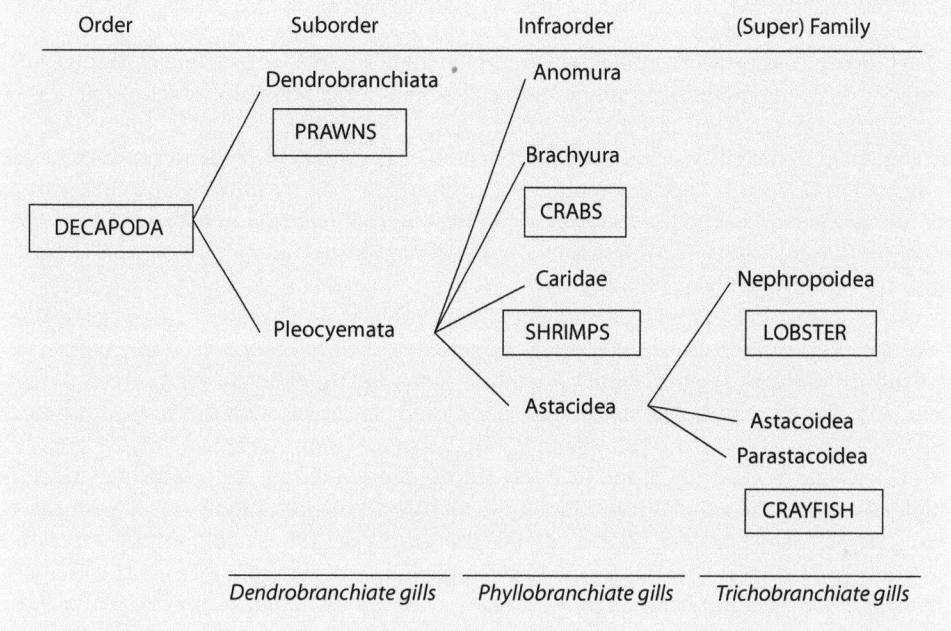

◘ Fig. 6.1 Overview of the different subgroups of decapod crustaceans discussed in this chapter. The nomenclature follows De Grave et al. (2009). To avoid confusion, members of the suborder Dendrobranchiata are referred to as "prawns" throughout the text, while "shrimps" solely refers to the infraorder Caridae

blue crab *Callinectes sapidus* and the Chinese mitten crab *Eriocheir sinensis*), this chapter aims to provide a broader review of the currently available literature on acid–base regulation also in the other major members of decapod crustaceans, namely, prawns (in this chapter referred to as synonym for penaeoid and sergestoid shrimp), Anomura, shrimp (caridean shrimp), lobster and crayfish (De Grave et al. 2009; ◘ Fig. 6.1).

6.2.1 Tissues Involved in Acid–Base Regulation

In crustaceans, anisosmotic extracellular regulation (AER), or the osmotic and ionic buffering of the extracellular fluid in order to maintain (acid–base) homeostasis, is believed to be mainly driven by the gills, antennal glands and gut (McNamara and Faria 2012). A tissue-specific inventory of epithelial membrane transporters then translates the changes of extracellular adjustments into the cell to ensure the intracellular maintenance of acid–base balance (Freire et al. 2008).

▪▪ Gills
Similar to fish and cephalopods (▶ see Chap. 11), the majority of the acid–base-relevant ion regulatory apparatus of decapod crustaceans is located in their gill epithelia (Henry et al. 2012; Larsen et al. 2014, and references therein). Not only are the gills involved in respiratory and acid–base physiology, but they are the major organs also for ion regulation

and ammonia excretion, therefore linking all of these regulatory processes (Freire et al. 2008; Henry et al. 2012).

All decapod crustaceans possess paired gills that are covered by a fine chitinous cuticle, lined by a single-layered epithelium and attached to a basal lamina (Freire et al. 2008). Depending on taxa, the number of paired gills, their location of attachment and the arrangement of the gill lamellae (phyllobranchiate, trichobranchiate, dendrobranchiate, as indicated in ◘ Fig. 6.1) vary substantially, providing more or less gill surface amplification for ion and gas exchange processes between the external (environment) and the internal medium (hemolymph). For further details, the reader is referred to the extensive descriptions by Taylor and Taylor (1992).

According to their different life strategies (i.e. primary habitat/habitat changes), gill epithelia of decapod crustaceans exhibit specific characteristics that can vary even within the respective sub-/infraorder or (super) family. Acid–base status and regulation in decapod crustaceans have been shown to be linked to external salinity and NaCl regulation (i.e. in the freshwater crayfish *Astacus leptodactylus* (Ehrenfeld 1974), *C. sapidus* (Henry and Cameron 1982) and *E. sinensis* (Whiteley et al. 2001), and therefore the tightness of the gill epithelium consequently might also affect the animals' capability for acid–base regulation. While the gill epithelia of strong hyper-regulators like *E. sinensis* (Weihrauch et al. 1999) and freshwater crayfish (Wheatly and Gannon 1995) represent a tight epithelium (conductance for ions <5 mS cm^{-2}), the epithelium of weak hyper-regulators like *C. maenas* (Weihrauch et al. 1999) and the American lobster *Homarus americanus* (Lucu and Towle 2010) is much leakier and allows for increased intercellular transport of ions (conductance $40–60$ mS cm^{-2}). Gills of osmoconforming crustaceans like *M. magister* (Hunter and Rudy 1975) or *Cancer pagurus* (Weihrauch et al. 1999) in contrast are highly permeable for ions (conductance > 200 mS cm^{-2}), and therefore these species are very limited in their capability to osmoregulate (Larsen et al. 2014).

Furthermore, specializations of gill epithelia can be seen at the ultrastructural level. Of the five different cell types found in decapod crustacean gill epithelia (thin cells, thick cells, flange cells, attenuated cells and pilaster cells; Freire et al. 2008), thin cells are generally believed to be associated with respiratory epithelia due to their low height ($1–5$ µM), a lack of extensive membrane folding and low number of mitochondria. Consequently, they have been considered to play an increased role in acid–base regulation rather than osmoregulation (Freire et al. 2008). Thin cells are found in all gills of osmoconforming crabs as well as the most anterior four to six pairs of gills of hyper-regulating crabs like *C. maenas* (Compere et al. 1989), *C. sapidus* (Copeland and Fitzjarrell 1968) and *E. sinensis* (Barra et al. 1983). Some thin cells were also observed in the gill epithelium of lobsters (Haond et al. 1998). In some hyper-regulating crabs like *C. maenas* (Compere et al. 1989) and *C. sapidus* (Copeland and Fitzjarrell 1968), thin cells are found to surround thick cells (also called ionocytes due to their supposed major role in ion transport) in the most posterior (osmoregulatory) pairs of gills, therefore indicating that acid–base regulatory properties might not be solely associated with the anterior gills in these species. To date, however, the direct site for acid–base regulation in euryhaline Brachyuran gills has not been confirmed, while osmoregulation has been demonstrated to be associated predominantly with the posterior gills (Henry et al. 2012) and ammonia with both anterior and posterior gills (Weihrauch et al. 1999; ▶ see Chap. 1).

The gill epithelia of lobsters (Haond et al. 1998), prawns (Bouaricha et al. 1994), shrimp (Freire and McNamara 1995) and freshwater crayfish (Morse et al. 1970) on the other hand are more homogeneous and possess so-called flange cells that exhibit features of both thin and thick cells of crabs and are therefore believed to incorporate both respiratory/acid–base and ion regulatory functions (Freire et al. 2008).

Even though most pilaster cells (resembling thin or thick cell criteria depending on their localization in the gill, Compere et al. 1989) in the epithelia of crabs and crayfish seem to inherit a mainly stabilizing function for the intra-lamellar septum, they are the exclusive sites for the vacuolar-type H^+-ATPase (V-(H^+)-ATPase) in *E. sinensis*, indicating an additional role for these cells in acid–base regulation in this species (Freire et al. 2008).

▪▪ Antennal Glands

Situated at the anterior end of the body at the base of the eyestalks, the paired antennal glands are mainly involved in the production (ultrafiltration) and ionic regulation of the urine to maintain extracellular water balance (Larsen et al. 2014). Therefore, they can be regarded as analogues of the nephron of the vertebrate kidney, the major acid–base regulatory organ in mammals and other terrestrial vertebrates (Weiner and Verlander 2013). Even though urine [Na^+], [K^+] and [Cl^-] are adjusted upon disturbance of (acid–base) homeostasis, antennal glands are believed to rather contribute to the regulation of divalent cations like Ca^{2+} and Mg^{2+} based on the respective clearance ratios (Wheatly 1985).

Only a few studies have investigated the role of antennal glands in acid–base regulation in decapod crustaceans. In freshwater-acclimated blue crabs *C. sapidus*, net urinary acid–base and ammonia efflux were negligible and did not change significantly when animals were exposed to hypercapnia (2 % pCO_2; Cameron and Batterton 1978a). In the Dungeness crab *Metacarcinus magister* acclimated to dilute salinity (~20 ppt), the increase in antennal gland-mediated HCO_3^- efflux resulted in an increased alkalinization of the urine, but was accompanied by an increase in HCO_3^- reabsorption over time, likely to assist in HCO_3^- accumulation in the hemolymph (Wheatly 1985). However, in respect to the overall base output in response to dilute salinity acclimation, the antennal gland of *M. magister* contributed to only 10 % at best (Wheatly 1985). Also in the freshwater-acclimated euryhaline crayfish *Pacifastacus leniusculus* exposed to hyperoxia, an initial extracellular acidosis resulted in an increase in HCO_3^- reabsorption from the urine to buffer hemolymph pH, but in parallel, an acidification of the urine was observed mainly due to increased ammonia (NH_4^+) excretion (Wheatly and Toop 1989). Similar to the observations of hyposaline-acclimated *M. magister*, however, net H^+ efflux accounted for only 10 % of the whole animal response in this crayfish. Interestingly, antennal glands of *P. leniusculus* show a significantly higher activity of carbonic anhydrase (CA), the enzyme involved in the hydration of CO_2 to form H_2CO_3 and subsequently dissociate to H^+ and HCO_3^-, compared to the gills (Wheatly and Henry 1987). When acclimated to hypersaline conditions, however, CA activity was progressively reduced with increased salinity (up to 80 % at ~25 ppt).

In conclusion, the existing data suggest an overall negligible involvement of antennal glands in acid–base regulation in decapod crustaceans.

▪▪ Gut and Gut Diverticula

Besides the respective adjustment of urine flow, gut-mediated fluid absorption and secretion of digestive fluid have been shown to be ion dependent in both hypo- and hyper-regulating crustaceans and likely help in the regulation of the hemolymph composition (Mantel and Farmer 1983). Accordingly, crustacean gut epithelia have been shown to possess Na^+/K^+-ATPase that in addition to their function in ion regulation might also promote the uptake of nutrients (Chu 1987; Chung and Lin 2006; Mantel and Farmer 1983). Even though being directly exposed to the environment and showing evidence for the capability to take up/excrete small ions like Na^+, Cl^-, K^+, Ca^{2+} and SO_4^- (Ahearn 1978;

Mantel and Farmer 1983), a potential role of the gut in acid–base regulation has not been investigated to date.

In addition to the gut, the presence of an electrogenic, likely apically situated $2Na^+:1H^+$-exchanger in the hepatopancreas of lobster and freshwater prawns (Ahearn et al. 1990, 1994) would provide an important key player for acid–base regulation in this tissue. Similar as for the gut, however, a direct involvement for the hepatopancreas in maintaining acid–base homeostasis has not been investigated to date. Clearly, future studies need to be performed in order to characterize the potential roles of gut and hepatopancreas in crustacean acid–base regulation.

6.2.2 Preadaptation Through Life Strategies?

The estimated 14,000 species of aquatic decapod crustaceans can be found in nearly all water bodies of the world. As described earlier, the capability of inhabiting specific water bodies mainly depends on the crustaceans' regulatory capacity due to the characteristics of the gill epithelium as well as the general permeability of their carapace. Acid–base regulatory capabilities are therefore likely correlated with life history, genetic predisposition and physiological plasticity (Melzner et al. 2009). As metazoans with a relatively high metabolic rate and level of activity, a high capacity to adjust body fluid pH (also dependent on the relatively large fluid volume) and relatively little expressed calcified structures compared to other marine calcifiers like corals, echinoderms and molluscs (decapod), crustaceans are believed to cope better with changes in their environment than other marine invertebrates (Wittmann and Pörtner 2013).

The following sections are intended to roughly characterize the basic life strategies for key species of the major decapod crustacean groups that will be discussed in more detail concerning their acid–base regulatory capabilities in the subsequent chapters.

▪▪ Prawns

While most Dendrobranchiata (Penaeidae) are euryhaline hyper-/hypo-osmoregulators and found in the marine environment (e.g. *(Marsu)Penaeus japonicus* (Cheng et al. 2013); *(Lito)Penaeus vannamei* (Liu et al. 2015)), some species of Sergestidae are found in freshwater. Many species inhabit deep (offshore) waters, while most Penaeidae are mainly found in shallow inshore tropical and subtropical waters and estuaries. Some prawn species are known to bury in mud substrates during the daytime (Tavares and Martin 2010), challenging their acid–base homeostasis as described below.

▪▪ Anomura

While king crabs (Lithodidea) like the southern king crab *Lithodes santolla* are subtidal species that can be found between 5 and 700 m depth in temperate waters (Urbina et al. 2013), hermit crabs (Paguroidea) like *Pagurus bernhardus* (De la Haye et al. 2011) and Porcelain crabs (Galatheoidea) like *Petrolisthes cinctipes* (Carter et al. 2013) are commonly found in the intertidal zone, potentially being trapped in rocky tide pools experiencing large spatial and temporal changes in abiotic parameters as discussed below.

▪▪ Crabs

With over 6700 species in 93 families, brachyuran crabs constitute to *ca.* 50 % of all decapod crustaceans (Ng et al. 2008). Accordingly, all imaginable life strategies and habitat uses are exhibited by this infraorder, including terrestrial species. Some of the most thoroughly investigated crabs are the osmoconforming Dungeness crab *M. magister*, the weak hyper-osmoregulating green crab *C. maenas* and the closely related blue crab *C. sapidus*, as well as the strong hyper-osmoregulating Chinese mitten crab *E. sinensis*. Like prawns and Anomura, they are oftentimes trapped in tide pools (Truchot and Duhamel-Jouve 1980) and some species bury in the sediment (Bellwood 2002).

▪▪ Shrimp

Besides brachyuran crabs (palaemonid), shrimps are the most diverse of the decapod groups with a great inter- and intraspecific variability in osmoregulatory capabilities. While most species can be found in freshwater and are strong hyper-regulators (i.e. genus *Macrobrachium*), some species inhabit estuarine (brackish) and even marine waters (i.e. genus *Palaemon*, *Palaemonetes*) and are hyper-/hypo-osmoregulators (Freire et al. 2003; McNamara and Faria 2012). Some shrimp species are associated with the intertidal zones and therefore more shallow waters (i.e. *Crangon crangon*, Almut and Bamber 2013) or are amphidromous and occupy those habitats during their early life stages (e.g. *Pandalus borealis*, Hammer and Pedersen 2013; *Macrobrachium olfersii*, Freire and McNamara 1995; McNamara and Lima 1997). Other shrimps are deep-water dwellers (e.g. *P. borealis*, Hammer and Pedersen 2013).

▪▪ Lobster

For the longest time, lobsters, especially the commercially important lobsters of the genus *Homarus* and the Norway lobster *Nephros norvegicus*, have traditionally been considered to be osmoconforming, stenohaline (salinity > 25 ppt) and limited to coastal and offshore habitats down to 700 m depth (Chapman 1980; Cooper and Uzmann 1980; Dall 1970). More recent research, however, identified them to be present also in estuarine and intertidal regions where they experience short-term fluctuation in salinity and other abiotic parameters (Charmantier et al. 2001). Additionally, the reproduction of lobsters seems to be dependent on higher water temperatures that lead to the animals' migration into different habitats and consequently, their exposure to different environmental conditions that potentially challenge acid–base homeostasis (Charmantier et al. 2001).

▪▪ Crayfish

Crayfish belong to the only decapod superfamily that is almost entirely found in freshwater (Reynolds et al. 2013). However, many crayfish species depend on the connection to the ocean in order to breed and therefore have a limited capability to osmo- and ion-regulate (Pequeux 1995). A remarkable number of crayfish build complex burrows in which they spend most of their life (Crandall and Buhay 2008; Guiasu 2002). Others are defined as stream or lake/pond/large river dwellers or are obligated cave dwellers (Crandall and Buhay 2008). Due to the very different chemistry of freshwater (low total osmolarity of ~4 mOsm/mainly $CaCO_3$ vs. 1000 mOsm/mainly NaCl in marine environments), ion regulation in crayfish is challenged and they maintain a lower, yet still hyper-osmotic extracellular ionic concentration and a lower carapace permeability for ions and water compared to marine decapod crustaceans (Wheatly and Gannon 1995).

6.2.3 Challenging Acid–Base Homeostasis

There are several factors that can challenge the acid–base regulatory machinery of animals. As (opportunistic) omnivores, all decapod crustaceans experience regular internal acid loads due to the *catabolism of proteins* and the resulting build-up of extracellular ammonia (mostly NH_4^+ at physiological pH, Weiner and Verlander 2013) that can affect extra- as well as intracellular acid–base homeostasis (Larsen et al. 2014). The response of decapod crustaceans upon *exercise (functional/internal hypoxia)* on the other hand results in a build-up of lactate and CO_2 in the hemolymph, therefore delivering H^+ and challenging extracellular acid–base regulation (Henry et al. 1994; see below).

Besides these intrinsic sources of acid–base disturbances, many of the mainly benthic aquatic decapods experience regular fluctuations of the abiotic parameters pH, pCO_2/pO_2, salinity and temperature in their surrounding environment. In intertidal zones, estuaries and water bodies like the Baltic Sea with restricted connection to the well-buffered open ocean, naturally recurrent elevated pCO_2 (*hypercapnia*, ~234 Pa vs. normal levels of 39 Pa) and changes in *pH* (7.5–8.2) and *temperature* (3.3–18.7 °C), as well as *salinity* (14.5–21.5 ppt), challenge acid–base homeostasis on a regular basis. Additionally, decapods in these shallow water environments are prone to be trapped in tide pools, where they experience even more drastic changes in all abiotic parameters (Truchot 1988), including extremely low levels of oxygen (*hypoxia*; Truchot and Duhamel-Jouve 1980). In extreme cases, these tide pools fall dry so that decapods are *exposed to air*. As mentioned above, many decapod crustacean species hide in burrows and caves or bury in the sediment to avoid predators (Larsen et al. 2014). With only limited water circulation around them, metabolic ammonia builds up around the animals, consequently exposing them to *high environmental ammonia* (HEA, Weihrauch et al. 2004a, b), another challenging factor for acid–base homeostasis mainly for osmoconforming crabs like *M. magister* (Martin et al. 2011). Furthermore, pH has been shown to decrease to as low as 6.5 already within the first few centimetres of sand and mud substrates, accompanied by elevated CO_2 levels of up to 1,600 Pa (Widdicombe et al. 2011).

Besides these naturally occurring challenges for acid–base homeostasis, global climate change and its impacts on acid–base regulation of decapod crustaceans and other invertebrates have become of greatest concern (Whiteley 2011). On the one hand, the anthropogenic increase of atmospheric pCO_2 and its oceanic uptake will result in a decrease of the surface ocean pH of up to 0.3 units by the year 2100 (corresponding to pCO_2 of 1000 ppm; IPCC 2013) and up to 1.4 units by the year 2300 (corresponding to a pCO_2 of 8000 ppm; Caldeira and Wickett 2005), a process termed ocean acidification (IPCC 2013). Even though crustaceans are predicted to be less sensitive to ocean acidification than other invertebrates, still one third of all investigated species in a current meta-analysis by Wittmann and Pörtner (2013) were negatively affected at an environmental pCO_2 of 851 and 1,370 µatm (scenario RCP8.5., Meinshausen et al. 2011). On the other hand, the increase of ocean surface temperature of up to 3 °C by the year 2100 as predicted by the Intergovernmental Panel on Climate Change (Collins et al. 2013) might pose an additional challenge for acid–base homeostasis in decapod crustaceans and even to shifts in whole ecosystem ecology and animal distributions (Walther et al. 2002). As a result, so-called dead zones and oxygen minimum zones (zones of depleted or low oxygen) are markedly increasing due to the anthropogenic pollution and the resulting increase in algal blooms and likely also due to a decrease in water circulation resulting from global warming (Mora et al. 2013). In combination, ocean acidification and warming negatively affected crustacean growth and potentially survival, but had no severe effects on calcification (Harvey et al. 2013).

6.3   Whole Animal Acid–Base Homeostasis and Regulation

6.3.1   Hemolymph Acid–Base Status

In decapod crustaceans, acid–base homeostasis on the whole animal level is described best by the carbonate system characteristics of the extracellular fluid. A disturbance of acid–base homeostasis can be of metabolic (shifts in aerobic/anaerobic metabolism and production of organic acids) or respiratory (shifts in respiratory CO_2) origin and leads to a decrease (acidosis) or increase (alkalosis) in hemolymph pH if not compensated for. Typically, these fluctuations are depicted in a Davenport diagram as shown in ◘ Fig. 6.2 (Davenport 1974).

Besides a small contribution of non-carbonate buffers like the respiratory pigments and other hemolymph proteins, adjustments in the hemolymph carbonate speciation allows for the buffering of the extracellular fluid upon an acid–base disturbance, according to following equation:

$$CO_2 + H_2O \leftrightarrow H_2CO_3 \leftrightarrow H^+ + HCO_3^-$$

Due to the low solubility of O_2 in water, aquatic animals need to establish a high flow rate of the external medium over their major gas exchange surfaces in order to ensure sufficient oxygenation. Aquatic decapod crustaceans are therefore very restricted in adjusting ventilation rates in order to regulate CO_2 flow. With a pK value of 6.2, extracellular CO_2 is mainly dissociated into H^+ and HCO_3^- at average physiological pH. Therefore, acid–base disturbances may be counteracted using primarily ion regulatory mechanisms at the gills (Henry et al. 2012; Truchot 1988).

◘ Tables 6.1, 6.2, 6.3, 6.4, 6.5, 6.6 and 6.7 give an overview of hemolymph acid–base characteristics of decapod crustaceans under control conditions, as well as changes

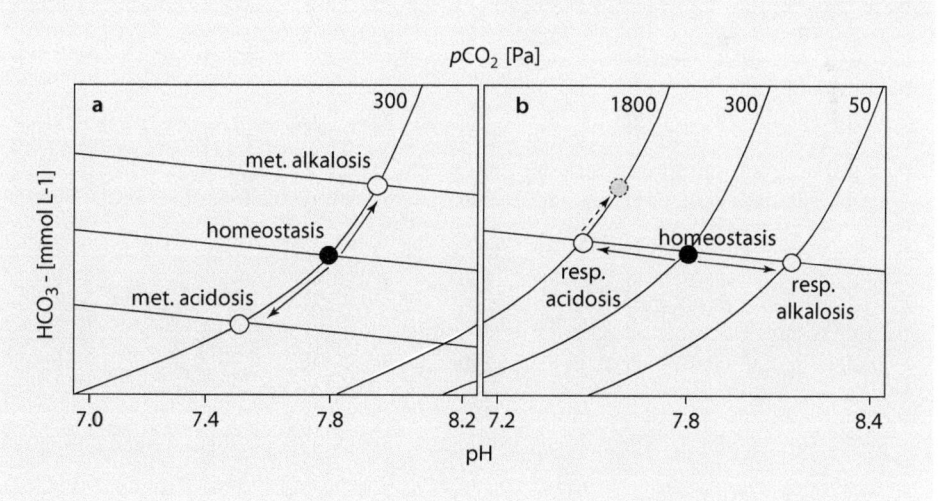

◘ **Fig. 6.2**   Davenport diagram for disturbances in acid–base homeostasis. (**a**) Metabolic and (**b**) respiratory components of acid–base disturbance. Indicated values are fitted roughly to represent an average decapod crustacean as listed in ◘ Tables 6.1, 6.2, 6.3, 6.4, 6.5, 6.6 and 6.7. Diagrams are reproduced according to Davenport (1974). *Filled circles*, acid–base homeostasis; *open circles*, status after disturbance. *Thin horizontal lines* indicate the non-carbonate buffer capacity of the hemolymph. *Thin curved lines* indicate CO_2 isopleths. The *grey circle* indicates a potential partial compensation by accumulation of HCO_3^- (see text for details)

Table 6.1 Physiological, whole animal response of decapod crustaceans upon exposure to air

Species	Group	References	Condition	Water salinity [ppt]	Water/air T [°C]	Hemolymph pH	Hemolymph pCO_2 [Pa]	Hemolymph HCO_3^- [mM]	Hemolymph TCO_2 [mM]	Lactate [mM]
Lithodes santolla	2	Urbina et al. (2013)	Control	30	12	7.65	–	–	–	–
			50 h air	30	12	6.78	–	–	–	–
Pachygrapsus crassipes	3	Burnett and McMahon (1987)	Control	32	23	7.8	306	–	5	1.0
			12 h air	N/A	23	7.7	1300	–	19	0.5
Eurytium albidigitum	3	Burnett and McMahon (1987)	Control	32	23	7.7	270	–	4	1.0
			8 h air	N/A	23	7.3	1500	–	9	0.5
Hemigrapsus nudus	3	Burnett and McMahon (1987)	Control	32	11	7.80	330	–	9	–
			4 h air	N/A	11	7.95	530	–	15	–
Neohelice granulata	3	Luquet and Ansaldo (1997)	Control	"BW"	20	7.82	280	–	–	–
			6 h air	N/A	20	7.75	310	–	–	–
Carcinus maenas	3	Simonik and Henry (2014)	Control	32	10	7.70	–	–	7.8	1.1
			6 h air	N/A	10	7.57	–	–	9.9	7.1

Species	Group	Reference	Condition	Salinity	Temp	pH				
Carcinus maenas	3	Truchot (1975b)	Control	35	15	7.82	150	–	3.86	–
			9 h air	N/A	15	7.59	680	–	9.67	–
			115 h air	N/A	15	7.79	590	–	13.79	–
Cancer productus	3	De Fur and McMahon (1984)	Control	32–35	10	7.84	320	–	10	0.9
			4 h air	N/A	10	7.62	510	–	13	9.0
Scylla serrata	3	Varley and Greenaway (1992)	Control	35	25	7.70	400	–	8	1.25
			6 h air	N/A	25	7.58	740	–	10	1.04
			7 days air	N/A	25	7.79	830	–	19	1.04
Homarus gammarus	5	Whiteley and Taylor (1990)	Control	32–35	15	7.8	130	4.1	–	0.41
			3 h air	32–35	10	7.6*	330*	6.6*	–	2.12*
			24 h air	32–35	20	7.8	800*	18.2*	–	3.24*
Homarus gammarus	5	Taylor and Whiteley (1989)	Control	35	15	7.78	400	9.0	–	1.8
			14 ha air	N/A	15	7.64*	1200	15.8	–	6.2
Austropotamobius pallipes	6	Taylor and Wheatly (1981)	Control	0	15	7.90	400	6.9	–	0.55
			4 h air	N/A	15	7.45*	1130*	6.9	–	8.28*
			24 h air	N/A	15	7.79	1040*	15.4*	–	0.57

Group definitions: (2) Anomura, (3) crabs, (5) lobster and (6) crayfish. Where available from the study, asterisks indicate significant differences from controls

Table 6.2 Physiological, whole animal response of decapod crustaceans upon changes in environmental pO_2 (hyper- or hypoxia)

Species	Group	References	Condition	Water pO_2 [kPa]	Water salinity [ppt]	Water T [°C]	Hemolymph pH	Hemolymph pCO_2 [Pa]	Hemolymph HCO_3^- [mM]	Hemolymph TCO_2 [mM]	Lactate [mM]
Carcinus maenas	3	Burnett and Johansen (1981)	Control	19.3 ?	35	10	7.85	450	–	13	–
			66 h	2.7–3.3	35	10	8.05*	170*	–	7*	–
			Control	19.3 ?	17	10	7.95**	425	–	15.9**	–
			22 h	2.7–3.3	17	10	8.29*	95*	–	6.2*	–
Carcinus maenas	3	Hill et al. (1991)	Control	21	30	10	7.86	240	6.52	8.2	1
			30 min	0	30	10	8.09*	190	–	8.1	3
			12 h	0	30	10	7.84	190	3.17*	5	18*
Palaemon elegans	4	Taylor and Spicer (1991)	Control	19.3 ?	32	10	7.79	–	–	5.4	0.55
			6 h mod	4	32	10	7.93*	–	–	1.9*	5.00*
			6 h sev	1.3	32	10	7.95*	–	–	5.3	0.73
Palaemon serratus	4	Taylor and Spicer 1991	Control	19.3 ?	32	10	7.8	–	–	5.1	0.6
			6 h mod	4	32	10	8.0*	–	–	1.5*	6.7*
			6 h sev	1.3	32	10	8.1*	–	–	4.9	1.2

Astacus leptodactylus	6	Dejours and Beekenkamp (1977)	Control	3.2	0	13	7.85	330	5.9	–	–
			44 days (hyper)	80	0	13	7.65	800	9	–	–
Pacifastacus leniusculus	6	Wheatly et al. (1991)	Control	–	0	12	7.83	330	7	–	–
			6 h (hyper)	67	0	12	7.55*	800	10*	–	–
			48 h (hyper)	67	0	12	7.73	730	13*	–	–
Austropotamobius pallipes	6	Wheatly and Taylor (1981)	Control	19.3	0	15	7.90	405	–	7.5	0.40
			24 h low	8.4	0	15	7.96	230	–	5.0	0.49
			24 h mod	4.8	0	15	7.98	195	–	4.0	0.97
Orconectes rusticus	6	Wilkes and McMahon (1982)	Control	16.5	0	15	7.78	480	–	5.84	–
			24 h	6–7.3	0	15	7.98*	140*	–	2.30*	–
			6 days	6–7.3	0	15	7.90*	270*	–	3.20*	–

Group definitions: (3) crabs, (4) shrimp and (6) crayfish. Asterisks indicate significant differences from controls, where available. Double asterisks indicate values significantly different between controls. Question marks indicate a value taken from a cross-reference within the paper

mod moderate, sev severe

Table 6.3 Physiological, whole animal response of decapod crustaceans upon changes in environmental pCO_2 (hypercapnia)

Species	Group	References	Condition	Water pCO_2 [kPa]	Water pH	Water salinity [ppt]	Water T [°C]	Hemolymph pH	Hemolymph pCO_2 [Pa]	Hemolymph HCO_3^- [mM]	Hemolymph TCO_2 [mM]
Callinectes sapidus	3	Henry et al. (1981)	Control	–	–	25	26	7.78	270	–	5
			48 h	2	–	25	26	7.61	2100	–	20
Callinectes sapidus	3	Cameron (1978)	Control	0.04	–	–	–	7.96	530	–	10
			15 min	1	–	–	–	7.65	1470	–	12
			44 h	1	–	–	–	7.80	1200	–	19
Carcinus maenas	3	Appelhans et al. (2012)	Control	0.066	8.06	15	13	7.82	380	8	–
			10 w mod	0.126	7.84	15	13	7.83	490	10	–
			10 w sev	0.351	7.36	15	13	7.81	550	12	–
Carcinus maenas	3	Fehsenfeld and Weihrauch (2013)	Control	0.054	7.7	10	14	7.9	263	6.6	–
			7 days	0.324	7.0	10	14	7.9	344	7.9	–
Carcinus maenas	3	Fehsenfeld and Weihrauch (2016a)	Control	0.09	7.76	32	13.2	7.87	156	5.5	–
			48 h	9.40	6.85	32	13.8	7.76*	688*	18.7*	–

Species	n	Reference	Condition								
Carcinus maenas	3	Truchot (1975a)	Control	0.04	7.8	35	17	7.79	180	3.9	–
			24 h	0.31	7.4	35	17	7.71	730	13.7	–
Necora puper	3	Spicer et al. (2007)	Control	0.02	7.98	34	15	7.90	190	6.6	–
			24 h	0.08	7.30	34	15	7.82	360	10.8	–
			24 h	1.11	6.70	34	15	7.97	1190	27.4	–
			24 h	6.04	6.05	34	15	7.59	4520	55.9	–
Necora puper	3	Small et al. (2010)	Control	0.074	7.85	35	17	7.73	190	2.63	–
			30 days sev	1.250	6.69	35	17	7.53	880	8.66	–
Metacarcinus magister	3	Hans et al. (2014)	Control	0.049	8.1	32	14	7.93	133	4.9	–
			7 days	0.330	7.4	32	14	8.01	402*	14.9*	–
Metacarcinus magister	3	Pane and Barry (2007)	Control	–	7.90	34	10	7.82	265	6	–
			90 min	1.013	7.08	34	10	7.42*	1000*	8*	–
			24 h	1.013	7.08	34	10	7.75	900*	18*	–
Pandalus borealis	4	Hammer and Pedersen (2013)	Control	0.052	8.1	35	7	7.76	250	5.1	–
			16 days	0.920	6.9	35	7	7.60	1000*	13.0*	–

(continued)

■ Table 6.3 (continued)

Species	Group	References	Condition	Water pCO_2 [kPa]	Water pH	Water salinity [ppt]	Water T [°C]	Hemolymph pH	Hemolymph pCO_2 [Pa]	Hemolymph HCO_3^- [mM]	Hemolymph TCO_2 [mM]
Palaemon elegans	4	Dissanayake et al. (2010)	Control	0.10	8.0	32	15	7.70	–	?	–
			14 days	0.31	7.5	32	15	7.55	–	Δ1.0	–
			30 days	0.38	7.5	32	15	7.80	–	Δ1.8	–
Palaemon serratus	4	Dissanayake et al. (2010)	Control	0.12	8.0	32	15	7.7	–	?	–
			14 days	0.32	7.5	32	15	7.7	–	Δ1.2	–
			30 days	0.37	7.5	32	15	8.2	–	Δ2.4	–

Group definitions: (3) crabs and (4) shrimp. Asterisks indicate significant differences from controls, where available. Question marks indicate unavailable data; accordingly, differences in hemolymph HCO_3^- are given as the difference in comparison to controls

w weeks, mod moderate, sev severe

◘ Table 6.4 Physiological, whole animal response of decapod crustaceans upon changes in environmental temperature

Species	References	Condition	Water T [°C]	Water salinity [ppt]	Hemo-lymph pH	Hemo-lymph pCO_2 [Pa]	Hemo-lymph HCO_3^- [mM]	Lactate [mM]
Carcinus maenas	Truchot (1973)	Control	15	20	7.75	335	6.6	–
		4 days	5	20	7.90	225	7.3	–
		4 days	10	20	7.75	280	6.5	–
		4 days	21	20	7.63	425	5.9	–
		4 days	26	20	7.53	545	5.5	–
Carcinus maenas	Howell et al. (1973)	Control	15	32	7.79	625	13.2	–
		7 days	5	32	8.07	305	16.7	–
		7 days	10	32	7.86	590	16.0	–
		7 days	20	32	7.77	560	10.2	–
Callinectes sapidus	Howell et al. (1973)	Control	15	32	7.77	320	5.9	–
		7 days	5	32	8.03	465	18.8	–
		7 days	10	32	7.80	425	9.2	–
Callinectes sapidus	Cameron and Batterton (1978a)	Control	10	0	7.94	280	10.35[a]	–
		7–14 days	27	0	7.74	400	7.60[a]	
Metacarcinus magister	McMahon et al. (1978)	Control	12	32	7.84	225	5.83	1.5
		10 days	7	32	7.91	235	7.42	1.4
		10 days	17	32	7.73	250	4.47	1.5

All studies have been conducted on brachyuran crabs
[a]Values are for total carbon

◻ Table 6.5 Physiological, whole animal response of decapod crustaceans upon environmental salinity

Species	Group	References	Condition	Water salinity [ppt]	Water pH	Water pCO_2/pO_2 [kPa]	Water T [°C]	Hemolymph pH	Hemolymph pCO_2 [Pa]	Hemolymph HCO_3^- [mM]
Callinectes sapidus	3	Henry and Cameron (1982)	Control	35	–	–	–	7.75	320	4.2
			24 h	12	–	–	–	7.84	400	9.0
Eriocheir sinensis	3	Truchot (1992)	Control	35	8.1	0.06	15	7.89	280	7.9
			3 days	0	8.4	0.06	15	8.02*	495*	16.0*
			Control	0	8.4	0.06	15	7.96	425	12.2
			24 h	35	8.1	0.06	15	7.96	240*	8.0*
Eriocheir sinensis	3	Whiteley et al. (2001)	Control	35	–	–	12	7.76	460	7.1
			6 h	12	–	–	12	7.50*	290	4.0
			24 h	12	–	–	12	7.94	310	9.0
Necora puber	3	Whiteley et al. (2001)	Control	35	–	–	12	7.82	300	7
			6 h	24	–	–	12	7.80	500	9
Carcinus maenas	3	Truchot (1981)	Control	35	8.19	0.039	15	7.85	215	5.5

Pacifastacus leniusculus	6	Wheatly and McMahon (1982)	48 h	12	7.85	0.040	15	8.08*	200	8.0*
			24 h	35	8.19	0.039	15	7.65*	190	3.0*
			Control	0	–	–	15	7.95	365	8.47
			48 h	8	–	–	15	8.00	180	5.73
			48 h	16	–	–	15	7.96	140	4.76*
			48 h	32	–	–	15	7.83*	90	2.83*

Group definitions: (3) crabs and (6) crayfish. Asterisks indicate significant differences from controls, where available

▪ **Table 6.6** Physiological, whole animal response of decapod crustaceans upon exercise (forced movement)

Species	Group	References	Condition	Water salinity [ppt]	Water T [°C]	Hemolymph pH	Hemolymph pCO_2 [Pa]	Hemolymph TCO_2 [mM]	Lactate [mM]
Metacarcinus magister	3	McDonald et al. (1979)	Control	27	8	7.94	230	8.3	0.69
			20 min	–	–	7.52	430	–	9.90
Callinectes sapidus	3	Booth et al. (1984)	Control	31	20	7.67	470	5.7	2
			30 min	31	20	7.24	800	3.6	15
Carcinus maenas	3	Hamilton and Houlihan (1992)	Control	31	15	7.83	210	5.2	3.9
				31	15	7.76*	250	4.3	9.7*
Homarus americanus	5	Rose et al. (1998)	Control	32	14	7.78	268	4.80	0.20
			30 min slow	32	14	7.64*	507*	5.51*	1.55*
			30 min fast	32	14	7.42*	603*	5.74*	1.29*

Group definitions: (3) crabs and (5) lobster. Asterisks indicate significant differences from controls, where available

■ Table 6.7 Physiological, whole animal response of decapod crustaceans upon a combination of different environmental stressors

Species	Group	References	Condition	Water pCO_2/pO_2 [kPa]	Water pH	Water T [°C]	Water salinity [ppt]	Hemolymph pH	Hemolymph pCO_2 [Pa]	Hemolymph HCO_3^- [mM]	Hemolymph TCO_2 [mM]	Lactate [mM]
Metapenaeus joyneri	1	Dissanayake and Ishimatsu (2011)	Control	0.04	8.2	15/20/25	32	7.65	–	–	–	–
			10 days HC	1.00	8.2	15	32	7.73	–	–	–	–
			10 days HC+T	1.00	8.2	20	32	7.79	–	–	–	–
			10 days HC+T	1.00	8.2	25	32	7.80 (2 d)	–	–	–	–
Hyas araneus	3	Harms et al. (2014)	Control	0.045	8.2	5	32	8.05	250	2.8	–	–
			10 weeks HC mod	0.100	7.8	5	32	8.00	350*	4.2	–	–
			10 weeks HC sev	0.190	7.6	5	32	7.75*	430*	4.0	–	–
			Control+T	0.037	8.2	10	34	7.98	210	1.4	–	–
			10 weeks mod HC+T	0.095	7.9	10	34	7.95	430*	5.1*	–	–

(continued)

Table 6.7 (continued)

Species	Group	References	Condition	Water pCO_2/pO_2 [kPa]	Water pH	Water T [°C]	Water salinity [ppt]	Hemolymph pH	Hemolymph pCO_2 [Pa]	Hemolymph HCO_3^- [mM]	Hemolymph TCO_2 [mM]	Lactate [mM]
Necora puber	3	Rastrick et al. (2014)	10 weeks sev HC+T	0.204	7.5	10	34	7.70*	430*	3.0	–	–
			(0) control	0.041	–	10	35	7.84	290	6.6	–	1.0
			(1) 2 weeks temp	0.041	–	15	35	7.89	300	7.1	–	1.0
			(2) 2 weeks HC	0.101	–	10	35	7.87	380	9.2*	–	1.0
			(3) 2 weeks HC+T	0.101	–	15	35	7.95	370	11.0*	–	1.0
			Control: 3 h air (0)	0.041	–	10	N/A	7.3**	1100**	7.0	–	8.0**
			3 h air of (1)	0.041	–	15	N/A	7.3	1600	12.0	–	7.5
			3 h air of (2)	0.101	–	10	N/A	7.5	1100	14.0	–	9.0
			3 h air of (3)	0.101	–	15	N/A	7.3	2100	14.0	–	10.0
Cancer productus	3	De Fur et al. (1980)	Control	18	–	10	32	7.89	280	–	9.0	–
			4 h air	–	–	10	N/A	7.62	1100	–	13.6	–

Chionoecetes tanneri	3	Pane and Barry (2007)	24 h hyperoxia	61–68	–	10	32	7.80	1100	–	20.0	–
			Control high O$_2$	–	7.85	3	34	7.8	240	5	–	–
			24 h HC, at high O$_2$	1.013	7.08	3.5	34	7.4*	535*	5	–	–
			Control low O$_2$	–	7.08	3	34	8.00	240	9.5	–	–
			24 h HC, at low O$_2$	1.013	7.08	3.5	34	7.65*	665*	11.5	–	–

Group definitions: (1) prawns and (3) crabs. Asterisks indicate significant differences, where available. Double asterisks indicate values signifcantly different between controls

HC hypercapnia, mod moderate, sev severe

induced by various stressors as described in ▶ Sect. 6.2.3 and below. Due to the vast number of publications on acid–base disturbances, we do not claim for the list to be complete, but rather tried to give representative examples for as many different species as possible. In case multiple studies were available for the same species, we attempted to include the most relevant publication(s) that was (were) comparable to the other studies. Data was partly extracted from graphs and values transformed into the units as depicted in the tables where applicable. If significantly different, values are given for the time point of maximum effect as well as from the end of the incubation period.

Under control conditions, all decapod crustaceans maintain their hemolymph pH typically between 7.7 and 8.0. Hemolymph pCO_2 levels, however, can vary quite substantially between crustacean species and are typically low (*ca.* 200–500 Pa) due to the almost complete dissociation to H^+ and HCO_3^- at physiological pH, but high enough likely to enable diffusion out of the body along the gradient between the extracellular fluid and the environment (*ca.* 40 Pa; Henry et al. 2012; Melzner et al. 2009). Prawns like the Japanese tiger prawn *Penaeus japonicus* seem to be an exception: These decapod crustaceans have a slightly lower than average hemolymph pH (7.5) and, accordingly, a higher pCO_2 (*ca.* 600 Pa; Cheng et al. 2013). Similar to hemolymph pCO_2, levels of hemolymph HCO_3^- have also been observed to vary between species and typically average between 4 and 9 mmol L^{-1}. As indicated in ◘ Tables 6.1, 6.2, 6.3, 6.4, 6.5, 6.6 and 6.7, control levels for hemolymph $[HCO_3^-]$ as high as 14 mmol L^{-1} have been observed in some studies, but these values have to be treated with caution as they may indicate that animals were in premoult rather than intermoult stages (▶ see Sect. 6.3.2).

▪▪ Exposure to Air (◘ Table 6.1)

The primary consequence for most aquatic decapod crustaceans of emerging from water is the collapse of their gills, physically impairing gas exchange processes. Many crabs are therefore retaining branchial water in their gill chambers likely to facilitate CO_2 diffusion (Burnett and McMahon 1987). Due to the higher solubility of O_2 in air compared to water, however, some decapod crustaceans like *Pachygrapsus crassipes* (Burnett and McMahon 1987) or *C. maenas* (Simonik and Henry 2014) are capable of extracting O_2 from the air and voluntarily move out of the water to offset acid–base disturbances resulting from other stressors like hypoxia (Wheatly and Taylor 1979).

Generally, exposure to air results in a pronounced respiratory acidosis with a pH drop of 0.1–0.2 units (in crabs) up to 0.5–0.7 units (in Anomura and crayfish), a two- to fivefold increase in hemolymph pCO_2 and a significant two- to threefold elevation of hemolymph HCO_3^- in all investigated decapod crustaceans (◘ Table 6.1).

It seems, however, that there are marked species-specific differences in compensating for the experienced acidosis. While some crabs (De Fur and McMahon 1984; Simonik and Henry 2014), lobsters (Taylor and Whiteley 1989; Whiteley and Taylor 1990) and crayfish (Taylor and Wheatly 1981) seem to switch partly to anaerobic metabolism and therefore experience an additional metabolic acidosis with a pronounced increase in hemolymph lactate levels, other crabs like *Eurytium albidigitum* (Luquet and Ansaldo 1997) and *Pachygrapsus crassipes* (Burnett and McMahon 1987) seem to undergo a metabolic depression (*E. albidigitum*) or maintain or even increase their aerobic metabolism (*P. crassipes*). Additionally, some crab species were observed to increase their strong ion difference (SID) likely via ion exchange processes at the gill in response to emersion (Burnett and McMahon 1987; Luquet and Ansaldo 1997; Truchot 1979), which is interpreted to help offset the experienced acidosis (Stewart 1978). Consequently, *C. maenas* (Truchot 1975b),

S. serrata (Varley and Greenaway 1992), *H. gammarus* (Whiteley and Taylor 1990), *A. pallipes* (Taylor and Wheatly 1981), *P. crassipes* and *H. nudus* (Burnett and McMahon 1987) are capable of fully compensating for the pH drop resulting from the experienced acidosis, while *E. albidigitum* is not (Burnett and McMahon 1987).

■ ■ **Hyper-/Hypoxia (☐ Table 6.2)**

Due to its low solubility in water compared to air, oxygen has to be considered one of the limiting factors in the aquatic environment (Dejours 1975). Hence, only subtle changes in water pO_2 result in immediate alterations of ventilation rates in aquatic decapod crustaceans in order to be able to maintain aerobic metabolism (Jouve-Duhamel and Truchot 1983; Truchot 1988). Consequently, hyperventilation as observed in moderate hypoxic conditions simultaneously leads to an increase in branchial CO_2 excretion and therefore respiratory alkalosis (elevated pH and lower HCO_3^-), while reduced ventilation in a (moderate) hyperoxic environment ultimately leads to accumulation of hemolymph pCO_2 and hence a respiratory acidosis (lower pH and higher HCO_3^-), as can be seen in ☐ Table 6.2. In shrimp and *C. maenas*, the increase in hemolymph lactate during moderate hypoxia and anoxia, respectively, indicates that in these decapod crustaceans, a metabolic component seems to be present that might explain the reduced levels of total carbon/HCO_3^-, but did not affect the actual increase in hemolymph pH (Taylor and Spicer 1991). Interestingly, when the same shrimp species *P. elegans* and *P. serratus* were exposed to a severe hypoxia (<2 Pa), lactate levels stayed constant and total carbon was not affected (Taylor and Spicer 1991).

■ ■ **Hypercapnia (☐ Table 6.3)**

In contrast to hypoxia, exposure to elevated environmental pCO_2 does not drive a ventilation response in decapod crustaceans due to its very similar solubility in air and water (Henry et al. 2012; Jouve-Duhamel and Truchot 1983). Nonetheless, exposure to hypercapnia leads to a respiratory acidosis marked by a rapid drop in hemolymph pH of up to 0.4 units and substantial increases in pCO_2 (two- to fourfold) in all investigated decapod crustaceans (☐ Table 6.3). Elevated extracellular pCO_2 is believed to be maintained in order to ensure an outwardly directed CO_2 gradient for the diffusion-based excretion of metabolic CO_2 (Melzner et al. 2009). Most decapod crustaceans are capable of fully compensating for the respiratory acidosis by accumulating HCO_3^- in their hemolymph to buffer excess protons, likely via active ion regulatory processes at the gill. While some species are capable of maintaining or even increasing their metabolic rate in response to hypercapnia (*C. maenas*, Appelhans pers. communication), others experience a metabolic depression (e.g. *M. magister*, Hans et al. 2014; *P. borealis*, Hammer and Pedersen 2013). Interestingly, green crabs *C. maenas* that are acclimated to full-strength seawater (32–35 ppt; Truchot 1975c; Fehsenfeld and Weihrauch 2016a) seemed to accumulate more CO_2 in their hemolymph than brackish-water acclimated specimen (Appelhans et al. 2012; Fehsenfeld and Weihrauch 2013). In contrast to brackish-water crabs, the resulting respiratory acidosis in the seawater-acclimated crabs was not fully compensated for after 24 h, and hemolymph pH decreased. This example indicates that acid–base and osmoregulation might indeed be closely linked in this species.

■ ■ **Temperature (☐ Table 6.4)**

It has been shown for poikilotherm animals such as decapod crustaceans that temperature correlates inversely with hemolymph pH in order to maintain extracellular H^+/OH^- ratios to ensure a constant net charge of proteins (Howell et al. 1973; Truchot 1973). In parallel, hemolymph pCO_2 seems to generally stay constant/increase only slightly with increasing temperature, while $[HCO_3^-]$ and/or total carbon (C_T) decreases more drastically. The authors of the respective studies (Cameron and Batterton 1978b; Truchot 1973) attributed the changes in C_T to active regulation of HCO_3^- via ion exchanges at the gills in order to compensate for the acidosis, rather than solution of the carapace or passive processes alone (Henry et al. 2012).

▪▪ Salinity (◘ Table 6.5)

Generally, acclimation to low salinity results in a metabolic alkalosis in all investigated decapod crustaceans, characterized by an increase in hemolymph pH at relatively stable pCO_2 and a significant increase in $[HCO_3^-]$. Conversely, when freshwater crayfish (Wheatly and McMahon 1982), freshwater *E. sinensis* (Truchot 1992) or brackish-water acclimated *C. maenas* (Truchot 1981) were acclimated to full-strength seawater, they developed a metabolic acidosis characterized by a decrease in hemolymph pH and HCO_3^-. In this case, however, an additional slight respiratory alkalosis (decrease in pCO_2) was observed in parallel, likely compensating for the respiratory acidosis. Throughout the time course of different salinity acclimations, species can exhibit specific alterations to this general pattern. An initial respiratory acidosis, for example, was observed in *C. maenas* upon acclimation to dilute salinity before switching into the expected metabolic alkalosis (Truchot 1981), and in *E. sinensis* a transient respiratory acidosis marked by a spontaneous drop in pH was only present at days 6 (Truchot 1992).

While Henry and Cameron (1982) attributed the observed increase in hemolymph $[HCO_3^-]$ following acclimation of *C. sapidus* to dilute salinity to the additionally observed change in the strong ion difference (SID), no equivalent observation was made in brackish-water acclimated *E. sinensis* in the study by Whiteley et al. (2001). In contrast, Truchot (1981, 1992) suggested metabolic adjustments correlated to isosmotic intracellular regulation in the cells to be responsible, resulting in either a measureable efflux of base or acid into the environment. These observations reveal the complex nature of acid–base disturbances upon different salinity acclimations, and consequently, the reasons for the observed metabolic alkalosis and acidosis are not yet fully understood.

▪▪ Exercise (◘ Table 6.6)

Hemolymph lactate levels are typically held lower than 1 mmol L^{-1} and negligible in undisturbed decapod crustaceans, but can increase more than tenfold in crabs that experience a metabolic acidosis due to exercise (forced movement). Furthermore, the experienced acidosis is characterized by an immediate drop in pH of up to 0.4 units and a twofold increase in hemolymph pCO_2, therefore also resembling characteristics of a respiratory acidosis. Interestingly, exercised lobsters seem to be able to avoid anaerobic metabolism during exercise for the most part and experience primarily a respiratory acidosis without the substantial rises in hemolymph lactate observed in other crustaceans (Rose et al. 1998).

In *M. magister* (McDonald et al. 1979), *C. maenas* and *C. sapidus* (Booth et al. 1984), the proton concentration in the hemolymph was observed to be lower than could be expected from the accumulated lactate, given that both are produced in equimolar quantities during glycolysis (Hochachka and Mommsen 1983). Due to the observed drastic increase in ammonia excretion, Booth et al. (1984) concluded that at least part of the protons are excreted as NH_4^+ via ion exchange processes at the gill epithelium.

▪▪ **Combined Stressors (▣ Table 6.7)**

As is clear by the previous sections, extracellular acid–base regulation in response to environmental disturbances can be quite complex despite some common principles. While the discussed studies isolated one stressor at a time and investigated its effects on the respective species' acid–base characteristics, environments are rarely that "simple" and a combination of simultaneous stressors seems much more likely, especially in tide pools (Truchot 1988) or in the face of ongoing global climate change (IPCC 2013).

For example, when combined with an increase in water temperature, hypercapnia resulted in a respiratory alkalosis in *N. puber* that was not observed when crabs were exposed to either one of the stressors alone (Rastrick et al. 2014). Even though the increase in HCO_3^- upon hypercapnia and high-temperature acclimation rendered the crabs more resistant to short periods of subsequent emersion, they still experienced the same magnitude of acidosis, and recovery from these stressors was significantly attenuated. In a different example, prior acclimation of green crabs to hypercapnia enabled them to avoid an uncompensated hypercapnic acidosis that was induced by low environmental alkalinity in normocapnic-acclimated animals (Truchot 1984). In *C. productus*, the respiratory acidosis usually observed following exposure to hyperoxia was not observed in crabs that were first exposed to air (De Fur et al. 1980). Finally, *P. joyneri* exposed to a combination of elevated temperature and hypercapnia no longer experienced an acidosis as to be expected by studies on other decapod crustaceans, but exhibited an alkalosis (Dissanayake and Ishimatsu 2011).

Even though studies on combined environmental stressors are rare, the existing data indicates alarming differences in the acid–base responses of decapod crustaceans in comparison to single-stressor studies. Therefore, it would be desirable for future research to focus on a more holistic and realistic approach.

6.3.2 Calcification, $CaCO_3$ and Moulting

Interestingly, as one of the most important physiological processes, growth in crustaceans is closely linked to the whole animal acid–base status and regulation. Due to their hard and inflexible exoskeleton, decapod crustaceans depend on a series of moults in order to grow. During the different pre-, post- and intermoult stages that compose the complex moult cycle (Mangum et al. 1985), the connectives between the living tissue and the extracellular cuticle are loosened and water uptake ensures the shedding of the old and the expansion of the new carapace. The exoskeleton contains the majority of the organismal $CaCO_3$ that is mobilized during the moult in order to soften this structure and either excreted into the environment or stored in gastroliths for the new exoskeleton (Ahearn et al. 2004). Generally, decapod crustaceans experience a pronounced premoult alkalosis (increase in hemolymph HCO_3^-) in order to compensate for a concomitant acidosis of mainly metabolic origin (increase in hemolymph lactate) after successful exuviation (Mangum et al. 1985). Interestingly, this HCO_3^- seems not to originate from mobilization of the exoskeletal stores. Even though an early study by Robertson (1960) detected a seemingly HCO_3^--correlated increase in hemolymph $[Ca^{2+}]$ and $[Mg^{2+}]$ in premoult *C. maenas*, later studies on the blue crab *C. sapidus* did not detect a change in hemolymph $[Ca^{2+}]$ but observed a decrease in $[Cl^-]$ instead, indicating a direct Cl^-/HCO_3^--exchange with the environment as the source of the extracellular HCO_3^- (Cameron and Wood 1985; Cameron 1978; Henry et al. 1981; Mangum et al. 1985). As a response to air exposure, however, crayfish (Wheatly and Gannon 1995), the anomuran porcelain crabs *Petrolisthes laevigatus* (Lagos and Cáceres 2008) and *Petrolisthes violaceus* (Vargas et al. 2010) and subpopulations of the brachyuran crab *Cyclograpsus cinereus* (Lagos et al. 2014) and *N. granulata* (Luquet and Ansaldo 1997) were able to mobilize exoskeletal Ca^{2+}/HCO_3^- stores in response to the acid–base disturbance.

Interestingly and in contrast to other invertebrate marine calcifiers like mussels (Beniash et al. 2010; Michaelidis et al. 2005) and corals (Langdon et al. 2000; ▶ Chap. 7), calcification of the carapace in response to hypercapnia (ocean acidification) seems to increase in the red rock cleaner shrimp *Lysmata californica* (Taylor et al. 2015), the prawns *(Lito)Penaeus occidentalis* and *P. monodon* (Wickins 1984), female Anomuran red king crab *Paralithodes camtschaticus* (Long et al. 2013), as well as the crab *C. sapidus*, the lobster *H. americanus* and the prawn *P. plebejus* (Ries et al. 2009). The robustness of the crustacean carapace is believed to be due to its increased amount of calcite, the less soluble form of $CaCO_3$ (Taylor et al. 2015), as well as its complete coverage with a relatively thick organic epicuticle (Ries et al. 2009), and the crustaceans' generally high capability for acid–base regulation. While an increase in calcification might sound advantageous, it potentially has negative effects on the crustaceans' moulting frequency (Wickins 1984) and crypsis/predator defence (Taylor et al. 2015). The metabolic investment and possible allocation of energy resources due to an increased calcification might also lead to other negative impacts in these crustaceans that might reduce the overall fitness (i.e. metabolic depression and reduced growth; Taylor et al. 2015). An increase in calcification as well as an increase in metabolic costs in premoult was already observed in very early life stages (zoea I larval stage) of the Anomuran red king crab *P. camtschaticus* (Long et al. 2013) and the brachyuran great spider crab *Hyas araneus* (Schiffer et al. 2013).

Only few decapod crustaceans like the velvet swimming crab *N. puber*, however, might also exhibit a dissolution of their exoskeleton in response to high levels of ocean acidification (Spicer et al. 2007). Also in late European lobster larvae, exposure to hypercapnia resulted in significantly lower carapace mass as well as less mineralization in response to hypercapnia (Arnold et al. 2009), as well as a delay in the first moult cycles (Keppel et al. 2012).

6.4   Gill Epithelial Acid–Base Regulation

6.4.1   Gill Epithelial Transporters Involved in Acid–Base Regulation

While the numerous studies on acid–base homeostasis in aquatic decapod crustaceans mainly have focussed on describing the whole animal acid–base status in response to diverse environmental stressors as described above, to date only a few investigations have commented on the actual regulatory mechanisms involved in these processes. These few studies indicate that the high acclimation potential of decapod crustaceans in response to environmental changes can be attributed mainly to ion exchange processes in the gill epithelium (see above). As the major site for osmoregulation and ammonia excretion (reviewed by Henry et al. 2012; Larsen et al. 2014; Weihrauch et al. 2004b), the gills possess many epithelial membrane transporters that are likely also involved in acid–base regulation. Indirect evidence was drawn from the observation that changes in hemolymph acid–base equivalents (H^+/HCO_3^-) were accompanied by changes in the strong ion difference (Na^+/Cl^-), when decapod crustaceans were exposed to air and dilute salinity (Burnett and McMahon 1987; Ehrenfeld 1974; Henry and Cameron 1982; Luquet and Ansaldo 1997; Truchot 1979) or when carbonic anhydrase, the enzyme responsible for the conversion of CO_2 into H_2CO_3 and subsequent dissociation to H^+ and HCO_3^-, was blocked (Burnett et al. 1981; Henry and Cameron 1983; Henry et al. 2003). While the crustacean gill epithelium has been subject to many investigations of membrane transporters involved in osmoregulation and ammonia excretion (for reviews see Henry et al. 2012; Larsen et al. 2014; Weihrauch et al. 2004a, b), hardly anything is known about the respective mechanisms for acid–base regulation. A recent set of gill perfusion experiments on anterior gills of seawater-acclimated *C. maenas*, however, shed some light on the gill transporter inventory potentially involved in branchial acid–base regulation and its linkage to ammonia regulation (see also section above) in this species (Fehsenfeld and Weihrauch 2016a).

 ◘ Figure 6.3a–d represents the current working models for osmoregulation, ammonia excretion and general acid–base regulation, as well as a new, more specific model for acid–base regulation and its overlap/link to ammonia excretion in the model organism *C. maenas*.

Trans-branchial active NaCl transport in moderate hyper-osmoregulators such as *C. maenas* (Riestenpatt et al. 1996) and *N. granulata* (Lucu and Siebers 1987; Onken et al. 2003) is fairly well characterized. As can be seen in ◘ Fig. 6.3a, basolateral Na^+/K^+-ATPase and Cl^--channels, as well as apical $Na^+/K^+/2Cl^-$-cotransporter supported by apical and basolateral K^+-channels, are key players in this osmoregulatory mechanism. A number of studies also indicated an alternative pathway for NaCl uptake via apical Na^+/H^+- and HCO_3^-/Cl^--exchangers, linked to the actions of a carbonic anhydrase (Henry et al. 2003; Lucu 1990; Onken et al. 2003; Tresguerres et al. 2008), therefore directly linking NaCl transport to the transport of acid–base equivalents. In *N. granulata*, however, a basolateral Na^+/H^+-exchanger seems to promote intracellular Na^+ uptake in exchange for H^+ rather than being situated apically (Tresguerres et al. 2008).

Overlapping with the model for NaCl transport, basolateral Na^+/K^+-ATPase and Cs^+/Ba^{2+}-sensitive K^+-channels have also been shown to be involved in ammonia excretion through the gills of *C. maenas* (Weihrauch et al. 1998, 2004a, b; ◘ Fig. 6.3b). In addition to the general Ba^{2+}-sensitive K^+-channels, a ZD7288-sensitive K^+-channel of the hyperpolarization activated cyclic nucleotide-gated potassium channel family (HCN) has recently been identified to contribute to NH_4^+ regulation over the gill epithelium of *C. maenas* (Fehsenfeld and Weihrauch 2016b).

Additionally, a cytoplasmic V-(H^+)-ATPase and a functional microtubule network have been hypothesized to promote ammonia excretion over the apical membrane via NH_3 trapping and transport in acidified vesicles in this species (Weihrauch et al. 2002), potentially linking ammonia excretion with acid–base regulation.

In comparison to the models for osmoregulation and ammonia excretion, the hypothetical model for general crustacean acid–base regulation after Freire et al. (2008) as seen in ◘ Fig. 6.3c is much more speculative. When considered in correlation with ammonia excretion as seen in ◘ Fig. 6.3d, however (▶ see also Sect. 6.5), potential pathways become more comprehensive. The most significant key player in *C. maenas*' branchial acid–base regulation (although not affecting ammonia excretion), as identified in a recent gill perfusion study applying pharmaceuticals to block specific transporters in *C. maenas* gills, was a potential basolateral Na^+/HCO_3^--cotransporter (Fehsenfeld and Weihrauch 2016a). A recently identified basolateral Na^+/HCO_3^--exchanger in the squid *Sepioteuthis lessoniana* (Hu et al. 2014, ▶ Chap. 11) has also been postulated to be important for acid–base regulation in the euryhaline crab *N. granulata* (Tresguerres et al. 2008).

A strictly apical distribution of V-(H⁺)-ATPase as hypothesized in the model of Freire et al. (2008) and depicted in ◘ Fig. 6.3c has only been identified in freshwater (and terrestrial) crustaceans (Tsai and Lin 2007), including the red crab *Dilocarcinus pagei* (Weihrauch et al. 2004a, b) and *E. sinensis* (Onken and Putzenlechner 1995; Tsai and Lin 2007), as well as many freshwater fish (Gilmour and Perry 2009), to generate an electrochemical gradient over the apical membrane to drive Na^+ uptake (Larsen et al. 2014; Weihrauch et al. 2001). While an apical V-(H⁺)-ATPase seems unlikely to be present in sea- and brackish-water acclimated decapod crustaceans for osmoregulatory purposes, an apical presence cannot be excluded to be involved in acid–base regulation. The pharmacological studies by Fehsenfeld and Weihrauch (2016a) on the isolated perfused gill and the immunohistochemical localization (Weihrauch et al. 2001), however, indicated a significant contribution of a cytoplasmic V-(H⁺)-ATPase that – together with the Rhesus-like protein – has been hypothesized to be involved in ammonia trapping in acidified vesicles as suggested by Weihrauch et al. (2002), therefore promoting the excretion of both, NH_3 and H^+.

Supporting the findings of the above-mentioned gill perfusion study in *C. maenas* gills (Fehsenfeld and Weihrauch 2016a), a study by Siebers et al. (1994), identified basolateral Na^+/K^+-ATPase (NKA) to be involved in branchial acid–base regulation. K^+-channels on the other hand provide a backflow of K^+ into the hemolymph. As mentioned earlier, both structures provide additional transport of NH_4^+ due to its similar size and charge compared to K^+ (Skou 1960; Lignon 1987; Weihrauch et al. 1998; Choe et al. 2000). Additionally, the HCN-like potassium channel recently identified to be involved in NH_4^+ movements over the gill epithelium of *C. maenas* as mentioned above has also been shown to be involved in branchial acid–base regulation in the respective study (Fehsenfeld and Weihrauch 2016b).

In fish, the Na^+/K^+-ATPase generates the electrochemical gradient over the basolateral membrane that is then the major driving force for the excretion of H^+ via apical Na^+/H^+-exchanger (NHE) in acid excretory epithelial cells (Choe et al. 2005; Edwards et al. 2002). While a potential electrogenic NHE ($2Na^+/1H^+$) has been identified to be present in crustacean gills (Shetlar and Towle 1989), an apical localization as suggested by studies on *Cancer antennarius* and *P. cinctipes* (Hunter and Kirschner 1986) and *C. maenas* (Weihrauch et al. 1998) is not clear to date due to the interference of the employed pharmaceuticals (amiloride) with the cuticle (Onken and Riestenpatt 2002; Weihrauch et al. 2002). A basolateral NHE that promotes H^+ excretion into the hemolymph, however, has been observed in *N. granulata* (Tresguerres et al. 2008). An additional cytoplasmic distribution of NHE that is potentially involved in both ammonia and proton excretion via vesicles is supported by phylogenetic analysis of diverse NHEs as conducted by Fehsenfeld and Weihrauch (2016a). Also in gills of *N. granulata*, Tresguerres et al. (2008) identified an apical Cl^-/HCO_3^--exchanger that would provide an apical exit of HCO_3^- as indicated in the proposed model (◘ Fig. 6.3d). In the gills of *C. maenas*, two isoforms of branchial carbonic anhydrase have been identified, a cytoplasmic and a membrane bound isoform (Boettcher et al. 1990; Serrano and Henry 2008). Supported by findings of the study of Siebers et al. (1994), carbonic anhydrase played a role in branchial acid–base regulation as observed in the inhibitor study by Fehsenfeld and Weihrauch (2016a).

The described epithelial transport processes closely resemble mechanisms that are observed in the mammalian kidney. While the crustacean epithelial ion regulatory mechanisms based on ◘ Fig. 6.3a has been compared to the thick ascending limb of the mammalian kidney in the past (Riestenpatt et al. 1996), the proposed new model for branchial acid–base regulation and its link to ammonia regulation as seen in ◘ Fig. 6.3d additionally resembles features of the mammalian kidney collecting duct (Weiner and Verlander 2013).

6.4.2 Genetic Responses to Acid–Base Disturbance

Two microarray and transcriptomic studies have identified changes in (mRNA) expression levels of gill epithelial transporters upon environmental disturbances that helped identify some of the candidate genes involved in acid–base regulation.

Interestingly, exposure to hypercapnia (400 Pa for 7 days) did not seem to elucidate a typical stress response in posterior gills of osmoregulating green crabs. Applying microarray and quantitative real-time experiments, Fehsenfeld et al. (2011) observed generally only subtle changes in mRNA expression levels among over 4400 genes in *C. maenas* and did not identify any changes in heat-shock proteins resembling direct indicators for stress. Instead, the data suggested an increased contribution of vesicular membrane transport, indicating that the proposed vesicular transport for active ammonia excretion (Weihrauch et al. 2002; ◘ Fig. 6.3b) might indeed contribute to $H^+(NH_4^+)$ excretion and therefore acid–base regulation in this species. Additionally, most of the annotated common ion transporters of the gill epithelium were not differentially expressed with the exception of a significant upregulated calcium-activated chloride channel and the downregulated Cl^-/HCO_3^- exchanger of the SLC 4 family, as well as a downregulated glycosyl-phosphatidylinositol-linked carbonic anhydrase VII. Interestingly, these genes were also affected by acclimation of green crabs to dilute salinity (Towle et al. 2011). One of the most downregulated transcripts in the hypercapnia/microarray study, the hippocampus abundant gene transcript or 1 (initially falsely annotated as a hyperpolarization activated nucleotide-gated potassium channel), has been confirmed to have significantly reduced mRNA expression levels in posterior gill 7 after 7 days of hypercapnia, as well as HEA-acclimation (Fehsenfeld and Weihrauch, unpublished data). A very similar response (downregulation) was observed in HCN as identified in the recent study of Fehsenfeld and Weihrauch (2016b). Both genes are therefore interesting novel candidate genes for further studies in respect to branchial acid–base regulation.

In a different study, changes in branchial mRNA expression levels of a number of important epithelial transporters were monitored by quantitative real-time PCR after acclimation of *C. maenas* to hypercapnia (Fehsenfeld and Weihrauch 2013). Similar to the results of the microarray study mentioned above, only subtle changes in mRNA expression were observed in individual gills, but the experiments indicated a role for the Rhesus-like protein, Na^+-K^+-ATPase and glycosyl-phosphatidylinositol-linked carbonic anhydrase VII in branchial acid–base regulation, as well as potentially the Na^+/H^+-exchanger and anion exchanger.

A different picture is generated in the branchial response of the great spider crab *Hyas araneus* upon exposure to different levels of environmental pCO_2 combined with varying temperatures (Harms et al. 2014). While Na^+/K^+-ATPase was upregulated following hypercapnia alone and hypercapnia in combination with temperature, mRNA levels of V-(H^+)-ATPase and carbonic anhydrase were only significantly elevated upon moderate and severe hypercapnia (Harms et al. 2014). Additionally, changes in genes involved in metabolism indicated an enhanced aerobic metabolism in response to moderate hypercapnia, while severe hypercapnia induced a metabolic depression. Specifically, decreased trehalose metabolism of the gills seems to be a common response of hypercapnia as well as temperature acclimation in *H. araneus*.

Similar to the response of *C. maenas* (Fehsenfeld et al. 2011), analysis of the Gene Ontology terms (GO-terms) indicated a restructuring of the gill epithelium and/or the cytoskeleton upon hypercapnia in *H. araneus* (Harms et al. 2014), a phenomenon that can also be observed upon acclimation to dilute salinity in posterior gills of *C. maenas* (Compere et al. 1989). In contrast to *C. maenas*, however, gill epithelia of *H. araneus* seem to undergo a pronounced stress response that includes the elevation of genes involved in intracellular oxidative stress defence, including a number of peroxidases.

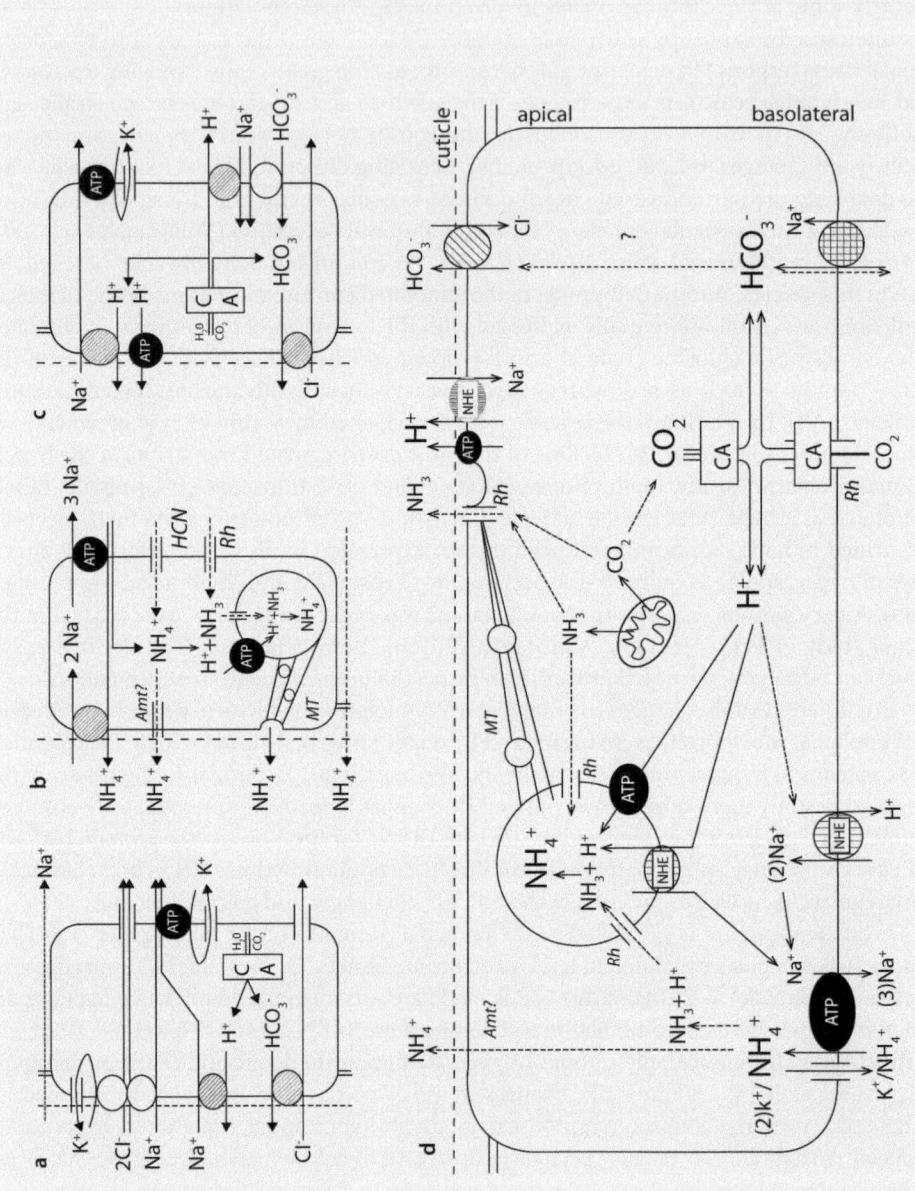

◼ **Fig. 6.3** Current working models for branchial ion regulatory processes in decapod crustaceans. (**a**) Osmoregulation, depicted after Onken et al. (2003); (**b**) ammonia excretion in the weak hyper-osmoregulator *C. maenas*, depicted after ▶ Chap. 1; and (**c**) general hypothetical model for acid–base regulation, depicted after Freire et al. (2008). (**d**) Depicts a recently developed model for acid–base regulation and its linkage with ammonia regulation in *C. maenas* after Fehsenfeld and Weihrauch (2016a). Potential overlaps in the transporter inventory of all models are basolateral Na^+/K^+-ATPase, basolateral K^+-channels (including a member of the HCN family, Fehsenfeld and Weihrauch 2016b) and potential apical Na^+/H^+-exchanger. (**a**, **c**, **d**) Also share a potential apical HCO_3^-/Cl^--exchanger. Light dashed lines to the left (**a–c**) and top (**d**) indicate the cuticle covering the apical membrane. Notice the different orientation of panel D. *Blue colour* indicates the involvement of the respective transporter only in acid–base, but not in ammonia excretion. *Bold* and *bigger letters* indicate proposed major routes for transport of respective acid-base equivalents. *Amt* ammonia transporter, *ATP* ATPase, *CA* carbonic anhydrase, *NHE* Na^+/H^+-exchanger, *MT* microtubule network, *Rh* Rhesus-like protein

6.5   Linking Acid–Base to Ammonia Regulation

Even though ammonia excretion in decapod crustaceans has been the focus of many studies, the potential importance of ammonia regulatory patterns in respect to acid–base regulation has not been acknowledged to date. Generally, ammonia exists in a pH-dependent equilibrium between the weak base NH_3 and its acidic form NH_4^+. With a pKa of 9.15, most ammonia is present as NH_4^+ at physiological pH (Weiner and Verlander 2013). Due to its physical properties, ammonia (and therefore ammonia excretion) might therefore very well contribute to acid–base homeostasis as an additional hemolymph buffer beside the carbonate system. Being the primary waste product of protein catabolism, NH_3/NH_4^+ levels are ultimately linked to the overall metabolic rate of the organism. As mentioned earlier, metabolic rates of decapod crustaceans are individually adjusted when experiencing external stressors that simultaneously also affect acid–base homeostasis. In response to hypoxia, for example, metabolism and hemolymph ammonia decreased significantly in *N. norvegicus* (Hagerman et al. 1990). A similar response was seen in *M. magister* when exposed to hypercapnia and included also a significant decrease in ammonia excretion rates (Hans et al. 2014). In *C. maenas* on the other hand, hemolymph ammonia as well as whole animal ammonia excretion increased significantly upon exposure to hypercapnia in both full-strength seawater and brackish-water acclimated specimen (Fehsenfeld and Weihrauch 2013; Fehsenfeld and Weihrauch 2016a). Interestingly, NH_4^+ excretion by individual gills of brackish-water acclimated *C. maenas* closely mirrored their H^+ excretion, indicating that acid excretion over the gill epithelium was mainly accomplished by NH_4^+ excretion (Fehsenfeld and Weihrauch 2013). Blocking basolateral V-(H^+)-ATPase, Na^+/K^+-ATPase and general K^+-channels (Fehsenfeld and Weihrauch 2016a), as well as the recently identified transporter HCN (Fehsenfeld and Weihrauch 2016b) by transporter-specific pharmaceuticals, simultaneously affected branchial NH_3/NH_4^+ excretion as well as the excretion of acid–base equivalents in this decapod crustacean.

Furthermore, as an identified key player in branchial ammonia excretion in crustaceans (Weihrauch et al. 2004b; Martin et al. 2011), Rhesus proteins have recently been strongly suggested to not only mediate NH_3 but to also act as CO_2 channels in human red blood cells (Endeward et al. 2008; Kustu and Inwood 2006; Musa-Aziz et al. 2009; Soupene et al. 2002, 2004) and fish gills (Perry et al. 2010). Interestingly, this protein is significantly downregulated in *C. maenas* in anterior gill 4 in response to hypercapnia (Fehsenfeld and Weihrauch 2013), as well as in posterior gills in response to both hypercapnia and high environmental ammonia (HEA; Fehsenfeld, pers. communication), but is significantly upregulated in HEA-acclimated *M. magister* (Martin et al. 2011), clearly indicating its role in acid–base regulation and providing a link to ammonia regulation in these decapod crustaceans.

6.6   Conclusion

Thanks to the numerous descriptive studies, we presently have a very good understanding of how various environmental factors influence acid–base homeostasis in aquatic decapod crustaceans. Accomplished mainly via adjustments of extracellular HCO_3^- concentrations and the correlated excretion of acid- and/or base equivalents, possibly connected to changes in the strong ion difference, crustaceans are capable of efficiently counteracting acid–base disturbances. Even though the collected data support a major role for the gill epithelium and the respective ion exchange processes in this organ in acid–base homeostasis, information on the actual mechanisms contributing its regulation are sparse. Studies on gill epithelial transporters involved in branchial acid–base regulation, however, deliver strong indications for a close link between ion and especially ammonia and acid–base regulation in the crustacean gill. The proposed mechanisms therefore resemble closely what is observed in the mammalian kidney, specifically in the thick ascending limb and the collecting duct. Future work needs to verify the localization of most of the respective proposed transporters in the crustacean gill epithelium and possibly other organs.

References

Ahearn GA (1978) Allosteric cotransport of sodium, chloride, and calcium by the intestine of freshwater prawns. J Membr Biol 42:281–300

Ahearn GA, Franco P, Clay LP (1990) Electrogenic 2 Na^+/1 H^+ exchange in crustaceans. J Membr Biol 116:215–226

Ahearn GA, Zhuang Z, Duerr J, Pennington V (1994) Role of the invertebrate electrogenic $2Na^+$/$1H^+$ antiporter in monovalent and divalent cation transport. J Exp Biol 196:319–335

Ahearn GA, Mandal PK, Mandal A (2004) Calcium regulation in crustaceans during the molt cycle: a review and update. Comp Biochem Physiol A 137:247–257

Almut G, Bamber S (2013) Behavioural responses of *Crangon crangon* (Crustacea, Decapoda) to reduced seawater pH following simulated leakage from sub-sea geological storage. J Environ Prot 4:61–67

Appelhans YS, Thomsen J, Pansch C et al (2012) Sour times: Seawater acidification effects on growth, feeding behaviour and acid–base status of *Asterias rubens* and *Carcinus maenas*. Mar Ecol Prog Ser 459: 85–98

Arnold KE, Findlay HS, Spicer JI et al (2009) Effect of CO_2-related acidification on aspects of the larval development of the European lobster, *Homarus gammarus* (L.). Biogeosciences 6:1747–1754

Barra J-A, Pequeux A, Humbert W (1983) A morphological study on gills of a crab acclimated to fresh water. Tissue Cell 15:583–596

Bellwood O (2002) The occurrence, mechanics and significance of burying behaviour in crabs (Crustacea: Brachyura). J Nat Hist 36:1223–1238

Beniash E, Ivanina A, Lieb NS et al (2010) Elevated level of carbon dioxide affects metabolism and shell formation in oysters *Crassostrea virginica*. Mar Ecol Prog Ser 419:95–108

Boettcher K, Siebers D, Becker W (1990) Localization of carbonic anhydrase in the gills of *Carcinus maenas*. Comp Biochem Physiol B 96:243–246

Booth CE, McMahon BR, De Fur PL, Wilkes PRH (1984) Acid–base regulation during exercise and recovery in the blue crab, *Callinectes sapidus*. Respir Physiol 58:359–376

Bouaricha N, Charmantier-Daures M, Thuet P et al (1994) Ontogeny of osmoregulatory structures in the shrimp *Penaeus japonicus* (Crustacea, Decapoda). Biol Bull 186:29–40

Burnett LE, Johansen K (1981) The role of branchial ventilation in hemolymph acid–base changes in the shore crab *Carcinus maenas* during hypoxia. J Comp Physiol 141:489–494

Burnett LE, McMahon BR (1987) Gas Exchange, hemolymph acid–base status, and the role of branchial water stores during air exposure in three littoral crab species. Physiol Zool 60:27–36

Burnett LE, Woodson PB, Rietow M, Vilicich VC (1981) Crab gill intra-epithelial carbonic anhydrase plays a major role in haemolymph CO_2 and chloride ion regulation. J Exp Biol 92:243–254

Caldeira K, Wickett ME (2005) Ocean model predictions of chemistry changes from carbon dioxide emissions to the atmosphere and ocean. J Geophys Res 110:1–12

Cameron JN (1978) Effects of hypercapnia on blood acid–base status, NaCl fluxes, and trans-gill potential in freshwater blue crabs, *Callinectes sapidus*. J Comp Physiol B 123:137–141

Cameron JN, Batterton CV (1978a) Antennal gland function in the freshwater blue crab, *Callinectes sapidus*: water, electrolyte, acid–base and ammonia excretion. J Comp Physiol B 123:143–148

Cameron JN, Batterton CV (1978b) Temperature and blood acid–base status in the blue crab, *Callinectes sapidus*. Respirin Physiol 35:101–110

Cameron JN, Wood CM (1985) Apparent H^+ excretion and CO_2 dynamics accompanying carapace mineralization in the blue crab (*Callinectes sapidus*) following moulting. J Exp Biol 114:181–196

Carter HA, Ceballos-Osuna L, Miller NA, Stillman JH (2013) Effects of ocean acidification on early life-history stages of the intertidal porcelain crab *Petrolisthes cinctipes*. J Exp Biol 216:1405–1411

Chapman CJ (1980) Ecology of juvenile and adult *Nephrops*. In: Cobb JS, Phillips BF (eds) The biology and management of lobsters. Academic Press, Inc., New York, p 1980

Charmantier G, Haond C, Lignot J-H, Charmantier-Daures M (2001) Ecophysiological adaptation to salinity throughout a life cycle: a review in homarid lobsters. J Exp Biol 204:967–977

Cheng SY, Shieh LW, Chen JC (2013) Changes in hemolymph oxyhemocyanin, acid–base balance, and electrolytes in *Marsupenaeus japonicus* under combined ammonia and nitrite stress. Aquat Toxicol 130–131:132–138

Choe H, Sackin H, Palmer LG (2000) Permeation properties of inward-rectifier potassium channels and their molecular determinants. J Gen Physiol 115:391–404

Choe KP, Kato A, Hirose S et al (2005) NHE3 in an ancestral vertebrate: primary sequence, distribution, localization, and function in gills. Am J Physiol Regul Integr Comp Physiol 289:R1520–R1534

Chu KH (1987) Sodium transport across the perfused midgut and hindgut of the blue crab, *Callinectes sapidus*: The possible role of the gut in crustacean osmoregulation. Comp Biochem Physiol A 87: 21–25

Chung KF, Lin HC (2006) Osmoregulation and Na, K-ATPase expression in osmoregulatory organs of *Scylla paramamosain*. Comp Biochem Physiol A 144:48–57

Collins M, Knutti R, Dufresne J-L et al (2013) Long-term climate change: projections, commitments and irreversibility. In: Stocker TF, Qin D, Plattner G-K et al (eds) Climate change 2013: the physical scienc basis. Contribution of working group I to the fifth assessment report of the intergovernmental panel of climate change. Cambridge University Press, Cambridge/New York

Compere P, Wanson S, Pequeux A et al (1989) Ultrastructural changes in the gill epithelium of the green crab *Carcinus maenas* in relation to the external salinity. Tissue Cell 21:299–318

Cooper RA, Uzmann JR (1980) Ecology of juvenile and adult *Homarus*. In: Cobb JS, Phillips BF (eds) The biology and management of lobsters. Academic Press, Inc., New York, pp 97–142

Copeland DE, Fitzjarrell AT (1968) The salt absorbing cells in the gills of the blue crab (*Callinectes sapidus* Rathbun) with notes on modified mitochondria. Zeitschrift fuer Zellforsch und Mikroskopische Anat 92:1–22

Crandall KA, Buhay JE (2008) Global diversity of crayfish (Astacidae, Cambaridae, and Parastacidae – Decapoda) in freshwater. Hydrobiologia 595:295–301

Dall W (1970) Osrnoregulation in the lobster *Homarus americanus*. J Fish Res Board Can 27:1123–1130

Davenport HW (1974) The ABC of acid base chemistry: The elements of physiological blood gas chemistry for medical students and physicians, 6th edn. The University of Chicago Press, Chicago

De Fur PL, McMahon BR (1984) Physiological compensation to short-term air exposure in Red rock crabs, *Cancer productus* Randall, from littoral and sublittoral habitats. II. Acid–base balance. Physiol Zool 57:151–160

De Fur PL, Wilkes PRH, McMahon BR (1980) Non-equilibrium acid–base status in *C. productus*: role of exo-skeletal carbonate buffers. Respir Physiol 42:247–261

De Fur PL, McMahon BR, Booth CE (1983) Analysis of hemolymph oxygen levled and acid–base status during emersion "in situ" in the red rock crab, *Cancer productus*. Biol Bull 165:582–590

De Grave S, Pentcheff ND, Ahyong ST et al (2009) A classification of living and fossil genera of decapod crustaceans. Raffles Bull Zool Suppl Ser 21:1–109

De la Haye KL, Spicer JI, Widdicombe S, Briffa M (2011) Reduced sea water pH disrupts resource assess-ment and decision making in the hermit crab *Pagurus bernhardus*. Anim Behav 82:495–501

Dejours P (1975) Principles of comparative respiratory physiology. Elsevier North Holland, New York

Dejours P, Beekenkamp H (1977) Crayfish respiration as a function of water oxygenation. Respir Physiol 30:241–251

Dissanayake A, Ishimatsu A (2011) Synergistic effects of elevated CO_2 and temperature on the metabolic scope and activity in a shallow-water coastal decapod (*Metapenaeus joyneri*; Crustacea: Penaeidae). ICES J Mar Sci 68:1147–1154

Dissanayake A, Clough R, Spicer JI, Jones MB (2010) Effects of hypercapnia on acid–base balance and osmo-/iono-regulation in prawns (Decapoda: Palaemonidae). Aquat Biol 11:27–36

Edwards SL, Donald JA, Toop T et al (2002) Immunolocalisation of sodium/proton exchanger-like proteins in the gills of elasmobranchs. Comp Biochem Physiol A 131:257–265

Ehrenfeld J (1974) Aspects of ionic transport mechanisms in crayfish *Astacus leptodactylus*. J Exp Biol 61:57–70

Endeward V, Cartron J-P, Ripoche P, Gros G (2008) RhAG protein of the Rhesus complex is a CO_2 channel in the human red cell membrane. FASEB J 22:64–73

Fehsenfeld S, Weihrauch D (2013) Differential acid–base regulation in various gills of the green crab *Carcinus maenas*: effects of elevated environmental pCO_2. Comp Biochem Physiol A 164:54–65

Fehsenfeld S, Weihrauch D (2016a) Mechanisms of acid–base regulation in seawater-acclimated green crabs (*Carcinus maenas*). Can J Zool 94:95–107

Fehsenfeld S, Weihrauch D (2016b) The role of an ancestral hyperpolarization-activated cyclic nucleotide-gated K+ channel in branchial acid–base regulation in the green crab, *Carcinus maenas*. J Exp Biol 219:1–10. doi:10.1017/CBO9781107415324.004

Fehsenfeld S, Kiko R, Appelhans Y et al (2011) Effects of elevated seawater pCO_2 on gene expression patterns in the gills of the green crab, *Carcinus maenas*. BMC Genomics 12:488

Freire CA, McNamara JC (1995) Fine structure of the gills of the fresh-water shrimp *Macrobrachium olfersii* (Decapoda): effect of acclimation to high salinity medium and evidence for involvement of the lamellar septum in ion uptake. J Crustac Biol 15:103–116

Freire CA, Cavassin F, Rodrigues EN et al (2003) Adaptive patterns of osmotic and ionic regulation, and the invasion of fresh water by the palaemonid shrimps. Comp Biochem Physiol A 136:771–778

Freire CA, Onken H, McNamara J (2008) A structure-function analysis of ion transport in crustacean gills and excretory organs. Comp Biochem Physiol A 151:272–304

Gilmour K, Perry S (2009) Carbonic anhydrase and acid–base regulation in fish. J Exp Biol 212:1647–1661

Guiasu RC (2002) Cambarus. In: Holdich DM (ed) Biology of Freshwater Crayfish. Blackwell Science Ltd, Oxford, pp 609–664

Hagerman L, Sondergaard T, Weile K et al (1990) Aspects of blood physiology and ammonia excretion in *Nephrops norvegicus* nuder hypoxia. Comp Biochem Physiol A 97:51–55

Hamilton NM, Houlihan DF (1992) Respiratory and circulatory adjustments during aquatic treadmill exercise in the European shore crab *Carcinus maenas*. J Exp Biol 162:37–54

Hammer KM, Pedersen SA (2013) Deep-water prawn *Pandalus borealis* displays a relatively high pH regulatory capacity in response to CO_2-induced acidosis. Mar Ecol Prog Ser 492:139–151

Hans S, Fehsenfeld S, Treberg JR, Weihrauch D (2014) Acid–base regulation in the Dungeness crab (*Metacarcinus magister*). Mar Biol 161:1179–7793

Haond C, Flik G, Charmantier G (1998) Confocal laser scanning and electron microscopical studies on osmoregulatory epithelia in the branchial cavity of the lobster *Homarus gammarus*. J Exp Biol 201:1817–1833

Harms L, Frickenhaus S, Schiffer M et al (2014) Gene expression profiling in gills of the great spider crab *Hyas araneus* in response to ocean acidification and warming. BMC Genomics 15:789

Harvey BP, Gwynn-Jones D, Moore PJ (2013) Meta-analysis reveals complex marine biological responses to the interactive effects of ocean acidification and warming. Ecol Evol 3:1016–1030

Henry RP, Cameron JN (1982) Acid–base balance in *Callinectes sapidus* during acclimation from high to low salinity. J Exp Biol 101:255–264

Henry RP, Cameron JN (1983) The role of carbonic anhydrase in respiration, ion regulation and acid–base balance in the aquatic crab *Callinectes sapidus* and the terrestrial crab *Gecarcinus lateralis*. J Exp Biol 103:205–223

Henry RP, Wheatly MG (1992) Interaction of respiration, ion regulation, and acid–base balance in the everyday life of aquatic crustaceans. Am Zool 32:407–416

Henry RP, Kormanik GA, Smatresk NJ, Cameron JN (1981) The role of $CaCO_3$ dissolution as a source of HCO_3^- for the buffering of hypercapnic acidosis in aquatic and terrestrial decapod crustaceans. J Exp Biol 94:269–274

Henry RP, Booth CE, Lallier FH, Walsh PJ (1994) Post-exercise lactate production and metabolism in three species of aquatic and terrestrial decapod crustaceans. J Exp Biol 186:215–234

Henry RP, Gehnrich S, Weihrauch D, Towle DW (2003) Salinity-mediated carbonic anhydrase induction in the gills of the euryhaline green crab, *Carcinus maenas*. Comp Biochem Physiol A 136:243–258

Henry RP, Lucu Č, Onken H, Weihrauch D (2012) Multiple functions of the crustacean gill: Osmotic/ionic regulation, acid–base balance, ammonia excretion, and bioaccumulation of toxic metals. Front Physiol 3:1–33

Hill AD, Taylor AC, Strang RHC (1991) Physiological and metabolic responses of the shore crab *Carcinus maenas* (L.) during environmental anoxia and subsequent recovery. J Exp Mar Bio Ecol 150:31–50

Hochachka PW, Mommsen TP (1983) Protons and anaerobiosis. Science 219:1391–1397

Howell BJ, Rahn H, Goodfellow D, Herreid C (1973) Acid–base regulation and temperature in selected invertebrates as a function of temperature. Integr Comp Biol 13:557–563

Hu MY, Guh Y-J, Stumpp M et al (2014) Branchial NH_4^+-dependent acid–base transport mechanisms and energy metabolism of squid (Sepioteuthis lessoniana) affected by seawater acidification. Front Zool 11:55

Hunter KC, Kirschner LB (1986) Sodium absorption coupled to ammonia excretion in osmoconforming marine invertebrates. Am J Physiol 251:R957–R962

Hunter KC, Rudy PPJ (1975) Osmotic and ionic regulation in the Dungeness crab, Cancer magister dana. Comp Biochem Physiol A 51A:439–447

IPCC (2013) Summary for policymakers. In: Stocke TF, Qin D, Plattner GK et al (eds) Climate change 2013: the physical science basis. Contribution of working group I to the fifth assessment report of the intergovernmental panel on climate change. Cambridge University Press, Cambridge/New York

Jouve-Duhamel A, Truchot J-P (1983) Ventilation in the shore crab Carcinus maenas (L.) as a function of ambient oxygen and carbon dioxide: field and laboratory studies. J Exp Mar Bio Ecol 70:281–296

Keppel EA, Scrosati RS, Courtenay SC (2012) Ocean acidification decreases growth and development in American lobster (Homarus americanus) larvae. J Northwest Atl Fish Sci 44:61–66

Kustu S, Inwood W (2006) Biological gas channels for NH_3 and CO_2: evidence that Rh (Rhesus) proteins are CO_2 channels. Transfus Clin Biol 13:103–110

Lagos ME, Cáceres CW (2008) Como afecta la exposición aérea el equilibrio ácido base de organismos móviles del intermareal: Petrolisthes laevigatus (Guérin, 1835) (Decapoda: Porcellanidae), como caso de estudio. Rev Biol Mar Oceanogr 43:591–598

Lagos M, Cáceres CW, Lardies MA (2014) Geographic variation in acid–base balance of the intertidal crustacean Cyclograpsus cinereus (Decapoda, Grapsidae) during air exposure. J Mar Biol Assoc U K 94: 159–165

Langdon C, Takahashi T, Sweeney C et al (2000) Effect of calcium carbonated saturation state on the calcification rate of an experimental coral reef. Global Biogeochem Cycles 14:639–654

Larsen EH, Deaton LE, Onken H et al (2014) Osmoregulation and excretion. Compr Physiol 4:405–573

Lignon JM (1987) Ionic permeabilities of the isolated gill cuticle of the shore crab Carcinus maenas. J Exp Biol 131:159–174

Liu M, Liu S, Hu Y, Pan L (2015) Cloning and expression analysis of two carbonic anhydrase genes in white shrimp Litopenaeus vannamei, induced by pH and salinity stresses. Aquaculture. doi:10.1016/j.aquaculture.2015.04.038

Long CW, Swiney KM, Foy RJ (2013) Effects of ocean acidification on the embryos and larvae of red king crab, Paralithodes camtschaticus. Mar Pollut Bull 69:38–47

Lucu Č (1990) Ionic regulatory mechanisms in crustacean gill epithelia. Comp Biochem Physiol A 97: 297–306

Lucu Č, Siebers D (1987) Linkage of Cl^- fluxes with ouabain sensitive Na/K exchange through Carcinus gill epithelia. Comp Biochem Physiol A 87:807–811

Luquet CM, Ansaldo M (1997) Acid–base balance and ionic regulation during emersion in the estuarine intertidal crab Chasmagnathus granulata Dana (Decapoda Grapsidae). Comp Biochem Physiol A 117:407–410

Mangum CP, McMahon BR, De Fur PL, Wheatly MG (1985) Gas exchange, acid–base balance, and the oxygen supply to the tissues during a molt of the blue crab Callinectes sapidus. J Crustac Biol 5:188–206

Mantel LH, Farmer LL (1983) Osmotic and ionic regulation. In: Mantel LH (ed) The Biology of Crustacea. Academic Press, Inc., New York, pp 53–161

Martin M, Fehsenfeld S, Sourial MM, Weihrauch D (2011) Effects of high environmental ammonia on branchial ammonia excretion rates and tissue Rh-protein mRNA expression levels in seawater acclimated Dungeness crab Metacarcinus magister. Comp Biochem Physiol A 160:267–277

McDonald DG, McMahon BR, Wood CM (1979) An analysis of acid–base disturbances in the haemolymph following strenuous activity in the Dungeness crab, Cancer magister. J Exp Biol 79:47–58

McMahon BR, Sinclair F, Hassall CD et al (1978) Ventilation and control of acid–base status during temperature acclimation in the crab, Cancer magister. J Comp Physiol B 128:109–116

McNamara JC, Faria SC (2012) Evolution of osmoregulatory patterns and gill ion transport mechanisms in the decapod Crustacea: a review. J Comp Physiol B 182:997–1014

McNamara JC, Lima AG (1997) The route of ion and water movements across the gill epithelium of the freshwater shrimp Macrobrachium olfersii (Decapoda, Palaemonidae): evidence from ultrastructural changes induced by acclimation to saline media. Biol Bull 192:321–331

Meinshausen M, Smith SJ, Calvin K et al (2011) The RCP greenhouse gas concentrations and their extensions from 1765 to 2300. Clim Change 109:213–241

Melzner F, Gutowska M, Langenbuch M et al (2009) Physiological basis for high CO_2 tolerance in marine ectothermic animals: pre-adaptation through lifestyle and ontogeny? Biogeosciences 6:2313–2331

Michaelidis B, Ouzounis C, Paleras A, Pörtner HO (2005) Effects of long-term moderate hypercapnia on acid–base balance and growth rate in marine mussels *Mytilus galloprovincialis*. Mar Ecol Prog Ser 293:109–118

Mora C, Wei CL, Rollo A et al (2013) Biotic and human vulnerability to projected changes in ocean biogeochemistry over the 21st century. PLoS Biol 11, e1001682

Morse HC, Harris PJ, Dornfeld EJ (1970) *Pacifastacus leniusculus*: Fine structure of arthrobranch with reference to active ion uptake. Trans Am Microsc Soc 89:12–27

Musa-Aziz R, Chen L-M, Pelletier MF, Boron WF (2009) Relative CO_2/NH_3 selectivities of AQP1, AQP4, AQP5, AmtB, and RhAG. Proc Natl Acad Sci U S A 106:5406–5411

Ng PKL, Davie PJF, Guinot D (2008) Systema brachyurorum : part I. An annotated checklist of extant brachyuran crabs of the world. Raffles Bull Zool 17:1–286

Onken H, Putzenlechner M (1995) A V-ATPase drives active, electrogenic and Na^+-independent Cl^- absorption across the gills of *Eriocheir sinensis*. J Exp Biol 198:767–774

Onken H, Riestenpatt S (2002) Ion transport across posterior gills of hyperosmoregulating shore crabs (*Carcinus maenas*): amiloride blocks the cuticular Na^+ conductance and induces current-noise. J Exp Biol 205:523–531

Onken H, Tresguerres M, Luquet CM (2003) Active NaCl absorption across posterior gills of hyperosmoregulating *Chasmagnathus granulatus*. J Exp Biol 206:1017–1023

Pane EF, Barry JP (2007) Extracellular acid–base regulation during short- term hypercapnia is effective in a shallow water crab, but ineffective in a deep- sea crab. Mar Ecol Prog Ser 334:1–9

Pequeux A (1995) Osmotic regulation in crustaceans. J Crustac Biol 15:1–60

Perry SF, Braun MH, Noland M et al (2010) Do zebrafish Rh proteins act as dual ammonia-CO_2 channels? J Exp Zool Part A 313(A):618–621

Rastrick SPS, Calosi P, Calder-Potts R et al (2014) Living in warmer, more acidic oceans retards physiological recovery from tidal emersion in the velvet swimming crab, *Necora puber*. J Exp Biol 217:2499–2508

Reynolds J, Souty-Grosset C, Richardson A (2013) Ecological roles of crayfish in freshwater and terrestrial habitats. Freshw Crayfish 19:197–218

Ries JB, Cohen AL, McCorkle DC (2009) Marine calcifiers exhibit mixed responses to CO_2-induced ocean acidification. Geology 37:1131–1134

Riestenpatt S, Onken H, Siebers D (1996) Active absorption of Na^+ and Cl^- across the gill epithelium of the shore crab *Carcinus maenas*: voltage-clamp and ion-flux studies. J Exp Biol 199:1545–1554

Riggs AF (1988) The Bohr effect. Annu Rev Physiol 50:181–204

Robertson JD (1960) Ionic regulation in the crab *Carcinus maenas* (L.) in relation to the moulting cycle. Comp Biochem Physiol 1:183–212

Rose RA, Wilkens JL, Walker RL (1998) The effects of walking on heart rate, ventilation rate and acid–base status in the lobster *Homarus americanus*. J Exp Biol 201:2601–2608

Schiffer M, Harms L, Pörtner HO et al (2013) Tolerance of *Hyas araneus* zoea I larvae to elevated seawater PCO_2 despite elevated metabolic costs. Mar Biol 160:1943–1953

Serrano L, Henry R (2008) Differential expression and induction of two carbonic anhydrase isoforms in the gills of the euryhaline green crab, *Carcinus maenas*, in response to low salinity. Comp Biochem Physiol D 3:186–193

Shetlar RE, Towle DW (1989) Electrogenic sodium-proton exchange in membrane vesicles from crab (*Carcinus maenas*) gill. Am J Physiol 257:R924–R931

Siebers D, Lucu Č, Böttcher K, Jürss K (1994) Regulation of pH in the isolated perfused gills of the shore crab *Carcinus maenas*. J Comp Physiol B 164:16–22

Simonik E, Henry RP (2014) Physiological responses to emersion in the intertidal green crab, *Carcinus maenas* (L.). Mar Freshw Behav Physiol 47:101–115

Skou JC (1960) Further investigations on a Mg^{++} +Na^+-activated adenosinetriphosphatase, possible related to the active, linked transport of Na^+ and K^+ across the nerve membrane. Biochim Biophys Acta 42:6–23

Small D, Calosi P, White D et al (2010) Impact of medium-term exposure to CO_2 enriched seawater on the physiological functions of the velvet swimming crab *Necora puber*. Aquat Biol 10:11–21

Somero GN (1986) Protons, osmolytes, and fitness of internal milieu for protein function. Am J Physiol 251:R197–R213

Soupene E, King N, Feild E et al (2002) Rhesus expression in a green alga is regulated by CO_2. Proc Natl Acad Sci U S A 99:7769–7773

Soupene E, Inwood W, Kustu S (2004) Lack of the Rhesus protein Rh1 impairs growth of the green alga *Chlamydomonas reinhardtii* at high CO_2. Proc Natl Acad Sci U S A 101:7787–7792

Spicer JI, Raffo A, Widdicombe S (2007) Influence of CO_2-related seawater acidification on extracellular acid–base balance in the velvet swimming crab *Necora puber*. Mar Biol 151:1117–1125

Stewart PA (1978) Independent and dependent variables of acid–base control. Respir Physiol 33:9–26

Tavares C, Martin JW (2010) Suborder dendrobranchiata bate, 1888. In: Schram FR, von Vaupel Klein JC, Forest J, Charmantier-Daures M (eds) Eucarida: Euphausiacea, Amphionidacea, and Decapoda (partim.). Treatise on zoology – anatomy, taxonomy, biology – The Crustacea. Koninklijke Brill NV, Leiden, pp 99–164

Taylor AC, Spicer JI (1991) Acid–base disturbances in the haemolymph of the prawns, *Palaemon elegans* (Rathke) and *P. serratus* (Pennant) (Crustacea: Decapoda) during exposure to hypoxia. Comp Biochem Physiol A 98:445–452

Taylor HH, Taylor EW (1992) Gills and lungs: The exchange of gases and ions. In: Harrison FW, Humes AG (eds) Microscopid anatomy of invertebrates, decapod Crustacea. Wiley-Liss, New York, pp 203–293

Taylor EW, Wheatly MG (1981) The effect of long-term aerial exposure on heart rate, ventilation respiratory gas exchange and acid–base status in the crayfish *Austropotamobius pallipes*. J Exp Biol 92:109–124

Taylor EW, Whiteley NM (1989) Oxygen transport and acid–base balance in the haemolymph of the lobster, *Homarus gammarus*, during aerial exposure and resubmersion. J Exp Biol 144:417–436

Taylor JRA, Gilleard JM, Allen MC, Deheyn DD (2015) Effects of CO_2-induced pH reduction on the exoskeleton structure and biophotonic properties of the shrimp *Lysmata californica*. Nat Sci Rep 5:10608

Towle D, Henry R, Terwilliger N (2011) Microarray-detected changes in gene expression in gills of green crabs (*Carcinus maenas*) upon dilution of environmental salinity. Comp Biochem Physiol D 6:115–125

Tresguerres M, Parks S, Sabatini S et al (2008) Regulation of ion transport by pH and [HCO_3^-] in isolated gills of the crab *Neohelice (Chasmagnathus) granulata*. Am J Physiol Regul Integr Comp Physiol 294:R1033–R1043

Truchot J-P (1973) Temperature and acid–base regulation in the shore crab *Carcinus maenas* (L.). Respir Physiol 17:11–20

Truchot J-P (1975a) Factors controlling the in vitro and in vivo oxygen affinity of the hemocyanin in the crab *Carcinus maenas* (L.). Respir Physiol 24:173–189

Truchot J-P (1975b) Blood acid–base changes during experimental emersion and reimmersion ot the intertidal crab *Carcinus maenas* (L.). Respir Physiol 23:351–360

Truchot J-P (1975c) Action de l'hypercapnie sur l'etat acide-base du sang chez le crabe *Carcinus maenas* (L.) (Crustace', De'capode). Comptes Rendus l'Académie des Sci 280:311–314

Truchot J-P (1979) Mechanisms of the compensation of blood respiratory acid–base disturbances in the shore crab, *Carcinus maenas* (L.). J Exp Biol 210:407–416

Truchot J-P (1981) The effect of water salinity and acid–base state on the blood acid–base balance in the euryhaline crab, *Carcinus maenas* (L.). Comp Biochem Physiol A 68:555–561

Truchot J-P (1984) Water carbonate alkalinity as a determinant of hemolymph acid–base balance in the shore crab, *Carcinus maenas*: a study at two different ambient PCO_2 and PO_2 levels. J Comp Physiol B 154:601–606

Truchot J-P (1988) Problems of acid–base balance in rapidly changing intertidal environments. Am Zool 28:55–64

Truchot J-P (1992) Acid–base changes on transfer between sea- and freshwater in the Chinese crab, *Eriocheir sinensis*. Respir Physiol 87:419–427

Truchot J-P, Duhamel-Jouve A (1980) Oxygen and carbon dioxide in the marine intertidal environment: diurnal and tidal changes in rockpools. Respir Physiol 39:241–254

Tsai J-R, Lin H-C (2007) V-type H^+-ATPase and Na^+, K^+-ATPase in the gills of 13 euryhaline crabs during salinity acclimation. J Exp Biol 210:620–627

Urbina MA, Paschke K, Gebauer P et al (2013) Physiological responses of the southern king crab, *Lithodes santolla* (Decapoda: Lithodidae), to aerial exposure. Comp Biochem Physiol A 166:538–545

Vargas M, Lagos ME, Contreras DA, Caceres CW (2010) Área de estructuras respiratorias y su efecto en la regulación del equilibrio ácido-base en dos especies de cangrejos porcelánidos intermareales, *Petrolisthes laevigatus* y *Petrolisthes violaceus*. Rev Biol Mar Oceanogr 45:245–253

Varley DG, Greenaway P (1992) The effect of emersion on haemolymph acid–base balance and oxygen levels in *Scylla serrata* Forskal (Brachyura: Portunidae). J Exp Mar Bio Ecol 163:1–12

Walther G-R, Post E, Convey P et al (2002) Ecological responses to recent climate change. Nature 416:389–395

Weihrauch D, Becker W, Postel U et al (1998) Active excretion of ammonia across the gills of the shore crab *Carcinus maenas* and its relation to osmoregulatory ion uptake. Comp Biochem Physiol B 168:364–376

Weihrauch D, Becker W, Postel U et al (1999) Potential of active excretion of ammonia in three different haline species of crabs. J Comp Physiol B 169:25–37

Weihrauch D, Ziegler A, Siebers D, Towle DW (2001) Molecular characterization of V-type H^+-ATPase (B-subunit) in gills of euryhaline crabs and its physiological role in osmoregulatory ion uptake. J Exp Biol 204:25–37

Weihrauch D, Ziegler A, Siebers D, Towle DW (2002) Active ammonia excretion across the gills of the green shore crab *Carcinus maenas*: participation of Na^+/K^+-ATPase, V-type H^+-ATPase and functional microtubules. J Exp Biol 205:2765–2775

Weihrauch D, McNamara JC, Towle DW, Onken H (2004a) Ion-motive ATPases and active, transbranchial NaCl uptake in the red freshwater crab, *Dilocarcinus pagei* (Decapoda, Trichodactylidae). J Exp Biol 207:4623–4631

Weihrauch D, Morris S, Towle DW (2004b) Ammonia excretion in aquatic and terrestrial crabs. J Exp Biol 207:4491–4504

Weiner ID, Verlander JW (2013) Renal ammonia metabolism and transport. Comp Physiol 3:201–220

Wheatly MG (1985) The role of the antennal gland in ion and acid–base regulation during hyposaline exposure of the Dungeness crab *Cancer magister* (Dana). J Comp Physiol B 155:445–454

Wheatly MG, Gannon AT (1995) Ion regulation in crayfish: freshwater adaptations and the problem of molting. Am J Zool 35:49–59

Wheatly MG, Henry RP (1987) Branchial and antennal gland Na^+/K^+-dependent ATPase and carbonic anhydrase activity during salinity acclimation of the euryhaline crayfish *Pacifastacus liniusculus*. J Exp Biol 133:73–86

Wheatly MG, McMahon BR (1982) Responses to hypersaline exposure in the euryhaline crayfish *Pacifastacus leniusculus*. J Exp Biol 99:425–445

Wheatly MG, Taylor EW (1979) Oxygen levels, acid–base status and heart rate during emersion of the shore crab *Carcinus maenas* (L.) into air. J Comp Physiol B 311:305–311

Wheatly MG, Taylor EW (1981) The effect of progressive hypoxia on heart rate, ventilation, respiratory gas exchange and acid–base status in the crayfish *Austropotamobius pallipes*. J Exp Biol 92:109–124

Wheatly MG, Toop T (1989) Physiological responses of the crayfish *Pacifastacus leniusculus* to environmental hyperoxia. J Exp Biol 143:53–70

Wheatly MG, Toop T, Morrison RT, Yow LC (1991) Physiological responses of the crayfish *Pacifasticus leniusculus* (Dana) to environmental hyperoxia. III. Intracellular acid–base balance. Physiol Zool 64:323–343

Whiteley NM (2011) Physiological and ecological responses of crustaceans to ocean acidification. Mar Ecol Prog Ser 430:257–271

Whiteley NM, Taylor EW (1990) The acid–base concequences of aerial exposure in the lobster, *Homarus gammarus* (L.) at 10 and 20C. J Therm Biol 15:47–56

Whiteley NM, Scott JL, Breeze SJ, McCann L (2001) Effects of water salinity on acid–base balance in decapod crustaceans. J Exp Biol 204:1003–1011

Wickins JF (1984) The effect of hypercapnic sea water on growth and mineralization in penaied prawns. Aquaculture 41:37–48

Widdicombe S, Spicer JI, Kitidis V (2011) Effects of ocean acidification on sediment fauna. In: Gattuso J-P, Hansson L (eds) Ocean Acidification. Oxford University Press, New York, pp 176–191

Wilkes PRH, McMahon BR (1982) Effect of maintained hypoxic exposure on the crayfish *Orconectes rusticus*. I. ventilatory, acid–base and cardiovascular adjustments. J Exp Biol 98:139–149

Wittmann AC, Pörtner H-O (2013) Sensitivities of extant animal taxa to ocean acidification. Nat Clim Chang 3:995–1001

Cell Biology of Reef-Building Corals: Ion Transport, Acid/Base Regulation, and Energy Metabolism

*Martin Tresguerres, Katie L. Barott, Megan E. Barron,
Dimitri D. Deheyn, David I. Kline, and Lauren B. Linsmayer*

© Springer International Publishing Switzerland 2017
D. Weihrauch, M. O'Donnell (eds.), *Acid-Base Balance and Nitrogen Excretion in Invertebrates*,
DOI 10.1007/978-3-319-39617-0_7

7.1   Summary

Coral reefs are built by colonial cnidarians that establish a symbiotic relationship with dinoflagellate algae of the genus *Symbiodinium*. The processes of photosynthesis, calcification, and general metabolism require the transport of diverse ions across several cellular membranes and generate waste products that induce acid/base and oxidative stress. This chapter reviews the current knowledge on coral cell biology with a focus on ion transport and acid/base regulation while also discussing related aspects of coral energy metabolism.

7.2   Introduction

Reef-building corals, also known as hermatypic or stony corals, are marine invertebrates from the phylum Cnidaria (Class Anthozoa, Subclass Hexacorallia, Order Scleractinia) (Bourne 1900) that build external calcium carbonate skeletons beneath their tissues. The reefs they build provide homes for a variety of other organisms, resulting in one of the most biodiverse ecosystems in the world, that provide ecosystem and economic services for hundreds of millions of people in tropical and subtropical areas (Burke et al. 2011).

Reef-building corals are colonial and establish an endosymbiotic relationship with dinoflagellates of the genus *Symbiodinium*. Reef-building corals fall into one of two clades, robust and complex, which diverged from each other between ~300 and ~400 million years ago (Romano and Cairns 2000; Romano and Palumbi 1996; Stolarski et al. 2011). Species belonging to the robust clade (e.g., *Stylophora*, *Pocillopora*, *Orbicella*) tend to have heavier calcified skeletons, while those from the complex clade (e.g., *Acropora*, *Porites*, *Siderastrea*) tend to have a more porous and less calcified skeleton. Another difference is the type of asexual growth mode used. Robust corals use intratentacular budding, when a parent corallite splits in the middle of the existing corallite into two daughter corallites. However, complex corals use extratentacular budding, when a parent corallite produces a daughter corallite external to the corallite wall (Wijsman-Best 1975). These differences in colony growth modes likely account for some of the skeletal differences between these two major clades of corals.

As the divergence between robust and complex clades occurred before corals developed skeletons, calcification likely evolved multiple independent times and therefore likely resulted in clade-specific or even species-specific mechanisms. This concept may well apply to other physiological processes, having important implications for coral biology and coral's responses to environmental stress, as not all coral species may respond similarly to the same stresses. Whenever possible, this book chapter will mention species- or clade-specific mechanisms. We advise the reader to consult the primary literature for specific details and to keep in mind that many of the mechanisms discussed may vary depending on species or clade, life history, geographic location, and local environmental conditions.

7.3   Coral Anatomy, Histology, and Cytology

A coral colony is comprised of multiple anemone-like polyps that are interconnected by a tissue layer overlying the skeletal plates that join the individual corallites, called the coenosarc, and internally by their gastrovascular cavity, called the coelenteron (◻ Fig. 7.1). Although the paradigm is that all polyps from a given coral colony are genetically identical clones, recent research has found that up to 50 % of coral colonies may be mosaics of different genotypes (Schweinsberg et al. 2015). Like all cnidarians, corals are diploblastic, meaning they have two germ layers, the ectoderm and the endoderm. During embryonic development, the blastula undergoes gastrulation and turns into a larva known as planulae, which result in the subdivision of both the ectoderm and endoderm into oral and aboral tissue layers (Babcock and Heyward 1986). This tissue organization is maintained after larval settlement and growth into juvenile and adult forms, with the oral ectoderm being in contact with seawater, the oral and aboral endoderm "sandwiched" in between and separated from each other by the coelenteron, and the aboral ectoderm (calicoblastic epithelium) sitting on top of the subcalicoblastic medium (SCM) and the skeleton (◻ Fig. 7.1b, c). The oral ectoderm and gastroderm, as well as their aboral counterparts, are interconnected by the mesoglea, an extracellular matrix comprised of connective protein fibers (mostly collagen), water, and some wandering amebocytes (Phillips 1963). ◻ Figure 7.1 shows a simplified diagram of coral anatomy. An example of some of the complexity that is not shown is that the polyp gastrodermal layers form mesenterial filaments with cells specialized for food digestion and reproduction (gonads), and the oral layers surrounding the polyp's mouth are modified into tentacles with abundant nematocysts for predatory food capture and defense (Galloway et al. 2006; Veron 1993). Additionally, areas of rapid growth and calcification may only have the oral and aboral ectodermal layers, with no gastrodermal tissue in between (Jokiel 2011a).

Each of the coral's tissue layers has several cell subtypes; however, the specific physiological function(s) of each cell subtype is not completely characterized. A "generic" coral oral ectoderm contains at least six cell subtypes: ciliated support cells with abundant microvilli, nematocysts, mucocytes, pigment cells, neurons, and epitheliomuscular cells (Goldberg 2002a) (◻ Fig. 7.1c). Ciliated cells are in direct contact with seawater and interact with the diverse microbiota present in the mucus and boundary layer immediately on top of the coral, with which they may exchange diverse substances including nitrogenous compounds, amino acids, and organic acids such as sulfur (Chimetto et al. 2008; Krediet et al. 2013; Raina et al. 2010). Corals may also have endosymbiotic cyanobacteria throughout the oral ectoderm and endoderm, some of which presumably aid in nitrogen fixation (Lesser et al. 2014). It has been argued that this interaction between coral and microbes is as essential as the symbiosis between coral and *Symbiodinium*, and the term "coral holobiont" has been coined to refer to the assemblage of diverse organisms living in close association with a coral colony (the cnidarian animal, *Symbiodinium* endosymbionts, microbes and viruses, endolithic algae, as well as fungi that might be present in between and underneath the coral tissues) (Rohwer et al. 2002). Since these microorganisms are fast evolving, they may provide at least some phenotypic plasticity for the coral holobiont to adapt to rapid environmental change (Rosenberg et al. 2007). However, like most other aspects of coral cell biology, the mechanisms behind these complex processes are largely unknown.

The most prominent cells in the oral gastroderm are the *Symbiodinium* algae and the cnidarian cells that host them; here we will use the term "symbiocyte" to refer to this host cell–symbiont cellular complex. The host cell tightly surrounds the endosymbiont, so its cytoplasm and organelles are only visible using high-magnification microscopy techniques. While most symbiocytes host a single *Symbiodinium*, they may occasionally host two or even

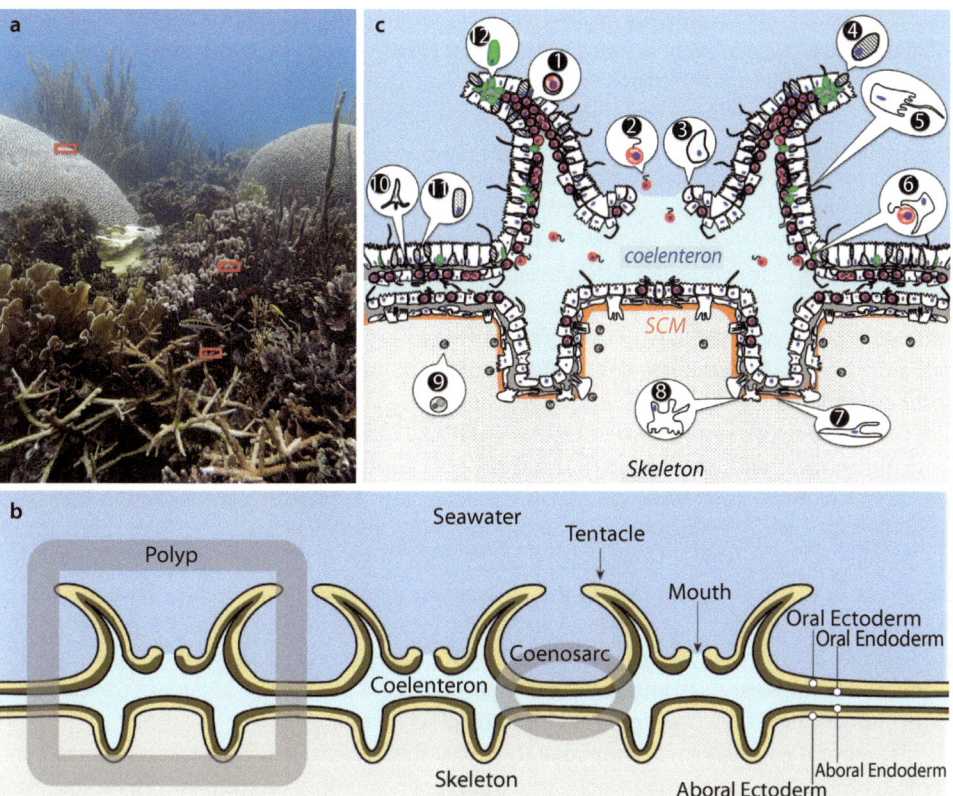

☐ **Fig. 7.1** Coral anatomy, histology, and cytology. (**a**) General view of a coral reef in Bocas del Toro, Panama. The areas enclosed in the *red boxes* are shown magnified as cartoons in b and c. (**b**) Diagram showing the different parts of a coral colony as well as the various coral tissue layers (aboral ectoderm = calicoblastic epithelium). (**c**) Coral cytology. (*1*) Symbiotic *Symbiodinium* in the symbiosome of a gastrodermal cell; (*2*) free-living *Symbiodinium*, with flagellum; (*3*) epitheliomuscular cells; (*4*) cnidocyte; (*5*) ciliated support cells; (*6*) gastrodermal cell in the process of phagocytize Symbiodinium; (*7*) calicoblastic cell; (*8*) desmocyte; (*9*) dead *Symbiodinium* encrusted in old skeleton; (*10*) neuron; (*11*) mucocyte; (*12*) pigment cell. *SCM* subcalicoblastic medium. Note: these are artistic renditions that do not necessarily reflect real cell sizes, morphology, relative proportion of the different cell types, or coral species-specific differences (Based on Allemand et al. (1998, 2011); Barott et al. (2015a), Goldberg (2001a, b, 2002a, b), Johnston (1980), Veron (1993))

three, which becomes more readily apparent in cell isolates (Barott et al. 2015b; Venn et al. 2009). Symbiocytes are morphologically and physiologically complex. *Symbiodinium* reside inside within the symbiosome, an intracellular space delimited by the host-derived symbiosome membrane that originates when algae are phagocytized from the gastrovascular cavity, which is later modified to sustain the various metabolic exchanges that make the symbiosis possible (Davy et al. 2012) (discussed in ▶ Sects. 7.6 and 7.7). Of particular importance, the symbiosome membrane must mediate the fluxes of dissolved inorganic carbon (DIC), nitrogenous compounds, and PO_4^{3-} from host cells to *Symbiodinium* and of photosynthates (e.g., organic compounds such as glycerol and glucose) in the opposite direction.

A given coral colony can host different strains or clades of *Symbiodinium*, and the specific strains and their relative abundances can dynamically change over time, potentially resulting in altered physiology and susceptibility or resistance to stress (Cunning et al. 2015a, b; Little et al. 2004). *Symbiodinium* give corals their typical brown/greenish color, but some corals are also biofluorescent under appropriate illumination conditions (Catala-Stucki 1959) due to fluorescent proteins (FPs). These FPs are produced in pigment cells in the oral layers and stored in granules, and they may also be transported to other cell types. FPs are present throughout the coral colony, including polyp tentacles, polyp wall, and in the coenosarc (FPs are discussed in ▶ Sect. 7.13).

The coral oral and aboral layers are separated by the gastrovascular cavity ("coelenteron"), which internally connects polyps within a colony. The fluid in the coelenteron is circulated throughout the colony by the action of flagella and/or cilia in gastrodermal cells, likely aided by peristaltic muscular contractions (Gladfelter 1983). Although the coelenteron opens to seawater *via* the polyps' mouths, the composition of the coelenteron fluid is quite different from that of seawater. For example, the concentrations of vitamin B_{12} (riboflavin), NO_3^-, NO_2^-, NH_4^+, and PO_4^{3-} can be between 30- and 2000-fold higher in the coelenteron, while bacterial counts can be 100-fold lower (Agostini et al. 2011). Similarly, the fluid in the coelenteron has distinct levels of dissolved oxygen, as well as CO_2, pH $[HCO_3^-]$, and $[CO_3^{2-}]$, which are different from both the surrounding seawater and the coral cells (Agostini et al. 2011; Furla et al. 2000). Furthermore, these parameters change dramatically due to photosynthesis, calcification, and food digestion, with pH fluctuations from 8.5 to 7.5 in *S. pistillata* (Furla et al. 2000) and 7.25 to 6.6 in *Galaxea fascicularis* (Agostini et al. 2011), with the higher values being in the light and the lower values in the dark. Due to the logarithmic nature of the pH scale, these pH values represent fluctuation in $[H^+]$ from ~3 to 30 nM in *S. pistillata* and from ~56 to 250 nM in *G. fascicularis*. The composition and chemistry of the coelenteron fluid is critical when considering calcification mechanisms and other aspects of coral physiology, including the effects of environmental stress such as ocean acidification on coral biology.

The aboral gastroderm is immediately below the coelenteron. This cell layer also contains symbiocytes, which may reach large numbers depending on the position on the coral colony (e.g., base vs. tip of a coral branch), the metabolic status, and the environmental conditions. The cytology and physiology of the aboral gastroderm is the least studied among all coral tissue layers. However, since this tissue layer lies in between the calicoblastic cells and the coelenteron, it likely serves essential functions in the exchange of DIC, Ca^{2+}, and H^+ for calcification. Similarly, the aboral gastroderm must be involved in mechanisms to transport sugars, fatty acids, and other molecules from the symbiocytes to the calicoblastic cells; however, these putative mechanisms remain unknown.

The aboral ectoderm is below the aboral gastroderm and immediately above the skeleton. This tissue layer contains calicoblastic cells and desmocytes. Both cell types seem to be responsible for building and maintaining the skeleton, and desmocytes additionally anchor the coral to the skeleton (described in detail in ▶ Sects. 7.8 and 7.9). The interface between calicoblastic cells and skeleton forms pockets filled with SCM, a gel-like fluid that is highly alkaline, hypersaturated with respect to aragonite, and contains a variety of proteins secreted by the coral (▶ see Sect. 7.10). The skeleton lies at the base of the coral colony and is made up of $CaCO_3$. The skeletal crystal structure of adult corals is almost exclusively aragonite (Wainwright 1964), but some of the first-formed elements in larval coral skeletons can be calcite (Vandermeulen and Watabe 1973). Traditionally, skeletal morphology was used for coral identification and classification. However, skeleton morphology is not a reliable indicator of coral evolutionarily history, as it can vary greatly due to environmental conditions. Current models on coral evolution are instead based on molecular phylogenetic analyses, which often do not match skeleton morphology (Fukami et al. 2004; Romano and Cairns 2000; Romano and Palumbi 1996; Stolarski et al. 2011). In addition, coral hybridization, especially during mass spawning events, makes molecular taxonomy challenging (van Oppen et al. 2001; Willis et al. 1997) and even raises the possibility of reticulate evolution (Diekmann et al. 2001; Odorico and Miller 1997).

Coral skeletons also include organic components, although these are much less abundant compared to aragonite (~0.1 % of the total skeletal dry weight), much more heterogeneous, and show vast species-specific differences (reviewed in Johnston (1980)). Organic components of the skeletal organic matrix (SOM) include proteins, lipids, and polysaccharides. Coral skeletons also have embedded dead host tissue and *Symbiodinium*, as well as micro- and macrosopic endolithic organisms such as cyanobacteria, bacteria, filamentous green algae, fungi, boring mollusks, and sponges. This diversity often complicates determining the various components of coral skeletons and their functions (reviewed in Johnston (1980)).

7.4 Physiological Challenges Associated with Photosynthesis and Calcification

The following chemical equilibrium equations are especially relevant for coral biology and are essential for understanding some of the concepts discussed in this chapter:

i $CO_2 + H_2O \Leftrightarrow H_2CO_3 \Leftrightarrow HCO_3^- + H^+ \Leftrightarrow CO_3^{2-} + 2H^+$ (DIC equilibria, important for A/B balance, movement of across biological membranes, and OA, among other processes)

ii $Ca^{2+} + 2HCO_3^- \Leftrightarrow CaCO_3 + 2H^+$ (relevant for calcification)

iii $6CO_2 + 12\ H_2O + Light \Leftrightarrow C_6H_{12}O_6 + 6O_2$ (photosynthesis equation)

The symbiosis between the coral host and *Symbiodinium* endosymbionts drives the productivity of reef ecosystems. Light-driven photosynthesis by *Symbiodinium* can provide the majority of the coral's energy needs (Muscatine 1990), allowing coral reefs to thrive in otherwise oligotrophic waters. The external skeleton is another hallmark of coral reefs, which requires the transport of both Ca^{2+} and DIC to the site of calcification. Photosynthesis and calcification have unique acid/base (A/B) requirements, and meeting these requirements can greatly disturb the A/B status of coral's internal and external microenvironments. However, the A/B requirements of photosynthesis and calcification are at odds, as photosynthesis requires an acidic pH to favor the speciation of DIC into CO_2 (the substrate for carbon fixation for ribulose-1,5-bisphosphate carboxylase/oxygenase (RuBisCo)), while calcification requires an alkaline pH to favor CO_3^{2-} formation (the "building blocks" for the skeleton). Similarly, photosynthesis consumes H^+, but calcification generates H^+, which, respectively, alkalize and acidify the surrounding microenvironment. In fact, diel and seasonal cycles of photosynthesis/respiration and calcification/dissolution of corals and associated reefs organisms can result in large swings in pH in the surrounding seawater, especially on shallow reef flats with long residence times where diel pH ranges can be 0.5–0.6 units and seasonal changes can be 0.7–0.8 pH units (Kline et al. 2015; Shaw et al. 2012; Silverman et al. 2012).

The sources of DIC for photosynthesis and calcification are the surrounding seawater and metabolically produced CO_2. However, once again there is no unifying concept on the relative contribution of each source, as available estimates depend on the coral species, environmental conditions, and experimental techniques utilized. All of the Ca^{2+} necessary for skeleton deposition is of external origin and therefore has to pass through four (if Ca^{2+} is taken up from seawater) or two (if it is taken up from the coelenteron or in the rapidly calcifying areas mentioned above) cell layers. Based on $^{45}Ca^{2+}$ influx and efflux kinetics, it has been suggested that all skeletal Ca^{2+} passes transcellularly through at least one coral tissue layer, presumably the calicoblastic epithelium (Tambutté et al. 1996). The proposed transcellular mechanisms include Ca^+ channels and plasma membrane Ca^+-ATPases (PMCA), as well as vesicular transport of DIC, Ca^{2+}, and skeletal organic matrix proteins (detailed in ▶ Sect. 7.9). An essential point to consider is that Ca^{2+} plays several crucial roles in cell homeostasis such as intracellular signaling and serving as a cofactor for various enzymes (e.g., Petersen and Petersen (1994), Friedman and Gasek (1995)). Additionally, free cytoplasmic Ca^{2+} would interact with the intracellular phosphate buffer system. Thus, the transcellular transport of Ca^{2+} likely involves some sort of Ca^{2+} sequestration within coral cells to prevent Ca^{2+} toxicity. Candidate mechanisms include vesicles or Ca^{2+}-binding proteins (cytosolic or intravesicular), which could be directionally trafficked across the cell toward the SCM.

Another physiological challenge for corals is that *Symbiodinium* photosynthesis produces copious amounts of O_2 as an end product. On the contrary, the biota and microbiota associated with corals can drive O_2 levels to hypoxic (or maybe even anoxic) levels at night, depending on density and water flow (Kühl et al. 1995; Ohde and van Woesik 1999; Shashar et al. 1993). These pronounced O_2 fluctuations have important implications for coral physiology in terms of their energy budget (▶ see Sect. 7.12).

The lack of coral-specific tools is a major disadvantage for studies on coral cellular physiology. For example, many previous studies proposed mechanisms based on the effects of pharmacological inhibitors for proteins from mammals, which might not be specific for the homologous coral proteins. Additionally, these types of pharmacological studies cannot determine the coral cell type or layer where the presumed target proteins are located, because drugs are dissolved in the surrounding seawater and therefore can reach multiple sites within a coral colony. Another hurdle is that drugs can form complexes with the various salts present in seawater, and therefore demonstrate different activity, or require different concentrations, compared to studies using mammalian physiological saline solutions. Recent advances in coral biology are rapidly expanding the experimental toolbox. For example, the sequencing of some coral genomes (Drake et al. 2014; Shinzato et al. 2011) has made it possible to develop coral-specific antibodies and perform shotgun transcriptomic and proteomic studies to identify candidate proteins involved in various responses based on differential regulation following stress. However, there is still the need for coral cell model systems that would allow genetic manipulations such as knocking down, silencing, or knocking in genes of interest to study the functions of encoded proteins. Elucidating the cellular mechanisms behind essential coral physiological processes is of crucial importance in coral research and will be critical for understanding and predicting coral responses to stress.

7.5   Acid/Base (A/B) Regulation

Unlike larger and more active animals such as vertebrate animals, mollusks, and crustaceans, corals do not have specialized organs to regulate the A/B status of their internal fluids. Thus, corals seem to rely on intracellular pH regulation (pHi) to achieve A/B homeostasis. In addition, corals maintain unique acidic and alkaline environments in the symbiosome space (Barott et al. 2015b) and in the SCM (Al-Horani et al. 2003; Venn et al. 2011), which, respectively, promote photosynthesis and calcification.

Tight pHi regulation is vital for all organisms as enzymes and most other cellular components are highly sensitive to pH. Universal challenges to pHi balance include metabolic activity and fluctuations in external pH (pHe). Additionally, photosynthesis and calcification in reef-building corals entail unique A/B conditions in microenvironments that are located only microns apart. In addition to standard "housekeeping" pHi, symbiocytes and calicoblastic cells must have developed specialized mechanisms to maintain extreme pH conditions in the symbiosome space and SCM, respectively, and to compensate for the pHi stress resulting from photosynthesis and calcification.

Like all cells, coral cells regulate pHi by passive buffering and by active transport of A/B relevant ions across cellular membranes. Information about the buffering capacity of cnidarian cells is limited to one study on isolated gastrodermal cells of *Anemonia viridis* (Laurent et al. 2014); their buffering capacity ranged from 20.8 to 43.8 mM/pH unit, similar to cells from other invertebrates. There were few differences between *Symbiodinium*-containing and *Symbiodinium*-free gastrodermal cells, at least in the dark (Laurent et al. 2013). Like all other animal cells (Putnam and Roos 1997; Roos and Boron 1981), active pHi regulation in corals likely involves Na^+/H^+ exchangers (NHEs, SLC9 family) and HCO_3^- transporters (e.g., SLC4 and SLC26 families). Homologous genes are present in the available cnidarian genomes, but they have not been functionally characterized yet. A detailed characterization of these transporters in terms of substrate specificity, kinetics, pharmacology, regulation, and potential cell-specific expression is an essential area of future research.

7.6   DIC Transport Across the Oral Epithelium

Photosynthesis relies heavily on seawater HCO_3^-, which requires ion-transporting proteins to transport it across the apical and basolateral membranes of the oral ectoderm cells and across the symbiocytes' membrane (Allemand et al. 1998). In addition, HCO_3^- could move across the paracellular pathway (i.e., in between cells) from seawater to the symbiocytes or be taken up by the symbiocytes from the coelenteron. In any case, DIC still has to move across the symbiocyte and symbiosome membranes, as well as across various *Symbiodinium* membranes, before finally reaching RuBisCo in the chloroplast stroma and pyrenoid.

The identity of the HCO_3^--transporter proteins has just begun to be elucidated. A recent paper suggests that Slc4-like transporters present in the apical membrane of oral ectodermal cells mediate HCO_3^- uptake from seawater in *A. yongei* (Barott et al. 2015a). Apically located Na^+/K^+-ATPase (NKA) might energize the uptake of HCO_3^- (and possibly other substances) from seawater (Barott et al. 2015a). However, apical NKA (Barott et al. 2015a) and the SLC4-like transporters present in *A. yongei* (◘ Fig. 7.2) were not detectable in *S. pistillata*. Once again, this highlights potential mechanistic differences between coral species and/or coral clades.

7.7   pHi Regulation in Gastrodermal Cells

Symbiodinium photosynthesis has a strong alkalinizing effect on their gastrodermal host cells. Exposure of isolated *S. pistillata* symbiocytes to light for 20 min induced a significant increase in pHi from ~7.10 to ~7.40 pH units (Venn et al. 2009). The alkalinizing effect is abolished by DCMU (Laurent et al. 2013), a specific blocker of plastoquinone that inhibits

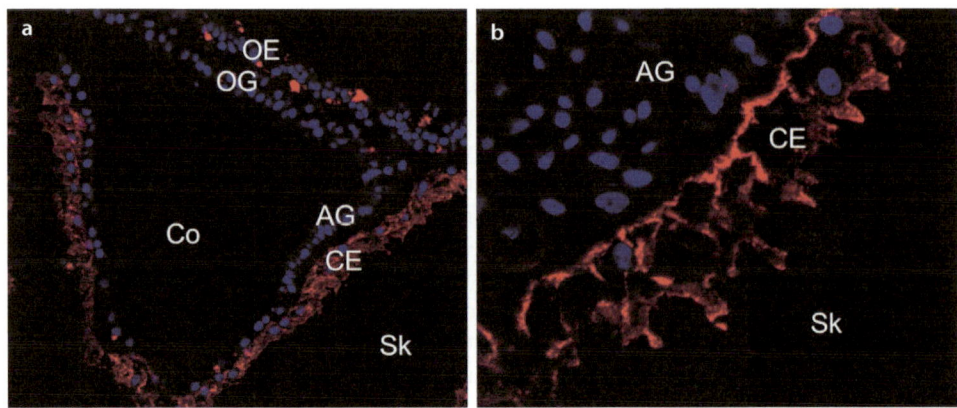

■ **Fig. 7.2** (**a**) 320x magnification; (**b**) m 1000x magnification. Localization of an SLC4-like protein in the coral *Stylophora pistillata* (*red*). Nuclei are stained in *blue*. *OE* oral ectoderm, *OG* oral gastroderm, *Co* coelenteron, *AG* aboral gastroderm, *CE* calicoblastic ectoderm, *Sk* skeleton (Methods followed those described in Barott et al. (2015a))

photosynthesis, confirming that the alkalinization was caused by *Symbiodinium* photosynthesis. Additionally, alkalinization occurred more quickly and reached higher values (maxing out at ~7.46) in cells exposed to irradiance levels of 300 μmol photons m^{-2}s^{-1} or higher (Laurent et al. 2013). Similar effects have been reported in isolated *P. damicornis* cells (Gibbin et al. 2014). Host cytoplasm alkalinization is caused by CO_2 removal by *Symbiodinium* photosynthesis and possibly by OH$^-$ secretion by *Symbiodinium* to the host cytoplasm (Allemand et al. 1998; Venn et al. 2009). Indeed, as photosynthesis draws down CO_2, HCO_3^- and H$^+$ are converted to CO_2 following the DIC equilibrium reaction, and the decline in [H$^+$] leads to alkalization. To prevent pHi from rising further, symbiocytes would need to excrete OH$^-$ and/or HCO_3^- or uptake H$^+$ and/or CO_2. For example, anemone endoderm cells may excrete OH$^-$ into the coelenteron upon light exposure (or uptake H$^+$) (Furla et al. 1998), and coral symbiocytes potentially transport HCO_3^- into the symbiosome as part of a carbon-concentrating mechanism (CCM) for photosynthesis (Barott et al. 2015b).

7.7.1 pH of the Symbiosome

In *S. pistillata* and *A. yongei*, the host-derived symbiosome membrane abundantly expresses vacuolar H$^+$-ATPase (VHA), a proton pump whose activity helps acidify the symbiosome to pH ~4 (Barott et al. 2015b). Acidification of the symbiosome space surrounding the zooxanthellae is part of a coral CCM that promotes *Symbiodinium* photosynthesis (Barott et al. 2015b). As *Stylophora* and *Acropora* belong to different coral clades, it is likely that this CCM is universal among coral species. The ultimate goal of the proposed CCM is to increase CO_2 to high-enough levels to sustain photosynthesis, which is required due to the low affinity of dinoflagellate RuBisCo for CO_2 over O_2 (Rowan et al. 1996). However, there are no known cellular mechanisms to actively transport gases such as CO_2, and at a pHi of ~ 7.10 (Venn et al. 2009), the most abundant form of DIC available inside coral gastrodermal cells would be in the form of HCO_3^-. Thus, DIC transport across biological membranes must happen by transporting H$^+$ and HCO_3^-, which can then dehydrate into CO_2 (and H_2O) in a reaction reversibly catalyzed by carbonic anhydrase (CA). Experiments using pharmacological drugs support the involvement of CAs, as inhibition of CA by Diamox reduces photosynthetic carbon assimilation by zooxanthellae by up to 85 % (Weis et al. 1989). However, CAs alone are not sufficient to transport DIC against its concentration gradient, which is a requirement for a CCM. This is where VHA becomes relevant by using energy from adenosine triphosphate (ATP) hydrolysis to transport H^+ against its electrochemical gradient ("pumping"), which presumably drives the transport of HCO_3^- into the symbiosome space by yet unidentified transporter proteins. Supporting this model, inhibition of VHA by bafilomycin resulted in an increase of the symbiosome pH by ~0.6 units (a ~70 % change in $[H^+]$) in *S. pistillata* isolated symbiocytes, as well as a reduction in O_2 production by ~30 % in *S. pistillata* colonies, and as high as an 80 % reduction in O_2 production in *A.yongei* colonies (Barott et al. 2015b). A potential additional source of H^+ transport into the symbiosome is the P-type H^+-ATPase, which is expressed by *Symbiodinium* only during symbiosis (Bertucci et al. 2010).

Elucidating the mechanisms of HCO_3^- transport across the symbiosome membrane is an important area of future research. Furthermore, once inside the symbiosome space, CO_2 still has to move across several other *Symbiodinium* membranes before reaching RuBisCo in the chloroplast stroma and pyrenoid; these mechanisms are also largely unknown. Finally, symbiosome acidification could help drive the transport of other molecules between host cells and *Symbiodinium*, as hypothesized in Barott et al. (2015b).

7.8 DIC Transport and pH Regulation in Calcifying Cells

The calicoblastic cells play a vital role in skeleton formation by promoting calcification through elevating pHe in the SCM, transporting DIC and Ca^{2+}, and excreting SOM proteins that promote $CaCO_3$ precipitation (reviewed in Cohen and McConnaughey (2003), Allemand et al. (2004), Johnston (1980)).

The pH of the SCM ranges between 8.1 and 8.4 pH units in the dark and between 8.7 and 9.3 pH units in the light, depending on the coral species and method used (Al-Horani et al. 2003; Venn et al. 2011). At those pH values, the calculated saturation state of aragonite is 20–30-fold hypersaturated in the light and 3–11-fold in the dark. However, the pHi of calicoblastic cells remains stable at ~7.4 in both light and dark conditions (Venn et al. 2011). This indicates that calicoblastic cells are capable of excreting HCO_3^- into the SCM against a steep electrochemical gradient and removing H^+ while also maintaining a stable pHi.

The sources and species of DIC used for calcification are not completely understood but may include CO_2, HCO_3^-, and CO_3^{2-} from seawater or metabolic origin that are ultimately transported and concentrated in the SCM (Allison et al. 2014). Based on immunolocalization data, members of the SLC4 family of HCO_3^- transporters may excrete HCO_3^- into the SCM both in *A. yongei* (Barott et al. 2015a) and *S. pistillata* (Zoccola et al. 2015). However, convincing functional data is not available due to the issues with drug specificity and site of action outlined in ▶ Sect. 7.2. The driving force for HCO_3^- secretion, and most likely many other transport processes, seems to be provided by NKA that is abundantly expressed in the basolateral membrane of calicoblastic cells from both *A. yongei* and *S. pistillata* (Barott et al. 2015a).

The contribution of metabolic CO_2 to the skeleton's aragonite has been estimated to be as high as 83 % (Erez 1978; Furla et al. 2000; Hughes et al. 2010). Metabolic CO_2 is chiefly produced during respiration in mitochondria, which can be abundantly present throughout coral tissues including calicoblastic cells (Barott et al. 2015b). This CO_2 could be rapidly hydrated into HCO_3^- by CA in the cell cytosol and exported into the SCM. The resulting H^+ would then be exported across the basolateral membrane into the mesoglea and eventually to the coelenteron. However, other studies have concluded that the respiration rate of corals is too low to significantly contribute DIC to the skeleton (e.g., Falkowski et al. (1984)).

7.9 Ca^{2+} Transport

Despite its central importance in coral biology, the mechanisms for Ca^{2+} transport in corals remain poorly understood. The route of Ca^{2+} transport from seawater to the coelenteron across the oral epithelium has been proposed to be either paracellular (Furla et al. 2000; Tambutté et al. 1996) or transcellular (Clode and Marshall 2002). A contribution of both routes, as well as of species-specific differences, is possible. Ectodermal cells in the oral and calicoblastic layers are connected by septate junctions (Barott et al. 2015a; Clode and Marshall 2002; Ganot et al. 2015; Johnston 1980), which limit the passage of ions and other molecules across the coral epithelium. Electrophysiology of coral microcolonies indicates that the coral epithelium has intermediate resistance, falling on the leaky end of vertebrate tight junctions' continuum but still "tighter" than leaky junctions. However, the relative contribution of the different coral tissue layers to coral epithelial resistance could not be differentiated due to methodological limitations (Tambutté et al. 2011).

There is evidence that all skeletal Ca^{2+} passes through at least one coral tissue layer (Tambutté et al. 1996). An L-type voltage-dependent Ca^{2+} channel was immunolocalized using heterologous antibodies in both the oral and aboral ectoderm of the coral *S. pistillata*, suggesting it functions to take up Ca^{2+} from seawater and export it into the SCM (Zoccola et al. 1999). In addition, pharmacological inhibitors of Ca^{2+} channels reduced calcification rates. However, due to their typically slow inactivation time and short mean open time, L-type Ca^{2+} channels are primarily involved in Ca^{2+} homeostasis and Ca^{2+} signaling in other organisms (Hosey and Lazdunski 1988). Although this does not rule out a role for L-type Ca^{2+} channels in coral calcification, it does highlight the need for studies on the kinetics and specificity of coral Ca^{2+} channels in order to better determine their physiological functions.

Plasma membrane Ca^{2+}-ATPases (PMCA) have been proposed to export Ca^{2+} across the apical membrane of calicoblastic cells in exchange for H^+ (Zoccola et al. 2004, Allemand et al. 2011, Davy et al. 2012, Cohen et al. 2009). Thus, PMCA could both deliver the Ca^{2+} needed for calcification and remove protons produced during $CaCO_3$ precipitation to maintain the alkaline pH of the SCM. However, this model is largely based on results from pharmacological inhibition, in this case with ruthenium red (Al-Horani et al. 2003; Ip et al. 1991; Marshall 1996). Although PMCA messenger RNA (mRNA) is present in calicoblastic cells, it is also abundantly expressed throughout other coral cell layers (Zoccola et al. 2004). Moreover, a recent immunolocalization study found PMCA protein in the cytoplasm and not the membrane of calicoblastic cells in *A. yongei* and *S. pistillata* (Barott et al. 2015a), further questioning the role of PMCA in Ca^{2+} excretion and H^+ removal. In *A. yongei*, PMCA protein was present throughout all coral tissue layers, with more intense localization in the cytoplasm of calicoblastic cells followed by the apical pole of cells in the oral ectoderm. PMCA protein was similarly present throughout all coral tissue layers of *S. pistillata*; however, it was most abundant in gastrodermal cells, more specifically in the cytoplasmic area facing the coelenteron.

Abundant vesicles have been observed in the calicoblastic cells of multiple species of corals, in some cases pinching in and out at the basolateral and apical membranes of calicoblastic cells (Barott et al. 2015a; Isa 1986; Johnston 1980). Furthermore, inhibitors of actin and tubulin polymerization reduced Ca^{2+} incorporation into the skeleton, suggesting trafficking of intracellular vesicles is required for calcification (Tambutté et al. 1996). However, these results could also be explained by effects on the trafficking of HCO_3^- or H^+ transporters (see (Tresguerres et al. 2006) for an example of transepithelial HCO_3^- secretion in shark gills).

7.10 Skeletal Organic Matrix (SOM)

In addition to $CaCO_3$ in the form of aragonite, coral skeletons have numerous proteins, lipids, and polysaccharides that are collectively referred as the skeletal organic matrix (SOM). A majority of the SOM components seem to be synthesized and secreted by calicoblastic cells (Puverel et al. 2004), although some precursors of the SOM may be synthesized by *Symbiodinium* or other coral tissues. Not surprisingly, significant differences in SOM composition exist between coral species (Johnston 1980; Puverel et al. 2005).

The SOM has been proposed to be especially important in the early mineralization phase, as "substrate, catalyst, or controlling agent in the chemical reactions that culminate in the deposition of new skeletal material" (Johnston 1980). Organic components present in deeper parts of the skeleton could be remnants from the early calcification process, or they could play additional structural roles such as modifying skeletal mechanical properties.

The SOM confers the skeleton with chemical properties that are different from pure $CaCO_3$; for example, some coral acid-rich proteins (CARPs) recently identified in the SOM of *S. pistillata* are able to spontaneously catalyze $CaCO_3$ precipitation in vitro, even at a seawater pH of 7.6 (Mass et al. 2013). Immunolabeling confirmed that at least four CARPs are present in calicoblastic cells and embedded in skeletal aragonite crystals but also in noncalcifying coral tissues suggesting roles other than calcification (Mass et al. 2014).

Another interesting protein found in the SOM is a secreted form of CA, hypothesized to play roles in calcification and A/B regulation in calicoblastic cells and the SCM (Moya et al. 2008). The SCM, which is the microenvironment where calcification occurs, clearly is vastly different from bulk seawater due to significant biological control by the coral tissues.

7.11 Light-Enhanced Calcification

The stimulatory effect of light on coral calcification was noted as early as the 1930s (Kawaguti 1937), subsequently demonstrated chemically by measuring the disappearance of Ca^{2+} from seawater (Kawaguti and Sakumoto 1948), and later corroborated by radioisotope studies showing incorporation of $^{45}Ca^{2+}$ from seawater into the coral skeleton (Goreau 1959; Goreau and Goreau 1959). This phenomenon, which was termed light-enhanced calcification (Chalker and Taylor 1975), was subsequently confirmed across multiple coral species (reviewed in Allemand et al. (1998), Allemand et al. (2011), Cohen and Holcomb (2009), Gattuso (1999), Johnston (1980)). Despite a few exceptions (e.g., Wijgerde et al. (2012, 2014)) (which might be explained by husbandry and/or illumination conditions), light-enhanced calcification seems to be a universal phenomenon among corals.

The physiological mechanisms supporting light-enhanced calcification are not completely understood, but all hypotheses proposed over the last 80 years are based on *Symbiodinium* photosynthetic activity facilitating calcification in one or multiple ways. Thus, the term "photosynthetically enhanced calcification" (Chalker and Taylor 1975) seems more appropriate. Some possible mechanisms for light-enhanced calcification include photosynthesis providing (a) carbohydrates as fuel to support the energetic cost of transporting DIC to the SCM and maintaining its alkaline pH, (b) O_2 to support aerobic metabolism, (c) a supply of SOM precursors, and (d) for the removal of calcification waste products such as CO_2, H^+, PO_4^{3-}, and NH_4^+ (reviewed in Allemand et al. (2011), Chalker and Taylor (1975), Johnston (1980)). Clearly, the ability to study any of these hypotheses requires the elucidation of the basic mechanisms for calcification, energy metabolism, and metabolic communication between coral host cells and *Symbiodinium*.

7.12   Coral Energy Metabolism

Symbiodinium translocate the majority of net-fixed carbon to the host, in some cases >95 % (Davies 1984; Falkowski et al. 1984; Muscatine et al. 1984). Under high light conditions, autotrophy from *Symbiodinium* photosynthesis can supply over 100 % of a coral's daily energy requirements (Falkowski et al. 1984). Most of the translocated carbon is immediately respired by the host, but a considerable amount (potentially as high as 50 %) is lost in coral mucus secretions (Crossland et al. 1980; Muscatine et al. 1984). The role of mucus is unknown but could be related to aiding in feeding; cleaning; staving off epiphytic, epizoic, and bacterial growth (Ducklow and Mitchell 1979; Muscatine 1973); protecting against damage (Benson et al. 1978); and getting rid of excess translocated carbon (Davies 1984). In any case, coral mucus release likely represents a significant source-dissolved organic carbon for reef microbial communities (Muscatine et al. 1984; Wild et al. 2004).

Originally, glycerol was believed to be the main organic compound supplied by *Symbiodinium* (Muscatine and Cernichiari 1969); however, more recent results indicate it is in fact glucose (Burriesci et al. 2012). Either way, both glycerol and glucose are nitrogen deficient and can sustain metabolic respiration but cannot be used to build new tissue (Davies 1984; Falkowski et al. 1984). Thus, corals must obtain nitrogenous compounds from heterotrophic feeding on zooplankton, bacteria, and sessile organisms, which they do using cnidocysts, mesenterial filament eversion and extrusion, and mucus trapping (Goreau et al. (1971) and references therein). Corals can also absorb PO_4^{3-}, amino acids, and other micronutrients directly from seawater (Goreau et al. 1971). The relative contribution of autotrophy and heterotrophy likely varies according to species and dynamic changes in environmental conditions such as light, *Symbiodinium* abundance and type, and nutrient availability.

Corals experience large daily variations in O_2 levels due to *Symbiodinium* photosynthesis during the day (which can increase O_2 concentrations in coral tissues and their diffusive boundary layer to over 4x higher than the surrounding seawater) and to respiration of reef organisms and microorganisms during the night that can result in very low O_2 levels (hypoxia) or even lack thereof (anoxia) (Kühl et al. 1995; Ohde and van Woesik 1999; Shashar et al. 1993). This implies that corals are able to dynamically switch between aerobic and anaerobic respiration; however, virtually nothing is known about coral anaerobic metabolic pathways and their relative contribution to the coral energy budget.

Because hypoxia and anoxia likely also occur during stressful conditions that reduce *Symbiodinium* abundance, the capacity to generate energy using anaerobic metabolic pathways might be especially important in determining a coral species' ability to survive mass-bleaching events.

7.13 Coral Fluorescence

Cnidarians, including corals, display biofluorescence under appropriate illumination conditions (Catala-Stucki 1959). Originally, coral fluorescence was proposed to play roles in UV protection and optimization of algal productivity (Ben-Zvi et al. 2015; Dawson 2007; Kawaguti 1969; Schlichter et al. 1994). As part of these studies, fluorescent mycosporine-like amino acids were discovered; however, these are found in the secreted mucus, and their fluorescence is relatively weak and labile. Subsequent research demonstrated that coral biofluorescence is largely due to endogenous fluorescent proteins (FPs).

The green fluorescence protein (GFP) was originally found associated with the luminous jellyfish *Aequorea victoria* (Shinomura et al. 1962) and was proposed to play a role in mimicry by changing the bioluminescence color of luciferase from blue (best for color dispersion in deeper waters) to green (best in shallow water) *via* Förster Resonance Energy Transfer (Ohmiya and Hirano 1996; Haddock et al. 2009). More recently, FPs were found in corals and sea anemones that do not produce bioluminescence (Matz et al. 1999; Wiedenmann et al. 2004). Although FPs are predominantly green, other colors have also evolved ranging from blue to red (Labas et al. 2002; Sabine et al. 2004; Alieva et al. 2008), and FPs of up to four colors have been found in a single coral (Kelmanson and Matz 2003). In addition, some proteins in this family, called chromoproteins, absorb but do not emit light (Dove et al. 1995, 2001). Altogether, FPs are one of the main pigments in cnidarians and give them a diversity of fluorescent colors (Oswald et al. 2007).

Adult corals express FPs throughout the colony, including in polyp tentacles, body wall, and coenosarc. The fluorescence intensity is usually stronger in the growing tips of corals, probably because the reduced *Symbiodinium* abundance in the coral tips results in less shading of FP fluorescence. FPs are also expressed in the eggs, larvae, and juveniles of many coral species (Kenkel et al. 2011; Roth et al. 2013); however, the specific fluorescence color may change throughout the coral life cycle.

The precise cellular localization of FPs remains a challenge (Leal et al. 2015). In many cnidarians, FPs occur largely in granules of coral pigments; however, FPs could also be transported across tissues in corals, as described in other organisms (Hanson and Kohler 2001; Schonknecht et al. 2008; McLean and Cooley 2013). FPs can represent up to 14 % of the soluble protein content in anthozoans and have slow decay lives with half-lives of about 20 days (Leutenegger et al. 2007). Furthermore, FPs can be accumulated at various stages of protein maturity, possibly building up as a nonfluorescent premature stage before maturing and becoming fluorescent following oxidation of the chromophore (Leutenegger et al. 2007). Thus, large amounts of FPs could be produced and stored in pigment cells but be readily available readily to other coral tissues where they might play different physiological functions.

Despite abundant information about FPs' spectral and biochemical properties in vitro, FP in vivo functions remain unclear. Since FPs can absorb shorter, potentially damaging high-energy wavelengths of light and transform them into longer, lower energy wavelengths that are beneficial for photosynthesis, two of the most likely FP functions are photoprotection (Roth et al. 2010; Salih et al. 2000; Smith et al. 2013) and photoenhancement of *Symbiodinium* photosynthesis (Dove et al. 2008; Kawaguti 1969; Roth 2014). Additionally, some FPs display antioxidant activity in vitro (Bou-Abdallah et al. 2006; Bomati et al. 2009), suggesting that they protect corals from reactive oxygen species generated as by-products of *Symbiodinium* photosynthesis in vivo and other sources of oxidative stress.

Fluorescence is dynamic in corals, being up- or downregulated as a function of life stage (Kenkel et al. 2011), presence of trematode parasites (Palmer et al. 2009), and various environmental conditions. For example, corals exposed to different intensities of light showed associated changes in FPs, with exposure to higher light levels resulting in larger FP abundance and greater fluorescence intensity (Roth et al. 2010). Similarly, a temperature shock (either cold or warm) induced a rapid decrease in fluorescence, which was detectable earlier than a RGB color change (Roth and Deheyn 2013; Roth et al. 2012). Fluorescence intensity returned to normal levels after corals adjusted to the cold shock or became more intense after continued exposure to warm water that induced bleaching; this was due to a loss of light shading by *Symbiodinium*. Similar declines in fluorescence level were observed under other stressors such as anoxia (Haas et al. 2014) and light intensity (Roth et al. 2010, 2013). Most likely, the decrease in fluorescence is related to the biochemical antioxidant property of GFP. Indeed, GFP otherwise is a very stable protein with a pKa between 5 and 6 that is sheltered from changes in the surrounding biochemical environment, except for its susceptibility to free radicals. These studies suggest that changes in fluorescence can be used as a proxy for physiological adjustments associated with oxidative stress in corals.

7.14   Potential Role of cAMP in Regulating Coral Physiology

The cyclic adenosine 3'5' monophosphate (cAMP) pathway can modulate virtually every aspect of cell biology by regulating the activity of target proteins *via* protein kinase A-dependent phosphorylation, exchange proteins activated by cAMP, and cyclic nucleotide-gated channels (Wong & Scott, 2004). The cAMP pathway seems to have a special significance in corals, as homogenates of *A. yongei* and *P. damicornis* have the highest cAMP production rates ever recorded for any organism: 17,000 pmol mg^{-1} min^{-1} and 30,000 pmol mg^{-1} min^{-1}, respectively (Barott et al. 2013). Furthermore, cAMP levels in *P. damicornis* fluctuate with the light/dark cycle, reaching their highest values during the day and the lowest at night (Barott et al. 2013). However, the physiological roles of cAMP in corals are virtually unknown.

Given the importance of A/B status on coral photosynthesis and calcification and the dynamic changes in A/B status corals experience during the diel cycle, corals must be able to sense A/B status to adjust and coordinate their physiology accordingly. A potential A/B sensor in corals is soluble adenylyl cyclase (sAC), an evolutionarily conserved HCO_3^--sensitive enzyme that produces cAMP (Buck et al., 1999; Chen et al., 2000). In vivo, sAC can potentially sense intra- and extracellular CO_2, pH, and HCO_3^- levels and trigger a variety of physiological responses *via* the cAMP pathway (reviewed in Tresguerres (2014), Tresguerres et al. (2010a, 2011, 2014)). Evidence for sAC presence in corals is available in genomic and transcriptomic databases from *A. digitifera* (Shinzato et al. 2011) and *S. pistillata* (Karako-Lampert et al. 2014); however, these putative sAC sequences have not been validated, and the activity of the encoded proteins has not been characterized.

This task, which is essential before characterizing sAC physiological roles in corals, is greatly complicated by the presence of multiple isoforms and splice variants, as previously described for mammals (Farrell et al. 2008; Geng et al. 2005). There is, however, strong biochemical evidence for sAC in corals: cAMP production in homogenates is stimulated by HCO_3^- in a dose–response manner, and the HCO_3^--stimulated cAMP activity is inhibited by KH7 (Barott et al. 2013), a pharmacological inhibitor of sAC in mammals (Hess et al. 2005) and sharks (Tresguerres et al. 2010b). Given the A/B physiology of corals, sAC is a potential regulator of coral calcification, photosynthesis, pHi and pHe, and metabolism, among other functions. Other potential A/B sensors found in other organisms include H^+-sensing G-protein-coupled receptors, transient receptor potential channels, and acid-sensing ion channels (reviewed in Tresguerres et al. (2010a)).

7.15 Effects of Ocean Acidification (OA) on Coral Biology

The oceans have absorbed over 25 % of the CO_2 emitted to the atmosphere since the beginning of the industrial revolution (Bindoff et al. 2007; Millero 2007; Sabine et al. 2004). Dissolved CO_2 reacts with water to create H_2CO_3, which in turn dissociates into H^+ (and HCO_3^-), which has resulted in a decline in surface open ocean pH of approximately 0.1 pH units. In addition to elevated CO_2 and reduced pH, OA reduces $[CO_3^{2-}]$ in seawater when H^+ combines with CO_3^{2-} to form HCO_3^-. This phenomenon has been termed ocean acidification (OA) (Caldeira and Wickett 2003; Feely et al. 2004).

The current atmospheric CO_2 concentration of 400 parts per million is the highest observed in the last 800,000 years (Lüthi et al. 2008) and possibly since the middle Pliocene (~3MYA). This rate of increase is likely 2–3 orders of magnitude faster than most natural CO_2 excursions known via geohistory (Beerling and Royer 2011; Pagani et al. 2010). The rate of OA will match the increase in atmospheric CO_2, raising concerns on whether marine organisms can cope. Calcifying organisms such as corals are believed to be particularly vulnerable to OA because reduced seawater $[CO_3^{2-}]$ might mean fewer "building blocks" available to build calcium carbonate skeletons (Kleypas and Langdon 2006) and reduced pH increases the dissolution of dead skeleton, reef framework, and sediments (Andersson and Gledhill 2013; Silverman et al. 2009).

The biological mechanisms described in previous sections of this book chapter indicate that OA impacts on corals are likely much more complex. Five characteristics of coral biology are particularly important when predicting the potential effects of OA on corals: (1) most of the coral skeleton is not directly exposed to seawater, but instead it is underneath four tissue layers and the coelenteron (▶ Sect. 7.2); (2) the DIC source for coral skeletons is predominantly CO_2 and HCO_3^- from seawater and metabolic origin, rather than CO_3^{2-} from seawater (▶ Sects. 7.5 and 7.8); (3) the actual site of coral calcification (the SCM) is highly alkaline, has elevated Ω_{ARG}, and SOM proteins that make it extremely different chemically from the surrounding seawater (▶ Sects. 7.8 and 7.10); (4) as a result of photosynthesis, respiration, calcification, and dissolution (▶ Sect. 7.3), corals induce daily changes in the pH and DIC status of their immediate environment that are much more pronounced than those predicted due to OA in the surface open ocean; and (5) elevated CO_2 levels associated with OA could actually promote *Symbiodinium* photosynthesis, which might mean more energy transferred to the corals (▶ Sects. 7.7 and 7.12).

The "proton flux" hypothesis proposes that the negative effect of OA on coral calcification is caused by the reduced H^+ gradient between coral tissues and seawater, which requires corals to spend more energy to secrete the excess H^+ generated during calcification (Jokiel 2011a, b, 2013; Ries 2011). This hypothesis could explain reduced calcification rates under OA from an energetic point of view. In addition, other studies (e.g., Castillo et al. (2014)) have suggested that increased *Symbiodinium* photosynthesis as a result of elevated CO_2 and subsequent translocation of carbon compounds to the host could offset the increased energetic demand of calcification.

It is also important to consider the experimental challenges and limitations associated with OA research on corals. To date, the majority of studies on the effects of OA on coral calcification have been manipulative experimental studies in aquarium or mesocosms in which the corals were removed from their natural ecosystem and placed under artificial light, seawater, nutrition, and flow conditions. Extrapolating these results to the ecosystem scale and different future scenarios becomes difficult because many variables impact corals synergistically with elevated CO_2 conditions, which often lead to nonlinear effects. For example, several studies suggest that a coral's response to OA is highly dependent on food supply (Cohen and Holcomb 2009; Langdon and Atkinson 2005), temperature (Anthony et al. 2008), and to natural diel and seasonal variability of the carbonate chemistry (Bates et al. 2010; Dufault et al. 2012; Kline et al. 2015; Price et al. 2012). Another limitation of aquarium and mesocosm studies is their duration, which typically ranges from days to only a few months and therefore cannot investigate long-term impacts. These points help explain why OA studies have reported negative, neutral, and even "positive" effects on corals depending on the species and experimental setup (reviewed in Kroeker et al. (2010)).

In situ studies are a critical complement to aquarium and mesocosm findings, as they can include much more natural conditions such as light, nutrition, and environmental variability. The two main types of in situ experimental methods that are commonly used to assess OA impacts are observational studies of natural systems with high-CO_2 levels such as vent/seep sites and upwelling areas, and semi-enclosed controlled CO_2 manipulative experiments. Studies in natural vent/seep and upwelling regions are beginning to provide data about the long-term impacts of OA on organisms, communities, and ecosystems (Fabricius et al. 2011; Hall-Spencer et al. 2008; Kroeker et al. 2013) and are one of the few approaches that can assess multiyear-long and potentially evolutionary responses to OA. However, there is no control over the carbonate chemistry, and there may be other unmeasured variables that influence the results (Andersson et al. 2015). Fully enclosed in situ studies have been previously used to measure reef metabolism (Yates and Halley 2003), and results across different coral reef zones suggest that reef flats would shift to net dissolution at pCO_2 levels between 470 and 1000 ppm (Yates and Halley 2006). However, the duration of this approach is limited because the enclosed community has major biogeochemical feedbacks to the seawater chemistry and waste products' buildup after 1–2 days. An emerging and highly promising approach is to perform controlled in situ manipulative experiments with semi-enclosed replicate Free Ocean Carbon Enrichment (FOCE) flumes (Gattuso et al. 2014; Kline et al. 2012; Marker et al. 2010). FOCE style experiments allow for controlled CO_2 manipulations that can be performed as a controlled offset in pH from ambient, with natural food, diel and seasonal carbonate chemistry variability and natural environmental conditions. Additionally FOCE style experiments make it possible to address critical questions that are unanswerable in aquariums or mesocosms such as the study of ecological interactions such as herbivory or competition and OA impacts on feeding and overall energy budgets.

In addition to impairing calcification, OA has been suggested to have negative effects on many other aspects of coral biology, including sperm flagellar motility (Morita et al. 2010), fertilization, larvae settlement (Albright et al. 2010) and metamorphosis (Nakamura et al. 2011), and energy metabolism (Kaniewska et al. 2012; Vidal-Dupiol et al. 2013). The mechanistic bases behind these effects are unknown but likely are related to A/B disturbances in the coral intracellular environment. However, when analyzing the relevance of this type of studies, one must carefully analyze the conditions used as commonly the experimental CO_2/pH levels are too extreme, are administered too rapidly, are applied as a constant level even if the corals are from a dynamic environment, or are applied over very short periods of time, which do not match realistic OA conditions. Similarly, conclusions based on transcriptomics studies must be taken with caution due to potential mismatches between mRNA abundance and protein abundance and enzymatic activity (Rocca et al. 2015), posttranslational modifications not detectable by transcriptomics, and lack of information about the specific cell types where mRNA changes take place.

7.16 Conclusions

Despite their relatively simple body plan and their basal position within the phylogenetic tree, corals are complex at both the cellular and molecular level. Corals possess several specialized cell types that achieve unique and essential physiological functions, for example, symbiotic metabolic exchanges with *Symbiodinium* or deposition of massive skeletons that provide the foundation for coral reefs. The cellular and molecular mechanisms behind these functions are mostly unknown; however, recent technological advances are making it possible to study coral cellular and molecular biology at unprecedented levels of detail. In addition to being a fascinating area of research in its own right, mechanistic information about coral cell biology is essential for better understanding and predicting coral-general and species-specific responses to ocean warming, acidification, pollution, and disease.

Acknowledgments Supported by NSF grant #1220641 (MT and DDD), NSF grant#1538495 (DIK and MT), and Alfred P. Sloan Research Fellowship #BR2013-103 (MT). Special thanks to Mr. Garfield Kwan for his help with ◘ Fig. 7.1b.

References

Agostini S, Suzuki Y, Higuchi T, Casareto BE, Yoshinaga K, Nakano Y, Fujimura H (2011) Biological and chemical characteristics of the coral gastric cavity. Coral Reefs 31:147–156

Al-Horani FA, Al-Moghrabi AM, De Beer D (2003) Microsensor study of photosynthesis and calcification in the scleractinian coral, *Galaxea fascicularis*: active internal carbon cycle. J Exp Mar Biol Ecol 288:1–15

Albright R, Mason B, Miller M, Langdon C (2010) Ocean acidification compromises recruitment success of the threatened Caribbean coral *Acropora palmata*. Proc Natl Acad Sci U S A 107:20400–20404

Alieva NO, Konzen KA, Field SF, Meleshkevitch EA, Hunt ME, Beltran-Ramirez V, Miller DJ, Wiedenmann J, Salih A, Matz MV (2008) Diversity and evolution of coral fluorescent proteins. PLoS One 3(7):e2680. doi:10.1371/journal.pone.0002680

Allemand D, Furla P, Bénazet-Tambutté S (1998) Mechanisms of carbon acquisition for endosymbiont photosynthesis in Anthozoa. Can J Bot 76:925–941

Allemand D, Ferrier-Pagès C, Furla P, Houlbrèque F, Puverel S, Reynaud S, Tambutté E, Tambutté S, Zoccola D (2004) Biomineralisation in reef-building corals: from molecular mechanisms to environmental control. Comptes Rendus Palevol 3:453–467

Allemand D, Tambutté E, Zoccola D, Tambutté S (2011) Coral calcification, cells to reefs. In: Dubinsky Z, Stambler N (eds) Coral reefs an ecosystem in transition. Springer, New York, pp 119–150

Allison N, Cohen I, Finch AA, Erez J, Tudhope AW, Edinburgh Ion Microprobe Facility (2014) Corals concentrate dissolved inorganic carbon to facilitate calcification. Nat Commun 5:5741

Andersson AJ, Gledhill D (2013) Ocean acidification and coral reefs: effects on breakdown, dissolution, and net ecosystem calcification. Ann Rev Mar Sci 5:321–348

Andersson AJ, Kline DI, Edmunds PJ, Archer SD, Bednaršek N, Carpenter RC, Chadsey M, Goldstein P, Grottoli AG, Hurst TP et al (2015) Understanding ocean acidification impacts on organismal to ecological scales. Oceanography 28:16–27

Anthony KR, Kline DI, Diaz-Pulido G, Dove S, Hoegh-Guldberg O (2008) Ocean acidification causes bleaching and productivity loss in coral reef builders. Proc Natl Acad Sci U S A 105:17442–17446

Babcock RC, Heyward AJ (1986) Larval development of certain gamete-spawning scleractinian corals. Coral Reefs 5:111–116

Barott KL, Helman Y, Haramaty L, Barron ME, Hess KC, Buck J, Levin LR, Tresguerres M (2013) High adenylyl cyclase activity and in vivo cAMP fluctuations in corals suggest central physiological role. Sci Rep 3:1–7

Barott KL, Perez SO, Linsmayer LB, Tresguerres M (2015a) Differential localization of ion transporters suggests distinct cellular mechanisms for calcification and photosynthesis between two coral species. Am J Physiol Regul Integr Comp Physiol 309:R235–R246

Barott KL, Venn AA, Perez SO, Tambutté S, Tresguerres M (2015b) Coral host cells acidify symbiotic algal microenvironment to promote photosynthesis. Proc Natl Acad Sci U S A 112:607–612

Bates NR, Amat A, Andersson A (2010) Feedbacks and responses of coral calcification on the Bermuda reef system to seasonal changes in biological processes and ocean acidification. Biogeosciences 7: 2509–2530

Beerling DJ, Royer DL (2011) Convergent Cenozoic CO_2 history. Nat Geosci 4:418–420

Ben-Zvi O, Eyal G, Loya Y (2015) Light-dependent fluorescence in the coral *Galaxea fascicularis*. Hydrobiologia 759:15–26

Benson AA, Patton JS, Abraham S (1978) Energy exchange in coral reef ecosystems. Atoll Res Bull 220:33–54

Bertucci A, Tambutté E, Tambutté S, Allemand D, Zoccola D (2010) Symbiosis-dependent gene expression in coral-dinoflagellate association: cloning and characterization of a P-type H^+–ATPase gene. Proc Biol Sci 277:87–95

Bindoff NL, Willebrand J, Artale V, Cazenave A, Gregory J, Gulev S (2007) *Observations: oceanic climate change and sea level*. In: Solomon S, Qin D, Manning M, Chen Z, Marquis M, Averyt K, Tignor M (eds) Climate change 2007: the physical science basis contribution of working group I to the fourth assessment report of the intergovernmental panel on climate change. Cambridge University Press, New York

Bomati EK, Manning G, Deheyn DD (2009) Amphioxus encodes the largest known family of green fluorescent proteins, which have diversified into distinct functional classes. BMC Evol Biol 9:1–11

Bou-Abdallah F, Chasteen ND, Lesser MP (2006) Quenching of superoxide radicals by green fluorescent protein. Biochim Biophys Acta Gen Subj 1760(11):1690–1695

Bourne GC (1900) The anthozoa. In: Lankester ER (ed) A treatise on zoology. Part II. The porifera and coelenterata. Adam & Charles Black, London, pp 1–84

Buck J, Sinclair ML, Schapal L, Cann MJ, Levin LR (1999) Cytosolic adenylyl cyclase defines a unique signaling molecule in mammals. Proc Natl Acad Sci 96:79–84

Burke L, Reytar K, Spalding M, Perry A (2011) Reefs at risk revisited. World Resources Institute, Washington, DC

Burriesci MS, Raab TK, Pringle JR (2012) Evidence that glucose is the major transferred metabolite in dinoflagellate-cnidarian symbiosis. J Exp Biol 215:3467–3477

Caldeira K, Wickett ME (2003) Oceanography: anthropogenic carbon and ocean pH. Nature 425:365

Castillo KD, Ries JB, Bruno JF, Westfield IT (2014) The reef-building coral *Siderastrea siderea* exhibits parabolic responses to ocean acidification and warming. Proc R Soc B Biol Sci 281:20141856

Catala-Stucki R (1959) Fluorescent effects from corals irradiated with ultra-violet rays. Nature 183:949

Chalker BE, Taylor DL (1975) Light-enhanced calcification, and the role of oxidative phosphorylation in calcification of the coral *Acropora cervicornis*. Proc R Soc B Biol Sci 190:323–331

Chen Y, Cann MJ, Litvin TN, Iourgenko V, Sinclair ML, Levin LR, Buck J (2000) Soluble adenylyl cyclase as an evolutionarily conserved bicarbonate sensor. Science 289:625–628

Chimetto LA, Brocchi M, Thompson CC, Martins RC, Ramos HR, Thompson FL (2008) Vibrios dominate as culturable nitrogen-fixing bacteria of the Brazilian coral *Mussismilia hispida*. Syst Appl Microbiol 31:312–319

Clode PL, Marshall AT (2002) Low temperature FESEM of the calcifying interface of a scleractinian coral. Tissue Cell 34:187–198

Cohen A, Holcomb M (2009) Why corals care about ocean acidification. Oceanography 22:118–127

Cohen AL, McConnaughey TA (2003) Geochemical perspectives on coral mineralization. Rev Mineral Geochem 54:151–187

Crossland CJ, Barnes DJ, Borowitzka MA (1980) Diurnal lipid and mucus production in the staghorn coral *Acropora acuminata*. Mar Biol 60:81–90

Cunning R, Silverstein RN, Baker AC (2015a) Investigating the causes and consequences of symbiont shuffling in a multi-partner reef coral symbiosis under environmental change. Proc Biol Sci 282:20141725

Cunning R, Vaughan N, Gillette P, Capo TR, Matté JL, Baker AC (2015b) Dynamic regulation of partner abundance mediates response of reef coral symbioses to environmental change. Ecology 96: 1411–1420

Davies PS (1984) The role of zooxanthellae in the nutritional energy requirements of *Pocillopora eydouxi*. Coral Reefs 2:181–186

Davy SK, Allemand D, Weis VM (2012) Cell biology of cnidarian-dinoflagellate symbiosis. Microbiol Mol Biol Rev 76:229–261

Dawson TL (2007) Light-harvesting and light-protecting pigments in simple life forms. Color Technol 123:129–142

Diekmann O, Bak R, Stam W, Olsen J (2001) Molecular genetic evidence for probable reticulate speciation in the coral genus Madracis from a Caribbean fringing reef slope. Mar Biol 139:221–233

Dove SG, Takabayashi M, Hoegh-Guldberg O (1995) Isolation and partial characterization of the pink and blue pigments of Pocilloporid and Acroporid corals. Biol Bull 189(3):288–297

Dove SG, Hoegh-Guldberg O, Ranganathan S (2001) Major colour patterns of reef-building corals are due to a family of GFP-like proteins. Coral Reefs 19(3):197–204

Dove SG, Lovell C, Fine M, Deckenback J, Hoegh-Guldberg O, Iglesias-Prieto R, Anthony KR (2008) Host pigments: potential facilitators of photosynthesis in coral symbioses. Plant Cell Environ 3:1523–1533

Drake JL, Mass T, Haramaty L, Zelzion E, Bhattacharya D, Falkowski PG (2014) Proteomic analysis of skeletal organic matrix from the stony coral *Stylophora pistillata*. Proc Natl Acad Sci 111:12728–12733

Ducklow HW, Mitchell R (1979) Composition of mucus released by coral reef coelenterates. Limnol Oceanogr 24:706–714

Dufault AM, Cumbo VR, Fan T-Y, Edmunds PJ (2012) Effects of diurnally oscillating pCO$_2$ on the calcification and survival of coral recruits. Proc Biol Sci 279:2951–2958

Erez J (1978) Vital effect on stable-isotope composition seen in foraminifera and coral skeletons. Nature 273:199–202

Fabricius KE, Langdon C, Uthike S, Humphrey C, Noonan S, Death G, Okazaki R, Muehllehner N, Glas MS, Lough JM (2011) Losers and winners in coral reefs acclimated to elevated carbon dioxide concentrations. Nat Clim Chang 1:165–169

Falkowski PG, Dubinsky Z, Muscatine L, Porter JW (1984) Light and bioenergetics of a symbiotic coral. BioScience 34:705–709

Farrell J, Ramos L, Tresguerres M, Kamenetsky M, Levin LR, Buck J (2008) Somatic 'soluble' Adenylyl Cyclase isoforms are unaffected in Sacy^{tm1Lex}/Sacy^{tm1Lex} 'knockout' mice. PLoS One 3:e3251

Feely RA, Sabine CL, Lee K, Berelson W, Kleypas J, Fabry VJ, Millero FJ (2004) Impact of anthropogenic CO$_2$ on the CaCO$_3$ system in the oceans. Science 305:362–366

Friedman PA, Gesek FA (1995) Cellular calcium transport in renal epithelia: measurement, mechanisms, and regulation. Physiol Rev 75:429–471

Fukami H, Budd AF, Paulay G, Solé-Cava A, Allen Chen C, Iwao K, Knowlton N (2004) Conventional taxonomy obscures deep divergence between Pacific and Atlantic corals. Nature 427:832–835

Furla P, Bénazet-Tambutté S, Jaubert J, Allemand D (1998) Functional polarity of the tentacle of the sea anemone *Anemonia viridis*: role in inorganic carbon acquisition. AJP: Regul Integr Comp Physiol 274:303–310

Furla P, Galgani I, Durand I, Allemand D (2000) Sources and mechanisms of inorganic carbon transport for coral calcification and photosynthesis. J Exp Biol 203:3445–3457

Galloway SB, Work TM, Bochsler VS, Harley RA, Kramarsky-Winters E, McLaughlin SM, Meteyer CU, Morado JF, Nicholson JH, Parnell PG et al (2006) Coral disease and health workshop: coral histopathology II. National Oceanic and Atmospheric Administration (NOAA), Silver Spring

Ganot P, Zoccola D, Tambutté E, Voolstra CR, Aranda M, Allemand D, Tambutté S (2015) Structural molecular components of septate junctions in cnidarians point to the origin of epithelial junctions in eukaryotes. Mol Biol Evol 32:44–62

Gattuso JP (1999) Photosynthesis and calcification at cellular, organismal and community levels in coral reefs: a review on interactions and control by carbonate chemistry. Am Zool 39:160–183

Gattuso JP, Hendriks IE, Brewer PG (2014) Free-ocean CO_2 enrichment (FOCE) systems: present status and future developments. Biogeosciences 11:4057–4075

Geng W, Wang Z, Zhang J, Reed BY, Pak CY, Moe OW (2005) Cloning and characterization of the human soluble adenylyl cyclase. AJP: Cell Physiol 288:C1305–C1316

Gibbin EM, Putnam HM, Davy SK, Gates RD (2014) Intracellular pH and its response to CO2-driven seawater acidification in symbiotic versus non-symbiotic coral cells. J Exp Biol 217:1963–1969

Gladfelter EH (1983) Circulation of fluids in the gastrovascular system of the reef coral *Acropora cervicornis*. Biol Bull 165:619–636

Goldberg WM (2001a) Acid polysaccharides in the skeletal matrix and calicoblastic epithelium of the stony coral *Mycetophyllia reesi*. Tissue Cell 33:376–387

Goldberg WM (2001b) Desmocytes in the calicoblastic epithelium of the stony coral *Mycetophyllia reesi* and their attachment to the skeleton. Tissue Cell 33:388–394

Goldberg WM (2002a) Feeding behavior, epidermal structure and mucus cytochemistry of the scleractinian *Mycetophyllia reesi*, a coral without tentacles. Tissue Cell 34:232–245

Goldberg WM (2002b) Gastrodermal structure and feeding responses in the scleractinian *Mycetophyllia reesi*, a coral with novel digestive filaments. Tissue Cell 34:246–261

Goreau TF (1959) The physiology of skeleton formation in corals. I. A method for measuring the rate of calcium deposition by corals under different conditions. Biol Bull 116:59–75

Goreau TF, Goreau NI (1959) The physiology of skeleton formation in corals. II. Calcium deposition by hermatypic corals under various conditions in the reef. Biol Bull 117:239–250

Goreau TF, Goreau NI, Yonge CM (1971) Reef corals: autotrophs or heterotrophs? Biol Bull 141:247–260

Haas AF, Smith JE, Thompson M, Deheyn DD (2014) Effects of reduced dissolved oxygen concentrations on physiology and fluorescence of hermatypic corals and benthic algae. Peer J 2:e235. doi:10.7717/peerj.235

Haddock SHD, Moline MA, Case JF (2009) Bioluminescence in the sea. 2:443–493. http://dx.doi.org/10.1146/annurev-marine-120308-081028

Hall-Spencer JM, Rodolfo-Metalpa R, Martin S, Ransome E, Fine M, Turner SM, Rowley SJ, Tedesco D, Buia M-C (2008) Volcanic carbon dioxide vents show ecosystem effects of ocean acidification. Nature 454:96–99

Hanson MR, Kohler RH (2001) GFP imaging: methodology and application to investigate cellular compartmentation in plants. J Exp Bot 52(356):529–539

Hess KC, Jones BH, Marquez B, Chen Y, Ord TS, Kamenetsky M, Miyamoto C, Zippin JH, Kopf GS, Suarez SS et al (2005) The "soluble" adenylyl cyclase in sperm mediates multiple signaling events required for fertilization. Dev Cell 9:249–259

Hosey MM, Lazdunski M (1988) Calcium channels: molecular pharmacology, structure and regulation. J Membr Biol 104:81–105

Hughes AD, Grottoli AG, Pease TK, Matsui Y (2010) Acquisition and assimilation of carbon in non-bleached and bleached corals. Mar Ecol Prog Ser 420:91–101

Ip YK, Lim ALL, Lim RW (1991) Some properties of calcium activated adenosine triphosphatase from the hermatypic coral *Galaxea fascicularis*. Mar Biol 111:191–197

Isa Y (1986) An electron microscope study on the mineralization of the skeleton of the staghorn coral *Acropora hebes*. Mar Biol 93:91–101

Johnston IS (1980) The ultrastructure of skeletogenesis in hermatypic corals. Int Rev Cytol 67:171–214

Jokiel P (2011a) The reef coral two compartment proton flux model: a new approach relating tissue-level physiological processes to gross corallum morphology. J Exp Mar Biol Ecol 409:1–12

Jokiel PL (2011b) Ocean acidification and control of reef coral calcification by boundary layer limitation of proton flux. Bull Mar Sci 87:639–657

Jokiel PL (2013) Coral reef calcification: carbonate, bicarbonate and proton flux under conditions of increasing ocean acidification. Proc Biol Sci 280:20130031

Kaniewska P, Campbell PR, Kline DI, Rodriguez-Lanetty M, Miller DJ, Dove S, Hoegh-Guldberg O (2012) Major cellular and physiological impacts of ocean acidification on a reef building coral. PLoS One 7:e34659

Karako-Lampert S, Zoccola D, Salmon-Divon M, Katzenellenbogen M, Tambutté S, Bertucci A, Hoegh-Guldberg O, Deleury E, Allemand D, Levy O (2014) Transcriptome analysis of the scleractinian coral *Stylophora pistillata*. PLoS One 9:e88615

Kawaguti S (1937) On the physiology of reef corals. II. The effect of light on color and form of reefs. Palao Trop Biol Stat Stud 2:199–208

Kawaguti S (1969) The effect of green fluorescent pigment on the productivity of the reef corals. Micronesica 5:313

Kawaguti S, Sakumoto D (1948) The effect of light on the calcium deposition of corals. Bul Oceanogr Inst Taiwan 4:65–70

Kelmanson IV, Matz MV (2003) Molecular basis and evolutionary origins of color diversity in great star coral *Montastraea cavernosa* (Scleractinia: Faviida). Mol Biol Evol 20(7):1125–1133

Kenkel CD, Traylor MR, Wiedenmann J, Salih A, Matz MV (2011) Fluorescence of coral larvae predicts their settlement response to crustose coralline algae and reflects stress. Proc Biol Sci 278:2691–2697

Kleypas J, Langdon C (2006) Coral reefs and changing seawater carbonate chemistry. In: Phinney JT, Hoegh-Guldberg O, Kleypas J, Skirving W, Strong A (eds) *Coral reefs and climate change science and management*. American Geophysical Union, Washington, DC

Kline DI, Teneva L, Hauri C, Schneider K, Miard T, Chai A, Marker M, Dunbar R, Caldeira K, Lazar B et al (2015) Six month in situ high-resolution carbonate chemistry and temperature study on a coral reef flat reveals asynchronous pH and temperature anomalies. PLoS One 10:e0127648

Kline DI, Teneva L, Schneider K, Miard T, Chai A, Marker M, Headley K, Opdyke B, Nash M, Valetich M et al (2012) A short-term in situ CO_2 enrichment experiment on Heron Island (GBR). Sci Rep 2:413

Krediet CJ, Ritchie KB, Paul VJ, Teplitski M (2013) Coral-associated micro-organisms and their roles in promoting coral health and thwarting diseases. Proc Biol Sci 280:20122328

Kroeker KJ, Kordas RL, Crim RN, Singh GG (2010) Meta-analysis reveals negative yet variable effects of ocean acidification on marine organisms. Ecol Lett 13:1419–1434

Kroeker KJ, Micheli F, Gambi MC (2013) Ocean acidification causes ecosystem shifts via altered competitive interactions. Nat Clim Chang 3:156–159

Kühl M, Cohen Y, Dalsgaard T, Jorgensen BB, Revbech NP (1995) Microenvironment and photosynthesis of zooxanthellae in scleractinian corals studied with microsensors for O_2, pH and light. Mar Ecol Prog Ser 117:159–172

Labas YA, Gurskaya NG, Yanushevich YG, Fradkov AF, Lukyanov KA, Lukyanov SA, Matz MV (2002) Diversity and evolution of the green fluorescent protein family. Proc Natl Acad Sci U S A 99(7):4256–4261

Langdon C, Atkinson MJ (2005) Effect of elevated pCO_2 on photosynthesis and calcification on corals and interactions with seasonal change in temperature/irradiance and nutrient enrichment. J Geophys Res 110:C09S07

Laurent J, Tambutté S, Tambutté E, Allemand D, Venn A (2013) The influence of photosynthesis on host intracellular pH in scleractinian corals. J Exp Biol 216:1398–1404

Laurent J, Venn A, Tambutté E, Ganot P, Allemand D, Tambutté S (2014) Regulation of intracellular pH in cnidarians: response to acidosis in *Anemonia viridis*. FEBS J 281:683–695

Leal MC, Jesus B, Ezequiel J, Calado R, Rocha RJM, Cartaxana P, Serodio J (2015) Concurrent imaging of chlorophyll fluorescence, Chlorophyll a content and green fluorescent proteins-like proteins of symbiotic cnidarians. Mar Ecol Evol Perspect 36(3):572–584

Lesser MP, Mazel CH, Gorbunov MY, Falkowski PG (2014) Discovery of symbiotic nitrogen-fixing cyanobacteria in corals. Science 305:1–5

Leutenegger A, D'Angelo C, Matz MV, Denzel A, Oswald F, Salih A, Nienhaus GU, Wiedenmann J (2007) It's cheap to be colorful. Anthozoans show a slow turnover of GFP-like proteins. FEBS J 274:2496–2505

Little AF, van Oppen MJ, Willis BL (2004) Flexibility in algal endosymbioses shapes growth in reef corals. Science 304:1492–1494

Lüthi D, Le Floch M, Bereiter B, Blunier T, Barnola J-M, Siegenthaler U, Raynaud D, Jouzel J, Fischer H, Kawamura K et al (2008) High-resolution carbon dioxide concentration record 650,000–800,000 years before present. Nature 453:379–382

Marker M, Kline DI, Kirkwood WJ, Headley K, Brewer PG, Peltzer ET, Miard T, Chai A, James M, Schneider K et al (2010) The coral proto free ocean carbon enrichment system (CP-FOCE): engineering and development, Proceedings of OCEANS IEEE 1–10

Matz MV, Fradkov AF, Labas YA, Savitsky AP, Zaraisky AG, Markelov ML, Lukyanov SA (1999) Fluorescent proteins from nonbioluminescent Anthozoa species. Nat Biotechnol 17:969–973

Marshall AT (1996) Calcification in hermatypic and ahermatypic corals. Science 271:637–639

Mass T, Drake JL, Haramaty L, Kim JD, Zelzion E, Bhattacharya D, Falkowski PG (2013) Cloning and characterization of four novel coral acid-rich proteins that precipitate carbonates *in vitro*. Curr Biol 23:1126–1131

Mass T, Drake JL, Peters EC, Jiang W, Falkowski PG (2014) Immunolocalization of skeletal matrix proteins in tissue and mineral of the coral *Stylophora pistillata*. Proc Natl Acad Sci U S A 111:12728–12733

McLean PF, Cooley L (2013) Protein equilibration through somatic ring canals in Drosophila. Science 340(6139):1445–1447

Millero FJ (2007) The marine inorganic carbon cycle. Chem Rev 107:308–341

Morita M, Suwa R, Iguchi A, Nakamura M, Shimada K, Sakai K, Suzuki A (2010) Ocean acidification reduces sperm flagellar motility in broadcast spawning reef invertebrates. Zygote 18:103–107

Moya A, Tambutté S, Bertucci A, Tambutté E, Lotto S, Vullo D, Supuran CT, Allemand D, Zoccola D (2008) Carbonic anhydrase in the scleractinian coral *Stylophora pistillata*: characterization, localization, and role in biomineralization. J Biol Chem 283:25475–25484

Muscatine L (1973) Nutrition of corals. In: Jones OA, Endean R (eds) Biology and geology of coral reefs, vol II, Biology I. Academic, New York, pp 77–115

Muscatine L (1990) The role of symbiotic algae in carbon and energy flux in reef corals. In: Dubinsky Z (ed) Ecosystems of the world. Elsevier, Amsterdam, pp 75–87

Muscatine L, Cernichiari E (1969) Assimilation of photosynthetic products of zooxanthellae by a reef coral. Biol Bull 137:506–523

Muscatine L, Falkowski PG, Porter JW, Dubinsky Z (1984) Fate of the photosynthetic fixed carbon in light- and shade-adapted colonies of the symbiotic coral *Stylophora pistillata*. Proc R Soc B Biol Sci 222:181–202

Nakamura M, Ohki S, Suzuki A, Sakai K (2011) Coral larvae under ocean acidification: survival, metabolism, and metamorphosis. PLoS One 6:e14521–e14527

Odorico DM, Miller DJ (1997) Variation in the ribosomal internal transcribed spacers and 5.8S rDNA among five species of *Acropora* (Cnidaria; Scleractinia): patterns of variation consistent with reticulate evolution. Mol Biol Evol 14:465–473

Ohde S, van Woesik R (1999) Carbon dioxide flux and metabolic processes of a coral reef, Okinawa. Bull Mar Sci 65:559–576

Ohmiya Y, Hirano T (1996) Shining the light: the mechanism of the bioluminescence reaction of calcium-binding photoproteins. Chem Biol 3:337–347

Oswald F, Schmitt F, Leutenegger A, Ivanchenko S, D'Angelo C, Salih A, Maslakova S, Bulina M, Schirmbeck R, Nienhaus GU, Matz MV, Wiedenmann J (2007) Contributions of host and symbiont pigments to the coloration of reef corals. FEBS J 274(4):1102–1109

Pagani M, Liu ZH, LaRiviere J, Ravelo AC (2010) High earth-system climate sensitivity determined from Pliocene carbon dioxide concentrations. Nat Geosci 3:27–30

Palmer CV, Roth MS, Gates RD (2009) Red fluorescent protein responsible for pigmentation in trematode-infected *Porites compressa* tissues. Biol Bull 216:68–74

Petersen OH, Petersen C (1994) Calcium and hormone action. Annu Rev Physiol 56:297–319

Phillips JH (1963) Immune mechanisms in the phylum Coelenterata. In: Dougherty EC (ed) The lower metazoa. University of California Press, Berkeley, pp 425–431

Price NN, Martz TR, Brainard RE, Smith JE (2012) Diel variability in seawater pH relates to calcification and benthic community structure on coral reefs. PLoS One 7:e43843

Putnam RW, Roos A (1997) Intracellular pH. In: Hoffman JF, Jamieson DJ (eds) Handbook of physiology. Oxford University Press, New York, pp 389–440

Puverel S, Tambutt E, Zoccola D, Domart-Coulon I, Bouchot A, Lotto SV, Allemand D, Tambutt S (2004) Antibodies against the organic matrix in scleractinians: a new tool to study coral biomineralization. Coral Reefs 24:149–156

Puverel S, Tambutté E, Pereira-Mouriès L, Zoccola D, Allemand D, Tambutté S (2005) Soluble organic matrix of two Scleractinian corals: partial and comparative analysis. Comp Biochem Physiol B Biochem Mol Biol 141:480–487

Raina J-B, Dinsdale EA, Willis BL, Bourne DG (2010) Do the organic sulfur compounds DMSP and DMS drive coral microbial associations? Trends Microbiol 18:101–108

Ries JB (2011) A physicochemical framework for interpreting the biological calcification response to CO_2-induced ocean acidification. Geochim Cosmochim Acta 75:4053–4064

Rocca JD, Hall EK, Lennon JT, Evans SE, Waldrop MP, Cotner JB, Nemergut DR, Graham EB, Wallenstein MD (2015) Relationships between protein-encoding gene abundance and corresponding process are commonly assumed yet rarely observed. ISME J 9:1693–1699

Rohwer F, Seguritan V, Azam F, Knowlton N (2002) Diversity and distribution of coral-associated bacteria. Mar Ecol Prog Ser 243:1–10

Romano SL, Cairns SD (2000) Molecular phylogenetic hypotheses for the evolution of scleractinian corals. Bull Mar Sci 67:1043–1068

Romano SL, Palumbi SR (1996) Evolution of scleractinian corals inferred from molecular systematics. Science 271:640–642

Roos A, Boron WF (1981) Intracellular pH. Physiol Rev 61:296–434

Rosenberg E, Koren O, Reshef L, Efrony R, Zilber-Rosenberg I (2007) The role of microorganisms in coral health, disease and evolution. Nat Rev Microbiol 5:355–362

Roth MS (2014) The engine of the reef: photobiology of the coral-algal symbiosis. Front Microbiol 5:422

Roth MS, Deheyn DD (2013) Effects of cold stress and heat stress on coral fluorescence in reef-building corals. Sci Rep 3:1421

Roth MS, Fan T-Y, Deheyn DD (2013) Life history changes in coral fluorescence and the effects of light intensity on larval physiology and settlement in Seriatopora hystrix. PLoS One 8:e59476

Roth MS, Goericke R, Deheyn DD (2012) Cold induces acute stress but heat is ultimately more deleterious for the reef-building coral *Acropora yongei*. Sci Rep 2:240

Roth MS, Latz MI, Goericke R, Deheyn DD (2010) Green fluorescent protein regulation in the coral *Acropora yongei* during photoacclimation. J Exp Biol 213:3644–3655

Rowan R, Whitney SM, Fowler A, Yellowlees D (1996) Rubisco in marine symbiotic dinoflagellates: form II enzymes in eukaryotic oxygenic phototrophs encoded by a nuclear multigene family. Plant Cell 8:539–553

Sabine CL, Feely RA, Gruber N, Key RM, Lee K, Bullister JL, Wanninkhof R, Wong CS, Wallace DW, Tilbrook B et al (2004) The oceanic sink for anthropogenic CO2. Science 305:367–371

Salih A, Larkum A, Cox G, Kühl M, Hoegh-Guldberg O (2000) Fluorescent pigments in corals are photoprotective. Nature 408:850–853

Schlichter D, Meier U, Fricke HW (1994) Improvement of photosynthesis in zooxanthellate corals by autofluorescent chromatophores. Oecologia 99:124–131

Schonknecht G, Brown JE, Verchot-Lubicz J (2008) Plasmodesmata transport of GFP alone or fused to potato virus X TGBp1 is diffusion driven. Protoplasma 232(3–4):143–152

Schweinsberg M, Weiss LC, Striewski S, Tollrian R, Lampert KP (2015) More than one genotype: how common is intracolonial genetic variability in scleractinian corals? Mol Ecol 24:2673–2685

Shashar N, Cohen Y, Loya Y (1993) Extreme diel fluctuations of oxygen in diffusive boundary layers surrounding stony corals. Biol Bull 185:455–461

Shaw EC, Mcneil BI, Tibrook B (2012) Impacts of ocean acidification in naturally variable coral reef flat ecosystems. J Geophys Res Oceans 117:C03038

Shinomura O, Johnson FH, Saiga Y (1962) Extraction, purification and properties of aequorin, a bioluminescent protein from the luminous hydromedusan, Aequorea. J Cell Comp Physiol 59:223–239

Shinzato C, Shoguchi E, Kawashima T, Hamada M, Hisata K, Tanaka M, Fujie M, Fujiwara M, Koyanagi R, Ikuta T et al (2011) Using the *Acropora digitifera* genome to understand coral responses to environmental change. Nature 476:320–323

Silverman J, Kline DI, Johnson L, Rivlin T, Schneider K, Erez J, Lazar B, Caldeira K (2012) Carbon turnover rates in the one tree Island reef: a 40-year perspective. Geophys Res Biogeosci 1127:G03023

Silverman J, Lazar B, Cao L, Caldeira K, Erez J (2009) Coral reefs may start dissolving when atmospheric CO_2 doubles. Geophys Res Lett 36:L05606

Smith EG, Angelo D, Salih C, Wiedenmann J (2013) Screening by coral green fluorescent protein (GFP)-like chromoproteins supports a role in photoprotection of zooxanthellae. Coral Reefs 32:463–474

Stolarski J, Kitahara MV, Miller DJ, Cairns SD, Mazur M, Meibom A (2011) The ancient evolutionary origins of Scleractinia revealed by azooxanthellate corals. BMC Evol Biol 11:316

Tambutté E, Allemand D, Mueller E, Jaubert J (1996) A compartmental approach to the mechanism of calcification in hermatypic corals. J Exp Biol 199:1029–1041

Tambutté E, Tambutté S, Segonds N, Zoccola D, Venn A, Erez J, Allemand D (2011) Calcein labelling and electrophysiology: insights on coral tissue permeability and calcification. Proc R Soc B BiolSci 279: 19–27

Tresguerres M (2014) sAC from aquatic organisms as a model to study the evolution of acid/base sensing. BBA Mol Basis Dis 1842:2629–2635

Tresguerres M, Barott KL, Barron ME, Roa JN (2014) Established and potential physiological roles of bicarbonate-sensing soluble adenylyl cyclase (sAC) in aquatic animals. J Exp Biol 217:663–672

Tresguerres M, Buck J, Levin LR (2010a) Physiological carbon dioxide, bicarbonate, and pH sensing. Pflugers Arch Eur J Physiol 460:953–964

Tresguerres M, Levin LR, Buck J (2011) Intracellular cAMP signaling by soluble adenylyl cyclase. Kidney Int 79:1277–1288

Tresguerres M, Parks SK, Katoh F, Goss GG (2006) Microtubule-dependent relocation of branchial V-H+– ATPase to the basolateral membrane in the Pacific spiny dogfish (*Squalus acanthias*): a role in base secretion. J Exp Biol 209:599–609

Tresguerres M, Parks SK, Salazar E, Levin LR, Goss GG, Buck J (2010b) Bicarbonate-sensing soluble adenylyl cyclase is an essential sensor for acid/base homeostasis. Proc Natl Acad Sci U S A 107:442–447

van Oppen MJ, McDonald BJ, Willis B, Miller DJ (2001) The evolutionary history of the coral genus *Acropora* (Scleractinia, Cnidaria) based on a mitochondrial and a nuclear marker: reticulation, incomplete lineage sorting, or morphological convergence? Mol Biol Evol 18:1315–1329

Vandermeulen JH, Watabe N (1973) Studies on reef corals. I. Skeleton formation by newly settled planula larva of *Pocillopora damicornis*. Mar Biol 23:47–57

Venn AA, Tambutté E, Lotto S, Zoccola D, Allemand D, Tambutté S (2009) Imaging intracellular pH in a reef coral and symbiotic anemone. Proc Natl Acad Sci U S A 106:16574–16579

Venn A, Tambutté E, Holcomb M, Allemand D, Tambutté S (2011) Live tissue imaging shows reef corals elevate pH under their calcifying tissue relative to seawater. PLoS One 6:e20013–e20019

Veron JE (1993) Corals of Australia and the Indo-Pacific, 2nd edn. University of Hawaii Press, Honolulu

Vidal-Dupiol J, Zoccola D, Tambutté E, Grunau C, Cosseau C, Smith KM, Freitag M, Dheilly NM, Allemand D, Tambutté S (2013) Genes related to Ion-transport and energy production are upregulated in response to CO_2-driven pH decrease in corals: new insights from transcriptome analysis. PLoS One 8: e58652

Wainwright SA (1964) Studies of the mineral phase of coral skeleton. Exp Cell Res 34:213–230

Weis VM, Smith GJ, Muscatine L (1989) A "CO_2 supply" mechanism in zooxanthellate cnidarians: role of carbonic anhydrase. Mar Biol 100:195–202

Wiedenmann J, Ivanchenko S, Oswald F, Nienhaus GU (2004) Identification of GFP-like proteins in nonbioluminescent, azooxanthellate anthozoa opens new perspectives for bioprospecting. Marine Biotechnol 6(3):270–277

Wijgerde T, Jurriaans S, Hoofd M, Verreth JA, Osinga R (2012) Oxygen and heterotrophy affect calcification of the scleractinian coral *Galaxea fascicularis*. PLoS One 7:e52702

Wijgerde T, Silva CIF, Scherders V, van Bleijswijk J, Osinga R (2014) Coral calcification under daily oxygen saturation and pH dynamics reveals the important role of oxygen. Biol Open 3:489–493

Wijsman-Best M (1975) Intra-and extratentacular budding in hermatypic reef corals. pp. 471–475

Wild C, Huettel M, Klueter A, Kremb SG, Rasheed MY, Jørgensen BB (2004) Coral mucus functions as an energy carrier and particle trap in the reef ecosystem. Nature 428:66–70

Willis BL, Babcock RC, Harrison PL, Wallace CC (1997) Experimental hybridization and breeding incompatibilities with the mating systems of mass spawning reef corals. Coral Reefs 16:S53–S65

Wong W, Scott JD (2004) AKAP signalling complexes: focal points in space and time. Nat Rev Mol Cell Biol 5:959–970

Yates KK, Halley RB (2003) Measuring coral reef community metabolism using new benthic chamber technology. Coral Reefs 22:247–255

Yates KK, Halley RB (2006) Diurnal variation in rates of calcification and carbonate sediment dissolution in Florida Bay. Estuar Coast 29:24–39

Zoccola D, Ganot P, Bertucci A, Caminiti-Segonds N, Techer N, Voolstra CR, Aranda M, Tambutté E, Allemand D, Casey JR et al (2015) Bicarbonate transporters in corals point towards a key step in the evolution of cnidarian calcification. Sci Rep 5:9983

Zoccola D, Tambutté E, Kulhanek E, Puverel S, Scimeca J-C, Allemand D, Tambutté S (2004) Molecular cloning and localization of a PMCA P-type calcium ATPase from the coral *Stylophora pistillata*. Biochim Biophys Acta 1663:117–126

Zoccola D, Tambutté E, Sénégas-Balas F, Michelis J-F, Failla J-P, Jaubert J, Allemand D (1999) Cloning of a calcium channel a1 subunit from the reef-building coral, *Stylophora pistillata*. Gene 227:157–167

Acid–Base Regulation in Insect Haemolymph

Philip G.D. Matthews

© Springer International Publishing Switzerland 2017
D. Weihrauch, M. O'Donnell (eds.), *Acid-Base Balance and Nitrogen Excretion in Invertebrates*,
DOI 10.1007/978-3-319-39617-0_8

8.1   Summary

Insects regulate the acid–base balance of their haemolymph by ventilation, buffering and active uptake and excretion of acid–base equivalents with their gut. Haemolymph pH varies widely across ontogeny and between species, varying from acidic (~6.4) to alkaline (~8.0). Terrestrial and aquatic insects are exposed to different acid–base challenges by virtue of their environment and have evolved different mechanisms to cope. Terrestrial insects use bicarbonate buffering and changes in ventilation to respond to metabolic acid loads, as well as clearing excess acid through excretion into the lumen of the hindgut. V-ATPase and Na^+/K^+-ATPase in the hindgut/Malpighian tubule complex generates the transmembrane electrochemical potential necessary to drive uptake and excretion of acid–base equivalents and ions. The activity of these transport processes is under hormonal control. While aquatic insects also use their hindgut for acid–base regulation, they possess additional ion-transporting chloride cells on parts of their cuticle that are in contact with the surrounding water. These cells provide an extra-renal pathway for regulating haemolymph pH and osmolality, relying on ion transport driven directly and indirectly by V- and Na^+/K^+-ATPases. While aquatic insects are among the most pH-tolerant animals on the planet, the mechanisms that allow them to tolerate chronic exposure to highly acidic or alkaline water across a wide range of ionic strengths are incompletely understood and require further investigation.

8.2   Introduction

The acid–base balance of an insect's body fluids results from the interplay between the partial pressure of carbon dioxide (PCO_2) in its body fluids, the buffering action of weak acids and the active pumping of ions across membranes within the insect as well as between the insect and its environment. By actively regulating these variables, insects can maintain their internal pH within a physiologically acceptable range. But the extraordinary diversity of insects and their distribution across both terrestrial and aquatic habitats make summarising their acid–base regulatory strategies no easy task. This diversity also means that the acid–base regulation of only a few insect species have received any particular attention, by virtue of being either economically important agricultural pests (e.g. *Schistocerca gregaria*, *Manduca sexta*) or vectors of disease (*Aedes* and *Anopheles*). While the mechanisms described from among these few exemplars likely apply to most insects, many new and surprising acid–base adaptations among the millions of unstudied insect species are yet to be discovered.

Insects regulate the acid–base balance of two separate extracellular compartments: the lumen of their gut and their haemolymph-filled haemocoel. This chapter reviews the acid–base balance within the insect's haemolymph. It considers the role of respiration and ion exchanging regions in the hindgut and Malpighian tubules in terrestrial insect acid–base balance strategies, comparing these physiological mechanisms with those utilised by aquatic insect species.

8.3   Acid–Base Regulation in Terrestrial Insects

8.3.1 Haemolymph Composition and pH

The two largest reserves of body water in an insect are located in its gut lumen and in its haemolymph. After the gut lumen, the haemolymph represents the second largest fraction of an insect's total body water. For example, it accounts for between 17 % and 21 % of the total body weight of the American cockroach (*Periplaneta americana*) (Wheeler 1963) but tends to be a higher fraction of the total body weight of larval insects, such as the death's-head hawkmoth caterpillar *Acherontia atropos* (34 % of body weight) (Wasserthal 1996). Haemolymph fills the open circulatory system, the haemocoel, and bathes the insect's tissues, supplying them with nutrients, hormones and ions, while acting as a sink for metabolic wastes. It also acts as a pH buffer and as a source and sink of acid–base equivalents. The acid–base balance of haemolymph is determined by three independent variables, namely, the strong ion difference (the difference between the sum of all cations and anions present in solution), the concentration and type of weak acids and the partial pressure of carbon dioxide (PCO_2) (Stewart 1983). Ideally, all three variables should be measured to provide a complete picture of the acid–base balance of a fluid. But as pH (or, more precisely, proton concentration $[H^+]$) is the variable that is regulated by physiological processes to maintain enzyme function, pH is the usual starting point for considering the acid–base properties of this fluid. An early, and still repeated, assertion is that insect haemolymph is acidic (Chapman 2013; Wigglesworth 1972). However, the data collated in ◘ Table 8.1 shows this to be incorrect. While lepidopteran and hymenopteran larvae appear to possess haemolymph that is uniformly acidic, measurements from other holometabolous insects show alkaline or near-neutral pH, while most of the hemimetabolous insects measured thus far maintain a haemolymph higher than pH 7 (◘ Table 8.1). The accuracy of some of the older measurements of insect haemolymph pH is questionable because measurements made in vitro used fluid pooled from several individuals. This approach potentially introduces errors due to changes in the composition of haemolymph, the clotting of protein and the loss of CO_2. More recent measurements of haemolymph pH, made in vivo by inserting a pH microelectrode or fibre-optic optode directly into the insect's haemocoel, have recorded pH values within a range that is comparable with in vitro measurements on similar insect species (e.g. Hetz and Wasserthal (1993), Matthews and White (2011)). Nonetheless, this data clearly shows that insects regulate their haemolymph pH across a very wide range of values. When considering the regulation of acid–base balance in insects, it is pertinent to note that nearly all haemolymph pH measurements are point measurements, with a few exceptions (Lettau et al. 1977; Matthews and White 2011; Hetz and Wasserthal 1993). Long-term recording from chronically implanted pH probes has shown that haemolymph pH often varies from minute to minute and may vary by more than 0.3 units across the course of a day (Lettau et al. 1977). Thus, at present, it is difficult to say how precisely insects regulate their extracellular pH environment, or whether closely defending intracellular pH is a more common strategy.

Since relatively few reliable measurements have been made on insect haemolymph, much work remains to be done to evaluate what the 'normal' haemolymph pH range is for

◨ **Table 8.1**   Summary of measured values of haemolymph pH from across eight orders of insects, with an emphasis on more recent measurements

Order	Species	pH	Method	References
Orthoptera	*Melanoplus bivittatus*	7.121	Electrode	Harrison (1988)
	Schistocerca gregaria	7.1	Electrode	Phillips et al. (1987)
	Paracinema tricolor	7.0	Optode	Groenewald et al. (2014)
	Schistocerca gregaria	7.28	Electrode	Stagg et al. (1991)
	Teleogryllus commodus	7.48[a] (fed) 6.76[a] (starved)	Electrode	Cooper and Vulcano (1997)
Blattodea	*Nauphoeta cinerea*	7.084	Electrode	Snyder et al. (1980)
	Nauphoeta cinerea	7.30[a]	Optode	Matthews and White (2011)
	Leucophaea maderae	6.9–7.24[a]	Electrode	Lettau et al. (1977)
Odonata	*Libellula julia*[b, c]	7.58[a]	Electrode	Rockwood and Coler (1991)
	Somatochlora cingulata[b, c]	7.60[a]	Electrode	Correa et al. (1985)
	Uropetala carovei[b, c]	8.2	Electrode	Bedford and Leader (1975)
Hemiptera	*Corixa dentipes*	7.04	Electrode?	Vangenechten et al. (1989)
	Corixa punctata	6.97	Electrode?	
Lepidoptera	*Bombyx mori*[d]	6.60	Electrode	Wyatt et al. (1956)
	Manduca sexta[b]	6.64[a]	Electrode	Moffett and Cummings (1994)
	Agapema galbina[d]	6.45	Electrode	Buck and Friedman (1958)
	Hyalophora cecropia[d]	6.52	Electrode	
	Acherontia atropos[b]	6.9	Electrode	Dow (1984)
	Lasiocampa quercus[b]	6.7	Electrode	
	Lichnoptera felina[b]	6.4	Electrode	
Diptera	*Gasterophilus intestinalis*[b]	6.8	Electrode	Levenbook (1950)
	Chironomus riparius[b, c]	7.2–7.3	Electrode	Jernelöv et al. (1981)
	Chironomus riparius[b, c]	7.7–8.0	ISME	Nguyen and Donini (2010)
	Rhynchosciara americana[b]	7.27	Electrode	Terra et al. (1974)

◼ **Table 8.1** (continued)

Order	Species	pH	Method	References
	Ochlerotatus taeniorhynchus[b, c]	7.7	Electrode	Clark et al. (2004)
	Aedes aegypti[b, c]	7.6	Electrode	
	Aedes dorsalis[b, c]	7.55–7.70	Electrode	Strange et al. (1982)
Coleoptera	*Leptinotarsa decemlineata*	6.53–6.74	Electrode	Pelletier and Clark (1992)
	Xylotrupes ulysses	7.0[a]	Optode	Matthews and White (2009)
Hymenoptera	*Neodiprion abietis*[b]	6.59	Electrode	Heimpel (1955)
	Neodiprion americanus[b]	6.54	Electrode	
	Neodiprion lecontei[b]	6.88	Electrode	
	Neodiprion sertifer[b]	6.81	Electrode	
	Neodiprion virginiana[b]	6.84	Electrode	
	Pikonema alaskensis[b]	6.55	Electrode	
	Hemichroa crocea[b]	6.69	Electrode	
	Pristiphora erichsonii[b]	6.64	Electrode	

[a]Indicates in vivo measurement
[b]Indicates larval or nymph/naiad stage
[c]Indicates aquatic habitat
[d]Indicates pupal stage

any given insect. This information is particularly crucial for studies involving isolated or cultured insect tissues. For example, while insect physiological saline is used as an artificial haemolymph in numerous *Drosophila* studies, the precise pH of these solutions varies. Some saline recipes recommend a pH between 6.6 and 6.9, similar to the pH measured in the third instar larvae (Echalier 1997), while other studies on larvae recommend a saline buffered to a pH of 7.0–7.2 (Badre et al. 2005; Stewart et al. 1994). Measurements of heartbeat in larva have been performed at a pH of 7.0 (Gu and Singh 1995), and adult *Drosophila* hearts have been measured at 7.2 (Papaefthimiou and Theophilidis 2001). While all of these values lie close to the values recorded from some other Diptera, the pH of adult *Drosophila* haemolymph still needs to be investigated. It remains to be seen how successfully these physiological salines recreate the normal acid–base environment of the haemolymph in vivo.

8.3.2 Haemolymph Buffering Capacity

Insects can respond to the addition of acid or base into their haemolymph by buffering or by active transport/excretion. Transporting acid–base equivalents between body compartments and/or the environment is a slower, long-term solution, while buffering immediately limits the extent of the pH shift. Buffering is achieved by the presence of weak acids and their conjugate bases acting as proton donors or accepters. A weak acid and its conjugate base buffer most effectively at the pH that coincides with its dissociation constant (pKa), the pH at which the concentration of weak acid, [HA], is the same as its conjugate base, [A⁻]. In biological systems, the most commonly occurring pH buffers are bicarbonate ions, proteins and phosphate (Truchot 1987). The importance of these buffers in maintaining the pH of an insect's haemolymph depends on their concentration in the haemolymph and the nature of the acid–base disturbance. Non-bicarbonate buffers donate and accept protons from non-volatile acids and bases, as well as from carbonic acid produced by the hydration of CO_2. In contrast, bicarbonate ions can only buffer metabolic and non-volatile acid disturbances, as shown by the reversible reaction:

$$CO_2 + H_2O \leftrightarrow H_2CO_3 \leftrightarrow HCO_3^- + H^+ \tag{8.1}$$

An increase in [H⁺] shifts this reaction from bicarbonate towards the production of CO_2. The volatile CO_2 can then be removed from the blood by gas exchange with the surrounding environment. The concentration of HCO_3^- in a fluid is partly determined by PCO_2 (in addition to [SID] and [HA]). The PCO_2 of a terrestrial vertebrate is around 4.7–6 kPa (Truchot 1987), while the PCO_2 of insect haemolymph is generally much lower, around 2 kPa in a cockroach (Matthews and White 2011) and between 1.5 and 2.19 kPa in the desert locust Schistocerca gregaria (Gulinson and Harrison 1996; Harrison et al. 1990). As a result, there is generally a lower concentration of bicarbonate available to act as a buffer. The bicarbonate buffering properties of insect blood have been most thoroughly investigated within the Orthoptera, with concentrations of haemolymph bicarbonate levels known from six species, varying between 5 and 13 mmol l⁻¹ (Gulinson and Harrison 1996; Harrison et al. 1990, 1995; Harrison 1988, 1989b; Krolikowski and Harrison 1996). In the normal physiological range of haemolymph pH, this equates to a bicarbonate buffer capacity of 12–28 mmol l⁻¹ pH unit⁻¹ (Harrison 2001). The role of the bicarbonate buffer system in insects has been demonstrated in the desert locust Schistocerca gregaria, which was injected with hydrochloric acid (HCl) to simulate a metabolic acidosis. The insect showed a reduction in haemolymph pH of 0.5 units, which caused [HCO_3^-] to drop by half and PCO_2 to double for the first 15 min following injection (Harrison et al. 1992). It was calculated that the bicarbonate buffer neutralised half of the added acid, liberating CO_2 which was completely restored to control levels 1 h after the injection. The remainder was buffered by non-bicarbonate buffers, and ultimately by transport and excretion processes associated with the gut (Harrison et al. 1992). The total buffer capacity of Schistocerca gregaria haemolymph at a pH of 7.31 is 33.6 mmol l⁻¹ pH unit⁻¹, attributed to bicarbonate (60 %), protein (30 %), organic and inorganic phosphate (9 %), with the remainder citrate and histidine (Harrison et al. 1990). One of the few non-orthopteran insects to have the buffer capacity of its haemolymph determined is the larva of the horse bot fly Gasterophilus intestinalis. The horse bot fly maggot is a specialist intestinal parasite that lives within the high-CO_2 environment of a horse's stomach, where it regularly experiences a PCO_2 in excess of 66 kPa (Levenbook 1950). Within this unusual environment, the pH buffer capacity of G. intestinalis haemolymph at its usual haemolymph pH of 6.8 is 47.0 mmol l⁻¹ pH unit⁻¹. This buffer capacity is again primarily due to bicarbonate (57.5 %), as well as proteins (30 %), succinate (6 %) and inorganic phosphate (5 %) (Levenbook 1950).

8.3.3 Regulation of Acid–Base Disturbance

Role of the Gut and Malpighian Tubules

Adding a strong acid or base into an insect, either from metabolic by-products, food, or, in the case of aquatic insects, directly from their environment, alters the acid–base balance of its haemolymph. As buffering can only ever limit the change in pH, other mechanisms must come into play to restore the original acid–base balance. This can only be achieved by the active transport of acid–base equivalents across epithelia, either between body fluid compartments within the insect, or between the insect and its environment. In terrestrial insects, this transport occurs between the haemolymph and gut lumen, across the walls of the gut and Malpighian tubules.

A generalised insect gut is divided into several distinct sections, beginning with the foregut, which is made up of the oesophagus, proventriculus, or gizzard, and crop. The midgut comprises the gastric caeca and ventriculus; and the hindgut is made up of the ileum, colon, rectum and anus (◘ Fig. 8.1). The pH of the gut lumen differs along most of its length from that of the surrounding haemolymph. The Malpighian tubules arise between the mid and hindgut regions and extend into the haemolymph, where they take up ions, metabolic wastes and water, excreting this primary urine from the haemolymph into the hindgut, where it is further modified before excretion. In an insect, this hindgut–Malpighian tubule complex is the main site of ion and water regulation and performs essentially the same functions as the vertebrate kidney, including acid–base regulation (Phillips 1981). A series of investigations into the desert locust *Schistocerca gregaria* provides the most complete picture of an insect's acid–base regulation and the central role played by the hindgut–Malpighian tubule complex. However, these studies have focused on the mechanisms used to remove excess acid, and no studies have yet examined the response of the locust to an alkaline challenge. A comprehensive review of this topic can be found in Phillips et al. (1994).

The Malpighian tubules produce the primary urine by actively taking up K^+ and Na^+ (but may preferentially secrete one or the other, depending on the insect species) from the haemolymph, as well as smaller amounts of Mg^{2+}, Cl^-, HCO_3^-, NH_4^+ and other haemolymph solutes (Maddrell and O'Donnell 1992). The transport of inorganic ions is driven by

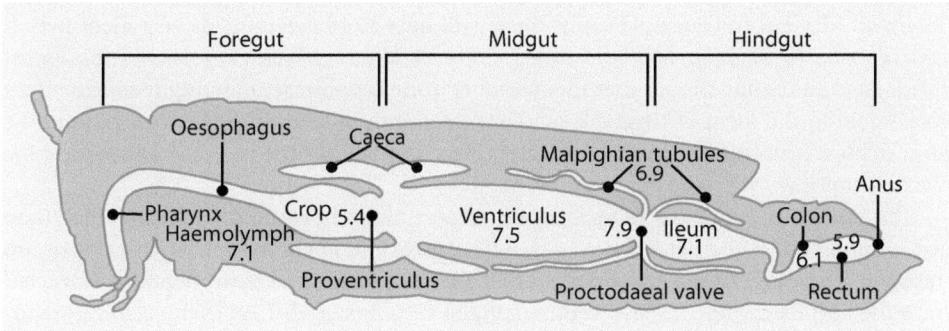

◘ **Fig. 8.1** Cross section of the locust. The pH of the haemolymph and the various sections of the gut are indicated in bold (Data from Thomson et al. (1988a); image adapted from Phillips (1964))

an electrogenic V-ATPase (O'Donnell and Maddrell 1995), which pumps protons into the tubule, keeping its pH around 0.5 units below that of the haemolymph (Stagg et al. 1991). This proton gradient energises the cation/H^+ antiporters, driving ion transport into the tubule lumen. The reduced pH also causes the PCO_2 in the fluid to increase to three times the haemolymph levels due to the titration of HCO_3^- to CO_2 (Phillips et al. 1994). But the V-ATPase proton pump is a minor contributor to total acid excretion, as most of the protons in the Malpighian tubules are recycled through the cation antiport system (Stagg et al. 1991). The osmotic gradient between the fluid in the tubules and the haemolymph drives the movement of water from the haemolymph into the tubules, flushing the primary urine into the ileum. Both the ileum and the rectum function to reabsorb Na^+, Cl^- and K^+ ions and water while excreting acid or base into the lumen using a range of pathways. In both segments, apical V-ATPase proton pumps actively secrete H^+ into the lumen, progressively lowering the luminal pH from the ileum to the rectum (◘ Fig. 8.1) (Thomson and Phillips 1992). The pH in the rectum is further reduced following the injection of a non-volatile acid, demonstrating that the increased excretion of acid into the rectum occurs to restore the acid–base balance of the haemolymph (Thomson et al. 1988a). Approximately 15 % of the injected acid load was accounted for by an increased transport of ammonium ions (NH_4^+) by the ileum and rectum (Harrison and Phillips 1992). Ammonia (NH_3) is produced primarily from the oxidation of proline, as well as other amino acids, before it is excreted into the gut lumen as NH_4^+ in exchange for the uptake of Na^+ into the haemolymph (Thomson et al. 1988b; Peach and Phillips 1991). It has been estimated that the rectum is responsible for approximately 60 % of all acid excretion in the locust, due to its large surface area ($\sim$0.62 cm^2) compared to the ileum, which contributes 30 % (0.4 cm^2), and the Malpighian tubules (less than 0.09 %) (Phillips et al. 1994).

While the acidification of the hindgut occurs predominantly by excretion of H^+ into the gut lumen rather than by the selective reabsorption of bicarbonate (Thomson et al. 1988a), both the ileum and rectum do appear to transport HCO_3^- from the gut into the haemolymph. Experiments on isolated epithelia have shown that these sections of the hindgut secrete a fluid from the gut lumen into the haemolymph that has a HCO_3^- concentration three times higher than that in lumen (Lechleitner et al. 1989). This absorption of bicarbonate by the locust hindgut appears to occur by a process that differs from most other animals, in that it does not rely on a Cl^-/HCO_3^- exchange system. The uptake of Cl^- from the gut lumen appears to be an active process coupled to the passive uptake of K^+ (Hanrahan and Phillips 1983; Phillips and Audsley 1995). Thus, the mechanisms used by the insect hindgut to transport bicarbonate still have not been conclusively identified. It also remains to be seen how the hindgut excretes excess base. Studies on the locust *Taeniopoda eques* has shown that this insect is quite capable of changing from excreting excess acid in the form of titratable acid and ammonium to excreting excess base in the form of bicarbonate, depending on its diet (Harrison and Kennedy 1994). But, again, the mechanisms have yet to be described in detail.

The secretion processes described above are under hormonal control, rather than being influenced directly by the acid–base status of the haemolymph, with rates of ion transport, secretion of H^+ and uptake of HCO_3^- responding to neuropeptides extracted from the corpus cardiacum and ventral ganglia (Phillips et al. 1998). The active component of the corpus cardiacum extract has been identified as a neuropeptide hormone called ion transport peptide (ITP). This hormone has been shown to stimulate Na^+, Cl^-, K^+ and fluid absorption by the ileum, while inhibiting H^+ secretion (Phillips et al. 1998). It is likely that ITP controls the Na^+, Cl^- and K^+ ion pumps by stimulating the intracellular production of cAMP (cyclic AMP or cyclic adenosine monophosphate). However, while cAMP has been found to inhibit acid secretion in the rectum of *S. gregaria,* it does not appear to have the same effect on the H^+ V-ATPase pumps in the ileum (Thomson et al. 1991), which appear to be inhibited by a different messenger molecule in response to ITP. While the V-ATPase transporters in locust rectum are inhibited by cAMP, a stimulatory effect has been recorded from other parts of the Malpighian tubule–hindgut complex. The V-ATPase in the Malpighian tubules is stimulated by cAMP, with studies on *Drosophila* showing a decrease of 0.4 pH units in the tubule fluid relative to the unstimulated fluid secretion (O'Donnell and Maddrell 1995).

Role of Respiration and Ventilation

The constant production of CO_2 by respiration constitutes the largest continuous influx of acid that most insects must contend with. Given that many flying insects are capable of increasing their rate of CO_2 production over 100-fold when they transition from rest to active flight, it is clear that they must possess effective mechanisms to deal with CO_2 and its potential impact on acid–base balance through carbonic acid production. As CO_2 is a volatile acid, releasing it to the atmosphere is the simplest mechanism to deal with this problem. While vertebrates transport CO_2 from their tissues to gas exchange organs in their blood, insects have a respiratory system that transports O_2 and CO_2 independently from their haemolymph: the tracheal system. Their tracheal system consists of a branching network of air-filled tubes that open to the atmosphere through spiracles, providing a high-conductance conduit between the haemolymph and tissues of the insect and the surrounding atmosphere. While the lack of respiratory pigments within the haemolymph limits the role it can play in transporting O_2 and CO_2 within the insect, it still functions as a significant buffer for respiratory CO_2 and a minor reserve of O_2. The high rate of diffusion and convection within the air-filled tracheal system and the large surface area of the finely branched tracheoles allow for the rapid removal of CO_2 from both tissues and haemolymph. By opening and closing their spiracles to regulate gas exchange between the atmosphere and their tracheal system, insects at rest maintain an internal PCO_2 between 1.8 and 3 kPa. This PCO_2 is considerably lower than that found in an endothermic vertebrate (4–5 kPa) but is similar to values recorded from terrestrial, air-breathing crabs (Wood and Randall 1981; Truchot 1987). Air-breathing vertebrates use their high internal PCO_2 and bicarbonate levels to rapidly neutralise the addition of a fixed acid. The high concentration of bicarbonate reacts with the acid to produce CO_2, which is then exhaled. The low CO_2 and bicarbonate levels reduce the extent to which insects may use ventilation to regulate their internal pH by expelling excess CO_2. Experiments on locusts partially support this view, as the injection of 50 µl of 0.5 mol l^{-1} hydrochloric acid (HCl) into the haemolymph of the locust *Romalea guttata* did not produce a significant increase in ventilation rate, despite significantly reducing pH (Gulinson and Harrison 1996). However, the injection of sodium bicarbonate ($NaHCO_3$) does cause ventilation frequency to increase. This increased gas exchange is unlikely to be a response to correct pH, since the treatment did not significantly alter the pH of the haemolymph, but it may be explained by a significant increase in the average intratracheal PCO_2 during the 10 min following the injection of $NaHCO_3$ (Gulinson and Harrison 1996).

These experiments examining the effect of haemolymph pH on ventilation raise some interesting questions regarding the nature of gas exchange control in insects. If respiratory pH regulation is important, then their respiratory chemoreceptors should respond to changes in pH as well as PCO_2, as they do in air-breathing vertebrates (Milsom 2002).

Using acids such as HCl to change pH did not alter the ventilation rate, but the addition of bicarbonate, and its liberation from the haemolymph as CO_2, did have this effect. Similarly, the spiracles of the housefly *Musca domestica* have been shown to open in response to increases in CO_2, and not to decreases in pH (Case 1957). However, this is not to say that insects do not use gas exchange to regulate the pH of their haemolymph. For example, vigorous hopping in the locust *Melanoplus bivittatus* at 35 °C was sufficient to cause the PCO_2 of its haemolymph to increase from approximately 2.9–5.2 kPa, and the pH of its haemolymph to decrease by 0.14 units (Harrison et al. 1991). While at rest after this exercise, the locust completely reversed the acidification of its haemolymph by employing a sustained period of elevated ventilation to remove the accumulated CO_2, thereby restoring the PCO_2. Thus, regulating ventilation to preserve a constant internal PCO_2 is sufficient to maintain a stable haemolymph pH.

While the stimulatory effect of CO_2 on ventilation is well documented, some studies have also found evidence that decreased haemolymph pH can produce a similar effect. For example, irrigating the central nervous system of the cockroach *Nauphoeta cinerea* with Ringer's solution, either equilibrated with a PCO_2 of 6.2 kPa or acidified to a pH of 6.97 with HCl, resulted in both cases in an equally significant rise in ventilation frequency (Snyder et al. 1980). Other cockroaches have displayed similar responses. The isolated nerve cord of *Blaberus craniifer* could be reversibly stimulated to produce spiking consistent with respiratory activity by exposing it to any of 13 different weakly dissociated acids in saline solution (Case 1961). However, the pH at which the different acids produced an effect varied, indicating that pH per se may not be the main stimulus.

Insects often experience microclimates that have CO_2 concentrations well above normal atmospheric levels. Habitats rich in decaying organic matter, such as dung pats, can have levels of CO_2 up to 20 kPa (Holter and Spangenberg 1997). These environments are challenging to acid–base regulation, as they can increase the PCO_2 of body fluids to dangerously high levels. By increasing their ventilation rate, *Nauphoeta cinerea* cockroaches are capable of maintaining a stable internal pH while breathing air with a PCO_2 of up to 1 kPa (Matthews and White 2011). Exposure to even higher levels of environmental hypercapnia results in a further increase in ventilation frequency. But once the ambient PO_2 exceeds the normal internal PCO_2 of the cockroach, their internal PCO_2 must necessarily increase, associated with a decrease in haemolymph pH. No amount of hyperventilation can restore the pH of the cockroach's haemolymph so long as the PCO_2 remains above the insect's normal PCO_2. The only way to restore the balance is to move the cockroach to a microhabitat with a lower CO_2 level. The locust *Schistocerca nitens* displays the same responses to short-term environmental hypercapnia, with both its abdominal ventilation rate and the PCO_2 of its haemolymph rising as haemolymph pH falls (Harrison 1989a). For some insects, escaping a hypercapnic microhabitat in favour of lower atmospheric CO_2 levels is not an option, and the acid–base effects of chronic hypercapnia must be dealt with. Whether insects in this situation are capable of producing a compensated respiratory acidosis by increasing bicarbonate reabsorption remains unknown.

In addition to regulating CO_2 removal, insects must also maintain O_2 uptake. Exposure to environmental hypoxia elicits hyperventilation, increasing the clearance rate of CO_2 from the insect's body fluids above rates of production. This decreases the internal PCO_2, shifting the bicarbonate/CO_2 equilibrium towards CO_2, which is then exhaled. The end result is a pH increase indicative of respiratory alkalosis. Exposure to a PO_2 of 5 kPa caused hyperventilation in the cockroach *Nauphoeta cinerea*, causing a rise in haemolymph pH of 0.34 units (Matthews and White 2011), and the same PO_2 caused pH to rise

0.15 units in the rhinoceros beetle *Xylotrupes ulysses* (unpublished data ☐ Fig. 8.2). The data from *X. ulysses* also shows the corrective response when returned to normoxic air. Once internal PO_2 has been restored, the beetle begins a breath-hold period that allows CO_2 to accumulate within its body fluids, restoring its haemolymph pH to pre-hyperventilation levels.

Effects of Discontinuous Gas Exchange

Some insects at rest breathe intermittently, alternately accumulating CO_2 while they hold their spiracles shut, then expelling the accumulated CO_2 in a burst or bout of ventilation. This pattern of gas exchange is best known from lepidopteran pupae, but also occurs in the Orthoptera, Blattodea, Coleoptera and Hymenoptera (Marais et al. 2005). The acid–base balance of the insect's haemolymph fluctuates during these discontinuous gas exchange cycles (DGCs), with in vivo measurement of haemolymph pH revealing periodic decreases of around 0.06 pH units below pH 6.74 in a butterfly pupa (Hetz and Wasserthal 1993) and 0.11 pH units below ~7.3 in a cockroach (Matthews and White 2011). In vitro measurements made on lubber grasshoppers (*Taeniopoda eques*) displaying DGCs by extracting haemolymph samples immediately preceding and following a CO_2 burst found DGCs caused even smaller changes in haemolymph pH of between 0.030 and 0.037 units (Harrison et al. 1995). The pH change observed during DGCs is due entirely to the accumulation and release of CO_2. An example of the relationship between intermittent gas exchange and pH fluctuation is shown in ☐ Fig. 8.2. Simultaneous record-

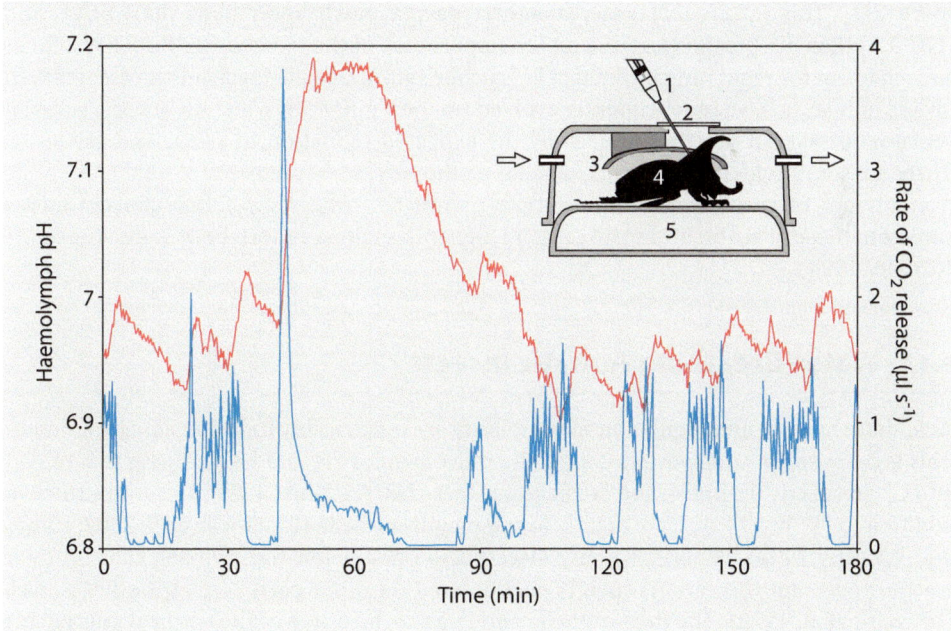

☐ **Fig. 8.2** Haemolymph pH (*red*) and exhaled CO_2 (*blue*) before, during and after exposure to hypoxia (5 % PO_2, grey rectangle) including an inset of experimental setup showing (*1*) syringe mounted optical pH probe, (*2*) respirometry chamber including wax seal around needle, (*3*) putty-filled restraint, (*4*) rhinoceros beetle, (*5*) temperature controlled water jacket maintained at 25 °C. Arrows indicate direction of airflow through the chamber to the CO_2 analyser

ing of haemolymph pH, using a fibre-optic pH optode implanted in the haemocoel of a rhinoceros beetle (*Xylotrupes ulysses*) in concert with the rate of CO_2 release from the insect measured using flow-through respirometry, clearly shows the decrease in haemolymph pH during periods where CO_2 is not released. The high intracellular buffering capacity estimated from whole body homogenates, relative to that of the haemolymph, suggests that pH changes associated with DGCs are insignificant within the insect's tissues (Bridges and Scheid 1982).

8.4   Acid–Base Regulation in Aquatic Insects

Terrestrial insects have reinvaded the aquatic environment numerous times throughout their evolutionary history, evolving either partially or completely aquatic lifecycles (Wootton 1988). Aquatic insects spend their lives in intimate contact with an aqueous environment with an ionic composition and pH that varies more or less from the internal fluids of the insect. This has two important consequences for acid–base regulation: First, it means that aquatic insects can excrete or take up acid–base equivalents directly from their surrounding environment across ion permeant epithelia. Second, because water has a far lower capacitance for O_2 than for CO_2 when compared with air (Dejours 1981), life in water can have a dramatic impact on an insect's internal PCO_2 and bicarbonate concentration. Aquatic animals must ventilate large volumes of water to extract sufficient O_2, and in the process, they quickly expel their respiratory CO_2 (Dejours 1988). In contrast, terrestrial animals can easily obtain O_2 from the air and so ventilate less and consequently retain more CO_2. This means that water-breathers have a much lower body fluid PCO_2 and $[HCO_3^-]$ than air-breathers, reducing the importance of the bicarbonate buffer system as a mechanism for regulating internal pH (Truchot 1987). Despite the fact that at least seven orders of insects have independently evolved tracheal gills with which to breathe water, it remains to be seen whether they show the expected reduction in PCO_2 and $[HCO_3^-]$. There are some evidences that this may not be the case. Measurements of total CO_2 in the haemolymph of a water-breathing damselfly nymph (*Austrolestes* sp.) show concentrations similar to those of air-breathing insects and higher than those of vertebrate water-breathers (Cooper 1994).

8.4.1 Water-Breathing Aquatic Insects

Acid–base balance and regulation of osmolarity are inextricably linked in aquatic animals. This is because ion transporters couple the movement of H^+ and HCO_3^- with that of Na^+ and Cl^-, respectively. In fish and crustaceans, these Na^+/H^+ and Cl^-/HCO_3^- antiporters, in addition to H^+ V-ATPases, are mainly found on gill epithelia (Claiborne et al. 2002; Henry and Wheatly 1992). Water-breathing insects also possess ion-transporting cells on their tracheal gills, but they are frequently encountered on other parts of their bodies as well (Wichard et al. 1972). The thin epithelia and large surface area of the tracheal gills enable them to function as ion exchange organs and to participate in acid–base regulation, in addition to their respiratory function. Chloride cells occur over the entire body surface of mayflies (Ephemeroptera) and stoneflies (Plecoptera) but occur in higher abundance on the surface of the tracheal gills (Komnick 1977). In comparison, the chloride cells of dragonflies and damselflies (Odonata) are restricted to the rectum, which is modified into a gas exchange organ (a rectal gill) in dragonflies (Wichard and Komnick 1974; Komnick and Achenbach 1979). In order to maintain a hypertonic haemolymph with respect to their freshwater environment, aquatic insects rely on the pumping action of chloride cells to counteract the constant passive loss of Na^+ and Cl^- ions into the surrounding water across permeable regions of their cuticle and expelled from their gut. However, exposure to acidic water has the potential to disrupt osmoregulation by interfering with the function of V-ATPase proton pumps, thereby preventing the Na^+/H^+ exchanger from functioning. There is evidence that many aquatic insects exposed to low pH succumb due to the disruption of Na^+ uptake and subsequent decrease in internal $[Na^+]$ due to ion efflux, rather than a loss of pH homeostasis. This has been observed in stoneflies (Lechleitner et al. 1985) and dragonflies (Rockwood and Coler 1991) which both show a significant decrease in haemolymph $[Na^+]$ following prolonged exposure to acidic water (pH 3 and 2.3, respectively). Exposing the stonefly nymph to alkaline water with a pH of 8.0 did not significantly change $[Na^+]$ (Lechleitner et al. 1985). Given that osmoregulation is a serious challenge in acidified water, it follows that an aquatic insect's ability to withstand highly acidic aquatic environments is enhanced by a low cuticular permeability (Havas and Advokaat 1995). This can be more easily achieved if the aquatic insect breathes not water, but air. Without the need for tracheal gills and the large permeable surface area they represent, air-breathing aquatic insects can potentially maintain a much lower cuticular permeability.

8.4.2 Air-Breathing Aquatic Insects

Many aquatic insects do not possess tracheal gills or breathe water, but remain air-breathers, either by maintaining contact with atmospheric air using a snorkel, by intermittently contacting the surface to refresh a bubble of atmospheric air carried on their body or by having a layer of air trapped permanently over their spiracles. For example, mosquito larva use a respiratory syphon to connect their tracheal system with the atmosphere while resting at the surface of the water, while many aquatic bugs (Hemiptera) and beetles (Coleoptera) carry a bubble of air on their body while diving. As a result, these insects do not bring water into direct contact with their thin respiratory epithelia and are likely to possess a PCO_2 and $[HCO_3^-]$ similar to terrestrial insects. But, again, this assumption remains untested. The absence of gills also reduces the degree of contact between the surrounding water and respiratory epithelia, limiting opportunities for both passive and active ion movement across the body wall. It is likely that this increases the importance of the gut in regulating acid–base balance and osmolality, as has been observed in the larvae of diving beetles (Komnick 1977). However, even air-breathing aquatic insects possess ion-transporting chloride cells on parts of their cuticle that remain in contact with the surrounding water, providing a direct pathway for ionic regulation with the aquatic environment in addition to the gut (Komnick 1977; Komnick and Schmitz 1977). Water boatmen (Corixidae) are aquatic hemipterans that carry a bubble of air on their ventral surface while underwater. While they do possess chloride cells on their head and legs (Komnick and Schmitz 1977), they lack tracheal gills and have very low rates of sodium efflux (Vangenechten et al. 1989; Witters et al. 1984). Consequently, they are highly tolerant of life in acidic water, showing insignificant decreases in haemolymph $[Na^+]$ in water with a pH as low as 3.0 (Needham 1990). Several species of corixids also show a high degree of tolerance to alkaline conditions, living in so-called athalassohaline lakes that are rich in carbonates and have very high pH (9–10). While the Malpighian tubules of these insects are capable of producing an alkaline primary urine when stimulated by cAMP (Szibbo and Scudder 1979), excretion of HCO_3^- in exchange for Cl^- and Na^+ for H^+ by the chloride cells are likely to play a central role in acid–base regulation (Cooper et al. 1987).

Dipteran larvae, including mosquitoes, midges and alkali flies show some of the most impressive acid–base regulation abilities in the animal kingdom. Of these insects, mosquito larvae are among the most capable pH regulators known. Larvae of both *Aedes aegypti* and *Ochlerotatus taeniorhynchus* are able to complete their development in water with a pH ranging from 4 to 11, and, more impressively, they can maintain the pH of their haemolymph to within 0.1 pH units or less across this range (Clark et al. 2004). This constancy of internal pH is reflected by the mosquito larva's ability to maintain internal osmolality in both very dilute (20 μmol l^{-1} NaCl) and hypersaline (>300 mmol l^{-1} NaCl) water (Garrett and Bradley 1984; Patrick et al. 2002). Rather than having chloride cells scattered over their entire cuticle, many dipteran larvae instead possess dedicated ion exchange organs called anal papillae. The anal papillae are finger-like projections of the cuticle that lie next to the anus and are involved with ion exchange rather than respiratory gas exchange. They are similar to gills in that they are also highly water permeable but afford a far smaller surface area for the passive movement of ions and water (Wigglesworth 1933). Restricting cutaneous ion and water transfer to a dedicated ion exchange organ allows the rest of the cuticle to be water and ion impermeable. The transport mechanisms used by mosquito larvae to achieve their acid–base and ionoregulatory feats are not yet completely understood, but the measurements of ion fluxes across the anal papillae and immunolocalization of the ion pumps on the anal papillae (Patrick et al. 2006) and along the alimentary canal (Okech et al. 2008) have greatly advanced our understanding of ion transport and pH regulation in these insects.

The presence of V-ATPase and Na^+/K^+ ATPase in the epithelium of anal papillae has been confirmed using immunolocalization techniques (Patrick et al. 2006) while the presence of a Cl^-/HCO_3^- transporter is currently only inferred from pharmacological experiments (Del Duca et al. 2011). Curiously, though, there is no evidence for the presence of a Na^+/H^+ ATPase on the papillae. The uptake of sodium into the papillae appears to be driven primarily by the electrical potential generated by apically located V-ATPase pumping protons into the surrounding water, assisted by a basal Na^+/K^+-ATPase (Patrick et al. 2006). It is not yet clear whether this V-ATPase is actively involved in pH regulation, and there are some lines of evidence that suggest otherwise. In particular, the observation that larvae of *A. aegypti* do not upregulate mitochondrial density in their papillae when reared in high or low pH suggests that this organ does not actively increase acid or base excretion in response to ambient conditions (Clark et al. 2007). Mitochondrial density is, however, upregulated in response to low salt levels (Sohal and Copeland 1966). Regardless, the presence of a V-ATPase demonstrates that the anal papillae are a site of active H^+ excretion. The movement of other ions across the anal papillae of *A. aegypti* have been measured by recording ion gradients adjacent to the epithelium using ion-selective microelectrodes. This has revealed a net influx of Na^+, Cl^- and K^+ and net efflux of acid and NH_4^+ (Donini and O'Donnell 2005). Unlike other animals that rely on the secretion of bicarbonate to drive Cl^- uptake, mosquitoes also appear to use a novel Cl^- channel that is stimulated by exposure to acidic conditions and by the inhibition of the apical V-ATPase. This Cl^- pathway has only been observed in insects and shows an enhanced rate of Cl^- uptake during exposure to low pH. It has been suggested that the Cl^- transport is driven by an electrodiffusive gradient that is enhanced by either a reduction in the rate of H^+

excretion increasing the electrical gradient for Cl^- entry through a Cl^- channel or by a steeper H^+ gradient driving uptake via a Cl^-/H^+ co-transporter (Patrick et al. 2002). But while the transport of H^+ is clearly required for ion uptake, there is little evidence supporting the anal papillae as significant organs of pH homeostasis. This leaves the mosquito's hindgut as the next most likely candidate.

The role that the mosquito larvae's hindgut–Malpighian tubule complex plays in the ionic and acid–base regulation of the haemolymph is currently incomplete. Previous studies examining the responses of A. aegypti to aquatic pH have found that drinking rates varied inversely with pH, being highest in acidic water (pH 4) and lowest in alkaline (pH 11) (Clark et al. 2007). From this study, it was concluded that acid excretion by the Malpighian tubules associated with increased fluid clearance through the hindgut and rectum was responsible for maintaining haemolymph pH homeostasis under acid conditions. Low fluid ingestion rates have also been documented for mosquitoes naturally adapted to alkaline, high carbonate water (Strange et al. 1982), indicating that both alkaline-tolerant and alkaline-adapted mosquito larvae share a similar response to high pH water. But the nature of this response and the associated processes used to maintain pH homeostasis in alkaline water are unknown. Immunolocalization of V-ATPases in the Malpighian tubules and posterior rectum of A. aegypti and Anopheles gambiae larvae (Okech et al. 2008; Patrick et al. 2006) and the observation that the rectum is acidified (pH < 6.2) in both acidic and alkaline water (Clark et al. 2007) indicate that acid is excreted through the hindgut. However, an acidified rectum raises an interesting issue in regard to pH regulation in Aedes dorsalis mosquito larvae inhabiting carbonate-rich hypersaline lakes. The first investigations into this species concluded that the rectum was responsible for excreting HCO_3^- to maintain a haemolymph pH of ~7.7 in alkaline water with a pH above 8.85 (Strange et al. 1982, 1984). But if this species is like other Aedes mosquito larvae, then they should maintain an acidic rectal lumen, regardless of their alkaline environment. Thus, if they were to regulate their internal pH in alkaline water by excreting bicarbonate into their acidic rectum, it would react with the acid and be converted to CO_2, subsequently diffusing into the surrounding water and producing a net loss of acid from the insect (Clark et al. 2007). Clearly, a great deal more research is required to produce a complete picture of the acid–base regulation of mosquitoes and aquatic insects in general.

8.4.3 A Comment on Mosquitoes and Midges

The non-biting midges (Chironomidae) and mosquitoes (Culicidae) both have aquatic larvae that are highly competent iono- and acid–base regulators, as well as being similar anatomically. However, they differ in their mode of respiration. Mosquito larvae are primarily air-breathers, using their respiratory syphon to access atmospheric oxygen. This abundant source of O_2 is further supplemented by the trans-cuticular diffusion of oxygen and carbon dioxide between the larvae and surrounding water, particularly across their thin anal papillae (Wigglesworth 1933). In contrast, the benthic chironomid larvae are exclusively water-breathers, taking up dissolved oxygen across their entire cuticle (Fox 1921). As chironomids not only lack access to the atmosphere but also live burrowed into stagnant sediment, they fill their haemolymph with high oxygen affinity haemoglobins (Weber 1980). While these respiratory proteins facilitate oxygen uptake, their ability to bind reversibly with protons also allows them to contribute to the pH buffering capacity of the haemolymph (Jernelöv et al. 1981). It has been suggested that populations of

Chironomus riparius living in acidic streams have elevated levels of haemoglobin precisely for this reason (Jernelöv et al. 1981). Furthermore, the reversible binding of H^+ by haemoglobin alters the oxygen affinity of the protein (i.e. induces a Bohr shift), such that oxygen delivery is altered by changes in their internal pH (Weber et al. 1985). As mosquito larvae rely on the same air-filled tracheal oxygen delivery system as other insects, their respiration is not linked to pH in this manner nor do they possess this additional pH buffer. Beyond this obvious difference, there may be yet more subtle divergences between the acid–base and ionoregulatory systems of mosquitoes and chironomids. Experiments using carbonic anhydrase inhibitors have revealed a surprising difference between how these insects take up Cl^-. The currently enigmatic Cl^- uptake mechanism observed among mosquitoes is insensitive to acetazolamide, indicating that bicarbonate originating from the hydration of CO_2 in the papillae is not involved in Cl^- transport (Patrick et al. 2002). In contrast, chironomids exposed to methazolamide show a dose-dependent decrease in Cl^- uptake, indicating that they couple the uptake of Cl^- across their anal papillae with HCO_3^- excretion (Nguyen and Donini 2010). The significance of these differences is currently unknown. But future efforts to compare and contrast the respiratory physiology, iono- and acid–base regulation of these two insects, and more broadly across the diverse infraorder Culicomorpha, have the potential to reveal much about how insects evolve different solutions in response to environmental challenges.

8.5   Future Directions

Our understanding of insect acid–base physiology continues to advance, with studies on the location and action of ion transporter proteins in the gut and anal papillae of mosquitoes and locusts further refining earlier observations of ion transport made on isolated epithelia. However, it is clear that while this research has provided a better framework for understanding osmoregulation and pH regulation, much remains to be done to understand how both aquatic and terrestrial insects regulate their extracellular acid–base balance in response to environmental and dietary challenges. In particular, while the processes used to excrete excess acid have been well investigated, there have been very few studies that specifically investigate how insects, aquatic or otherwise, eliminate excess base. It also remains to be determined why some aquatic insects are capable of regulating internal pH homeostasis in the face of highly acid or alkaline water, while other species have such a low sensitivity to these conditions.

References

Badre NH, Martin ME, Cooper RL (2005) The physiological and behavioral effects of carbon dioxide on *Drosophila melanogaster* larvae. Comp Biochem Physiol A Mol Integr Physiol 140(3):363–376

Bedford JJ, Leader JP (1975) Haemolymph composition in the larva of the dragonfly, *Uropetala carovei*. J Entomol Ser A Gen Entomol 50(1):1–7

Bridges CR, Scheid P (1982) Buffering and CO_2 dissociation of body fluids in the pupa of the silkworm moth, *Hyalaphora cecropia*. Respir Physiol 48(2):183–197

Buck J, Friedman S (1958) Cyclic CO_2 release in diapausing pupae III: CO_2 capacity of the blood: Carbonic anhydrase. J Insect Physiol 2(1):52–60

Case JF (1957) Differentiation of the effects of pH and CO_2 on spiracular function of insects. J Cell Comp Physiol 49(1):103–113

Case JF (1961) Effects of acids on an isolated insect respiratory center. Biol Bull 121(2):385

Chapman RF (2013) The insects: structure and function, 5th edn. Cambridge University Press, Cambridge

Claiborne JB, Edwards SL, Morrison-Shetlar AI (2002) Acid–base regulation in fishes: cellular and molecular mechanisms. J Exp Zool 293(3):302–319. doi:10.1002/jez.10125

Clark TM, Flis BJ, Remold SK (2004) pH tolerances and regulatory abilities of freshwater and euryhaline Aedine mosquito larvae. J Exp Biol 207(13):2297–2304. doi:10.1242/jeb.01021

Clark TM, Vieira MAL, Huegel KL, Flury D, Carper M (2007) Strategies for regulation of hemolymph pH in acidic and alkaline water by the larval mosquito Aedes aegypti (L.) (Diptera; Culicidae). J Exp Biol 210(24):4359–4367. doi:10.1242/jeb.010694

Cooper PD (1994) Mechanisms of hemolymph acid-base regulation in aquatic insects. Physiol Zool 67(1):29–53

Cooper PD, Vulcano R (1997) Regulation of pH in the digestive system of the cricket, Teleogryllus commodus Walker. J Insect Physiol 43(5):495–499. doi:http://dx.doi.org/10.1016/S0022-1910(97)85495-9

Cooper PD, Scudder GGE, Quamme GA (1987) Ion and CO_2 regulation in the freshwater water boatman, Cenocorixa blaisdelli (Hung.) (Hemiptera, Corixidae). Physiol Zool 60(4):465–471. doi:10.2307/30157908

Correa M, Coler RA, Yin C-M (1985) Changes in oxygen consumption and nitrogen metabolism in the dragonfly Somatochlora cingulata exposed to aluminum in acid waters. Hydrobiologia 121(2):151–156. doi:10.1007/bf00008718

Dejours P (1981) Principles of comparative respiratory physiology, 2nd edn. Elsevier/North-Holland Biomedical Press, Amsterdam

Dejours P (1988) Respiration in water and air. Elsevier Science Publishers B.V, Amsterdam

Del Duca O, Nasirian A, Galperin V, Donini A (2011) Pharmacological characterisation of apical Na^+ and Cl^- transport mechanisms of the anal papillae in the larval mosquito Aedes aegypti. J Exp Biol 214(23):3992–3999. doi:10.1242/jeb.063719

Donini A, O'Donnell MJ (2005) Analysis of Na^+, Cl^-, K^+, H^+ and NH_4^+ concentration gradients adjacent to the surface of anal papillae of the mosquito Aedes aegypti: application of self-referencing ion-selective microelectrodes. J Exp Biol 208(4):603–610. doi:10.1242/jeb.01422

Dow JA (1984) Extremely high pH in biological systems: a model for carbonate transport. Am J Physiol Regul Integr Comp Physiol 246(4):R633–R636

Echalier G (1997) Drosophila cells in culture. Academic, San Diego

Fox HM (1921) Methods of studying the respiratory exchange in small aquatic organisms, with particular reference to the use of flagellates as an indicator for oxygen consumption. J Gen Physiol 3(5):565–573

Garrett M, Bradley TJ (1984) The pattern of osmotic regulation in larvae of the mosquito Culiseta inornata. J Exp Biol 113(1):133–141

Groenewald B, Chown SL, Terblanche JS (2014) A hierarchy of factors influence discontinuous gas exchange in the grasshopper Paracinema tricolor (Orthoptera: Acrididae). J Exp Biol 217(19):3407–3415. doi:10.1242/jeb.102814

Gu G-G, Singh S (1995) Pharmacological analysis of heartbeat in Drosophila. J Neurobiol 28(3):269–280. doi:10.1002/neu.480280302

Gulinson SL, Harrison JF (1996) Control of resting ventilation rate in grasshoppers. J Exp Biol 199(2):379–389

Hanrahan JW, Phillips JE (1983) Cellular mechanisms and control of KCl absorption in insect hindgut. J Exp Biol 106(1):71–89

Harrison JM (1988) Temperature effects on haemolymph acid-base status in vivo and in vitro in the two-striped grasshopper Melanoplus bivittatus. J Exp Biol 140(1):421–435

Harrison JF (1989a) Ventilatory frequency and haemolymph acid-base status during short-term hypercapnia in the locust, Schistocerca nitens. J Insect Physiol 35(11):809–814

Harrison JM (1989b) Temperature effects on intra- and extracellular acid-base status in the American locust, Schistocerca nitens. J Comp Physiol B-Biochem Syst Environ Physiol 158(6):763–770

Harrison JF (2001) Insect acid-base physiology. Annu Rev Entomol 46:221–250

Harrison JF, Kennedy MJ (1994) In-vivo studies of the acid-base physiology of grasshoppers: The effect of feeding state on acid-base and nitrogen excretion. Physiological Zoology 67(1):120–141

Harrison JF, Phillips JE (1992) Recovery from acute haemolymph acidosis in unfed locusts: II. Role of ammonium and titratable acid excretion. J Exp Biol 165(1):97–110

Harrison JF, Wong CJH, Phillips JE (1990) Haemolymph buffering in the locust Schistocerca gregaria. J Exp Biol 154(1):573–579

Harrison JF, Phillips JE, Gleeson TT (1991) Activity physiology of the two-striped grasshopper, Melanoplus bivittatus: Gas exchange, hemolymph acid-base status, lactate production, and the effect of temperature. Physiol Zool 64(2):451–472. doi:10.2307/30158185

Harrison JF, Wong CJ, Phillips JE (1992) Recovery from acute haemolymph acidosis in unfed locusts: I. Acid transfer to the alimentary lumen is the dominant mechanism. J Exp Biol 165(1):85–96

Harrison JF, Hadley NF, Quinlan MC (1995) Acid-base status and spiracular control during discontinuous ventilation in grasshoppers. J Exp Biol 198(8):1755–1763

Havas M, Advokaat E (1995) Can sodium regulation be used to predict the relative acid-sensitivity of various life-stages and different species of aquatic fauna? Water Air Soil Pollut 85(2):865–870. doi:10.1007/BF004/6938

Heimpel AM (1955) The pH in the gut and blood of the larch sawfly, Pristiphora erichsonii (Htg.), and other insects with reference to the pathogenicity of Bacillus cereus Fr. and Fr. Can J Zool 33(2):99–106

Henry RP, Wheatly MG (1992) Interaction of respiration, ion regulation, and acid-base balance in the everyday life of aquatic crustaceans. Am Zool 32(3):407–416

Hetz SK, Wasserthal LT (1993) Miniaturized pH-sensitive glass-electrodes for the continuous recording of hemolymph pH in resting butterfly pupae. Verhandlungen Der Deutschen Zoologischen Gesellschaft 86:92

Holter P, Spangenberg A (1997) Oxygen uptake in coprophilous beetles (Aphodius, Geotrupes, Sphaeridium) at low oxygen and high carbon dioxide concentrations. Physiol Entomol 22(4):339–343

Jernelöv A, Nagell B, Svenson A (1981) Adaptation to an acid environment in Chironomus riparius (Diptera, Chironomidae) from Smoking Hills, NWT, Canada. Holarct Ecol 4(2):116–119. doi:10.2307/3682527

Komnick H (1977) Chloride cells and chloride epithelia of aquatic insects. Int Rev Cytol 49:285–329

Komnick H, Achenbach U (1979) Comparative biochemical, histochemical and autoradiographic studies of Na$^+$/K$^+$-ATPase in the rectum of dragonfly larvae (Odonata, Aeshnidae). Eur J Cell Biol 20(1):92–100

Komnick H, Schmitz M (1977) Cutane Chloridaufnahme aus hypo-osmotischer Konzentration durch die Chloridzellen von Corixa punctata. J Insect Physiol 23(2):165–173. doi:http://dx.doi.org/10.1016/0022-1910(77)90026-9

Krolikowski K, Harrison J (1996) Haemolymph acid-base status, tracheal gas levels and the control of post-exercise ventilation rate in grasshoppers. J Exp Biol 199(2):391–399

Lechleitner RA, Cherry DS, Cairns J Jr, Stetler DA (1985) Ionoregulatory and toxicological responses of stonefly nymphs (Plecoptera) to acidic and alkaline pH. Arch Environ Contam Toxicol 14(2):179–185. doi:10.1007/bf01055609

Lechleitner RA, Audsley N, Phillips JE (1989) Composition of fluid transported by locust ileum: influence of natural stimulants and luminal ion ratios. Can J Zool 67(11):2662–2668. doi:10.1139/z89-376

Lettau J, Foster WA, Harker JE, Treherne JE (1977) Diel changes in potassium activity in the haemolymph of the cockroach Leucophaea maderae. J Exp Biol 71(1):171–186

Levenbook L (1950) The physiology of carbon dioxide transport in insect blood: Part III. The buffer capacity of Gastrophilus blood. J Exp Biol 27(2):184–191

Maddrell SH, O'Donnell MJ (1992) Insect Malpighian tubules: V-ATPase action in ion and fluid transport. J Exp Biol 172(1):417–429

Marais E, Klok CJ, Terblanche JS, Chown SL (2005) Insect gas exchange patterns: a phylogenetic perspective. J Exp Biol 208(23):4495–4507

Matthews PG, White CR (2009) Rhinoceros beetles modulate spiracular opening to regulate haemolymph pH. Comp Biochem Physiol A Mol Integr Physiol 153(2, Suppl 1):S101–S102

Matthews PGD, White CR (2011) Regulation of gas exchange and haemolymph pH in the cockroach Nauphoeta cinerea. J Exp Biol 214(18):3062–3073. doi:10.1242/jeb.053991

Milsom WK (2002) Phylogeny of CO$_2$/H$^+$ chemoreception in vertebrates. Respir Physiol Neurobiol 131(1–2):29–41. doi:http://dx.doi.org/10.1016/S1569-9048(02)00035-6

Moffett D, Cummings S (1994) Transepithelial potential and alkalization in an in situ preparation of tobacco hornworm (Manduca sexta) midgut. J Exp Biol 194(1):341–345

Needham KM (1990) Specific and seasonal variation in survival and sodium balance at low pH in five species of waterboatmen (hemiptera: Corixidae), Masters Thesis, University of British Columbia

Nguyen H, Donini A (2010) Larvae of the midge Chironomus riparius possess two distinct mechanisms for ionoregulation in response to ion-poor conditions. Am J Physiol Regul Integr Comp Physiol 299(3):R762–R773. doi:10.1152/ajpregu.00745.2009

O'Donnell MJ, Maddrell SH (1995) Fluid reabsorption and ion transport by the lower Malpighian tubules of adult female Drosophila. J Exp Biol 198(8):1647–1653

Okech BA, Boudko DY, Linser PJ, Harvey WR (2008) Cationic pathway of pH regulation in larvae of Anopheles gambiae. J Exp Biol 211(6):957–968. doi:10.1242/jeb.012021

Papaefthimiou C, Theophilidis G (2001) An in vitro method for recording the electrical activity of the isolated heart of the adult *Drosophila melanogaster*. In Vitro Cell Dev Biol Anim 37(7):445–449. doi:10.2307/4295243

Patrick ML, Gonzalez RJ, Wood CM, Wilson RW, Bradley TJ, Val AL (2002) The characterization of ion regulation in Amazonian mosquito larvae: evidence of phenotypic plasticity, population-based disparity, and novel mechanisms of ion uptake. Physiol Biochem Zool 75(3):223–236. doi:10.1086/342002

Patrick ML, Aimanova K, Sanders HR, Gill SS (2006) P-type Na^+/K^+-ATPase and V-type H^+-ATPase expression patterns in the osmoregulatory organs of larval and adult mosquito *Aedes aegypti*. J Exp Biol 209(23):4638–4651. doi:10.1242/jeb.02551

Peach JL, Phillips JE (1991) Metabolic support of chloride-dependent short-circuit current across the locust (*Schistocerca gregaria*) ileum. J Insect Physiol 37(4):255–260. doi:http://dx.doi.org/10.1016/0022-1910(91)90059-9

Pelletier Y, Clark CL (1992) The haemolymph plasma composition of adults, pupae, and larvae of the colorado potato beetle, *Leptinotarsa decemlinaeta* (Say), and development of physiological saline solution. Can Entomol 124(6):945–949

Phillips JE (1964) Rectal absorption in the desert locust, *Schistocerca Gregaria* Forskål I Water. J Exp Biol 41(1):15–38

Phillips JE (1981) Comparative physiology of insect renal function. Am J Physiol Regul Integr Comp Physiol 241(5):R241–R257

Phillips JE, Audsley N (1995) Neuropeptide control of ion and fluid transport across locust hindgut. Am Zool 35(6):503–514. doi:10.2307/3884156

Phillips JE, Hanrahan J, Chamberlin M, Thomson B (1987) Mechanisms and control of reabsorption in insect hindgut. In: Evans PD, Wigglesworth VB (eds) Advances in insect physiology, vol 19. Academic Press, pp 329–422. doi:http://dx.doi.org/10.1016/S0065-2806(08)60103-4

Phillips JE, Thomson RB, Audsley N, Peach JL, Stagg AP (1994) Mechanisms of acid-base transport and control in locust excretory system. Physiol Zool 67(1):95–119

Phillips JE, Meredith J, Audsley N, Richardson N, Macins A, Ring M (1998) Locust ion transport peptide (ITP): a putative hormone controlling water and ionic balance in terrestrial insects. Am Zool 38(3):461–470

Rockwood JP, Coler RA (1991) The effect of aluminum in soft water at low pH on water balance and hemolymph ionic and acid-base regulation in the dragonfly *Libellula julia* Uhler. Hydrobiologia 215(3):243–250. doi:10.1007/bf00764859

Snyder GK, Ungerman G, Breed M (1980) Effects of hypoxia, hypercapnia, and pH on ventilation rate in *Nauphoeta cinerea*. J Insect Physiol 26(10):699–702

Sohal RS, Copeland E (1966) Ultrastructural variations in the anal papillae of *Aedes aegypti* (L.) at different environmental salinities. J Insect Physiol 12(4):429–434. doi:http://dx.doi.org/10.1016/0022-1910(66)90006-0

Stagg AP, Harrison JF, Phillips JE (1991) Acid-base variables in Malpighian tubule secretion and response to acidosis. J Exp Biol 159(1):433–447

Stewart PA (1983) Modern quantitative acid-base chemistry. Can J Physiol Pharmacol 61(12):1444–1461

Stewart BA, Atwood HL, Renger JJ, Wang J, Wu CF (1994) Improved stability of *Drosophila* larval neuromuscular preparations in haemolymph-like physiological solutions. J Comp Physiol A 175(2):179–191. doi:10.1007/BF00215114

Strange K, Phillips JE, Quamme GA (1982) Active HCO_3^- secretion in the rectal salt gland of a mosquito larva inhabiting $NaHCO_3$-CO_3 lakes. J Exp Biol 101(1):171–186

Strange K, Phillips JE, Quamme GA (1984) Mechanisms of CO_2 transport in rectal salt gland of *Aedes*. II. Site of Cl^--HCO_3^- exchange. Am J Physiol 246(5):R735–R740

Szibbo CM, Scudder GGE (1979) Secretory activity of the segmented Malpighian tubules of *Cenocorixa bifida* (Hung.) (Hemiptera, Corixidae). J Insect Physiol 25(12):931–937. doi:http://dx.doi.org/10.1016/0022-1910(79)90105-7

Terra WR, De Bianchi AG, Lara FJS (1974) Physical properties and chemical composition of the hemolymph of *Rhynchosciara americana* (Diptera) larvae. Comp Biochem Physiol B Comp Biochem 47(1):117–129. doi:http://dx.doi.org/10.1016/0305-0491(74)90096-0

Thomson RB, Phillips JE (1992) Electrogenic proton secretion in the hindgut of the desert locust, *Schistocerca gregaria*. J Membr Biol 125(2):133–154. doi:10.1007/BF00233353

Thomson RB, Speight JD, Phillips JE (1988a) Rectal acid secretion in the desert locust, *Schistocerca gregaria*. J Insect Physiol 34(9):829–837. doi:http://dx.doi.org/10.1016/0022-1910(88)90116-3

Thomson RB, Thomson JM, Phillips JE (1988b) NH_4^+ transport in acid-secreting insect epithelium. Am J Physiol Regul Integr Comp Physiol 254(2):R348–R356

Thomson RB, Audsley N, Phillips JE (1991) Acid-base transport and control in locust hindgut: artefacts caused by short-circuit current. J Exp Biol 155(1):455–467

Truchot JP (1987) Comparative aspects of extracellular acid-base balance, vol 20. Zoophysiology, Springer-Verlag, Berlin

Vangenechten JHD, Witters II, Vanderborght OLJ (1989) Laboratory studies on invertebrate survival and physiology in acid waters. In: Morris R, Taylor EW, Brown DJA, Brown JA (eds) Acid toxicity and aquatic animals. Cambridge University Press, Cambridge, pp 153–169

Wasserthal LT (1996) Interaction of circulation and tracheal ventilation in holometabolous insects. Adv Insect Physiol 26:297–351

Weber RE (1980) Functions of invertebrate hemoglobins with special reference to adaptations to environmental hypoxia. Am Zool 20(1):79–101

Weber RE, Braunitzer G, Kleinschmidt T (1985) Functional multiplicity and structural correlations in the hemoglobin system of larvae of *Chironomus thummi thummi* (Insecta, Diptera): Hb components CTT I, CTT IIβ, CTT III, CTT IV, CTT VI, CTT VIIB, CTT IX and CTT X. Comp Biochem Physiol B 80(4):747–753

Wheeler RE (1963) Studies on the total haemocyte count and haemolymph volume in *Periplaneta americana* (L.) with special reference to the last moulting cycle. J Insect Physiol 9(2):223–235

Wichard W, Komnick H (1974) Fine structure and function of the rectal chloride epithelia of damselfly larvae. J Insect Physiol 20(8):1611–1621. doi:http://dx.doi.org/10.1016/0022-1910(74)90090-0

Wichard W, Komnick H, Abel JH Jr (1972) Typology of ephemerid chloride cells. Z Zellforsch 132(4):533–551. doi:10.1007/bf00306640

Wigglesworth VB (1933) The function of the anal gills of the mosquito larva. J Exp Biol 10(1):16–26

Wigglesworth VB (1972) The principles of insect physiology, 7th edn. Chapman and Hall, London

Witters H, Vangenechten JHD, Van Puymbroeck S, Vanderborght OLJ (1984) Interference of aluminium and pH on the Na-influx in an aquatic insect *Corixa punctata* (Illig.). Bull Environ Contam Toxicol 32(1):575–579. doi:10.1007/BF01607540

Wood CM, Randall DJ (1981) Haemolymph gas transport, acid-base regulation, and anaerobic metabolism during exercise in the land crab (*Cardisoma carnifex*). J Exp Zool 218(1):23–35. doi:10.1002/jez.1402180104

Wootton RJ (1988) The historical ecology of aquatic insects: an overview. Palaeogeogr Palaeoclimatol Palaeoecol 62(1–4):477–492. doi:http://dx.doi.org/10.1016/0031-0182(88)90068-5

Wyatt GR, Loughheed TC, Wyatt SS (1956) The chemistry of insect hemolymph: organic components of the hemolymph of the silkworm, *Bombyx mori*, and two other species. J Gen Physiol 39(6):853–868. doi:10.1085/jgp.39.6.853

Acid–Base Loops in Insect Larvae with Extremely Alkaline Midgut Regions

Horst Onken and David F. Moffett

© Springer International Publishing Switzerland 2017
D. Weihrauch, M. O'Donnell (eds.), *Acid-Base Balance and Nitrogen Excretion in Invertebrates*,
DOI 10.1007/978-3-319-39617-0_9

9.1   Benefits of Extreme pH in Digestive Systems

Digestive compartments with extremely acidic pH values occur in many vertebrates, generating a luminal pH of 1–3 in the stomach. The low pH denatures protein, facilitating its breakdown by digestive enzymes. The proteases secreted in the stomach are adapted to function in a low-pH environment. Moreover, extremely acidic pH in vertebrate stomachs serves as a barrier to avoid, or at least limit, invasion of the downstream digestive tract by possibly harmful microorganisms.

Digestive compartments with extremely alkaline pH values of 10–12 occur in many larvae of endopterygote insects of the orders Coleoptera, Diptera, Trichoptera, and Lepidoptera (cf. Clark 1999). Again, the extreme pH denatures protein, facilitating its breakdown, and digestive enzymes that are active in this midgut region have pH optima in the alkaline pH range (Eguchi et al., 1990). The strongly alkaline medium is thought to dissociate tannin-protein complexes, resulting in increased ability to assimilate protein from a plant-based diet (Berenbaum 1980). Although not yet studied in greater detail, first results indicate that the alkaline midgut region may act as a barrier for microorganisms (Onken et al. 2014).

9.2   Acid–Base Loops Generate Compartments of Extreme pH

In principle, a digestive compartment with extreme pH is generated by a loop of acid and base equivalents. In the epithelium that lines the compartment of extreme pH, a conjugated acid–base pair (HA $\leftrightarrows$ H$^+$ + A$^-$) is generated. Protons and bases are then transported together with appropriate counter ions in opposite directions, one into the lumen of the digestive tract and the other to the blood or hemolymph. This process can generate a region or compartment of extreme pH, if the pair of acid–base plus counter ion accumulated in the lumen of the digestive tract is strong (highly dissociated). Downstream, another epithelium generates acid–base transport to reunite the acid–base pair, reestablishing a near-neutral pH in the digestive tract.

In vertebrates, parietal cells in gastric glands use carbonic anhydrase to produce H$^+$ and HCO$_3^-$ from CO$_2$ and H$_2$O. As H$^+$ is pumped by apical K$^+$/H$^+$-ATPases into the stomach lumen, HCO$_3^-$ is exchanged for Cl$^-$ across the basal membrane. The overall process results in secretion of the strong acid HCl into the stomach, and NaHCO$_3$ alkalinizes the venous blood leaving the stomach. Pancreatic duct cells close the acid–base loop by taking up sodium bicarbonate from the blood and actively secreting it to the lumen which ultimately neutralizes the acidic gastric chyme in the duodenum (◘ see Fig. 9.1a).

In larval insects an anterior midgut region alkalinizes the midgut content, and a posterior midgut region reestablishes a midgut pH close to neutral (◘ see Fig. 9.1b). In the following, we will review in greater detail the current knowledge about the acid–base loop that generates an extremely alkaline pH in the midgut of larval insects; we will point out steps that are not yet uncovered or understood, and we will present possible hypotheses to fill these knowledge gaps in future experimental work. If the acid–base loops are not carefully balanced, the hemolymph pH may change, requiring adjustment by other systems of organismal pH regulation (▶ Chap. 8).

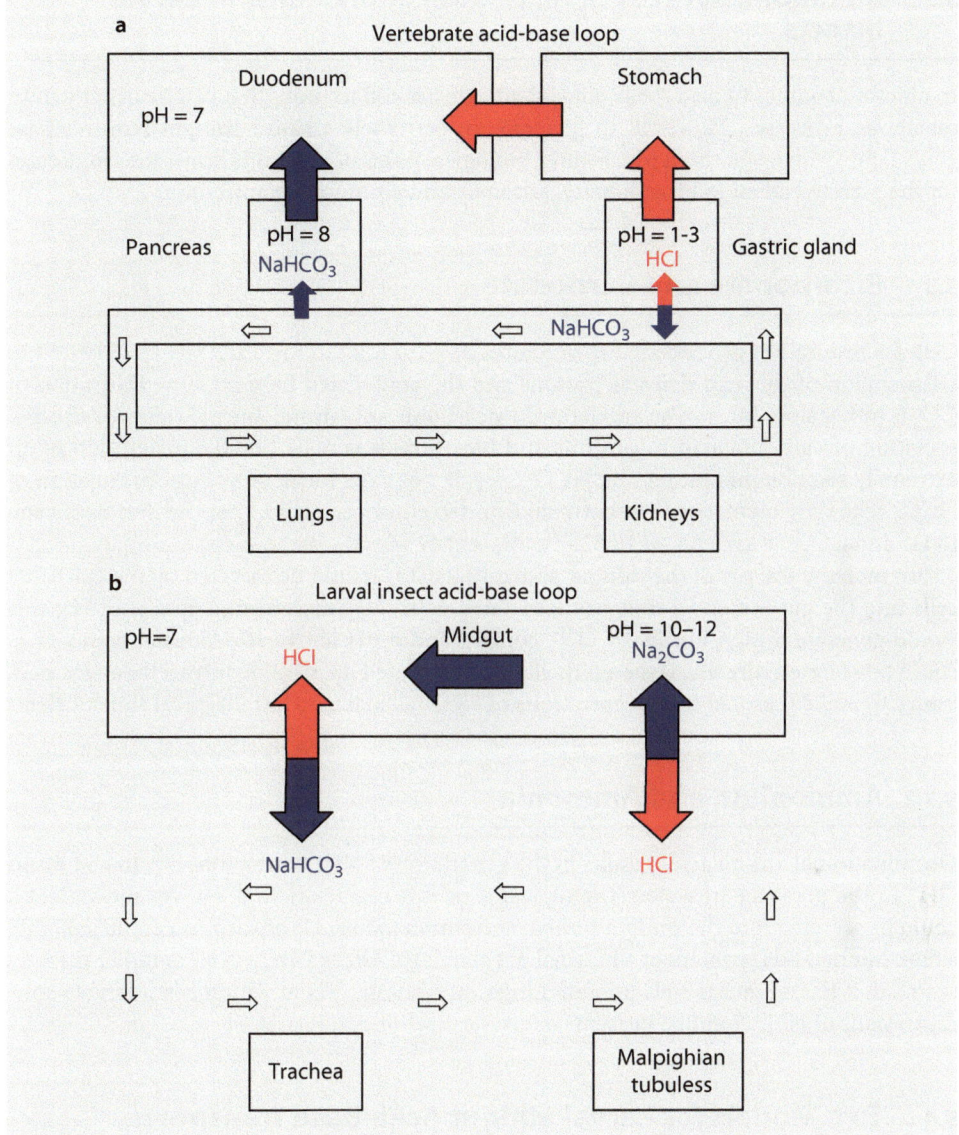

□ Fig. 9.1 Acid–base loops in digestive systems of vertebrates (**a**) and of larval insects with strongly alkaline midgut sections (**b**). Separation and reunification of acid and base are shown. Temporary disequilibria may result in pH changes of the blood or hemolymph that are balanced by the organs involved in pH regulation: kidneys and lungs in vertebrates and Malpighian tubules and trachea in insects

9.3   Metabolic Sources of Acid–Base Equivalents in Larval Insects

In animals, sources of acid–base equivalents are limited to those that can be generated in metabolic processes. However, to generate an extremely alkaline midgut requires base equivalents with a very high pH. From a biological point of view, this limits the candidates for the secreted alkali to bicarbonate/carbonate and ammonium/ammonia.

9.3.1   Bicarbonate and Carbonate

Cellular respiration produces CO_2 which reacts with H_2O to form the weak acid H_2CO_3. Dissociation of the acid delivers protons and the conjugated base HCO_3^-. Hydration of CO_2 is rather slow, but can be accelerated by carbonic anhydrase. The pK value for the dissociation of carbonic acid in protons and bicarbonate is 6.35. For the generation of an extremely alkaline midgut region, HCO_3^- seems not to be ideal, because accumulation of HCO_3^- can only increase pH to between 8 and 9. However, HCO_3^- can further dissociate to H^+ and CO_3^{2-} with a pK of 10.33.

To increase the pH of the midgut above 10, HCO_3^- could be secreted by the epithelial cells into the gut lumen. If this was associated with the simultaneous uptake of H^+, this would generate a high luminal CO_3^{2-} content and a pH above 10. Alternatively, CO_3^{2-} could be secreted directly, if the epithelial cells tolerated an alkaline intracellular pH necessary to make reasonable concentrations of CO_3^{2-} available to a transapical transporter.

9.3.2   Ammonium and Ammonia

Deamination of amino acids results in the generation of NH_3 which converts to NH_4^+ and OH^- in the presence of water. The pK value of this dissociation is 9.4. As above, NH_4^+ could be secreted into the midgut lumen, and simultaneous H^+ absorption could generate a high luminal NH_3 content at a luminal pH above 10. Alternatively, NH_3 could be directly secreted, if the epithelial cells tolerated high intracellular pH at which considerable concentrations of NH_3 could be present.

9.4   Two Models for Larval Midgut Acid–Base Transport: Lepidopterans and Dipterans

Insects can have midgut regions with acidic, neutral, or alkaline pH values, and their digestive enzymes have corresponding pH optima. Only the larvae of many endopterygote insects of the orders Coleoptera, Diptera, Trichoptera, and Lepidoptera have extremely alkaline midgut pH (cf. Clark 1999). Most studies of acid–base transport in the midgut of larval insects have been performed on lepidopterans and dipterans. The tobacco hornworm (*Manduca sexta*) is the major model organism for acid–base transport in lepidopterans, whereas larval yellow-fever mosquitoes (*Aedes aegypti*) serve the same role among dipterans. The results obtained with these two groups indicate marked differences, and we will discuss the two groups sequentially in the following sections.

a   b

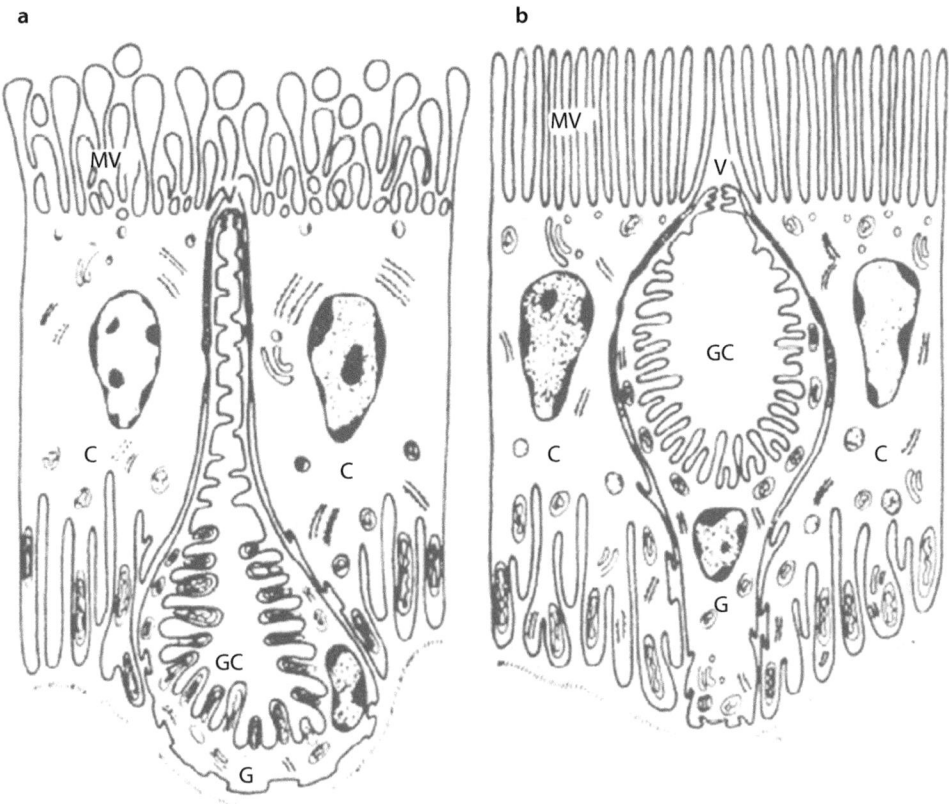

◘ Fig. 9.2 Half-schematic representation of anterior-middle (**a**) and posterior (**b**) midgut of larval lepidopterans. Luminal membranes face upward. *C* columnar cell cytoplasm, *G* goblet cell cytoplasm, *GC* goblet cavity, *MV* columnar cell microvilli, *V* goblet valve (From Moffett and Koch (1992) after Cioffi (1979))

9.4.1 Acid–Base Loop in Lepidopterans

The alkaline midgut of a number of lepidopteran larvae was first reported by Waterhouse (1949), followed by numerous reports for more lepidopteran species (for a review see Berenbaum 1980). The pH profile of the larval midgut of four species of lepidopteran larvae was measured in detail by Dow (1984). The pH of the slightly acidic food bolus rises sharply in the anterior midgut to 9.5–11, reaches a maximum (pH 10.5–12) in the middle of the midgut, and then declines slightly to about pH 10 in the posterior midgut. The gut content is reacidified in the hindgut (see Moffett 1994) to produce feces with a pH similar to the food.

The structure of the midgut shows macroscopically folding patterns and cytological features that allow three regions to be distinguished: anterior, middle, and posterior (Cioffi and Harvey 1981). In all three regions, the midgut epithelium consists of two cell types (◘ see Fig. 9.2). Columnar cells are characterized by a brush border. Goblet cells are distributed among the columnar cells in a ratio of about 1:3. Unlike the mucus-secreting goblet cells from vertebrate intestines, lepidopteran goblet cells are characterized by a goblet cavity of vacuolar origin that opens to the midgut lumen through a valve. The membrane of the goblet cavity shows many microvilli. The goblet cells differ considerably along the midgut. In the anterior and middle regions, the goblet cavity is large and stretches almost through the entire cell. Each microvillus contains a mitochondrion like a finger in a glove. In the posterior midgut, the goblet cavity of the cells is clearly smaller, and mitochondria are at the bases of the microvilli rather than inserted into them.

From a transport perspective, the lepidopteran midgut was first explored for its high capacity of active potassium secretion. The isolated epithelium generated transepithelial voltages above 100 mV (lumen positive), and the short-circuit current almost entirely reflected active potassium secretion (rev. by Harvey and Zerahn 1972). The first mechanism proposed for this potassium secretion was passive entry through potassium channels into goblet cells from a hemolymph of high potassium concentration and active transport across the membrane of the goblet cavity by an electrogenic potassium pump associated with portasomes in this membrane (Harvey et al. 1981).

Dow (1984) was the first to propose that potassium secretion and strong alkalinization are coupled processes. Active potassium secretion was thought to be accompanied by HCO_3^- generated from carbon dioxide and water and catalyzed by carbonic anhydrase (Turbeck and Foder 1970). Excess luminal protons were thought to be driven from the goblet cavity back into the cell by the large voltage generated by the electrogenic potassium pump, resulting in secretion of potassium carbonate and an extremely alkaline pH. However, it was later demonstrated (Moffett and Cummings 1994) that the pH gradient across the epithelium in quasi intact larvae was not matched by the transepithelial voltage, demanding an active process of H^+ removal from the goblet cavity.

Finally, the portasomes in the membrane of the goblet cavities were identified as V-type proton pumps (Klein et al. 1991) that energize K^+/H^+ exchangers in goblet cavity membranes (Wieczorek et al. 1991); and K^+/amino acid symporters in apical membranes of columnar cells. Based on measurements of electrochemical gradients, Moffett and Koch (1992) pointed out that the V-ATPase could only drive the exchanger, if it were electrogenic, exchanging more than one proton for each K^+ ion. Azuma et al. (1995) identified the $K^+:H^+$ ratio as 1:2. At this stage, luminal alkalinization in lepidopteran midgut is thought to be generated by the mechanism shown in ◘ Fig. 9.3 (cf. Wieczorek et al. 1999): V-ATPases pump protons into the goblet cavity, charging the membrane to a voltage above 200 mV (cell negative). This energizes the exchanger to secrete potassium and absorb double the number of protons, resulting in deprotonation of carbonic acid to carbonate and in net secretion of K_2CO_3, generating an extremely alkaline luminal pH. In the isolated midgut, it could be shown that bafilomycin and high doses of amiloride inhibit the potassium secretion measured as short-circuit current (Schirmanns and Zeiske 1994).

As elegant and as straightforward as this mechanism appears, there are still considerable knowledge gaps and inconsistencies that need to be resolved in the future.

1. Most of the results that lead to the above-described mechanism were obtained with isolated midgut epithelia or with membrane vesicles. However, isolated epithelia are only capable of generating acid–base fluxes of 8–9 µequiv cm^{-2} h^{-1} (Chamberlin 1990a, b; Coddington and Chamberlin 1999), barely 10–20 % of the acid–base flux expected *in vivo* (Clark et al. 1998). Semi-intact midgut preparations were able to generate *in vivo*-like acid–base fluxes and produce and maintain high luminal pH of 11 (Clark et al. 1998), whereas isolated midguts generated a luminal pH of only 7.8 in the microenvironment of apical folds (Dow and O'Donnell 1990). These observations evidently demand a search of a regulatory factor that stimulates acid–base transport

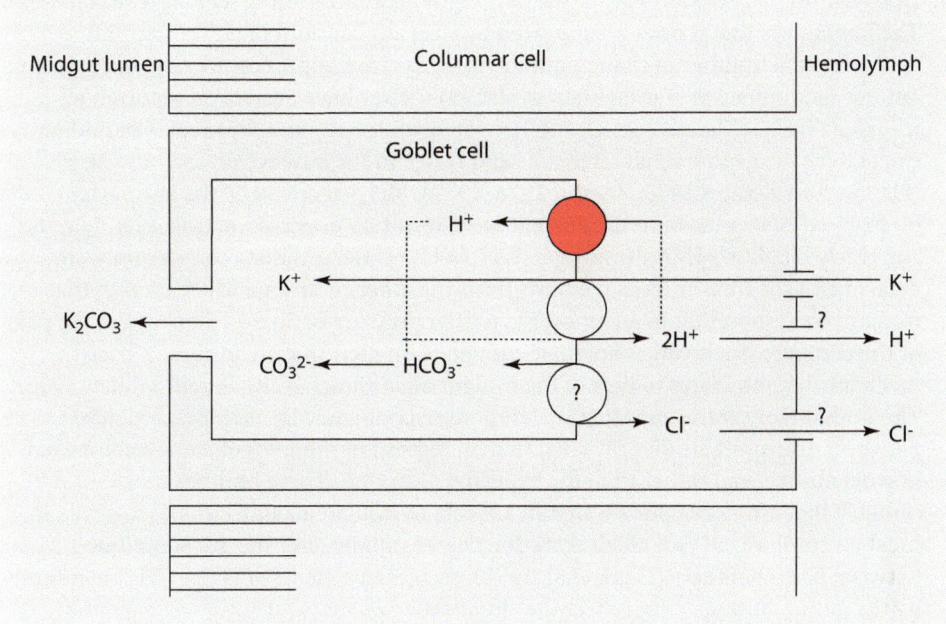

□ **Fig. 9.3** Current model of alkalinization in the midgut of lepidopteran larvae

to *in vivo*-like quantities. Moreover, they raise doubt about the above mechanism of alkalinization in which acid–base transport and alkalinization are directly coupled. How can it then be that isolated midgut preparations maintain very high rates of potassium transport, but almost completely lose their capability to transport acid–base equivalents? Is there a factor that in some way couples/uncouples potassium secretion and acid–base transport?

2. The V-ATPase-driven $K^+/2H^+$ exchange has so far only been demonstrated in experiments with membrane vesicles (Wieczorek et al. 1991; Azuma et al. 1995), but its molecular entity has not been identified.

3. The lepidopteran midgut contains high carbonic anhydrase activity (Turbeck and Foder 1970). In the anterior and middle midgut, carbonic anhydrase is associated with goblet cells, whereas the enzyme is associated with columnar cells in the posterior midgut (Ridgway and Moffett 1986). If carbonic anhydrase is located inside of the goblet cells, HCO_3^- would be formed intracellularly, and a transporter would be required to secrete it to the goblet cavity. To fill this knowledge gap, carbonic anhydrase localization with higher resolution and/or identification of a HCO_3^- transporter is required.

4. How do protons leave the goblet cells to enter the hemolymph? To our knowledge, the necessary pathway of protons across the basal membrane to the hemolymph has never been explored.

5. The above model assumes that the valve of the goblet cavity is open to allow transfer of K^+ and CO_3^{2-} from the goblet cavity to the midgut lumen. However, it seems that most valves in the isolated midgut are closed (Moffett et al 1995). Most interestingly, valves could be opened with cytochalasin that interferes with actin-based cell motility.

6. Does ammonia secretion play a role in midgut alkalinization in lepidopteran larvae? The hypothesis that it does is supported by a certain amount of circumstantial evidence. In a number of tissues, amiloride-sensitive antiporters are not very selective among alkali metal ions (Soleiman et al. 1991). They have even been reported to transport NH_4^+ (Thomson et al. 1988). Moreover, the isolated lepidopteran midgut can actively transport all alkali metals and NH_4^+ in the absence of Ca^{2+} and Mg^{2+} (Harvey and Zerahn 1972; Zerahn 1971, 1977). NH_4^+ secreted by the antiporter could be stripped of its proton in the goblet cavity, resulting in accumulation of NH_3 at the extremely alkaline pH in the midgut. NH_3/NH_4^+ concentrations of 2–3 mm were determined for fresh midgut contents from the anterior and middle region of the midgut, corresponding to an ammonia partial pressure of 80.6 ± 16 mmHg at the pH of the contents. Such values are high enough to suggest that an ammonia-based mechanism contributes to part of the midgut alkalinization (Koch and Moffett 1996). The midgut concentrations of NH_3/NH_4^+ were confirmed by Weihrauch (2006). However, transport studies revealed that all regions of the midgut show a somewhat low net absorption, rather than the expected secretion. These findings make it unlikely that ammonia plays a significant role in midgut alkalinization. However, it must be emphasized that alkali secretion ceases rapidly after the gut is mounted between half-chambers (Clark et al. 1998); so measurements of $NH_3-NH_4^+$ handling by the gut *in vitro* may not reflect the situation *in vivo*.

The alkaline midgut contents are moved by peristaltic activity to posterior regions within the midgut. The contents reach a maximal pH in the middle midgut, and then the pH drops slightly in the posterior midgut. It could be that the somewhat lower pH in the posterior midgut is based on a lower rate of alkalinization, or on a higher permeability of the epithelium for acid–base equivalents (which would partly collapse the generated gradient). It is noteworthy to mention that carbonic anhydrase is not located in goblet cells but on the apical border of columnar cells in the posterior midgut. This location would be consistent with the promotion of equilibration of luminal CO_3^{2-} with secreted acid for recovery as CO_2, even though present evidence from *in situ* studies supports net alkali secretion by all segments of the midgut (Moffett 1994; Moffett et al. 1996).

In contrast to the midgut, the content of the hindgut (ileum and rectum) is slightly acidic, indicating that the greater part of reacidification of the gut contents occurs here (Moffett 1994). Unfortunately, the lepidopteran hindgut has not yet been intensively studied with respect to acid–base transport, and the best model tissue at present is that of the hindgut of *Schistocerca gregaria* (Thomson et al. 1988; Thomson and Phillips 1992). In preliminary studies (Newberry and Moffett, unpublished), the isolated ileum secreted acid equivalents at a rate of about 1 nmol cm^{-2} min^{-1} and was able to lower the pH of lightly buffered saline to 5.5. It can be assumed that either net acid secretion or absorption of K^+ and/or HCO_3^-/CO_3^{2-} is involved. Rates of acid secretion by the isolated ileum appear to greatly underestimate that of the ileum *in vivo*, even with theophylline stimulation.

To close the acid–base loop, either the acid absorbed from the midgut needs to return to the hindgut or the base absorbed from the hindgut needs to return to the midgut. In the simplest case, this circuit would be completed by the flow of hemolymph. Phenol red dye injected into the hemocoel of the penultimate segment of intact *Manduca sexta* larvae reached the anteriormost segment at about the same time as it filled the posteriormost segment, indicating a movement of dye in the hemolymph from posterior to anterior (Moffett 1994). This would suggest that the acid–base loop is closed, at least in part, by hemolymph moving from the hindgut to the midgut.

The same study (Moffett 1994) proposed that about 10 % of fluid, K^+ ,and acid–base equivalents recycled from posterior to anterior would be returned to the vicinity of the midgut via the Malpighian tubules. The physical relationship of the Malpighian tubules to the gut varies considerably among insect orders, but in lepidopteran larvae the relationship, as first described by Irvine (1969), is so specific and so close that it is impossible not to believe that it is of functional importance. The initial segment of the tubule is tightly associated with the rectum, forming a cryptonephral complex (Ramsay 1976). The tubules then leave this complex anteriorly to form a plexus around the ileum. After leaving the ileum, each tubule proceeds anteriorly across the surface of the midgut, turns posteriorly at the junction of middle and anterior midgut to form a loop, and proceeds posteriorly to drain into the gut at the junction of the midgut and ileum. The anteriorward projection of the Malpighian tubules could have only one rationale – to deliver to the midgut something collected from the distal gut. Logically, the possibilities are alkali and K^+. By the same reasoning, the descending segments should operate to take solutes away from the vicinity of the midgut, and the logical possibilities are acid equivalents, including uric acid and NH_4^+.

Phenol red injected into the hemocoel rapidly passes into both the gut and the Malpighian tubules by active processes. Phenol red has two color transitions – from yellow to scarlet at 6.7 and from scarlet to magenta at 8.0. Dissection of an animal that had been injected with phenol red reveals a comprehensive picture of pH values throughout the animal and its gut contents and Malpighian tubular fluid (◙ Fig. 9.4a). The picture shows that initially acidic tubular fluid becomes strongly alkaline as it passes through the ileal

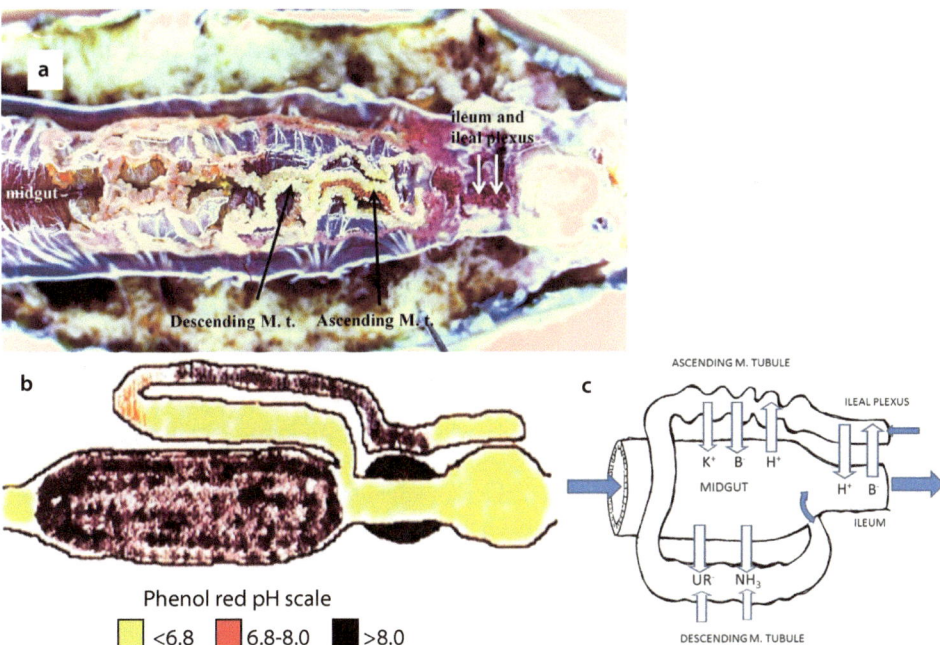

◙ **Fig. 9.4** pH values, acid–base loops, and nitrogen excretion mechanisms in the gut–Malpighian tubule complex in *Manduca sexta* larvae. (**a**) *Phenol red* distribution in gut and Malpighian tubules of a fifth-instar *M. sexta* at 30 min postinjection; *yellow color* indicates pH < 6.8; scarlet indicates 6.8 < pH > 8.0; magenta indicates pH > 8.0. (**b**) Simplified representation of the indicator colors observed in experiments like that shown in ◙ Fig. 9.4a. (**c**) Summary of known and hypothesized solute movements in the gut–Malpighian tubule complex, including recycling loops for K^+, H^+, and base (B^-) and excretion of NH_3 by pH trapping and urate (UR) by precipitation of the protonated acid

plexus, whereas at the same time, the gut contents become strongly acidic. The alkaline tubular fluid arrives in the neighborhood of the gut via the ascending segments. In attenuated tubule segments, some of the K^+ of the primary secretion has been replaced by Na^+ by the time the fluid reaches the end of the ascending segment (Irvine 1969); this K^+ could be regarded as being delivered to the midgut whether it arrives via the ascending segment or has been extracted by an earlier segment of the tubule and arrives by posterior-to-anterior movement of hemolymph. Acidification of the tubular fluid is measureable by the time the urine enters the descending segments (◉ see Fig. 9.4a, b), and the tubular fluid remains acidic for the remainder of its course to the ileum. In principle, this process can be expected to remove excess H^+ from the vicinity of the midgut, deliver HCO_3^- from the tubular fluid to the midgut's basal surface, and remove NH_3 from it, in the same way that filtered bicarbonate is recovered by conversion to diffusible CO_2 in the mammalian proximal tubule and pH trapping of ammonium is carried out by the mammalian distal tubule.

The midgut and the medial Malpighian tubules both express a transport mechanism (termed the type A pump by Nijout 1975) that accepts organic anions, including urate and hippurate. The effect of urate secretion by the descending segment of the Malpighian tubules with simultaneous acid secretion would be to cause some urate ion to be precipitated, and indeed this precipitate is seen as a white slurry in the tubular fluid. Urate secreted by the midgut would be expected to remain in the anionic form until the gut contents mixed with acidic urine in the ileum. In late fifth-instar larvae, the uric acid, presumably derived from both Malphigian tubule and midgut secretion, appears as a visible coating of the fecal pellets (Buckner et al. 1980). The parallel operation of these two organ systems scavenge uric acid from the hemolymph and deliver it to the lumen of the distal gut, achieving an excretion rate for uric acid of about 16 μmol hour^{-1} for a fifth-instar animal (Buckner et al. 1980). The effectiveness of this process is dependent on recycling of acid and base equivalents between the midgut and the ileum and rectum. A summary of known and hypothesized solute movements, including recycling loops for K^+, H^+, and base (B^-) and trapping of NH_3 as NH_4^+ and uric acid as precipitate, is shown in ◉ Fig. 9.4c (▶ Chap. 4).

9.4.2 Acid–Base Loop in Dipterans

Senior-White (1926) and Ramsay (1950) were the first to recognize an alkaline midgut in larval mosquitoes. Later, the midgut alkalinity was better quantified (Dadd 1975; Stiles and Paschke 1980), and other dipteran larvae with alkaline midguts were described (Lacey and Federici 1979; Terra et al. 1979; Undeen 1979; Martin et al. 1980; Espinoza-Fuentes et al. 1984; Shanbhag and Tripathi 2005). The more detailed studies of acid–base transport in the midgut of dipterans have mainly been performed on larval mosquitoes. According to Dadd (1975), the pH of the midgut lumen rises sharply from 7.5 to 10 in the anterior ventriculus. In the posterior ventriculus, the pH gradually returns to neutral values.

In larval mosquitoes, the midgut consists of four sections (◉ see Fig. 9.5). The cardia, a short midgut section that is involved in secreting the peritrophic membrane, and the gastric ceca, eight somewhat spherical, lateral extensions, are located in the thorax. From the gastric ceca in the thorax to the hindgut in the sixth abdominal segment, the midgut consists of a single tube-like structure, the ventriculus, also called the "stomach" in the mosquito literature (Clements 1992). Some authors use "anterior and posterior midgut" when referring to the ventriculus. The ventriculus consists of histologically different anterior

Fig. 9.5 Larval yellow-fever mosquito (*Aedes aegypti*) and the isolated gut, showing the anterior (Ant) and posterior (Post) regions of the ventriculus. Cardia and ceca can be seen anterior of the ventriculus. Malpighian tubules and hindgut are posterior of the ventriculus. The entire ventriculus is approximately 5 mm in length (From Clark et al. 1999)

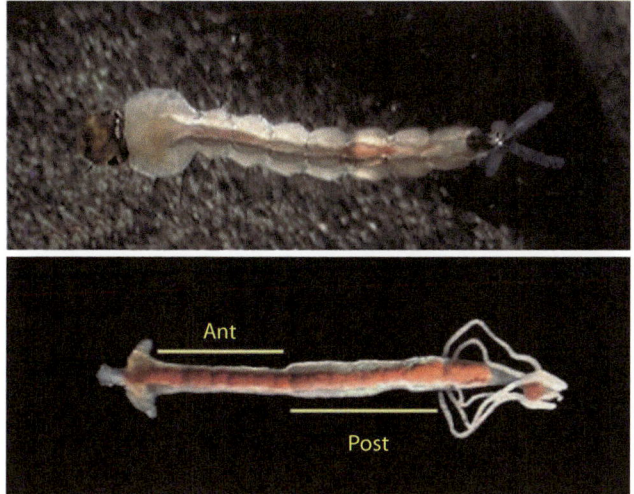

and posterior regions (Wigglesworth 1942; Zhuang et al. 1999) separated by a transitional region (Clark et al. 2005). The epithelial cells in the anterior ventriculus are characterized by many short (0.4 µm; Clark et al. 2005) microvilli and a basal membrane with intensive infoldings (Zhuang et al. 1999). In the transitional region, the short microvilli grow continuously in length (from approximately 1 µm in the anterior part of the transitional region to about 4 µm in the posterior transitional region; Clark et al. 2005). The posterior ventriculus is characterized by cells with half-dome-like apical membranes and long microvilli (3–4 µm; Zhuang et al. 1999; Clark et al. 2005). No goblet cells were detected in the midgut of larval mosquitoes.

Isolated anterior ventriculus preparations bathed with mosquito saline alkalinize the lumen (Onken et al. 2008). Serotonin stimulates and accelerates this alkalinization. In the presence of serotonin, the luminal pH rises from 7 to 10 within a few minutes. When perfused and bathed with mosquito saline, such preparations generate a lumen-negative transepithelial voltage that is also stimulated with submicromolar concentrations of serotonin (Clark et al. 1999). The electrical potential profile of the epithelial cells was measured (Clark et al. 2000). After stimulation with serotonin, the average basal membrane voltage was above 100 mV (cell negative). In these experiments, the apical membrane was slightly depolarized, and the transepithelial voltage was about 8 mV (lumen negative). Much higher transepithelial voltages have been recorded (Clark et al. 1999; Onken et al. 2004a, 2008), indicating that either the basal membrane voltage can be much higher or that the apical membrane can depolarize more. The apical membrane was determined to carry about 90 % of the transcellular electrical resistance (Clark et al. 2000). In the presence of serotonin, the epithelium generated an average short-circuit current of approximately 450 µA cm^{-2} (Onken et al. 2006). If this current reflected only acid–base transport, it would generate an acid–base transport rate of about 17 µmol cm^{-2} h^{-1}. The transepithelial conductance amounted to about 20 mS cm^{-2} at a paracellular leak conductance of 1 mS cm^{-2}. These values and the high transepithelial voltages characterize the epithelium as tight. Acid and chloride fluxes were also determined from measurements of H$^+$ and Cl$^-$ concentration gradients at the basal membrane of semi-intact larvae, using ion-selective microelectrodes (Boudko et al. 2001a). The highest H$^+$ fluxes in the anterior ventriculus were 0.02 µmol cm^{-2} h^{-1}. Chloride fluxes were significantly higher with an average of about 12 µmol cm^{-2}

h^{-1}. These results suggest that the H^+ fluxes detected with this technique should not be taken as true acid–base fluxes generated by the tissue. It may well be that the hemolymph buffering capacity reduces the measurable pH gradients on which the method is based.

Portasomes identified as V-ATPases are located in basal membranes of the anterior ventriculus of larval yellow-fever mosquitoes (Zhuang et al. 1999; Patrick et al. 2006). Evidently, pumping protons from the cytoplasm to the hemolymph is consistent with an epithelium that alkalinizes the lumen. Boudko et al. (2001b) demonstrated that basal membranes of the anterior ventriculus of semi-intact yellow-fever mosquitoes generate pH gradients that are consistent with H^+ absorption in this midgut region. These gradients were reduced in the presence of bafilomycin, a specific inhibitor of V-ATPases. The lumen-negative transepithelial voltage of isolated and perfused preparations of the anterior ventriculus was completely inhibited with concanamycin, another specific inhibitor of V-ATPases (Onken et al. 2004a). Alkalinization to pH values of 10 was also measured with isolated and perfused anterior ventriculus preparations (Onken et al. 2008). It was stimulated with submicromolar concentrations of serotonin and inhibited with concanamycin. Addition of serotonin raised the intracellular pH from 6.9 to 7.6 (Onken et al. 2009a), indicating that serotonin stimulates proton pumping across the basal membrane of epithelial cells in the anterior ventriculus. All these results demonstrate that V-ATPases generate the driving force for acid absorption/base secretion in the anterior ventriculus of larval *Aedes aegypti*. V-ATPases were also identified in basal membranes of the anterior ventriculus of *Anopheles gambiae* (Okech et al. 2008), indicating that these proton pumps serve the same role in other mosquito species.

Na^+/K^+-ATPase was localized in basal membranes of the very first part of the anterior ventriculus of larval *Aedes aegypti* (Patrick et al. 2006). This is where this ion pump is usually expressed in epithelial cells. However, in the major part of the anterior ventriculus, the Na^+/K^+-ATPase was localized in the apical membrane (Patrick et al. 2006). Similar findings were obtained with larval *Anopheles gambiae* (Okech et al. 2008). The unusual expression of Na^+/K^+-ATPase in apical membranes was suspicious and suggested that it may be of importance for luminal alkalinization. However, exposing the midgut lumen to ouabain, both in intact larvae, and using perfused anterior ventriculus preparations did not affect alkalinization (Onken et al. 2009b). Thus, the role of apical Na^+/K^+-ATPase in this tissue remains obscure.

Boudko et al. (2001a) observed Cl^- concentration gradients on the hemolymph side of basal membranes of the anterior ventriculus epithelial cells. The gradients for H^+ (see above) and Cl^- allow H^+ and Cl^- fluxes to be calculated. These fluxes were not equimolar, but still suggested that the tissue is absorbing HCl. The Cl^- fluxes were reduced by addition of DIDS to the hemolymph-side medium of the semi-intact preparation. On the basis of their results, the authors presented a hypothetical model in which V-ATPases and Cl^- channels generate HCl absorption across the basal membrane of anterior ventriculus cells. This interpretation was supported by effects of hemolymph-side DIDS and DPC on the transepithelial voltage of isolated anterior ventriculus preparations (Onken et al. 2004a).

In many tissues, carbonic anhydrase (CA) accelerates production of H^+ and HCO_3^- from CO_2 and H_2O and supports transport of acid–base equivalents. Boudko et al. (2001a) and Corena et al. (2002) observed that alkalinization and fluxes of H^+ and Cl^- in the anterior midgut of living larvae or semi-intact larvae were inhibited with acetazolamide or methazolamide, potent inhibitors of CA. On the other hand, CA activity was not at all detectable in the anterior ventriculus, and Hansson's histochemistry for the localization of CA stained only the posterior ventriculus, but not the anterior ventriculus (Corena et al. 2002). Moreover, acetazolamide or methazolamide did neither affect the transepithelial

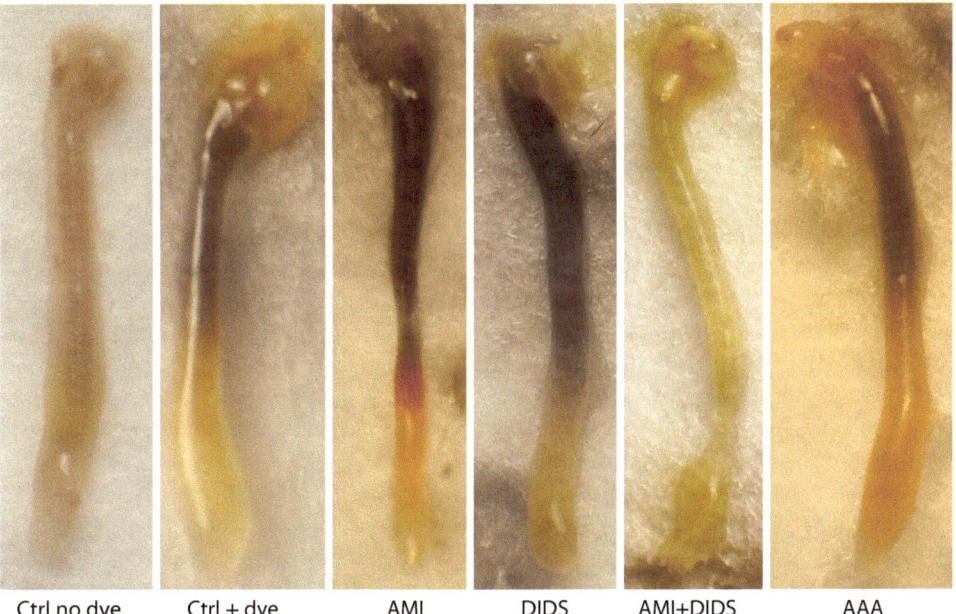

| Ctrl no dye | Ctrl + dye | AMI | DIDS | AMI+DIDS | AAA |

☐ **Fig. 9.6** Alkalinization in the anterior ventriculus of larval yellow-fever mosquitoes (*Aedes aegypti*). Larvae were fed with cellulose 2 h before the experiment in order to make the color of the indicator better visible. Larvae were then exposed to thymol blue (0.04 %, pH 9.2) and fed with cellulose for at least 1 h. During this phase drugs (amiloride, DIDS, amiloride + DIDS, acetazolamide each at a concentration of 1 mM) were present or not (control). Larvae (10–15 per group) were inspected under the microscope, and some midguts (3-7) were isolated and photographed. The *yellow color* in ceca and posterior ventriculus is consistent with a neutral pH. Midgut alkalinization was detected as *dark blue* color in the anterior ventriculus in all cases, but in the simultaneous presence of amiloride and DIDS

voltage nor alkalinization in isolated anterior ventriculus preparations (Onken et al. 2004a, 2008), and we are unable to reproduce effects of acetazolamide on living larvae (☐ see Fig. 9.6). Corena et al. (2004) demonstrated that the anterior ventriculus of larval anopheline mosquitoes has high CA content, whereas larval culicine mosquitoes, including *Aedes aegypti*, have low or no CA content in the anterior ventriculus. Linser et al. (2009) analyzed different types of CA in the midgut of larval *Anopheles gambiae* and relate them to their potential role in midgut acid–base transport. In summary, the results seem to show that CA may play different roles in larval midgut acid–base transport of different mosquito species. For larval *Aedes aegypti*, however, the results seem to indicate that CA is not involved in the luminal alkalinization in the anterior ventriculus.

Na⁺-free solution or amiloride in the hemolymph-side bathing medium of isolated and perfused anterior ventriculus preparations resulted in marked reductions of the transepithelial voltage and the blockade of luminal alkalinization (Onken et al. 2004a, 2008). The authors interpreted this effect as a primary inhibition of Na^+/H^+ exchangers in the basal membrane that secondarily affects the apical transporters involved in luminal alkalinization. Basal Na^+/H^+ exchangers could contribute to H^+ absorption and satisfy a Na^+ dependence of transporters in the apical membrane. Basal Na^+/H^+ exchangers (NHE3) were indeed identified in the basal membrane of the anterior ventriculus (Pullikuth et al. 2006).

A number of luminal manipulations were performed with isolated and perfused anterior ventriculus preparations in order to characterize the apical transporters involved in luminal alkalinization in larval *Aedes aegypti* (Onken et al. 2008, 2009b; Izeirovski et al. 2009). As mentioned above, luminal ouabain did not alter transepithelial voltage or alkalinization (Onken et al. 2009b). Na^+-dependent amino acid absorption was suggested by voltage changes induced by addition and removal of amino acids in the presence and absence of luminal Na^+ (Izeirovski et al. 2009). Such amino acid transporters were indeed identified in the anterior ventriculus of *Anopheles gambiae* with immunohistochemical techniques (Okech et al. 2008). On the other hand, neither luminal DIDS nor luminal amiloride prevented the tissue from alkalinization (Onken et al. 2008). Luminal alkalinization was even observed in the absence of Cl^-, which seemed to rule out electroneutral Cl^-–HCO_3^- exchangers as a pathway for base secretion. Only in the presence of luminal amiloride and DIDS together was alkalinization stopped in isolated ventriculus preparations and in living larvae (◘ see Fig. 9.6). This result indicated that two transporters –one sensitive to amiloride and the other sensitive to DIDS – are responsible for apical acid–base transport. Inhibition of only one of the two transporters apparently does not prevent the tissue from alkalinizing the luminal medium.

The best candidate for the amiloride-sensitive transporter is an electrogenic Na^+–$2H^+$ exchanger (NHA). It could make use of the transmembrane electrical gradient to absorb H^+ from the lumen. Such a transporter would also benefit from the high intracellular pH value of 8.6 in the presence of luminal pH 10 (Onken et al. 2009a). It could also explain the Na^+ dependence of apical acid–base transport suggested by the results with hemolymph-side amiloride or Na^+-free saline (see above). Such an NHA was indeed identified in larval *Anopheles gambiae* where it is expressed in intracellular vesicles and in the apical membrane of the anterior ventriculus (Rheault et al. 2007; Okech et al. 2008). The *Aedes aegypti* genome seems to contain an NHA which shows about 75 % identity to the NHA of *Anopheles gambiae* (Rheault et al. 2007).

There are multiple DIDS-sensitive anion transporters. Cl^-/HCO_3^- exchangers could provide a pathway for absorption of Cl^- from the lumen that is then transported across the basal membrane to the hemolymph (see above). Moreover, this assumption is consistent with the measured concentrations of Cl^- and HCO_3^-/CO_3^{2-} in the lumen (low Cl^-, high HCO_3^-/CO_3^{2-}), tissue (high Cl^- and HCO_3^-/CO_3^{2-}), and hemolymph (high Cl^-, low HCO_3^-/CO_3^{2-}) of the animals (Boudko et al. 2001a). Such Cl^-/HCO_3^- exchangers would rely on the driving force developed by the HCO_3^- concentration gradient, and, therefore, on the conversion of HCO_3^- to CO_3^{2-} by the H^+ absorption via the putative Na^+/$2H^+$ exchangers. Alternatively, Na^+/$2HCO_3^-$ symporters could be the DIDS-sensitive apical transporters. Such transporters could use the HCO_3^- concentration gradient and the membrane voltage as driving force. At an intracellular pH of 8.6 (as it was measured when the tissue was exposed to a luminal pH of 10; Onken et al. 2009a), it could even be that a fraction of base is secreted as CO_3^{2-}. In both assumptions, the stoichiometry is not correct, because the system would not secrete enough Na^+ for the secreted CO_3^{2-}. However, it is very likely that additional Na^+ is secreted via the paracellular pathways driven by the lumen-negative transepithelial voltage.

The overall mechanism of luminal alkalinization in the anterior ventriculus of larval *Aedes aegypti* based on the above-described findings and considerations is shown in the hypothetical model of ◘ Fig. 9.7. Nevertheless, some important questions remain to be addressed in the future.

1. The apical localization of Na^+/K^+-ATPase has been demonstrated in the largest part of the anterior ventriculus (Patrick et al. 2006). Apical localization of this ATPase is unusual in epithelia and could relate to the strong alkalinization. However, alkalinization was observed in living larvae and isolated anterior ventriculus in the presence of

☐ Fig. 9.7 Current transport model for alkalinization in the anterior ventriculus of larval yellow-fever mosquitoes (*Aedes aegypti*). (For details see text)

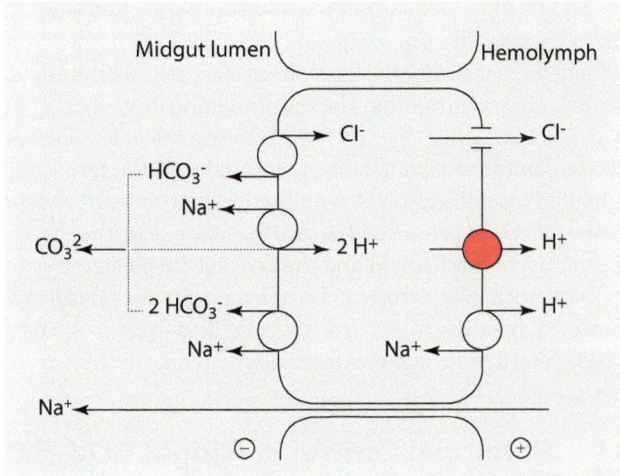

ouabain (Onken et al. 2009b). If not required for alkalinization, it is still of interest to further investigate the role of Na⁺/K⁺-ATPase in this midgut section.

2. Apical transporters must be verified in molecular and physiological approaches, including their influence on apical membrane voltage. Electrogenic $Na^+/2H^+$ exchange and $Na^+/2HCO_3^-$ (or Na^+/CO_3^{2-}) exchange should depolarize the apical membrane, which would be consistent with the electrical potential profile measured by Clark et al. (2000). Amiloride and DIDS should result in hyperpolarization of the apical membrane voltage and reduction of V_{te}.

3. Two findings raise the suspicion that alkalinization in the anterior ventriculus may involve secretion of NH_3. It was observed that luminal alkalinization requires the presence of amino acids in the hemolymph-side bath (Izeirovski et al. 2009). This indicates the energy metabolism in the epithelial cells of the anterior ventriculus depends on amino acids. At the alkaline intracellular pH, a considerable quantity of the produced nitrogenous waste could be present as NH_3 and diffuse across the apical membrane into the lumen to produce $(NH_4)_2CO_3$.

In the midgut, the food is enclosed in the peritrophic membrane. In the laboratory, food was observed to pass through the ventriculus from anterior to posterior in about 30 min when larvae had access to food (cf. Clements 1992). However, this time may be much longer at the lower food supply of a natural environment. Interestingly, the fluid in the ectoperitrophic space between the peritrophic membrane and the midgut epithelial cells moves from posterior to anterior and is apparently absorbed in the ceca (Ramsay 1950, 1951; Volkmann and Peters 1989a, b; Wigglesworth 1933). Consequently, the situation is somewhat more complex than shown in ☐ Fig. 9.1b, because alkaline midgut content moves to the posterior midgut, whereas alkaline ectoperitrophic fluid moves anterior to the ceca. The luminal pH becomes neutral in the midgut sections anterior and posterior of the anterior ventriculus. Accordingly, the epithelia of the gastric ceca and the posterior ventriculus must acidify the lumen. The two tissues show similarities in the presence of certain transporters. V-ATPase has been localized in these tissues in apical membranes (Zhuang et al. 1999; Patrick et al. 2006), whereas Na⁺/K⁺-ATPase and Na⁺/H⁺ exchangers (NHE3) were detected in basal membranes (Patrick et al. 2006; Pullikuth et al. 2006). Carbonic anhydrase activity is high in the posterior ventriculus and in the ceca of *Aedes aegypti* (Corena et al. 2002).

So far, functional studies have been performed only with the posterior ventriculus of *Aedes aegypti*. The transepithelial voltage is lumen positive and can be stimulated with serotonin (Clark et al. 1999). Jagadeshwaran et al. (2010) studied V_{te} and lumen reacidification in the posterior ventriculus. The results demonstrate that V_{te} and reacidification are stimulated with serotonin and depend on functioning apical V-ATPases. Other drugs (DIDS, amiloride, acetazolamide, ouabain, and others) reduced the transepithelial voltage, but none of them inhibited reacidification. A hypothetical mechanism for acid–base transport in the posterior ventriculus was proposed (◘ see Fig. 9.8), but further experiments are required to uncover the complete mechanism and the possible involvement of further transporters.

Not much is known about hemolymph circulation in larval mosquitoes (cf. Clements 1992). However, it seems to be clear that the acid–base loop must be closed via the hemolymph. Malpighian tubules and the tracheal system may well be involved in hemolymph pH regulation.

9.5   Hormonal Control of Midgut Acid–Base Transport

9.5.1 Lepidopteran Midgut

The midgut of *Manduca sexta* loses most of its alkalinization capacity after isolation. This observation demonstrates that alkalinization is affected by a stimulatory factor *in vivo*. Chamberlin (1989) observed that the transepithelial voltage was significantly stimulated by the use of a more complete saline, containing a number of amino acids and metabolites. When this saline was used in acid–base flux experiments (Chamberlin 1990a, b; Coddington and Chamberlin 1999), the highest flux rate was 8.9 µequiv cm^{-2} h^{-1} in the isolated anterior midgut. The acid–base fluxes were not different under open-circuit and short-circuit conditions. The fluxes were approximately 10 % of the short-circuit current (Coddington and Chamberlin 1999) and between 10 and 20 % of the rate *in vivo* (Clark et al. 1998). Using isolated midgut in Ussing chambers, Clark et al. (1998) observed an initial rate of 14 µequiv cm^{-2} h^{-1}, but the rate dropped to about 3 µequiv cm^{-2} h^{-1} within the first 30 min after removing the midgut from the animals. The transepithelial voltage was high during this time, indicating that acid–base flux and voltage (and short-circuit

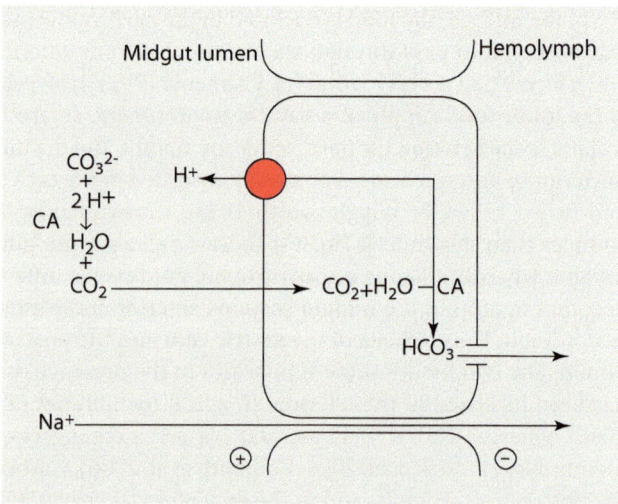

◘ **Fig. 9.8** Current transport model for reacidification in the posterior ventriculus of larval yellow-fever mosquitoes (*Aedes aegypti*). (For details see text)

current) are not correlated in such preparations. Clark et al. (1998) tried a number of manipulations to stabilize or increase acid–base fluxes, including addition of hemolymph, $KHCO_3$, macerated or sonicated head, central nervous system, ventral nerve cord, midgut, or Malpighian tubules, and use of 5 % CO_2 instead of pure oxygen. None of these manipulations reproduced the high acid–base transport rates observed *in vivo*. Clark et al. (1998) also established a cannulated *in situ* midgut preparation that actually showed high acid–base fluxes and reached luminal pH values of 11. In order to find the alkalinizing factor, body parts were removed step by step. Surprisingly, even the cannulated midgut alone was able to generate high acid–base flux rates and alkalinize the midgut lumen, indicating that the alkalinizing factor is produced by the midgut itself.

It seems evident that V-ATPase, the only ion-motive ATPase in plasma membranes of the midgut, is driving luminal alkalinization in larval *Manduca sexta*. Therefore, it is of importance to mention that mechanisms of V-ATPase regulation have been described and reviewed (Merzendorfer et al. 1997; Wieczorek et al. 2000, 2003, 2009).

In order to verify the proposed mechanism of luminal alkalinization in larval lepidopterans (see above), either the stimulating factor must be found and applied to the isolated midgut or the alkalinizing cannulated *in situ* preparation must be used in future experiments.

Reacidification in the hindgut of lepidopterans has barely been studied at all (see above). Consequently, nothing is known about the mechanisms and their regulation.

9.5.2 Dipteran Midgut

Serotonin evidently plays a major role in the regulation of midgut activities in larval yellow-fever mosquitoes. Submicromolar concentrations of serotonin stimulate the transepithelial voltage in anterior and posterior ventriculus (Clark et al. 1999). Serotonin also stimulates alkalinization in the anterior ventriculus (Onken et al. 2008) and reacidification in the posterior ventriculus (Jagadeshwaran et al. 2010). Finally, the drug also stimulates peristaltic activity of the midgut (Onken et al. 2004b). Moffett and Moffett (2005) showed that the gut of larval *Aedes aegypti* receives extensive serotonergic input from central neurons that project through axon terminals and varicosities to all parts of the midgut.

In the anterior ventriculus, serotonin increases the cellular electromotive force, indicating V-ATPase stimulation in the basal membrane (Onken et al. 2006). Simultaneously, serotonin reduces the transcellular resistance (Onken et al. 2006). Because about 90 % of the transcellular resistance is associated with the apical membrane (Clark et al. 2000), the decrease of the transcellular resistance induced by serotonin is most likely based on activation of transporters in the apical membrane. Serotonin induces an increase of intracellular Ca^{2+} (Moffett et al. 2012). The effect is not observed in the absence of extracellular Ca^{2+}, indicating that the signaling pathway involves Ca^{2+} channels in the basal membrane. Five putative serotonin receptors were identified in the anterior ventriculus of larval *Aedes aegypti*, including $5HT_2$-like receptors which could be expected to initiate the observed Ca^{2+} signaling pathway (Moffett et al. 2012). Also cyclic AMP stimulated alkalinization in the anterior ventriculus (Moffett et al. 2012). The effect depended on the extracellular Ca^{2+}. Cytochalasin inhibited the stimulation by serotonin, indicating the involvement of actin-based intracellular motility like vesicle fusion (Moffett et al. 2012). It seems that serotonin induces Ca^{2+}- and cyclic AMP-dependent fusion of vesicles containing V-ATPase with the basal membrane, while inducing vesicles containing NHA (see Rheault et al. 2007; Okech et al. 2008) and anion transporters to fuse with the apical membrane. This scenario is summarized in a hypothetical model in ◘ Fig. 9.9a.

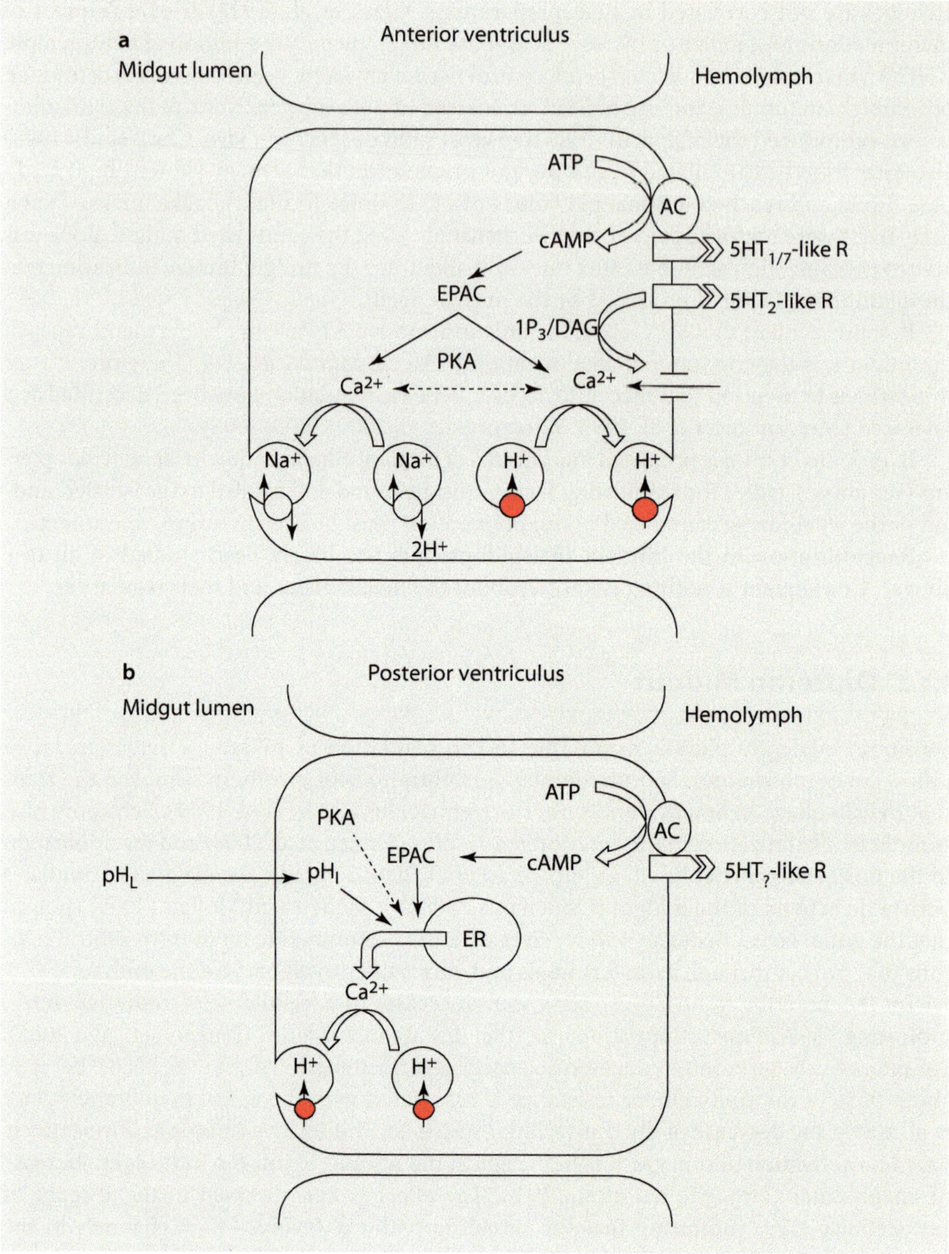

◻ Fig. 9.9 Summary and interpretation of results relating to the regulation of alkalinization in the anteririor ventriculus (**a**) and reacidification in the posterior ventriculus (**b**) by serotonin (Modified after Moffett et al. (2012)). *PKA* protein kinase A, *EPAC* guanine nucleotide exchange factor

In the posterior ventriculus, serotonin also causes an increase in intracellular Ca^{2+} (Moffett et al. 2012). However, the source for Ca^{2+} is the endoplasmic reticulum in this tissue. Like in the anterior ventriculus, cyclic AMP also mimicked the effects of serotonin. Serotonin receptors have not yet been identified in the posterior ventriculus. Most interestingly, alkalinization of the luminal medium resulted in intracellular acidification and an increase of intracellular Ca^{2+} (Moffett et al. 2012), indicating that the alkaline midgut lumen itself could serve as a signal to stimulate reacidification in the posterior ventriculus. The regulatory scenario in the posterior ventriculus is summarized in ◐ Fig. 9.9b.

Whereas serotonin stimulates acid–base transport and peristaltic activity, a number of peptides were observed to have inhibitory effects on transport and/or peristalsis (Onken et al. 2004b).

9.6   Future Perspectives

Reviewing the current knowledge of acid–base loops in larval insects clearly demonstrates that considerable knowledge gaps still exist in the understanding of acid–base transport mechanisms, how they are regulated and balanced in the whole animal. A number of such points have been highlighted above. Modern molecular and microscopic techniques should be able to identify and localize transporters and components of signal transduction cascades. However, a complete understanding of acid–base loops requires to return to the tissue and organism level to verify the presence and importance of the identified components.

Acknowledgments This paper is dedicated to the collaborators of our work on acid–base transport and its regulation in larval insects: Monica G. Bassous, P. Bower, Thomas M. Clark, Christopher Corbo, S. A. Cummings, Hana Davis, Greg G. Goss, Alan R. Koch, Sejmir Izeirovski, Urmila Jagadeshwaran, Margarita Javoroncov, Stacia B. Moffett, P. Newberry, Scott K. Parks, Malay Patel, Zeping Wang, and R. Woods.

Our work has been funded by the NSF (IBN-0091208) and NIH (1R01AI063463-01A2). We thank Marc Klowden for a longtime supply with mosquito eggs, before we started our own colonies.

References

Azuma M, Harvey WR, Wieczorek H (1995) Stoichiometry of K^+/H^+ antiport helps to explain extracellular pH 11 in a model epithelium. FEBS Lett 361:153–156

Berenbaum M (1980) Adaptive significance of midgut pH in larval Lepidoptera. Am Nat 115:138–146

Boudko DY, Moroz LL, Harvey WR, Linser PJ (2001a) Alkalinization by chloride/bicarbonate pathway in larval mosquito midgut. Proc Natl Acad Sci U S A 98:15354–15359

Boudko DY, Moroz LL, Linser PJ, Trimarchi JR, Smith PJS, Harvey WR (2001b) In situ analysis of pH gradients in mosquito larvae using non-invasive, self-referencing, pH-sensitive microelectrodes. J Exp Biol 204:691–699

Buckner JS, Caldwell JM, Reinecke JP (1980) Uric acid excretion in larval Manduca sexta. J Insect Physiol 26:7–12

Chamberlin ME (1989) Metabolic stimulation of transepithelial potential difference across the midgut of the tobacco hornworm (Manduca sexta). J Exp Biol 114:295–311

Chamberlin ME (1990a) Ion transport across the midgut of the tobacco hornworm (Manduca sexta). J Exp Biol 150:425–442

Chamberlin ME (1990b) Luminal alkalinization by the isolated midgut of the tobacco hornworm (Manduca sexta). J Exp Biol 150:467–471

Cioffi M (1979) The morphology and fine structure of the larval midgut of a moth (*Manduca sexta*) in relation to active ion transport. Tissue Cell 11:467–479

Cioffi M, Harvey WR (1981) Comparison of potassium transport in three structurally distinct regions of the insect midgut. J Exp Biol 91:103–106

Clark TM (1999) Evolution and adaptive significance of larval midgut alkalinization in the insect superorder Mecopterida. J Chem Ecol 25:1945–1960

Clark TM, Koch AR, Moffett DF (1998) Alkalization by *Manduca sexta* anterior midgut in vitro: requirements and characteristics. Comp Biochem Physiol 121A:181–187

Clark TM, Koch A, Moffett DF (1999) The anterior and posterior 'stomach' regions of larval *Aedes aegypti* midgut: regional specialization of ion transport and stimulation by 5-hydroxytryptamine. J Exp Biol 202:247–252

Clark TM, Koch A, Moffett DF (2000) The electrical properties of the anterior stomach of the larval mosquito (*Aedes aegypti*). J Exp Biol 203:1093–1101

Clark TM, Hutchinson MJ, Huegel KL, Moffett SB, Moffett DF (2005) Additional morphological and physiological heterogeneity within the midgut of larval *Aedes aegypti* (Diptera: Culicidae) revealed by histology, electrophysiology, and effects of *Bacillus thuringiensis* endotoxin. Tissue and Cell 37:457–468

Clements AN (1992) The biology of mosquitoes. Chapman and Hall, London

Coddington EJ, Chamberlin ME (1999) Acid/base transport across the midgut of the tobacco hornworm *Manduca sexta*. J Insect Physiol 45:493–500

Corena MDP, Seron TJ, Lehman HK, Ochrietor JD, Kohn A, Tu C, Linser PJ (2002) Carbonic anhydrase in the midgut of larval *Aedes aegypti*: cloning, localization and inhibition. J Exp Biol 205:591–602

Corena MDP, Fiedler MM, VanEkeris L, Tu C, Silverman DN, Linser PJ (2004) Alkalization of larval mosquito midgut and the role of carbonic anhydrase in different species of mosquitoes. Comp Biochem Physiol 137C:207–225

Dadd RH (1975) Alkalinity within the midgut of mosquito larvae with alkaline-active digestive enzymes. J Insect Physiol 21:1847–1853

Dow JAT (1984) Extremely high pH in biological systems: a model for carbonate transport. Am J Physiol 246:R633–R635

Dow JAT, O'Donnell MJ (1990) Reversible alkalinization by *Manduca sexta* midgut. J Exp Biol 150:247–256

Espinoza-Fuentes FP, Ferreira C, Terra WR (1984) Spatial organization of digestion in the larval and imaginal stages of the sciarid fly *Trichosia pubescens*. Insect Biochem 14:631–638

Harvey WR, Zerahn K (1972) Active transport of potassium and other alkali metals by the isolated midgut of the silkworm. In: Bronner F, Kleinzeller A (eds) Current topics in membranes and transport, vol III. Academic Press, New York, pp 367–410

Harvey WR, Cioffi M, Wolfersberger MG (1981) Portasomes as coupling factors in active ion transport and oxidative phosphorylation. Am Zool 21:775–791

Irvine HB (1969) Sodium and potassium secretion by isolated insect Malpighian tubules. Am J Physiol 217:1520–1527

Izeirovski S, Moffett SB, Moffett DF, Onken H (2009) The anterior midgut of larval yellow fever mosquitoes (*Aedes aegypti*): effects of amino acids, dicarboxylic acids, and glucose on the transepithelial voltage and strong luminal alkalinization. J Exp Zool 311A:719–726

Jagadeshwaran U, Onken H, Moffett SB, Moffett DF (2010) Cellular mechanisms of acid secretion in the posterior midgut of the larval mosquito (*Aedes aegypti*). J Exp Biol 213:295–300

Klein U, Löffelman G, Wieczorek H (1991) The midgut as a model system for insect K^+-transporting epithelia: immunocytochemical localization of a vacuolar-type H^+ pump. J Exp Biol 161:61–75

Koch AR, Moffett DF (1996) Ammonia is a factor in alkalinization of midgut contents by *Manduca sexta* larvae (Lepidoptera: Sphingidae). In: Proceeding of XX international congress of entomology Florence, Italy 05–031

Lacey LA, Federici BA (1979) Pathogenesis and midgut histopathology of *Bacillus thuringiensis* in *Simulium vitlatum* (Diptera: Simuliidae). J Invert Pathol 33:171–182

Linser PJ, Smith KE, Seron TJ, Oviedo MN (2009) Carbonic anhydrases and anion transport in mosquito midgut pH regulation. J Exp Biol 212:1662–1671

Martin MM, Martin JS, Kukor JJ, Merritt RW (1980) The digestion of protein and carbohydrate by the stream detritivore, *Tipula abdominalis* (Diptera, Tipulidae). Oecologia 46:360–364

Merzendorfer M, Gräf R, Huss M, Harvey WR, Wieczorek H (1997) Regulation of proton-translocating V-ATPases. J Exp Biol 200:225–235

Moffett DF (1994) Recycling of K^+, acid-base equivalents and fluid between gut and hemolymph in lepidopteran larvae. Physiol Zool 67:68–81

Moffett DF, Cummings SA (1994) Transepithelial potential and alkalinization in an *in situ* preparation of tobacco hornworm (*Manduca sexta*) midgut. J Exp Biol 194:341–345

Moffett DF, Koch A (1992) The insect goblet cell – a problem in functional cytoarchitecture. NIPS 7:19–23

Moffett SB, Moffett DF (2005) Comparison of immunoreactivity to serotonin, FMRFamide and SCPb in the gut and visceral nervous system of larvae, pupae and adults of the yellow fever mosquito *Aedes aegypti*. J Insect Science 5:20

Moffett DF, Koch AR, Woods R (1995) Electrophysiology of K^+ transport by midgut epithelium of lepidopteran insect larvae III. Goblet valve patency. J Exp Biol 198:2103–2113

Moffett DF, Koch AR, Bower B, Newberry P (1996) Acid-base relations between gut, Malpighian tubules and hemolymph in Lepidopteran larvae: some pieces of the puzzle. In: Proceedings of XX international congress of entomology 05–030

Moffett DF, Jagadeshwaran U, Wang Z, Davis HM, Onken H, Goss GG (2012) Signaling by intracellular Ca^{2+} and H^+ in larval mosquito (*Aedes aegypti*) midgut epithelium in response to serosal serotonin and lumen pH. J Insect Physiol 58:506–512

Nijhout HF (1975) Excretory role of the midgut in larvae of the tobacco hornworm. J Exp Biol 62:221–230

Okech BA, Boudko DY, Linser PJ, Harvey WR (2008) Cationic pathway of pH regulation in larvae of *Anopheles gambiae*. J Exp Biol 211:957–968

Onken H, Moffett SB, Moffett DF (2004a) The transepithelial voltage of the isolated anterior stomach of mosquito larvae (*Aedes aegypti*): pharmacological characterization of the serotonin-stimulated cells. J Exp Biol 207:1779–1787

Onken H, Moffett SB, Moffett DF (2004b) The anterior stomach of larval mosquitoes (*Aedes aegypti*): effects of neuropeptides on transepithelial ion transport and muscular motility. J Exp Biol 207:3731–3739

Onken H, Moffett SB, Moffett DF (2006) The isolated anterior stomach of larval mosquitoes (*Aedes aegypti*): voltage-clamp measurements with a tubular epithelium. Comp Biochem Physiol 143A:24–34

Onken H, Moffett SB, Moffett DF (2008) Alkalization in the isolated and perfused anterior stomach of larval *Aedes aegypti*. J Insect Science 8:46–66

Onken H, Parks SK, Goss GG, Moffett DF (2009a) Serotonin-induced high intracellular pH aids in alkali secretion in the anterior midgut of larval yellow fever mosquito *Aedes aegypti* L. J Exp Biol 212:2571–2578

Onken H, Patel M, Javoroncov M, Izeirovski S, Moffett SB, Moffett DF (2009b) Strong alkalinization in the anterior midgut of larval yellow fever mosquitoes (*Aedes aegypti*): involvement of luminal Na^+/K^+-ATPase. J Exp Zool 311A:155–161

Onken H, Bassous MG, Moffett DF, Corbo C (2014) The alkaline anterior midgut of larval mosquitoes as a barrier for microorganisms. FASEB Journal 28:1100.5

Patrick ML, Aimanova K, Sanders HR, Gill SS (2006) P-type Na^+/K^+-ATPase and V-type H^+-ATPase expression patterns in the osmoregulatory organs of larval and adult mosquito *Aedes aegypti*. J Exp Biol 209:4638–4651

Pullikuth AK, Aimanova K, Kang'ethe W, Sanders HR, Gill SS (2006) Molecular characterization of sodium/proton exchanger 3 (NHE3) from the yellow fever vector, *Aedes aegypti*. J Exp Biol 209:3529–3544

Ramsay JA (1950) Osmotic regulation in mosquito larvae. J Exp Biol 27:145–157

Ramsay JA (1951) Osmotic regulation in mosquito larvae: the role of the malpighian tubules. J Exp Biol 28:62–73

Ramsay JA (1976) The rectal complex in the larvae of Lepidoptera. Proc R Soc London B 271:203–226

Rheault MR, Okech BA, Keen SBW, Miller MM, Meleshkevitch EA, Linser PJ, Boudko DY, Harvey WR (2007) Molecular cloning, phylogeny and localization of AgNHA1: the first Na^+/H^+ antiporter (NHA) from a metazoan, *Anopheles gambiae*. J Exp Biol 210:3848–3861

Ridgway RL, Moffett DF (1986) Regional differences in the histochemical localization of carbonic anhydrase in the midgut of tobacco hornworm (*Manduca sexta*). J Exp Zool 237:407–412

Schirmanns K, Zeiske W (1994) An investigation of the midgut K^+ pump of the tobacco hornworm (*Manduca sexta*) using specific inhibitors and amphotericin B. J Exp Biol 188:191–204

Senior-White R (1926) Physical factors in mosquito ecology. Bull Entomol Res 16:187–248

Shanbhag S, Tripathi S (2005) Electrogenic H^+ transport and pH gradients generated by a V-H^+-ATPase in the isolated perfused larval *Drosophila* midgut. J Membrane Biol 206:61–72

Soleiman M, Lesoine GA, Bergman JA, Aronson PS (1991) Cation specificity and modes of the $Na^+:CO_3^{2-}:HCO_3^-$ cotransporter in renal basolateral membrane vesicles. J Biol Chem 266:8706–8710

Stiles B, Paschke JD (1980) Midgut pH in different instars of three *Aedes* mosquito species and the relation between pH and susceptibility of larvae to a nuclear polyhedrosis virus. J Invert Pathol 35:58–64

Terra WR, Ferreira C, De Bianchi AG (1979) Distribution of digestive enzymes among the endo- and ecto-peritrophic spaces and midgut cells of *Rhynchosciara* and its physiological significance. J Insect Physiol 25:487–494

Thomson RB, Phillips JE (1992) Electrogenic proton secretion in the hindgut of the desert locust, *Schistocerca gregaria*. J Membr Biol 125:133–154

Thomson RB, Thomson JM, Phillips JE (1988) NH_4^+ transport in acid-secreting insect epithelium. Am J Physiol 254:R348–R356

Turbeck BO, Fodor B (1970) Studies on a carbonic anhydrase from the midgut epithelium of larvae of Lepidoptera. Biochim Biophys Acta 212:139–149

Undeen AH (1979) Simuliid larval midgut pH and its implications for control. Mosq News 39:391–393

Volkmann A, Peters W (1989a) Investigation on the midgut caeca of mosquito larva. I. Fine structure. Tissue Cell 21:243–251

Volkmann A, Peters W (1989b) Investigations on the midgut caeca of mosquito larvae II. Functional aspects. Tissue Cell 21:253–261

Waterhouse DF (1949) The hydrogen ion concentration in the alimentary canal of larval and adult Lepidoptera. Aust J Sci Res, Ser B, Biol Sci 132:428–437

Weihrauch D (2006) Active ammonia absorption in the midgut of the tobacco hornworm *Manduca sexta* L.: transport studies and mRNA expression analysis of a Rhesus-like ammonia transporter. *Insect Biochem. Mol Biol* 36:808–821

Wieczorek H, Putzenlechner M, Zeiske W, Klein U (1991) A vacuolar-type proton pump energizes K^+/H^+ antiport in an animal plasma membrane. J Biol Chem 266:15340–15347

Wieczorek H, Brown D, Grinstein S, Ehrenfeld J, Harvey WR (1999) Animal plasma membrane energization by proton-motive V-ATPases. Bioessays 21:637–648

Wieczorek H, Grüber G, Harvey WR, Huss M, Merzendorfer H, Zeiske W (2000) Structure and regulation of insect plasma membrane H^+ V-ATPase. J Exp Biol 203:127–135

Wieczorek H, Huss M, Merzendorfer H, Reineke S, Vitavska O, Zeiske W (2003) The insect plasma membrane H^+ V-ATPase: intra-, inter-, and supramolecular Aspects. J Bioenerg Biomembr 35:359–366

Wieczorek H, Beyenbach KW, Huss M, Vitavska O (2009) Vacuolar-type proton pumps in insect epithelia. J Exp Biol 212:1611–1619

Wigglesworth VB (1933) The function of the anal gills of the mosquito larvae. J Exp Biol 10:16–26

Wigglesworth VB (1942) The storage of protein, fat, glycogen and uric acid in the fat body and other tissues of mosquito larvae. J Exp Biol 19:56–77

Zerahn K (1971) Active transport of the alkali metals by the isolated midgut of *Hyalophora cecropia*. Phil Trans R Soc London B 262:315–321

Zerahn K (1977) Potassium transport in insect midgut. In: Gupta BL, Moreton RB, Oschman JL, Wall BJ (eds) Transport of ions and water in animals. Academic, New York/London, pp 381–401

Zhuang Z, Linser PJ, Harvey WR (1999) Antibody to H^+ V-ATPase subunit E colocalizes with portasomes in alkaline larval midgut of a freshwater mosquito (*Aedes aegypti* L.). J Exp Biol 202:2449–2460

pH Regulation and Excretion in Echinoderms

Meike Stumpp and Marian Y. Hu

© Springer International Publishing Switzerland 2017
D. Weihrauch, M. O'Donnell (eds.), *Acid-Base Balance and Nitrogen Excretion in Invertebrates*,
DOI 10.1007/978-3-319-39617-0_10

10.1 Adult Echinoderms

10.1.1 Extracellular Acid–Base Regulation

Echinoderms consist of five classes: Crinoidea (sea lilies), Asteroidea (sea stars), Ophiuroidea (brittle stars), Echinoidea (sea urchins), and Holothuroidea (sea cucumbers). Most of them are key stone species in many marine ecosystems and especially sea urchins and sea cucumbers are economically important organisms in many countries.

Adult echinoderms are characterized by their pentameric symmetry and an ambulacrarian vascular system (◻ see Fig. 10.1 for an echinoid body plan). Some of the classes have heavily calcified tests, like sea urchins, brittle stars, and crinoids. Sea stars and sea cucumbers have calcified spines or test plates and are not as heavily calcified. All echinoderms have a so-called coelomic fluid, their blood derivate with coelomic cells that resemble higher animals' immune cells. The composition of the coelomic fluid is similar to that of seawater and is prone to acid–base disturbances when the animals encounter hypercapnic conditions or emersion. The coelomic fluid can be relatively easily sampled from the animals, and examinations regarding the extracellular acid–base status in echinoderms were primarily conducted on this fluid.

Among the five classes of echinoderms, there is data on extracellular acid–base status on four of them (no data on crinoids). Of these four classes of echinoderms, only Echinoidea and Ophiuroidea were reported to actively compensate for acid–base disturbances indicated by an accumulation of bicarbonate in response to elevated pCO_2 (Stumpp et al. 2012b; Calosi et al. 2013; Hu et al. 2014). The effects of CO_2-induced seawater acidification on echinoderms have lately received considerable attention, as calcifying species were predicted to be particularly sensitive to ocean acidification. The maintenance of calcification is directly linked to acid–base regulatory capacities making this physiological process a key candidate to address sensitivities in echinoderms (Ries et al. 2009). Furthermore,

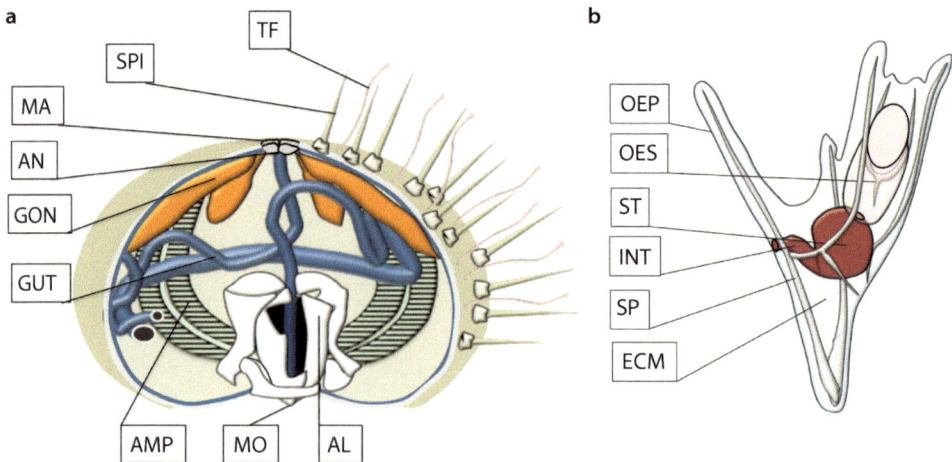

◻ **Fig. 10.1** Body plan of an adult sea urchin (**a**) and an echinopluteus larvae (**b**). Aristotle's lantern (*AL*); mouth (*MO*); ampullae (*AMP*); gut (*GUT*); gonads (*GON*); anus (*AN*); madrepore plate (*MA*); spines (*SPI*); tube feet (*TF*); extracellular matrix (*ECM*); spicule (*SP*); intestine (*INT*); stomach (*ST*); esophagus (*OES*); outer epithelium (*OEP*)

echinoderms are key stone species in many marine habitats, and vulnerability of these organisms to near-future climate change could have severe repercussions on populations and ecosystem stabilities.

Echinoidea

Sea urchins of the species *Strongylocentrotus droebachiensis* exposed to 1400 and 3800 µatm CO_2 were able to fully or partially compensate extracellular pH (pH_e) disturbances within 7–10 days of exposure to elevated pCO_2 (Dupont and Thorndyke 2012; Stumpp et al. 2012b) and were able to sustain this compensation for over 45 days. The maximum bicarbonate accumulation recorded in this species was 2.5 mmol L^{-1} above their control levels of 4.0–5.3 mmol L^{-1} HCO_3^-. Interestingly, the control levels of $[HCO_3^-]_e$ were already higher than that of seawater (around 2 mmol L^{-1}). Sea urchins of the species *Arbacia lixula* and *Paracentrotus lividus* also accumulated bicarbonate up to 3.5 mmol L^{-1} (from 2 mmol L^{-1} control) and 5–5.1 mmol L^{-1} (3.6–3.8 mmol L^{-1} control levels) in response to 1500 µatm and 2500 µatm CO_2 (Calosi et al. 2013; Collard et al. 2014b). With this increase in bicarbonate, both species were able to fully or partially compensate induced extracellular acid–base disturbances (Catarino et al. 2012). *Tripneustes ventricosus* accumulated up to a $[HCO_3^-]_e$ of 5.0 mmol L^{-1} (from 2.9 mmol L^{-1} control levels) in response to 2000 µatm CO_2 (Collard et al. 2014b). *Psammechinus miliaris* accumulated 0.5 mmol L^{-1} HCO_3^- from 1.8 mmol L^{-1} to 2.3 within 7 days of exposure to 2400 µatm CO_2 (Miles et al. 2007). These examples highlight that at least some sea urchins have a higher potential to actively regulate their extracellular pH than previously thought.

Nevertheless, there are also examples for sea urchins that do not compensate pH_e disturbances: The cidaroid pencil urchin *Eucidaris tribuloides* was not able to increase extracellular bicarbonate levels within 32 days of exposure to 2000 µatm CO_2 (Collard et al. 2014b). Data on irregular and/or infaunal sea urchins such as sand dollars or sea mice are still missing and may offer more conclusive patterns regarding acid–base physiology in different groups of echinoids.

In general, it seems that noncidaroid, regular sea urchins are able to accumulate up to 2.5 mmol L^{-1} bicarbonate in order to counteract acid–base disturbances. The mechanisms underlying bicarbonate accumulation, however, are insufficiently understood. Based on coelomic Mg^{2+} and Ca^{2+} concentration measurements in response to elevated pCO_2, an accumulation based on test dissolution was suggested (Miles et al. 2007). Contrary, bicarbonate concentrations in the coelomic fluid of control animals is already significantly higher than seawater in most studied sea urchins (e.g., 2–3 mmol L^{-1} above seawater $[HCO_3^-]_{SW}$ of 2–3 mmol L^{-1}) and favors the hypothesis, that already under control conditions an active HCO_3^- accumulation mechanism is employed. While temporary test dissolution may still contribute to the initial compensation phase in sea urchins, it seems unlikely that test dissolution is solely responsible for the observed reestablishment of pH_e homeostasis.

There are two tissues that could contribute to bicarbonate accumulation: the peritoneal epithelium that separates the sea urchins test from the coelomic fluid and the intestinal tissue. The peritoneal epithelium, though being tight for molecules larger than 0.3 kDa, is no barrier for ions and small molecules, such as HCO_3^- (Holtmann et al. 2013). Therefore, the organ that most likely contributes to the observed bicarbonate accumulation is the digestive system. The intestine prevents loss of bicarbonate ions by establishing a tight barrier for HCO_3^- ions (Holtmann et al. 2013). However, the mechanisms and a potential function of the intestine in acid–base regulation remain unexplored.

Ophiuroidea, Asteroidea, and Holothuroidea

There is only one study reporting extracellular acid–base status for an ophiuroid. The infaunal brittle star *Amphiura filiformis* has extracellular bicarbonate levels of 4–6 mmol L^{-1} under control conditions (compared to 2 mmol L^{-1} in seawater) and increased extracellular bicarbonate levels up to 8 mmol L^{-1} when exposed to 1500 and 3200 µatm CO_2 (Hu et al. 2014). The compensation reaction happened within 24 h after onset of exposure and resulted in a full (1500 µatm) and partial (3200 µatm) recovery of pH_e. *A. filiformis* was able to sustain high bicarbonate levels during 14 days along the experimental period (Hu et al. 2014). Interestingly, transcript levels for V-type H^+-ATPase were elevated in disk tissue, and not in the arm tissue, in response to CO_2 stress indicating that proton transport might also be an important mechanism for extra- and/or intracellular pH compensation.

The sea star *Asterias rubens* did neither accumulate bicarbonate in response to elevated pCO_2 (1200–3500 µatm, (Appelhans et al. 2012)) nor were they demonstrated to compensate acid–base disturbances in their coelomic fluid (Collard et al. 2014b) over an experimental duration ranging from 14 days to 10 weeks. The sea star *Leptasterias polaris* did also not compensate pHe disturbances during 7 days of exposure to 1275 µatm CO_2 (Dupont and Thorndyke 2012).

The sea cucumbers *Holothuria scabra* and *Holothuria parva* did not significantly accumulate bicarbonate to compensate their pHe in response to hypercapnic conditions (2000–3500 µatm) (Collard et al. 2014a) in the experimental period of 6–12 days that may, however, be too short an interval to initiate a compensation reaction. $[HCO_3^-]_e$ was still slightly higher (between 2.7 and 3.0 mmol L^{-1}) than in seawater (between 2.2 and 2.7 mmol L^{-1}).

These observations indicate that the ability to compensate for acid–base disturbances via active accumulation of HCO_3^- in body fluids is not a universal feature in all echinoderms but seems to be restricted to Echinoidea and Ophiuroidea. The fact that most holothurians and asteroids are less calcified than echinoids and ophiuoroids can lead to speculations that acid–base regulatory abilities are correlated to the degree of calcification in echinoderms, with the exception of the pencil urchin *E. tribuloides*, which is a heavily calcified sea urchin even among sea urchins. In order to test this hypothesis, a comparative approach using a larger range of species and a standardized experimental design would be needed. Moreover, additional studies on the mechanistic basis of pHe regulation in echinoderms will represent an important future task to better understand the mechanisms of acid–base transport across gut epithelia in a so far unexplored animal group.

10.1.2 Nitrogen Excretion

It has been proposed that excretion of protons may be facilitated by the excretion of ammonia (for mechanistic details on the cellular level, see chapters on crustaceans and cephalopods). There are only a few studies examining the ammonia excretion rates in response to hypercapnia in echinoderms (Stumpp et al. 2012b; Hu et al. 2014).

It seems that in all studied echinoderm species, the compensation reaction for CO_2-induced acid–base disturbances is accompanied by a significant increase in NH_3/NH_4^+ excretion rates. For example, *Strongylocentrotus droebachiensis* increased ammonia excretion rates by 70 % in response to 2800 µatm CO_2 (Stumpp et al. 2012b). *Amphiura filiformis* significantly excreted 65 % more ammonia under elevated pCO_2 conditions (3400 and 6600 µatm CO_2) (Hu et al. 2014). Species, e.g., *Holothuria scabra* and *Holothuria parva*, with uncompensated perivisceral fluids did not increase their ammonia excretion in response to hypercapnic conditions (Collard et al. 2014a).

This observation suggests that pH_e regulatory mechanisms are linked to ammonia excretion as a mechanism to facilitate bicarbonate formation/accumulation and/or proton (equivalent) excretion. A potent mechanism to excrete protons is essential in species that are capable of actively accumulating HCO_3^- as the formation of HCO_3^- from CO_2 and H_2O is ultimately associated with the formation of H^+ that needs to be secreted from the animals. As the present body of knowledge regarding species-specific NH_3/NH_4^+ secretion characteristics and mechanisms for echinoderms are scarce, definite conclusions cannot be drawn from the present literature data.

10.2 Larval Echinoderms

Echinoderms generally reproduce by free-swimming larvae – with very few exceptions of brooding species. Larvae are either lecithotrophic or in most of the cases planktotrophic. Depending on the class, you find larvae with calcified endoskeletons (sea urchins and brittle stars) or larvae that are not calcified until the adult rudiment is formed (sea stars and sea cucumbers). The calcified structures, if present, are surrounded by a cytoplasmic sheet of primary mesenchyme cells, which is again embedded in a vast extracellular space between the epidermis and digestive tract. The extracellular space of echinoderm larvae is characterized by extensive almost cell-free gelatinous material occupying the primary body cavity and surrounding the calcifying primary mesenchyme cells (PMCs) in echinopluteus larvae (Strathmann 1989; Crawford 1990). The gel-like compounds of the primary body cavity are extremely flexible and the larval body is able to rebound to its original shape following muscle contraction (e.g., during swallowing of food particles). Thus, it is a highly energy-saving system for larval movement, allowing for large larvae with relatively little cellular material, a situation not unlike that in many Scyphomedusae (Strathmann 1989; Crawford 1990). The larval calcium carbonate structures, if present, combined with the gelatinous material support the larval morphology and orientation of the larvae in the water column (Pennington and Strathmann 1990). ◘ See Fig. 10.1b for a larval body plan of an echinopluteus.

Thus, there are three morphological structures/compartments that are of interest for acid–base regulation: intracellular pH regulation in calcifying primary mesenchyme cells, the extracellular space between the ectoderm and endoderm, and the extracellular/luminal space of the digestive system. While there is genetic data available of whole larvae exposed to hypercapnic conditions, mechanistic data on acid–base regulation in echinoderms is, again, scarce and is within the echinoderms limited to larvae from sea urchins.

10.2.1 Intracellular Acid–Base Regulation in Calcifying Primary Mesenchyme Cells

Echinoderm larvae are considered to be particularly sensitive to seawater acidification and to the connected changes in calcium carbonate saturation state of seawater (Ω_{Cal}) since their internal skeleton is composed of high-magnesium calcite, the most soluble form of $CaCO_3$ (Beniash et al. 1997; Raz et al. 2003). However, sea urchin larvae are able to maintain calcification rates (when corrected for developmental or growth delay) even during exposure to unrealistically high CO_2 partial pressures (0.35 kPa CO_2, pH 7.25, Ω_{Cal} 1.02) (Martin et al. 2011). This indicates that the calcification machinery has substantial capacities to compensate for $CaCO_3$ dissolution due to CO_2-induced reductions in seawater pH. The calcification process in sea urchin larvae has been relatively well investigated (e.g., Nakano et al. 1963; Sikes et al. 1981; Wilt 2002; Raz et al. 2003; Politi et al. 2008) and primary mesenchyme cells (PMCs) were identified as the cell type mediating calcification in sea urchin larvae. These cells are located within the extracellular matrix of the primary body cavity and form a syncytium covering the spicules of the pluteus larvae. This syncytial sheath covers the entire surface of the spicules and communicates with the extracellular environment of the primary body cavity (Decker et al. 1987; Beniash et al. 1997). Recent studies using a pharmacological approach demonstrated that several loop diuretics including azosemide, bumetanide, and furosemide could significantly inhibit spicule formation at concentrations that significantly inhibit the $Na^+/K^+/2Cl^-$ symporter (NKCC) (Basse et al. 2015). Using histological analyses it has been concluded that the impaired formation of spicules was due to an inhibition of cytoplasmic cord formation and maintenance in PMCs.

The high-magnesium calcitic spicules are formed through the production of a transient amorphous calcium carbonate (ACC) phase within vesicles in PMCs which is subsequently released into the spicular cavity (Beniash et al. 1997; Raz et al. 2003; Politi et al. 2004, 2008). To fuel the calcification process, bicarbonate is absorbed from the seawater (40 %) as well as generated from metabolic CO_2 (60 %) (Sikes et al. 1981). On the other hand, Ca^{2+} is exclusively obtained directly from the seawater (Nakano et al. 1963). Although the general principle of calcification is well understood, mechanistic information regarding transporters that facilitate Ca^{2+} or HCO_3^- uptake in PMCs is still limited. Several pharmacological studies suggested that Ca^{2+} channels and pumps are key players in the provision of Ca^{2+} for spicule formation (Hwang and Lennarz 1993; Gunaratne and Vacquier 2007). In order to remove protons generated during $CaCO_3$ precipitation, PMCs must possess efficient acid–base regulatory properties (McConnaughey and Whelan 1996; Buitenhuis et al. 1999). A recent study applied the intracellular pH dye acid (2,7-bis (2 carboxyethyl)-5-(and-6)-carboxy fluorescein (BCECF)) in order to investigate pH regulatory mechanisms in sea urchin PMCs (Stumpp et al. 2012a). Using pulses of NH_3/NH_4^+ solution, it has been demonstrated that PMCs have significant pH_i regulatory abilities. PMCs have a control pH_i of 6.9 and 20 mmol L^{-1} NH_3/NH_4^+ solution induced an intracellular alkalinization leading to an increase of pH_i by approximately 0.7 units. Washout of the NH_3/NH_4^+ solution induced an acidosis of approximately 0.7 pH units below the control value, followed by a pH recovery to control levels within 15 min. Rates of pH_i recovery after the NH_3/NH_4^+ pulse were significant but weak compared to recovery times reported for strong ion regulatory cells of vertebrates (Boron and De Weer 1976; Bleich et al. 1995; Hasselblatt et al. 2000; Boron 2004) and similar to those reported for crayfish neurons (~20 min, (Moody 1981)). These experiments indicated that PMCs that are involved in skeletogenesis in sea urchins have a significant acid–base regulatory capacity.

The rate of pH compensation following an NH_3/NH_4^+-induced acidosis was further investigated in the presence of low-Na^+ and low-HCO_3^- artificial seawater solutions in order to test the potential role of these ions in mediating pH_i. These experiments clearly indicated that the compensation reaction after NH_3/NH_4^+-induced acidosis in PMCs is highly dependent on Na^+ and HCO_3^- transport and suggest the involvement of Na^+-dependent HCO_3^- import mechanisms during pH recovery. These are an indirect sink of ATP mediated through secondary active transporters (e.g., transporters of the SLC4 family, including Na^+/HCO_3^- cotransporters and Na^+-dependent Cl^-/HCO_3^- exchangers) that are driven through ATP-dependent ion pumps such as the Na^+/K^+-ATPase or the V-type H^+-ATPase (Boron 2004). The existence of a pH regulatory machinery in PMCs of sea urchins is underlined by the identification of genes coding for ion transporters including Na^+/K^+-ATPase (NKA), Na^+/HCO_3^- cotransporters (NBC), H^+-ATPases (HA), and Na^+/H^+-exchangers (NHE) from the sea urchin PMC transcriptome (Zhu et al. 2001). However, since protons are generated during the intracellular calcification process, the identification and characterization of proton secretion mechanisms involving transporters such as the V-type H^+-ATPase or $Na^+/H+$ exchangers represent an important future task to better understand the mechanisms of calcification in echinoid PMCs.

10.2.2 Extracellular Acid–Base Regulation

Primary Body Cavity

Microelectrode and microfluorescence measurements of extracellular pH (pH_e) within the primary body cavity of *S. droebachiensis* pluteus larvae revealed that the fluid of the primary body cavity pH conforms to the surrounding seawater (Stumpp et al. 2012a). Given that the fluid composition of the primary body cavity is similar to seawater, this indicates that the internal pCO_2 within the animals is similar to that of seawater pCO_2. Usually, extracellular pCO_2 of adult, heterotrophic metazoans is at least 100 Pa higher than the seawater pCO_2 in order to establish a CO_2 diffusion gradient away from the organism (Melzner et al. 2009). The result would imply that sea urchin larvae do not have an outward directed CO_2 diffusion gradient due to metabolic CO_2 production. During acute exposure of larvae to elevated seawater pCO_2, pH_e decreased linearly with pH_{SW} indicating that the epidermis is extremely leaky to ions and small molecules. A transepithelial KCl^- potential of 0 mV of the ectodermal epithelium also suggests that the epithelium is not a specialized transporting epithelium (Stumpp et al. 2012a). In fact, diffusion experiments using different sizes of BCECF-substituted dextrans demonstrated that the ectodermal epithelium forms no barrier for molecules up to 10 kDa if sufficient time for diffusion is allowed (60 min). Molecules up to 4 kDa penetrate the epidermis within 15 min (Stumpp et al. 2012a). Gases and small ions can therefore be expected to be in equilibrium with seawater leading to good conditions for calcification at present conditions, but on the other hand being sensitive to environmental variability in pH or future ocean acidification.

Stomach Lumen

Sea urchin larvae have a tripartite gut consisting of a foregut (esophagus), midgut (stomach), and a hindgut (intestine). Food particles (e.g., microalgae) are collected and concentrated in the esophagus and then swallowed by opening of the anterior cardiac sphincter. Food is processed in the stomach and absorbed in the stomach as well as the intestine. While the pH of the primary body cavity is conformed with seawater, the luminal pH_e in the stomach of *S. droebachiensis* increased after feeding commenced from pH_e 8.9 at 5 days post fertilization to pH_e 9.5 at 9 days post fertilization (Stumpp et al. 2013) and up to pH 9.0 in *S. purpuratus* (Stumpp et al. 2015). Alkaline gastric pHs are known to occur in a range of insects and their larvae such as lepidopterans (Dow 1984) and dipteran larvae (Boudko et al. 2001). Among marine species alkaline digestive conditions were additionally found in the marine copepod *Calanus helgolandicus* (Pond et al. 1995) as well as in tornaria larvae of the hemichordate *Ptychodera flava* – reaching a pH of 10.5 in their stomach lumen (Stumpp et al. 2015). Alkaline digestive conditions may represent an evolutionary advantage for marine herbivorous larvae as alkaline conditions may facilitate breakdown of plant and algal proteins (Felton and Duffey 1991). Moreover, alkaline conditions higher than pH 9.5 in larval digestive systems can be regarded as an efficient defense mechanism against environmental pathogens because most marine protists and viruses are killed or growth inhibited by alkaline conditions exceeding pH 9.5 (Pedersen and Hansen 2003; Jończyk et al. 2011). However, these unique digestive features were proposed to be a key process that makes these larvae sensitive to climate change and the associated reductions in seawater pH (ocean acidification) by impairing digestion and exerting additional energetic costs to the larvae in future oceans (Stumpp et al. 2013) (for details, see next paragraph).

Chronic exposure of *S. droebachiensis* larvae to elevated pCO_2 resulted in a lower stomach lumen pH_e at 7 and 9 days post fertilization in two CO_2 treatments (1015 and 2460 µatm pCO_2). The data suggest that the environmental CO_2 induced acidification is exceeding the larval capacity of fully alkalinize the stomach lumen to control levels. The drop in stomach pH_e of 0.25 and 0.4 pH units at the high pCO_2 level (2460 µatm at 7 and 9 dpf, respectively) is, however, much smaller than the environmental pH_{SW} drop of 0.67 pH units, possibly indicating an activation of compensatory mechanisms. Acute acidification experiments down to pH 7 in *S. purpuratus* resulted in a partial compensation reaction of the gastric pH within 3 h. This compensation reaction was likely mediated by cells of the stomach epithelium.

In contrast to the thin squamous epidermal cell layers, the echinopluteus' stomach epithelium consists of simple columnar cells (of 10–15 µm height) with microvilli extending into the stomach lumen (Burke 1981). At least two cell types are present in the stomach epithelium. Type I cells are characterized by a variety of different vacuoles and are expected to function in absorption and storage of nutrients as well as the secretion of digestive enzymes. Type II cells are assumed to phagocytose and digest whole algal cells (Burke 1981).

Immunohistochemical and in situ hybridization using antibodies and probes against specific ion transporters could demonstrate that ion pumps and transporters including Na^+/K^+-ATPase (NKA) and V-type H^+-ATPase (VHA) are highly expressed in specialized epidermal cells of the stomach and intestine. The colocalization of NKA, VHA, and Na^+H^+ exchangers revealed the subcellular localization of these proteins in stomach cells. While NKA and VHA are localized in the luminal membrane positive Na^+/H^+ exchanger (NHE), immunoreactivity was mainly found in the basolateral membrane (Stumpp et al. 2015). The mechanistic basis of gastric alkalization was further examined using H^+-selective microelectrodes in combination with specific inhibitors for acid–base relevant ion transporters. These studies underlined the histological findings by demonstrating that gastric alkalization in sea urchin larvae is inhibited by ouabain and bafilomycin, specific inhibitors for the Na^+/K^+-ATPase (NKA) and the V-type H^+-ATPase (VHA), respectively. In addition, gastric alkalization was decreased by treatments with ethyl-isopropyl amiloride (EIPA, a specific inhibitor of Na^+-dependent H^+ exchangers (NHEs)) as well as artificial seawater with reduced Na^+ (5 mmol L^{-1}) and K^+ (0 mmol L^{-1}) concentrations.

Gene expression analyses of genes encoding acid–base relevant transporters in pluteus upon challenge of the gastric pH regulatory machinery with acidified seawater identified genes coding for the transporters mediating gastric alkalinization. During acute (1–3 h) challenge of the gastric pH regulatory machinery by exposing larvae to acidified conditions

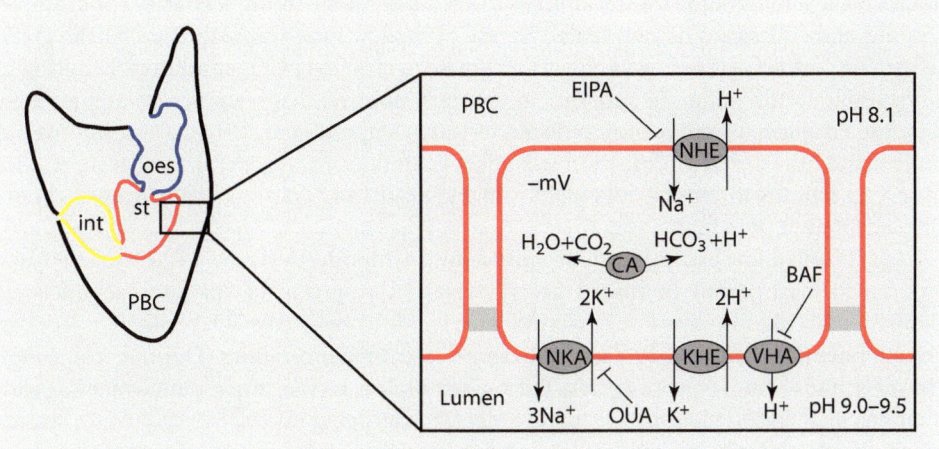

☑ **Fig. 10.2** First working model for gastric alkalinization in sea urchin pluteus larvae. The stomach epithelium is rich in ion transporters that are capable of regulating gastric pH. The current model suggests that gastric alkalinization is achieved by a basolateral (facing the primary body cavity, PBC) NHE that is fueled by a Na$^+$/K$^+$-ATPase located in the luminal membrane. Protons that are excreted via NHE derive from the hydration of metabolic CO$_2$ via cytosolic carbonic anhydrases (CA) as well as from a luminal K$^+$/H$^+$ antiport mechanism energized by luminal V-type H$^+$-ATPase (VHA) (modified after Stumpp et al. 2015).

(pH 7.0), ion transport genes, including NKA, VHA, NHE3, and a KHE, were significantly up-regulated when compared to larvae kept under control conditions. Accordingly a first working model for gastric alkalinization in sea urchin larvae has been proposed (Stumpp et al. 2015). Gastric alkalinization in the sea urchin stomach is mainly generated through cationic exchange processes, leading to a net export of protons from the stomach lumen (summarized in ☑ Fig. 10.2). NKA located in luminal membranes creates an electrochemical gradient, and a low intracellular [Na$^+$] that drives basolateral excretion of H$^+$ via NHE in the the primary body cavity (PBC). On the luminal side, H$^+$ is imported via a coupling of VHA and a putative electrogenic K$^+$/H$^+$ antiporter. A similar cationic-based alkalinization mechanism driven by the VHA coupled to secondary active transport processes including K$^+$/H$^+$ antiport has been proposed for the midgut alkalinization in lepidopteran insect larvae (Wieczorek et al. 1991). The employment of NHE localized in membranes facing the PBC that mediate excretion of H$^+$ in exchange for Na$^+$ represents an energetically favorable mechanism due to high external [Na$^+$] ($\approx$460 mmol L^{-1}) compared to low intracellular [Na$^+$] ($\approx$30 mmol L^{-1}) (Robertson 1949).

10.3 Effects of Near-Future Ocean Acidification on Echinoderms

Effects of simulated ocean acidification experiments on adult echinoderms generally included reduced growth and feeding rates, decreased fertility (measured as number of eggs per female or gonad growth), and occasionally reduced metabolic rates (Appelhans et al. 2012; Dupont and Thorndyke 2012; Stumpp et al. 2012b). This most likely leads to alterations in the energy budget that may be due to elevated costs for cellular and extracellular pH homeostasis as described in sea urchins and brittle stars. However, most studies employed experimental time frames between a few days and a few months, possibly not allowing enough time for the animals to acclimate to the future scenario. For example, Dupont and colleagues demonstrated in one of the few long-term studies that the green sea urchin S. *droebachiensis* was able to acclimate to elevated pCO_2 conditions resulting in comparable fertility after 16 months. In contrast, animals that were acclimated for only 4 months displayed significantly reduced fertility (Dupont et al. 2012). Thus, in order to understand how ocean acidification will affect echinoderms in general, long-term studies of several months to years would yield valuable results in regard to acclimation and possibly adaptation potential.

Apart from a few exceptions with nonfeeding echinoderm larvae, i.e., *Crossaster papposus*, which displayed increased growth rates in response to seawater acidification (Dupont et al. 2010b), exposure of echinoderm larvae to elevated pCO_2 resulted in reduced growth rates (Dupont et al. 2010a) and in some cases high mortalities (Dupont et al. 2008). However, most studies were conducted on sea urchin larvae, since pluteus larvae with their calcified endoskeleton were readily identified as being especially sensitive to seawater acidification. Only two studies looked at noncalcifying, feeding marine larvae such as bipinnaria larvae of sea stars. Interestingly, despite the lack of calcified structures, exposure of sea star larvae to environmentally realistic increases in CO_2 levels could also demonstrate slowed larval growth (Gonzalez-Bernat et al. 2012; Byrne et al. 2013). This clearly indicates that other physiological processes than calcification may cause sensitivities to seawater acidification in marine larval stages.

Based on the findings of alkaline digestive systems in sea urchin larvae (and also hemichordate larvae (Stumpp et al. 2015)) and their sensitivity to lowered pH (Stumpp et al. 2013), the hypothesis of gastric pH homeostasis being a dominant factor determining sensitivity in echinoderm larvae arises. Apart from potential elevated costs for intracellular and/or extracellular pH regulation itself (Stumpp et al. 2012a, 2015), a lowered stomach pH might have negative consequences for larval energy acquisition. Digestive enzymes secreted into the gut to hydrolyze nutrients usually exhibit specific pH optima at which the enzymes are most efficient. Sea urchin larval total protease activity exhibited a pH optimum at pH 11 with a steady increase in activity from pH 8. A potential digestive pH drop of 0.3 pH units would thus translate into 8 % less digestive efficiency. The digestive potential to digest algal cells measured in *in vivo* digestion assays supports the *in vitro* total protease measurements by demonstrating a significantly lower digestive potential in sea urchin larvae raised under future elevated pCO_2 conditions (Stumpp et al. 2013). Decreased digestive potential together with increased acid–base regulatory costs would inevitably result in modifications of the larval energy budget and could potentially explain the observed reduced growth rates in sea urchin plutei, as well as in noncalcifying but feeding echinoderm larvae, such as sea star larvae.

10.4 Conclusions

The present body of knowledge regarding pH regulatory abilities in echinoderms demonstrated that despite their relatively basal position among deuterostomes, these animals have evolved significant pH regulatory abilities. Although echinoderms were demonstrated to exhibit extracellular pH regulatory abilities, the organs and mechanisms involved remain unexplored. Few studies indicated that the pH compensatory mechanism involves accumulation of buffers (e.g., HCO_3^-) as well as the excretion of proton equivalents such as NH_4^+. Although extracellular pH regulatory abilities were found in a range of adult echinoids, their larval stages lack extracellular pH regulatory abilities. However, the regulation of an alkaline gastric pH homeostasis seems to play an important role in digestive function, which could be identified as an "Achilles' heel" in the context of CO_2-induced seawater acidification.

The fact that echinoderms are an evolutionary ancient group, that has also survived acidification events during geological timescales, suggests that this animal group has the ability to successfully cope with environmental pH/pCO_2 fluctuations. In this context special attention should be dedicated to the ability of these animals to acclimate and to genetically or epigenetically adapt to changing environmental conditions. As larval stages were identified as the weakest ontogenetic link, these pelagic life stages deserve special attention. Thus, in a first step, a better understanding regarding the physiological processes in echinoderm larval stages are needed in order to better understand animal–environment interactions in this ecologically and economically important group.

Bibliography

Appelhans JS, Thomsen J, Pansch C, Melzner F, Wahl M (2012) Sour times: seawater acidification effects on growth, feeding behaviour and acid-base status of *Asterias rubens* and *Carcinus maenas*. Mar Ecol Prog Ser 459:85–97

Basse WC, Gutowska MA, Findeisen U, Stumpp M, Dupont ST, Jackson DJ, Himmerkus N, Melzner F, Bleich M (2015) A sea urchin Na^+K^+ 2Cl cotransporter is involved in the maintenance of calcification-relevant cytoplasmic cords in *Strongylocentrotus droebachiensis* larvae. Comp Biochem Physiol A Mol Integr Physiol 187:184–192

Beniash E, Aizenberg J, Addadi L, Weiner S (1997) Amorphous calcium carbonate transforms into calcite during sea urchin larval spicule growth. Proc R Soc Lond B 264:461–465

Bleich M, Köttgen M, Schlatter E, Greger R (1995) Effect of NH_4^+/NH_3 on cytosolic pH and the K^+ channels of freshly isolated cells from the thick ascending limb of Henle's loop. Pflugers Arch 429:345–354

Boron WF (2004) Regulation of intracellular pH. Adv Physiol Educ 28:160–179

Boron WF, De Weer P (1976) Intracellular pH transients in squid giant axons caused by CO_2, NH_3, and metabolic inhibitors. J Gen Physiol 67:91–112

Boudko DY, Moroz LL, Harvey WR, Linser PJ (2001) Alkalization by chloride/bicarbonate pathway in larval mosquito midgut. Proc Natl Acad Sci U S A 98:15354–15359

Buitenhuis ET, de Baar HJW, Veldhuis MJW (1999) Photosynthesis and calcification by *Emiliania huxleyi* (Prymnesiophyceae) as a function of inorganic carbon species. J Phycol 35:949–959

Burke DR (1981) Structure of the digestive tract of pluteus larva of *Dendraster excentricus* (Echinodermata: Echinoidea). Zoomorph 98:209–225

Byrne M, Gonzalez-Bernat M, Doo S, Foo S, Soars N, Lamare M (2013) Effects of ocean warming and acidification on embryos and non-calcifying larvae of the invasive sea star *Patiriella regularis*. Mar Ecol Prog Ser 473:235–246

Calosi P, Rastrick SPS, Graziano M, Thomas SC, Baggini C, Carter HA, Hall-Spencer JM, Milazzo M, Spicer JI (2013) Distribution of sea urchins living near shallow water CO_2 vents is dependent upon species acid-base and ion-regulatory abilities. Mar Poll Bull 30:470–484

Catarino A, Bauwens M, Dubois P (2012) Acid-base balance and metabolic response of the sea urchin *Paracentrotus lividus* to different seawater pH and temperatures. Environ Sci Poll Res 19:2344–2355

Collard M, Eeckhaut I, Dehairs F, Dubois P (2014a) Acid-base physiology response to ocean acidification of two ecologically and economically important holothuroids from contrasting habitats, *Holothuria scabra* and *Holothuria parva*. Environ Sci Poll Res 21:13602–13614

Collard M, Srey A, Dehairs F, Dubois P (2014b) Euechinoidea and Cidaroidea respond differently to ocean acidification. Comp Biochem Physiol A 174:45–55

Crawford BJ (1990) Changes in the arrangement of the extracellular matrix, larval shape, and mesenchyme cell migration during asteroid larval development. J Morphol 206:147–161

Decker GL, Morrill JKB, Lennarz WJ (1987) Characterization of sea urchin primary mesenchyme cells and spicules during biomineralization, in vitro. Development 101:297–312

Dow JA (1984) Extremely high pH in biological systems: a model for carbonate transport. Am J Physiol 246:R633–R636

Dupont S, Ortega-Martinez O, Thorndyke MC (2010a) Impact of near-future ocean acidification on echinoderms. Ecotoxicology 19:449–462

Dupont ST, Dorey N, Stumpp M, Melzner F, Thorndyke MC (2012) Long-term and trans-life-cycle effects of exposure to ocean acidification in the green sea urchin *Strongylocentrotus droebachiensis*. Mar Biol. Volume 160, Issue 8, pp 1835–1843. doi: 10.1007/s00227-012-1921-x

Dupont ST, Havenhand JN, Thorndyke W, Peck L, Thorndyke MC (2008) Near-future level of CO2-driven ocean acidification radically affects larval survival and development in the brittlestar Ophiothrix fragilis. MEPS, Volume373, pp285-294

Dupont ST, Lundve B, Thorndyke MC (2010b) Near future ocean acidification increases growth rate of the lecithotrophic larvae and juveniles of the sea star *Crossaster papposus*. J Exp Zool 314B:1–8

Dupont ST, Thorndyke MC (2012) Relationship between CO_2-driven changes in extracellular acid-base balance and cellular immune response in two polar echinoderm species. J Exp Mar Biol Ecol 424–425:32–37

Felton GW, Duffey SS (1991) Reassessment of the role of gut alkalinity and detergency in insect herbivory. J Chem Ecol 17:1821–1836

Gonzalez-Bernat MJ, Lamare M, Barker M (2012) Effects of reduced seawater pH on fertilisation, embryogenesis and larval development in the Antarctic seastar *Odontaster validus*. Polar Biol 36:235–247

Gunaratne HJ, Vacquier VD (2007) Sequence, annotation and developmental expression of the sea urchin Ca^{2+}-ATPase family. Gene 397:67–75

Hasselblatt P, Warth R, Schulz-Baldes A, Greger R, Bleich M (2000) pH regulation in isolated in vitro perfused rat colonic crypts. Pflugers Arch 441:118–124

Holtmann WC, Stumpp M, Gutowska MA, Syré S, Himmerkus N, Melzner F, Bleich M (2013) Maintenance of coelomic fluid pH in sea urchins exposed to elevated CO_2: the role of body cavity epithelia and stereom dissolution. Mar Biol 160:2631–2645

Hu MY, Casties I, Stumpp M, Ortega-Martinez O, Dupont ST (2014) Energy metabolism and regeneration are impaired by seawater acidification in the infaunal brittlestar *Amphiura filiformis*. J Exp Biol 217:2411–2421

Hwang SP, Lennarz WJ (1993) Studies on the cellular pathway involved in assembly of the embryonic sea urchin spicule. Exp Cell Res 205:383–387

Jońezyk E, Klak M, Międzybrodzi R, Górski A (2011) The influence of external factors on bacteriophages-review. Folia Microbiol 56:191–200

Martin S, Richier S, Pedrotti M-L, Dupont S, Castejon C, Gerakis Y, Kerros M-E, Oberhänsli F, Teyssié J-L, Jeffree R, Gattuso J-P (2011) Early development and molecular plasticity in the Mediterranean sea urchin *Paracentrotus lividus* exposed to CO_2-driven acidification. J Exp Biol 214:1357–1368

McConnaughey TA, Whelan JF (1996) Calcification generates protons for nutrient and bicarbonate uptake. Earth Sci Rev 41:95–117

Melzner F, Gutowska MA, Langenbuch M, Dupont S, Lucassen M, Thorndyke MC, Bleich M, Pörtner HO (2009) Physiological basis for high CO_2 tolerance in marine ectothermic animals: pre-adaptation through lifestyle and ontogeny. Biogeosci 6:2313–2331

Miles H, Widdicombe S, Spicer JI, Hall-Spencer JM (2007) Effects of anthropogenic seawater acidification on acid-base balance in the sea urchin *Psammechinus miliaris*. Mar Poll Bull 54:89–96

Moody WJ (1981) The ionic mechanism of intracellular pH regulation in crayfish neurones. J Physiol 316:293–308

Nakano E, Okazaki K, Iwamatsu T (1963) Accumulation of radioactive calcium in the larvae of the sea urchin *Pseudecentrotus depressus*. Biol Bull 125:125–136

Pedersen MF, Hansen PJ (2003) Effects of high pH on the growth and survival of six marine heterotrophic protists. Mar Ecol Prog Ser 260:33–41

Pennington JT, Strathmann RR (1990) Consequences of the calcite skeletons of planktonic echinoderm larvae for orientation, swimming, and shape. Biol Bull 179:121–133

Politi Y, Arad T, Klein E, Weiner S, Addadi L (2004) Sea urchin spine calcite forms via a transient amorphous calcium carbonate phase. Science 306:1161–1164

Politi Y, Metzler RA, Abrecht M, Gilbert B, Wilt FH, Sagi I, Addadi L, Weiner S, Gilbert PUPA (2008) Transformation mechanisms of amorphous calcium carbonate into calcite in the sea urchin larval spicule. Proc Natl Acad Sci U S A 105:17362–17366

Pond DW, Harris RP, Brownlee C (1995) A microinjection technique using a pH-sensitive dye to determine the gut pH of *Calanus helgolandicus*. Mar Biol 123:75–79

Raz S, Hamilton PC, Wilt FH, Weiner S, Addadi L (2003) The transient phase of amorphous calcium carbonate in sea urchin larval spicules: the involvement of proteins and magnesium ions in its formation and stabilization. Adv Funct Mater 13:480–486

Ries JB, Cohen AL, McCorkle DC (2009) Marine calcifiers exhibit mixed responses to CO_2-induced ocean acidification. Geology 37:1131–1134

Robertson JD (1949) Ionic regulation in some marine invertebrates. J Exp Biol 26:182–200

Sikes CS, Okazaki K, Fink RD (1981) Respiratory CO_2 and the supply of inorganic carbon for calcification of sea urchin embryos. Comp Biochem Physiol A 70:285–291

Strathmann RR (1989) Existence and functions of a gel filled primary body cavity in development of echinoderms and hemichordates. Biol Bull 176:25–31

Stumpp M, Hu MY, Casties I, Saborowski R, Bleich M, Melzner F, Dupont S (2013) Digestion in sea urchin larvae impaired under ocean acidification. Nat Climate Change 3:1044–1049

Stumpp M, Hu MY, Melzner F, Gutowska MA, Dorey N, Himmerkus N, Holtmann WC, Dupont ST, Thorndyke MC, Bleich M (2012a) Acidified seawater impacts sea urchin larvae pH regulatory systems relevant for calcification. Proc Natl Acad Sci U S A 109:18192–18197

Stumpp M, Hu MY, Tseng Y-C, Guh YJ, Chen YC, Yu JK, Su YH, Hwang PP (2015) Evolution of extreme stomach pH in bilateria inferred from gastric alkalization mechanisms in basal deuterostomes. Sci Rep 5:1–9

Stumpp M, Trübenbach K, Brennecke D, Hu MY, Melzner F (2012b) Resource allocation and extracellular acid-base status in the sea urchin *Strongylocentrotus droebachiensis* in response to CO_2 induced seawater acidification. Aquat Toxicol 110–111:194–207

Wieczorek H, Putzenlechner M, Zeiske W, Klein U (1991) A vacuolar-type proton pump energizes K^+/H^+ antiport in an animal plasma membrane. J Biol Chem 266:15340–15347

Wilt FH (2002) Biomineralization of the spicules of sea urchin embryos. Zool Sci 19:253–261

Zhu X, Mahairas G, Illies M, Cameron RA, Davidson EH, Ettensohn CA (2001) A large-scale analysis of mRNAs expressed by primary mesenchyme cells of the sea urchin embryo. Development 128: 2615–2627

Acid–Base Regulation and Ammonia Excretion in Cephalopods: An Ontogenetic Overview

Marian Hu and Yung-Che Tseng

© Springer International Publishing Switzerland 2017
D. Weihrauch, M. O'Donnell (eds.), *Acid-Base Balance and Nitrogen Excretion in Invertebrates*,
DOI 10.1007/978-3-319-39617-0_11

11.1 Summary

Among invertebrates cephalopods have evolved a high degree of behavioral and physiological complexity that is comparable to that found in vertebrates. This high-performance lifestyle, including well-developed sensory and locomotory abilities, is directly associated with high energetic costs that require efficient metabolic and regulatory pathways to maintain body homeostasis. Amino acid catabolism is the major energy source in cephalopods with ammonia being the major metabolic end product. Furthermore strong fluctuations in extracellular pH during exercise have led to the presence of ion-regulatory epithelia that highly express a conserved set of ion transporters to mediate extracellular acid–base homeostasis and excretion of nitrogenous waste products.

Cephalopods are oviparous with the embryo developing inside a protective egg capsule that creates an extreme microenvironment in terms of low pH and oxygen concentrations. In order to tolerate these environmental stressors already embryonic stages possess specialized ion-regulatory cells located on the yolk epithelium and integument. This microenvironment of the developing embryo is particularly interesting in the context of CO_2-induced seawater acidification as increases in environmental pCO_2 (hypercapnia) are additive to the already high pCO_2 inside the egg. In several cephalopod species environmental hypercapnia causes a decrease in developmental rates, probably associated to energy reallocations toward compensatory processes (pH regulation) and/or metabolic depression.

Adults of important representatives of coleoid cephalopods, including squid cuttlefish and octopus, utilize branchial epithelia as a major site for gas exchange, acid–base regulation, and excretion. The multiple functions of the cephalopod gill are achieved by a complex folding pattern, generating a large surface area, beneficial for gas exchange and a spatial separation of ion-regulatory epithelia. The concave inner epithelium of the squid and cuttlefish gill expresses important ion transporters and pumps and creates a semitubular structure that shows many parallel features to vertebrate excretory organs. The branchial ion-regulatory machinery responses to CO_2-incuced pH disturbances by an upregulation of ion transporters, followed by a compensation of the extracellular acidosis. This compensatory process is accompanied by elevations in blood HCO_3^- levels and increased proton or proton equivalent excretion. Enhanced NH_4^+ excretion and upregulation of the Rhesus protein during acid–base disturbances suggest that excretion of NH_4^+ as an proton equivalent and acid–base regulation are linked. Based on morphological, functional, and molecular studies, an acid-trapping mechanism for NH_4^+ is proposed in the semitubular environment of the decapod cephalopod gill. Despite effective acid–base-regulatory capacities, cephalopods respond with different sensitivities toward sea water acidification. While sluggish species like cuttlefish are more tolerant, active pelagic squids seem to be more vulnerable during prolonged exposure to strong (pH 7.4) acidification levels. Energetic limitations are discussed as a potential factor for higher sensitivities toward acid–base disturbances in fast-swimming open water squids. Although cuttlefish were demonstrated to be relatively tolerant during long-term exposure to acidified conditions, chronic elevations in blood HCO_3^- levels are probably the cause for hypercalcification of the internal cuttlebone. This internalized molluscan shell allows the animal to control its buoyancy and changes in density may affect swimming behavior and energetics. This chapter summarizes the recent advances in understanding transepithelial acid–base regulatory and excretory mechanisms in cephalopods and highlights the importance to better understand these physiological aspects in an environmental context.

11.2 Cephalopod Biology: An Introduction to Extracellular pH Regulation and Excretion

Cephalopods belong to the phylum Mollusca and evolved during the Cambrian explosion 500 million years ago (O'Dor and Webber 1986; Smith and Caron 2010). Cephalopods are classified into subgroups including extinct Belemnites, Ammonites, and extant subclasses like the Coleoidea and Nautiloidea. While the Nautiloidea are represented by *Nautilus* spp. and *Allonautilus* spp., the coleoid cephalopods can be divided into octopuses, squid, and cuttlefish. Cephalopods, and coleoid cephalopods in particular, can be characterized by a high degree of mobility and vertebrate-like sensory abilities which are believed to be a result from the competition with fish for similar resources in the marine environment. For example, cephalopods exhibit a high degree of cephalization and a complex neural system with very efficient sensory organs such as lens eyes, chemoreceptors, and balance receptors (Budelmann et al. 1997; Hu et al. 2009). However, in contrast to fish, cephalopods, including squid, cuttlefish, and octopods, use jet propulsion to generate high swimming speeds. In comparison to the undulatory swimming movements of fish, this less efficient swimming mode is associated with high energetic costs which constrain energetic capacities in active pelagic cephalopods (Wells and O'Dor 1991; O'Dor 2002). Another physiological feature that sets energetic limitations in cephalopods is found in oxygen transport efficiencies of the extracellular oxygen transporting pigment, hemocyanin. Despite a considerable evolutionary refinement of oxygen transport by hemocyanins within the cephalopoda, this respiratory pigment can only carry about half the oxygen of the cellular hemoglobin of vertebrates (Brix et al. 1989). Thus, in order to optimize the transport efficiency of this pigment, cephalopod hemocyanins usually have a large Bohr effect increasing the pigments' pH sensitivity (Pörtner 1990). In order to protect highly pH-sensitive gas transport mechanisms, cephalopods, including squid and cuttlefish, have evolved efficient mechanisms to regulate extracellular acid–base homeostasis (Pörtner 1994; Pörtner and Zielinski 1998). Despite their low osmo-regulatory abilities which restrict cephalopods to the marine environment, these organisms have evolved vertebrate-like epithelia controlling gas exchange and extracellular pH homeostasis and excretion of nitrogenous waste products (Potts 1965; Wells and Wells 1982; Wells 1990; Hu et al. 2015). A well-developed intra- and extracellular acid–base regulatory machinery is particularly important for active cephalopod species as jetting and fast swimming can lead to strong CO_2-induced temporal acid–base disturbances (Pörtner et al. 1991).

Active pelagic squids have the highest metabolic rates among all marine species due to their less efficient swimming mode, and most cephalopods have a relatively short lifespan usually not exceeding 1–2 years (Llpiński 2010). Accordingly, cephalopods have very high growth rates and can reach body weights of several kilograms within 1 year. This "live fast and die young" lifestyle is predominantly fueled through protein metabolism, and ammonia is the main end product (Boucher-Rodoni and Mangold 1988, 1989, 1994). In principle, ammonia in its uncharged gaseous form as NH_3 can passively diffuse across biological membranes, whereas the charged ionic form, NH_4^+, cannot cross cell membranes, and transport of this molecule requires active transport mechanisms that facilitate the passage across biological membranes (Wright and Wood 2009). At physiological pH of approximately 7.2–7.8, 95–99 % of the total ammonium occurs in the hydrated form as NH_4^+. High concentrations of ammonia are toxic for organisms as they can have severe detrimental effects on the central nervous system and can lead to intra- as well as extracellular acid–base disturbances (Albrecht 2007). Studies on ammonia tolerance demonstrated lethal concentrations 50 (LC50) values for ammonia in the μM range for most aquatic organisms including fish, crustaceans, and cephalopods (Miller et al. 1990; Randall and Tsui 2002; Camargo and Alonso 2006). Pelagic squids were demonstrated to be particularly sensitive to elevated seawater NH_4^+ concentrations >10 µM (Hanlon 1990). Accordingly, besides substantial acid–base regulatory mechanisms, cephalopods have also evolved potent ammonia excretion pathways to control intra- and extracellular NH_3/NH_4^+ homeostasis. Some mid-water cephalopods (e.g., Cranchiidae) store NH_4^+ in the mM range in coelomic chambers, while other ammoniacal species store it in specialized vacuoles within muscle tissues to support neutral buoyancy (Seibel et al. 2004). Although exchange of Na^+ against NH_4^+ can serve to improve neutral buoyancy, most shallow-water cephalopods actively secrete ammonia via branchial epithelia (Potts 1965; Schipp et al. 1979; Boucher-Rodoni and Mangold 1988, 1994). In cephalopods, the paired ctenidia (gills), branchial heart appendages, and renal appendages represent the major excretory organs which are connected by blood vessels to create a directed blood flow. Blood returning from the body through the vena cava enters the lateral vena cava and is distributed to the renal appendages from where the urine is secreted into the renal sac. After passage of the renal appendages, the blood flow reaches the branchial heart by passing the branchial heart appendages. From here the blood is pumped through the gill via the first-order afferent vessels and leaves the gill via the first-order efferent vessels from where it enters the systemic heart (for a schematic overview, see Potts 1965). From here the oxygenated blood is pumped back into the body. While the gills are mainly involved in epithelial ion exchange processes and gas exchange, the renal appendages and branchial heart appendages are mainly involved in urine formation (Schipp and von Boletzky 1976; Donaubauer 1981). The urine of most cephalopods was described to be acidic (pH 5–6) and rich in ammonia (NH_4^+), whereas other nitrogenic compounds like nitrate or nitrite were absent (Emanuel and Martin 1956). As cephalopods produce acidic urine, it can be assumed that similar to what is observed in the vertebrate and mammalian kidney, ammonia excretion and pH regulation are linked processes.

11.3   Early Life Stages

11.3.1 The Oviparous Development

Cephalopods have a direct developmental mode which takes place inside a protective egg capsule. Growth and development is fueled through maternally provided yolk reserves predominantly located in the external yolk sac (Portmann 1926; von Boletzki 1987a). After hatch the juvenile often termed "paralarvae" spends the first days in the marine environment by using up their internal energy reserves before starting to actively capture prey organisms (e.g., crustacean larvae). The embryonic development inside an egg capsule protects the embryo from biotic (e.g., predation) and abiotic stressors (e.g., osmo-protection) by creating a physical barrier between the developing embryo and the external medium (Charmantier and Charmantier-Daures 2001). On the opposite site, this protective capsule creates a diffusion barrier for gases such as O_2, CO_2 and NH_3 but also ions including NH_4^+. This results in steep gradients of O_2, CO_2 and NH_4^+ between the perivitelline fluid (PVF) and the surrounding seawater, which increase during development as a result of increased metabolic rates of the embryo (◘ Fig. 11.1a). Thus cephalopod early life stages are exposed to low-oxygen (hypoxia), high-CO_2 (hypercapnia), and high-ammonia concentrations during their early life history that have led to the evolution of special physiological adaptations. For example, it has been observed that the eggs of many cephalopod species undergo a swelling process during development,

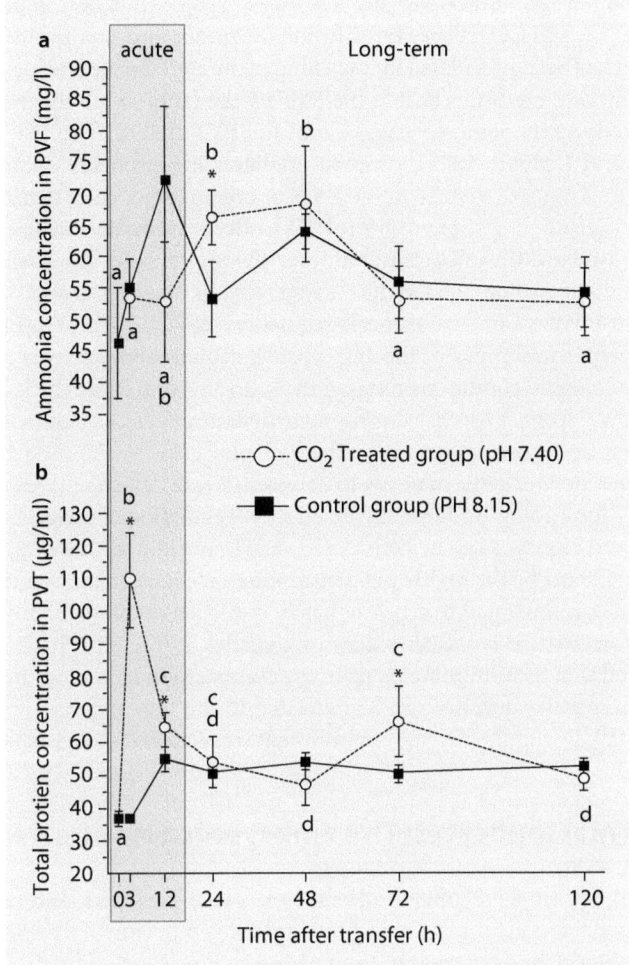

☐ **Fig. 11.1** Total ammonia and protein concentrations in the perivitelline fluid (PVF) of developing squid embryos under control and acidified conditions. NH_4^+ levels in the PVF of developing squid (ca. 50–70 mg l–1) are significantly higher than those of seawater (<0.05 mg l^{-1}) (**a**). NH_4^+ levels remained constant along the developmental period of 120 h and did not significantly differ between CO_2 treatments. Note: NH_4^+ levels in the mM range were detected in the PVF using an Ammonia Assay Kit (Sigma Aldrich) based on the reaction of ammonia with α-ketoglutaric acid (KGA) and NADPH in the presence of L-glutamate dehydrogenase (GIDH). This reaction forms L-glutamate and oxidized nicotinamide adenine dinucleotide phosphate (NADP$^+$). The decrease in absorbance is detected at 340 nm, due to the oxidation of NADPH that is proportional to the ammonia concentration. The measured NH_4^+ levels are surprisingly high (mM range) and we cannot exclude that the reaction is affected by the high concentration of proteins in the PVF. Protein levels determined in the PVF of squid embryos ranged between 40 and 55 mM under control conditions (**b**). Animals kept under acidified conditions had elevated PVF protein concentrations of 110 mmol l^{-1} in the PVF at the time point of 3 h. The remaining time points (except for 72 h) did not show any differences among different CO_2 treatments. Values are presented as mean ± SD. Letters denote significant differences along the incubation period of 120 h and asterisks indicate significant differences among CO_2 treatments (Student's t-test, $p < 0.05$, $n = 6$–8)

leading to an enlarged egg surface area and a reduced capsule thickness in order to maintain sufficient gas (e.g., O_2 and CO_2) fluxes by diffusion (Seymour and Bradford 1994; Cronin and Seymour 2000). It has been speculated that swelling of the egg capsule is achieved by the secretion of osmotically active substances into the PVF by the embryo but the exact mechanisms remain unexplored. As the ionic composition of the PVF is similar to that of the seawater (De Leersnyder and Lemaire 1972), osmotic gradients are probably established by other osmotically active substances such as amino acids or organic acids (e.g., taurine). Preliminary experiments demonstrated that exposure to CO_2-induced seawater acidification increased egg swelling to optimize diffusion permeabilities of the egg capsule. The swelling is accompanied by an accumulation of proteins in the PVF suggesting that amino acids are probably the major component involved in PVF osmotic regulation (❏ Fig. 11.1b). Despite such mechanisms to optimize permeabilities of the egg capsule, cephalopod embryos are still exposed to increasing hypercapnic conditions ranging from 0.13 to 0.41 kPa which correspond to a pH drop in the PVF from 7.7 to 7.2 during their oviparous phase (Gutowska and Melzner 2009). These hypercapnic conditions within the PVF may lead to acid–base disturbances that require acid–base regulatory mechanisms to counteract this developing respiratory acidosis already at an early time point. Additionally, an effective regulation of extracellular pH homeostasis in cephalopod embryos can be considered very important as these early stages already rely on oxygen transport by the highly pH-sensitive extracellular hemocyanins (Decleir et al. 1971). It has been demonstrated that before hatch, cuttlefish embryos highly express a third hemocyanin isoform which is less abundant in juveniles and adults (Strobel et al. 2012). It has been suggested that these modifications in the composition of hemocyanin isoforms and the occurrence of hemocyanin isoform 3 are related to the very different abiotic conditions between the egg (PVF) and the marine environment (seawater) (Strobel et al. 2012).

11.3.2 Epidermal Ionocytes: The Embryonic Site of pH Regulation

The external yolk sac of the cephalopod embryo is well perfused, and pulsatory movements of the yolk drive a current of blood through the embryonic circulatory system that consists of large blood lacunae mainly in the head region (Portmann 1926; von Boletzki 1987b). Immunohistological, ultrastructural, electrophysiological, and gene expression studies demonstrated that the yolk epithelium represents an ion-regulatory site in cephalopod embryos (Hu et al. 2011a, 2013; Bonnaud et al. 2013). Similar to the situation in teleosts, Na^+/K^+-ATPase (NKA)-rich cells that are scattered over the yolk epithelium are equipped with a pH regulatory machinery to mediate extracellular acid–base homeostasis (Hu et al. 2013). Besides the high concentration of basolateral NKA, these epidermal ionocytes found in decapod cephalopods including cuttlefish and squid can be characterized by basolateral V-type H^+-ATPase (VHA) and apical Na^+/H^+-exchanger3 (NHE3). The presence of cation and particularly H^+ transporting proteins suggests a central role of these ionocytes in mediating pH homeostasis. Furthermore, the epidermal ionocytes of cephalopod embryos can be positively stained by Concanavalin A and MitoTracker similar to epidermal ionocytes in teleost larvae (Hu et al. 2011a). High concentrations of mitochondria close to the basolateral membrane were also confirmed by ultrastructural analyses underlining the high-energy turnover of this cell type (Hu et al. 2013). These cellular features are very similar to those found in teleosts and allow the adoption of term "mitochondria-rich cells (MRCs)" in the context of epidermal ionocytes of cephalopod early life stages.

Electrophysiological examinations using H^+-selective microelectrodes demonstrated that cephalopod embryos have outward-directed H^+ gradients on their body surface, indicating a net excretion of protons. While the skin, especially in the head and tentacle region, is characterized by a strong positive proton gradient ($\Delta[H^+]$ 4 µM) from the animal, the cephalopod yolk epithelium is characterized by a comparatively lower outward-directed proton gradient ($\Delta[H^+]$ 0.5 µM). Although the yolk sac organ is characterized by a lower efflux of protons, one has to consider that this organ exhibits a relatively large surface area compared to the head and the mantle. Especially in earlier stages with a decreasing body-surface-to-yolk-surface ratio, the yolk will represent an even more significant site for proton secretion and thus pH regulation. Pharmacological studies using 5-N-ethyl-N-isopropyl-amiloride (EIPA), a specific inhibitor for Na^+/H^+ exchangers, could demonstrate that proton gradients decrease in a dose-dependent manner when embryos were treated with 10 and 1 µM of this compound (Hu et al. 2011a). These findings demonstrated that NHEs are important players in transepithelial H^+ secretion of squid embryos.

In vertebrates, the process of epidermal development, including the differentiation of ionocytes, relies on tightly regulated signaling pathways controlled by the nonneural (bone morphogenetic protein (BMPs)) and the neural ectoderm (including the chordin and noggin) and is evolutionarily conserved (De Robertis and Kuroda 2004; Moreau and Leclerc 2004). In response to stimulation by BMPs, downstream targets of transcription factors, including $\Delta Np63$, are activated in epidermal stem cells, inducing the subsequent process of terminal differentiation (Hsiao et al. 2007; Jänicke et al. 2007). Recently a partial sequence of $\Delta Np63$ transcription factor was cloned from the cuttlefish *Sepia pharaonis* indicating that similar regulatory pathways as those found in vertebrates also exist in lophotrochozoans. The stem cell marker $\Delta Np63$ is expressed in the integument of embryonic stages of *S. pharaonis*, indicating a role of this regulatory pathway in the differentiation of epidermal cells in cephalopods (Hu et al. 2015). Thus, future studies addressing the regulatory mechanisms of ionocyte differentiation in cephalopods will represent an important task to explore the regulatory pathways that control the differentiation of epidermal ionocytes in molluscs.

11.3.3 Effects of Seawater Acidification on Early Developmental Stages: Acid–Base Regulation, Metabolism, and Development

Environmental hypercapnia is a stressor for marine species that has lately received considerable attention in the context of climate change (Pörtner et al. 2004; Fabry et al. 2008; Melzner et al. 2009). Anthropogenic CO_2 emissions are predicted to lead to a rise in surface ocean pCO_2 from 0.04 kPa up to 0.08–0.14 kPa within this century (Cao and Caldeira 2008). The resulting increased hydration of CO_2 to H_2CO_3 and its subsequent dissociation to H^+ and HCO_3^- changes seawater chemistry causing a drop in ocean pH, a process termed ocean acidification (OA). Oftentimes, early life stages have been identified as the most vulnerable link when a species is confronted with seawater acidification (Kroeker et al. 2013). This has also been demonstrated for several cephalopod species, including cuttlefish (*S. officinalis*) and squid (*Sepioteuthis lessoniana*, *Loligo vulgaris*, and *Doryteuthis pealeii*), with larval stages being more sensitive than juveniles and adults (Hu et al. 2011b, 2013; Kaplan et al. 2013; Rosa et al. 2014). CO_2-induced seawater acidification has been demonstrated to elicit reductions in developmental rates (indicated by the embryo's wet mass) in squid and cuttlefish embryonic stages by 16 % (*S. officinalis*) and 28 % (*S. lessoniana*) at an ambient pCO_2 of 0.4 kPa (pH 7.3–7.4) compared to animals kept under control conditions at 0.04 kPa (pH 8.1) (Hu et al. 2011b, 2013). In *S. officinalis* no effects on growth and development were observed for moderate hypercapnia levels of 0.14 kPa pCO_2 or below (Hu et al. 2011b; Dorey et al. 2013). Interestingly, in this species reductions in growth and development were less pronounced using the dry or wet mass but were more evident when using the mantle length of the developing embryo (Hu et al. 2011b). This suggests that elevated pCO_2 affects the size, in particular length of the cuttlefish embryo, but to a lesser extent the total mass. In this respect it needs to be mentioned that the aragonitic cuttlebone contributes approximately 4 % to the total body mass (Gutowska et al. 2010b; Dorey et al. 2013). As hypercalcification (for more details refer to section *Calcification*) of the cuttlebone upon environmental hypercapnia significantly increased the weight of this internal calcified structure (Gutowska et al. 2010b), it is likely that reductions in body mass are partly compensated by a heavier cuttlebone.

On the one hand, the higher sensitivity in cephalopod early life stages may be explained by lower pH regulatory capacities, but on the other hand, the fact that these organisms develop in a naturally acidified environment may exert an additional challenge to early life stages. CO_2 perturbation experiments demonstrated that environmental hypercapnia is additive to the already high pCO_2 inside the egg capsule of cuttlefish and squid embryos to maintain a sufficient diffusion gradient of CO_2 (Hu et al. 2011b, 2013; Dorey et al. 2013; Rosa et al. 2013). The maintenance of this diffusion gradient of approximately 0.2–0.3 kPa is essential in order to excrete CO_2 from the PVF. For example, under control conditions (0.04 kPa pCO_2), the PVF pCO_2 of squid (*S. lessoniana*) and cuttlefish (*S. officinalis*) is about 0.2–0.3 kPa (pH 7.4) in late-stage embryos and increases to 0.4–0.5 kPa (pH 7.1) when exposed to environmental hypercapnia (0.14 kPa) (Hu et al. 2011b, 2013; Dorey et al. 2013; Rosa et al. 2013). As mentioned earlier, besides pH compensatory mechanisms, the cuttlefish embryo seems to be capable of actively controlling the swelling process of the egg enabling a fine-tuning of the egg capsule's diffusion permeability. In this respect two studies (Lacoue-Labarthe et al. 2009; Dorey et al. 2013) have demonstrated an increased swelling of *S. officinalis* eggs exposed to hypercapnic conditions of 0.14 kPa, a process that could improve CO_2 permeability and reduce partial pressure gradients during environmental hypercapnia. In fact, osmolarity of the PVF in squid (*S. lessoniana*) increased upon exposure to acidified conditions (◘ Fig. 11.2a). However, measurements of major ions including Na^+, Cl^+, and K^+ in the PVF of squid exposed to control (pH 8.15) and acidified (pH 7.40) conditions could not detect any substantial differences in ionic composition (◘ Fig. 11.1b–d). As mentioned earlier, proteins secreted by the embryo are probably the most important compound contributing to PVF osmolarity (◘ Fig. 11.1b). However, besides the modulation of permeabilities across the egg capsule, cephalopod embryos can actively compensate for hypercapnia-induced acid–base disturbances by activation of their epithelial acid–base regulatory machinery (Hu et al. 2013). Gene expression analysis performed on whole embryos and yolk epithelia demonstrated that gene candidates relevant for acid–base regulation including *NHE3*, *VHA*, *Rhesus protein (RhP)*, and Na^+/HCO_3^- cotransporter (*NBC*) significantly increased transcript expression levels in response to acidified conditions compared to animals kept under control conditions. This indicates that the acid–base regulatory machinery of squid early life stages responds to increased environmental hypercapnia by potentially increasing regulatory capacities. Using scanning ion selective electrode technique (SIET), it has been demonstrated that during environmental hypercapnia an increase in transcript abundance of acid–base transporters is accompanied by higher H^+ secretion rates in squid (*S. lessoniana*)

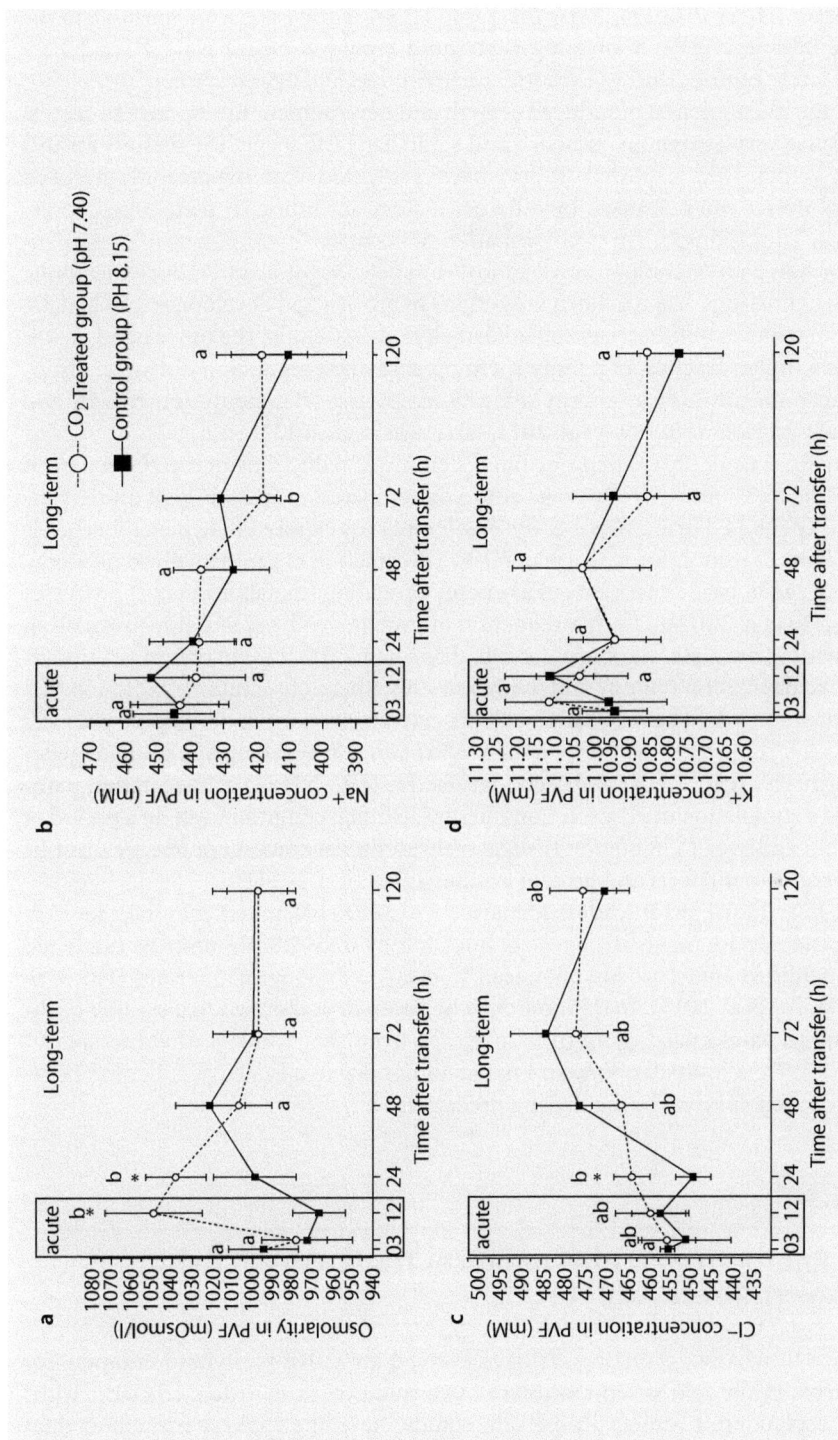

❏ **Fig. 11.2** Concentration of major ions in the perivitelline fluid (PVF) of developing squid embryos under control and acidified conditions. Results indicate an initial increase in PVF osmolarity upon hypercapnic exposure (**a**). The PVF concentrations of ions including Na^+ (**b**), Cl^- (**c**), and K^+ (**d**) do not reveal any strong differences between control and CO_2-treated animals. Only for Cl^- a significant (two-way ANOVA, $p < 0.05$, $n = 6$–8) increase has been measured at the time point of 24 h. Values are presented as mean ± SD. Letters denote significant differences along the incubation period of 120 h and asterisks indicate significant differences among CO_2 treatments (two-way ANOVA, $p < 0.05$, $n = 6$–8)

embryonic stages (Hu et al. 2013). These increased H$^+$ secretion rates were sensitive to the specific NHE-inhibitor EIPA, indicating that squid embryos utilize Na$^+$/H$^+$ exchange pathways to secrete protons during exposure to environmental hypercapnia.

Although the phenomenon of reduced growth and development in response to seawater acidification is very evident in cuttlefish and squid early life stages, the reasons for this phenomenon remain less explored. It has been suggested that hypercapnia-induced developmental delays can be caused by different factors, including (i) metabolic depression, (ii) energy allocations, or (iii) a combination of both.

On the one hand, an uncompensated acidosis has been suggested to induce metabolic depression and, thus, hypercapnia-born reductions in growth and development in aquatic organisms (Langenbuch and Pörtner 2002; Pörtner et al. 2004). On the other hand, it was proposed that a higher fraction of energy spent on acid–base regulation during environmental hypercapnia can lead to energy allocations leaving less energy for growth and developmental processes (Stumpp et al. 2011, 2012; Hu et al. 2013).

A study by Rosa et al. (2013) demonstrated significant reductions in metabolic rate at pH 7.5 and 22 °C in *S. officinalis* late-stage embryos compared to animals kept under control conditions (pH 8.1). Furthermore, *S. officinalis* embryos exposed to hypercapnia levels of 0.4 kPa for 5 weeks responded with a general downregulation of genes involved in energy consuming (e.g., acid–base genes) as well as energy providing (metabolic genes) processes in gill tissues (Hu et al. 2011b). Such a reduction of protein synthesis and downregulation of energy providing processes has been described as a general feature of organisms undergoing metabolic depression (Guppy and Withers 1999). These observations indicated that metabolic depression might be associated with observed reductions in developmental rates in response to environmental hypercapnia in *S. officinalis*. However, upregulation of genes involved in primary- (e.g., VHA) and secondary-active (e.g., NHE3) acid-secretion pathways as well as a stimulation of H$^+$ excretion rates by acidified conditions (along a period of 5 days) in squid embryos (*S. lessoniana*) suggest that a higher fraction of energy must be spend on pH regulatory efforts and protein synthesis (Hu et al. 2013).

Although near-future acidification scenarios (−0.3–0.4 pH units) may not severely affect embryonic development, stronger acidification by−0.6−0.8 pH units as predicted for naturally acidified environments may lead to delayed developmental rates (Hu et al. 2011b, 2013; Dorey et al. 2013). In this context it has been hypothesized that smaller cephalopod hatchlings would need more time to escape from the "window of vulnerability" (Pecl et al. 2004). Thus, a smaller body size in hatchlings due to elevated pCO$_2$ could negatively influence their survival by facilitating predation.

11.4   Adults

11.4.1 The Energetic and Physiological Trade-Offs of Being an Invertebrate Athlete

In convergence to teleosts, cephalopods have evolved an active lifestyle to compete for similar resources in the marine environment. As a trade-off to their less efficient swimming mode, cephalopods, especially pelagic squids, have the highest metabolic rates among marine high-power taxa, including tuna and sharks (Rosa and Seibel 2008). This leads to strong temporal fluctuations in body fluid pCO$_2$ between routine swimming and exercise, which can lead to an acidosis if not actively compensated for (Heisler 1986).

Thus, cephalopods have evolved efficient mechanisms to control extracellular acid–base homeostasis and to protect gas transport via highly pH-sensitive hemocyanins. Moreover, their active lifestyle is exclusively fueled through amino acid metabolism leading to an extensive production of NH_4^+ which needs to be secreted across excretory epithelia. In convergence to fish (Evans et al. 2005) and crustaceans (Henry et al. 2012), the cephalopod gill constitutes the organ with the largest surface area being in direct exchange with the surrounding seawater and thus covers functions of gas exchange, pH regulation, and excretion.

11.4.2 Structure and Function of the Cephalopod Gill

Similar to fish (Evans et al. 2005) and decapod crustaceans (Henry et al. 2012), the cephalopod gill is also one of the most important organs for ion-regulatory processes in juveniles and adults (Evans et al. 2005; Henry et al. 2012; Hu et al. 2015). During larval development of squid and cuttlefish that takes place inside an protective egg capsule, rudimentary gill structures occur at stage 20 (for staging system refer to (Arnold 1965)) and significantly differentiate over the course of embryonic development as well as after hatching (Arnold 1965; Schipp et al. 1979; Hu et al. 2010). These studies demonstrated that the cephalopod gill changes morphologically within several days after hatching to become a highly folded epithelium with a high density of vesicles and mitochondria and a high concentration of a cytosolic type of carbonic anhydrase (cCA) and NKA (Schipp et al. 1979; Hu et al. 2010). Furthermore, determinations of enzyme activities and immunohistological stainings demonstrated that NKA expression and activity steeply increase toward hatching in late-stage (29–30) embryos. This differentiation indicates that gas exchange and ion-regulatory capacity might be fully activated only after leaving the protective egg capsule (Hu et al. 2011a).

The cephalopod gill is a highly folded epithelium consisting of first- to third-order lamellae and vessels. The third-order gill lamellae consist of two epithelial layers lining a blood sinus. The lamellae itself is folded and creates inner (concave) and outer (convex) foldings which cover different functions including gas exchange and active ion transport. While the outer convex epithelium is thin and mainly involved in gas exchange, the inner concave epithelium is thicker and rich in mitochondria and has high concentrations of proteins that are involved in transepithelial ion transport. In squid and cuttlefish, this concave, inner epithelium creates a semitubular space. The gills are attached to the mantle wall laterally and are expanded during inhalation, when the mantle expands. The ventilatory current goes into the center of the gill and passes the second order gill lamellae in countercurrent to the blood flow with oxygen extraction efficiencies ranging between 50 and 80 % (Wells and Wells 1982; Wells 1990).

Besides its function in regulating ion (e.g., acid–base) homeostasis, the gill has been also identified as an important site for the elimination of nitrogenous waste products. For example, despite the existence of other potential excretory organs such as branchial hearts or renal appendages (equivalent to the vertebrate kidneys), gills were proposed to represent the most important site for NH_4^+ excretion in cephalopods (Potts 1965). High concentrations of Na^+/K^+-ATPase localized in basolateral membranes of branchial epithelia of cephalopods underline its predominant role in active ion transport but also excretory functions as this enzyme can also accept NH_4^+ as a substrate in competition with K^+ ions (Schipp et al. 1979; Hu et al. 2011b).

11.4.3 Extracellular Acid–Base Regulation in Cephalopods

In response to environmental acidification, cephalopods can quickly restore blood pH involving active regulatory mechanisms. Besides the accumulation of HCO_3^- to buffer the excess of protons, recent studies demonstrated that this compensation reaction is accompanied by an increased excretion of protons as well (Hu et al. 2011a, 2013). Furthermore, acidified conditions stimulate the expression of VHA and NHE3 in branchial epithelia underlining the importance of H^+ secretion mechanisms during the compensation of an extracellular acidosis (Hu et al. 2014). The predominant expression of acid–base transporters in branchial epithelia reveals this organ to represent a major site of acid–base regulation.

Cephalopods have branchial ion-regulatory epithelia, which are equipped with ion transporters beneficial for coping with acid–base disturbances (for branchial acid–base regulation, see also ▶ Chap. 6). Using degenerate primers designed against highly conserved regions of the NKA, VHA, NHE3, carbonic anhydrase (CA), and Na^+/HCO_3^- cotransporters (NBC_e), partial DNA sequences encoding these enzymes were identified in squid and cuttlefish (Hu et al. 2011b, 2014). In situ hybridization and immunohistochemical methods with antibodies specifically designed against acid–base transporters (NKA, VHA, NHE3, and Rhesus protein (RhP)) of squid could reveal the subcellular localization of these enzymes in cephalopod ion-regulatory epithelia (Hu et al. 2011b, 2014). NHE3, for example, which is an essential player for proton secretion in vertebrates (Wagner et al. 2003; Yan et al. 2007; Watanabe et al. 2008; Lin et al. 2012), is also expressed in apical membranes of epidermal ionocytes and gill epithelia of squid (*S. lessoniana*) and cuttlefish (*S. officinalis*) (Hu et al. 2011a, 2013, 2014). Proton selective electrodes in combination with the application of the specific inhibitor for NHE proteins, EIPA, demonstrated that at least 60 % of H^+ secretion across the surface of epidermal ionocytes of embryonic squids is mediated through Na^+/H^+ antiport (Hu et al. 2011a). The involvement of NHEs instead of VHA to mediate apical proton secretion in marine species is thermodynamically favored by the strong Na^+ gradient between cytosol (30 mM) and seawater (470 mM) (Robertson 1949; Potts 1994). It was hypothesized that NHE-based proton secretion may represent a fundamental pathway in marine organisms which however needs further confirmation by testing more marine species including crustaceans, echinoderms, annelids, and mollusks. Besides the secretion of protons, cephalopods including squid and cuttlefish were demonstrated to actively accumulate extracellular HCO_3^- to stabilize blood pH during an acidosis (Gutowska et al. 2010a; Hu et al. 2014). In vertebrate systems, the basolateral resorption of HCO_3^- into the blood is achieved by various transporters of the SLC4 family, including anion exchangers, and Na^+/HCO_3^- cotransporters (Romero et al. 2004; Lee et al. 2011). In this context, an electrogenic Na^+/HCO_3^- cotransporter (NBC_e) and a Na^+-dependent Cl^-/HCO_3^- exchanger (NDCBE) were cloned from the squid giant axon that are two potential candidates for the transport of HCO_3^- ions in cephalopod ion-regulatory epithelia (Virkki et al. 2002; Piermarini et al. 2006). While NDCBE had low expression levels in branchial epithelia but higher RNA levels in neurons, the electrogenic NBC was highly expressed in gill epithelia and expression of this gene was stimulated upon hypercapnic exposure (Piermarini et al. 2006; Hu et al. 2011b, 2014). Another important enzyme group in the context of acid–base balance is CAs that catalyze the formation of H_2CO_3 from water and CO_2 and its subsequent dissociation to H^+ and HCO_3^- (Lindskog et al. 1971). Besides their function in vertebrate systems (Evans et al. 2005; Hwang and Lee 2007; Gilmour and Perry 2009), CAs were also demonstrated to play an important role in the acid–base regulatory mechanisms of aquatic invertebrates (e.g., crustaceans) (Henry 1984, 1988). To date, studies addressing the role of CAs in mediating acid–base homeostasis in cephalopods are still scarce. Earlier studies by Schipp and colleagues (Schipp et al. 1979) located a cytosolic carbonic anhydrase (cCA) in the transporting epithelium of the cuttlefish gill using histochemical methods. Using degenerate primers, a partial sequence of a cCA isoform was cloned from cuttlefish and squid gills (Hu et al. 2011b). This gene also responded with increased transcript levels in gill epithelia of squid (*S. lessoniana*) upon acute (<24 h) exposure to moderate hypercapnia ($pCO_2 = 0.14$ kPa, pH 7.7) (Hu et al. 2014). According to results of this study, a first working model for the cephalopod branchial acid–base regulatory machinery during hypercapnic acclimation was suggested (Hu et al. 2014). As shown by Hu et al. (2014), in response to a hypercapnia-induced acidosis, the drop in blood pH is accompanied by an increase in pCO_2 in order to maintain a sufficient outward-directed CO_2 gradient. CO_2 can easily diffuse across biological membranes, and its subsequent hydration in the cytosol, catalyzed by a cytosolic CA, leads to the formation of HCO_3^- ions and protons. The protons generated in this process are exported across the apical membrane via NHE3, while the bicarbonate absorbed across the basolateral membrane, to buffer the excess of protons. The secretion of protons via apical NHE3 is driven by the high environmental [Na^+], while basolateral HCO_3^- absorption has been suggested to be achieved by an electrogenic Na^+/HCO_3^- cotransporter driven by the intracellular formation of HCO_3^- catalyzed by CA and a negative membrane potential created by the basolateral NKA.

11.4.4 Branchial Ammonia Excretion Mechanisms

Gill epithelia of cephalopods including cuttlefish and octopods were also suggested to represent the major site for the excretion of nitrogenous waste products (Potts 1965; Schipp et al. 1979). An efficient NH_4^+ regulatory machinery is probably very important for cephalopods as these animals are highly ammonotelic, exclusively fueling their active lifestyle through protein metabolism (Boucher-Rodoni and Mangold 1988, 1994). In vertebrate systems, Rhesus glycoproteins including Rhbg and Rhcg were identified as central players to facilitate the export of ammonia (NH_3) in concert with proton secretion mechanisms leading to a NH_4^+ trapping mechanism in the acidified boundary layer of excretory cells (Shih et al. 2008; Wright and Wood 2009; Gruswitz et al. 2010; Larsen et al. 2014). Just lately Rhesus proteins (RhP) were also found and characterized in a range of invertebrates including crustaceans and insects (Weihrauch 2006; Weihrauch et al. 2009; Martin et al. 2011). Also in cephalopods one RhP has been cloned, which has the highest degree of identity to RhPs from the decapod crustaceans *Carcinus maenas* (50 %) (AAK50057.2) and *Metacarcinus magister* (56 %) (AEA41159.1). Phylogenetic analyses based on amino acid identities demonstrated that the invertebrate specific RhP1 clade clearly separates from vertebrate Rh proteins including RhBG, RhCG, RhAG clades, as well as RhP2 and Rh30 clades (Hu et al. 2013).

In various cephalopod species, the largest fraction of ammonia is produced by amino acid metabolism. Some mid-water cephalopods (e.g., cranchiidae) store NH_4^+ in the mM range in coelomic chambers (e.g., cranchiidae), while other ammoniacal species store it in specialized vacuoles within muscle tissues to support neutral buoyancy via unexplored transport mechanisms (Seibel et al. 2004). Although exchange of Na^+ against NH_4^+ in specialized body compartments can serve to improve neutral buoyancy, most shallow-water cephalopods actively secrete ammonia via branchial epithelia (Potts 1965;

Boucher-Rodoni and Mangold 1994). For shallow-water octopods, it was hypothesized that ammonia is excreted as ammonia (NH_3) accompanied with an excretion of protons to form the ammonium ion (NH_4^+) (Potts 1965). The apical localization of RhP in cephalopod gill epithelia (Hu et al. 2014) supports earlier studies suggesting that the cephalopod gill is the major site for NH_4^+ excretion. Moreover, colocalization of NHE3 and RhP and a coordinated upregulation of these genes during a hypercapnic challenge strongly suggests an interplay of these two proteins, supporting the hypothesis of an acid-trapping mechanism of NH_4^+ in the semitubular space of the third-order gill lamellae similar to that demonstrated for diverse vertebrate excretory organs including skin, gill, and kidneys (Wright and Wood 2009; Liu et al. 2013; Weiner and Verlander 2013). For example, in acid-secreting intercalated cells of the collecting duct of the mammalian kidney, acid secretion is coupled to the export of NH_3 that is trapped as ammonium ions (NH_4^+) in the acidic urine (Gruswitz et al. 2010). Here, the basolateral localization of Rhbg and Rhcg is believed to facilitate the entry of the de-protonized ammonia (NH_3) from the blood into the cell where the majority of NH_3 is protonized to the ammonium ion at a physiological pH of approximately 7.2. Apical excretion and subsequent trapping of NH_4^+ in the urinary space of the collecting duct are achieved by apical Rhcg and proton pumps like the VHA and the H^+/K^+-ATPase (Wagner et al. 2011). Interestingly, the gills of coleoid cephalopods show striking morphological and functional similarities to the collecting duct of the mammalian kidney. The third-order lamellae of the cephalopod gill form a semitubular structure creating a luminal space into which NH_4^+ is secreted by the interplay of RhP and NHE3 (Hu et al. 2014). At the basolateral membrane, the entry of the ammonium ion (NH_4^+) is achieved by the NKA that can also accept NH_4^+ as a substrate (Skou 1957). Here it remains to be investigated if the microenvironment within the semitubular space of the third-order lamellae is also acidified, similar to the mammalian collecting duct. Accordingly, a coupling of NH_3/NH_4^+ excretion and pH regulatory mechanism can be proposed for cephalopods, which also represents a fundamental pathway of acid–base regulation in vertebrate systems (Boron 2006; Wu et al. 2010).

Recent studies using perfused gills of *Octopus vulgaris* opened new venues to study epithelial transport in cephalopod gill epithelia. These studies provided the first direct evidence for the coupling of NH_4^+ excretion and acid–base regulation in the cephalopod gill. A simulated acidosis by lowering the pH by addition of HCL in perfusion solutions to pH 7.2 (bath solution same ionic composition as perfusion solution, pH 8.1) led to an increase in apical NH_4^+ release into the bath by the gill, accompanied by a restoration of the perfusate pH to 7.6 after gill passage (Hu et al., personal communication). These experiments further demonstrated that the perfused cephalopod gill is not only capable of active NH_4^+ secretion but also involved in maintaining blood NH_4^+ levels at a concentration of approximately 250–300 µM. For example, when the octopus gill was perfused with a 150 µM NH_4^+ solution, the perfusate after gill passage had elevated NH_4^+ concentration of approximately 260 µM. In contrast to terrestrial vertebrates that have blood NH_4^+ concentrations of approximately 30 µM (Meijer et al. 1990), aquatic species are generally characterized by extracellular NH_4^+ concentrations in the range of 100–600 µM (Potts 1965; Wright 1995; Fehsenfeld and Weihrauch 2013). In the light of these, first insights demonstrated a direct link between pH regulation and NH_4^+ excretion in a marine mollusk. This is particularly important, as ammonia-based proton secretion mechanisms are found in a wide range of aquatic organisms and may represent an evolutionary ancient mechanism to cope with CO_2 induced acid–base disturbances.

11.4.5 Effects of CO_2-Induced Seawater Acidification on Cephalopods

Energetics and Acid–Base Regulation

Environmental hypercapnia alters positive diffusion gradients of CO_2 from the animal to the seawater and consequently affects the acid–base physiology of all water breathing animals, as intra- and extracellular CO_2 concentrations must increase to maintain a sufficient diffusion gradient to excrete metabolic CO_2 (Melzner et al. 2009). Among a range of marine species, some have been identified as more sensitive (e.g., less active calcifying species such as corals or echinoderms), whereas others (many active species such as adult fish and cephalopods (Fabry et al. 2008; Gutowska et al. 2008; Dupont et al. 2010)) can tolerate relatively high CO_2 concentrations over long exposure times. It was hypothesized that the degree of sensitivity is directly linked to a species' ability to compensate for acid–base disturbances by actively accumulating bicarbonate and eliminating protons from their body fluids (Melzner et al. 2009).

Initially, different hypotheses regarding the tolerance of cephalopods towards hypercapnia have been proposed based on their physiological features. On the one hand, high Bohr coefficients of the cephalopod hemocyanins were suggested to be a critical physiological factor that would make cephalopods particularly sensitive to environmental pH fluctuations (Pörtner et al. 2004). On the other hand, substantial extracellular acid–base regulatory capabilities as found in most cephalopod species represent a common feature that was hypothesized to make ectothermic marine animals robust to seawater acidification (Melzner et al. 2009). However, metabolic data from the cuttlefish *S. officinalis* (acute, 24 h exposure to 0.6 kPa pCO_2) and the squids *Dosidicus gigas* (acute, 1 h exposure to 0.1 kPa pCO_2) and *S. lessoniana* (acute; 20 h medium term; 168 h 0.16 and 0.41 kPa pCO_2) indicated that these animals respond very differently towards environmental hypercapnia (Gutowska et al. 2008; Rosa and Seibel 2008; Hu et al. 2014). Although these studies partly used unrealistically high pCO_2 levels that are beyond those predicted for the coming century (usually < 0.1 kPa), they indicated a higher sensitivity in active pelagic squids compared to cuttlefish that have a benthic lifestyle. Accordingly, it has been suggested that different lifestyles and energetic limitations can limit the ability to mobilize energy resources to fuel acid–base compensatory processes in different cephalopod species (Hu et al. 2014). Different physiological responses towards seawater acidification among different cephalopod species are also reflected in extracellular acid–base regulatory features. *S. officinalis* has control blood HCO_3^- levels of 3.4 mM and a partial compensation of pHe via HCO_3^- accumulation during exposure to environmental hypercapnia (0.6 kPa pCO_2; pH 7.1) (Gutowska et al. 2010a). In the same study, it was suggested that a partial compensation of 0.2 pH units below control levels is sufficient to achieve sufficient gas transport via the blood pigment hemocyanin under acidified conditions in this less active cephalopod species. However, for the squid *S. lessoniana*, a full compensation of extracellular pH was evident after 20 h exposure to acidified conditions (pH 7.3) (Hu et al. 2014). It has been suggested that more sluggish cephalopod species like cuttlefish and octopods may not rely on pH-dependent oxygen transport to the same extent as more active pelagic squid species (Pörtner 1990, 1994). If this is the case, different pH sensitivities of hemocyanins may represent a key factor that can lead to differential sensitivities to seawater acidification among cephalopod species. Interestingly, blood HCO_3^- levels in the cuttlefish *S. officinalis* increased by approximately

7.5 mM within 48 h in response to 0.6 kPa CO_2 exposure, whereas those in squid were only increased by 2 mM when exposed to a similar acidification level (Gutowska et al. 2010a; Hu et al. 2014). This indicates that different pH buffering/regulatory mechanisms, including nonbicarbonate buffering and H^+ extrusion mechanisms, exist among cephalopods. Nonbicarbonate buffer values determined for squid species ranged between 4.7 mmol l^{-1} pH unit ^{1} (*S. lessoniana*), 5 mmol l^{-1} pH unit^{-1} (*Illex illecebrosus*), and 5.8 mmol l^{-1} pH unit^{-1} (*Loligo pealei*), whereas those determined for cuttlefish, *S. officinalis*, were 10 mmol l^{-1} pH unit^{-1} (Lykkeboe and Johansen 1982; Pörtner 1990) indicating an even lower HCO_3^--independent buffering potential in squid species. According to these observations, it can be suggested that control of extracellular pH in squids is dominated by efficient H^+ extrusion mechanisms and to a lesser extent to HCO_3^- buffering mechanisms. This stands in contrast to the very pronounced HCO_3^- buffer reaction in cuttlefish, which also depends on H^+ secretion to a certain degree, but allows for a energetically less-expensive long-term acclimation (maintenance of elevated blood HCO_3^- levels *vs.* continuous elimination of protons) to acidified conditions.

The evolution of these different physiological features may be a result from the long-term selection pressure exerted on these organisms to occupy different (e.g., benthic *vs.* pelagic) ecological niches in the marine environment. In contrast to most pelagic squids, the cuttlefish *S. officinalis* is preadapted to environmental fluctuations in environmental pH as well as moderate hypercapnia which naturally occur in many coastal habitats (Feely et al. 2008; Melzner et al. 2013). However, more studies using cephalopod species with different lifestyles and energetic requirements are needed in order to provide more conclusive answers regarding the energetics of acid–base regulation in cephalopods that may strongly determine the degree of sensitivity towards CO_2-induced sea water acidification.

Calcification

Although many cephalopods including squid and octopuses have lost their molluscan shell, cuttlefish like *S. officinalis* have retained and internalized their shell, which forms the cuttlebone in the Sepiidae. This gas-filled cuttlebone serves these animals to control their buoyancy by actively pumping water in or out of the aragonitic chambers through the generation of osmotic gradients (Denton and Gilpin-Brown 1961). These features enable *S. officinalis* to have an energetically favorable swimming mode compared to most pelagic squids which are negatively buoyant, but restricts their life to shallow, benthic habitats (O'Dor 2002). This gas-filled chamber limits the vertical distribution range of sepiids due to the potential implosion of the cuttlebone when exceeding the tolerable hydrostatic pressure (Sherrard 2000). Several studies conducted in the context of ocean acidification demonstrated that calcification of the cuttlebone was impacted by environmental hypercapnia. ^{45}Ca uptake increased and lamellar spacing decreased in response to environmental hypercapnia leading to a higher density of the chambered aragonitic cuttlebone in CO_2-treated animals (Gutowska et al. 2010b; Dorey et al. 2013). The phenomenon of hypercalcification of internal $CaCO_3$ structures in response to environmental hypercapnia has also been observed for fish otoliths, cuttlefish statoliths, and crustacean carapaces (Checkley et al. 2009; Lacoue-Labarthe et al. 2009; Ries et al. 2009) and was speculated to be the consequence of a chronic elevation of HCO_3^- levels in body fluids (Checkley et al. 2009; Gutowska et al. 2010b). The aragonitic statoliths of cephalopods are paired organs located in the statocysts, a structure which is filled with endolymph and located inside the head cartilage. The statocysts serve as gravitation, balance, and acceleration receptors by using the denser statolith attached to mechano-sensitive hair cells that transduce nanometer scale displacements into neuronal signal (Budelmann 1988). It has been demonstrated that exposure to environmentally relevant pCO_2 elevations induced malformed statolith with a significantly reduced surface area and altered crystal structure in squid *(Doryteuthis pealeii)* paralarvae (Kaplan et al. 2013). The reasons for these malformations remain unclear, but may also be related to chronically altered acid–base equilibria in cephalopod early life stages.

Abnormally formed cuttlebones and statoliths may negatively influence swimming behavior, prey capture, and metabolism. Preliminary studies demonstrated that embryonic development under hypercapnic conditions (63 days, 0.4 kPa CO_2) evoked a reduced ability to catch prey and to launch attacks against prey organisms (Maneja et al. 2011). Thus, future studies addressing the complex behavioral patterns of cephalopods raised under different seawater acidification scenarios will represent a fruitful research field in order to clarify the question regarding hypercalcification-born behavioral defects under acidified conditions.

11.5   Conclusion and Future Research Perspectives

Recent advances in understanding transepithelial ion transport in epithelia of cephalopods demonstrated that these organisms evolved sophisticated pH regulatory mechanisms to control extracellular pH homeostasis. However, although acid–base regulatory abilities were investigated in great detail during the last 40 years, the mechanistic basis for acid–base regulation remained relatively unclear. Substantial acid–base regulatory capabilities as found in other active marine ectotherms including fish and crustaceans are also a physiological feature of many cephalopod species. In convergence to fish and crustaceans, the gill of cephalopods has been identified as a central organ mediating acid–base balance. However, excretory organs including renal appendages and branchial heart appendages need to be considered as additional sites that are contributing to the overall acid–base homeostasis in these animals. Accordingly, it can be hypothesized that there must be a certain coordination of excretory and pH regulatory functions between different organs that are probably controlled by endocrine mechanisms. Thus, future studies using recently established perfusion techniques in different cephalopod excretory organs in combination with recently identified hormones (e.g., octapressin) will help to better understand the regulatory pathways that control excretion and pH homeostasis in cephalopods. Furthermore, preliminary studies indicated that similar to the situation in vertebrates also cephalopods have evolved sophisticated regulatory pathways that control the differentiation of epidermal cells, including ionocytes. Future studies using histological and functional markers for epidermal ionocytes will allow a precise and efficient identification of key regulatory pathways that control cellular differentiation in cephalopods. The comparison of evolutionary distant groups, such as lophotrochozoans, arthropods, basal deuterostomes, and chordates, will create an exciting research direction to better understand the evolution of regulatory pathways responsible for the proper differentiation of ion-regulatory epithelia.

Even though many evolutionary conserved features similar to ion-regulatory epithelia in vertebrates and mammals are also present in cephalopods, certain differences do exist largely due to an exclusively marine and highly ammonotelic lifestyle of cephalopods. For example, the predominant role of NHEs in apical H^+ secretion can be regarded as an adaptation to marine conditions as the utilization of Na^+/H^+ exchange mechanisms is thermodynamically favored by the strong Na^+ gradient between cytosol (30 mM) and seawater (470 mM) (◘ Fig. 11.3). Interestingly, the V-type H^+-ATPase (VHA) as a predominant player of H^+ secretion in apical membranes of regulatory epithelia in vertebrate systems seems to have an alternative function in acid–base regulatory processes in cephalopods.

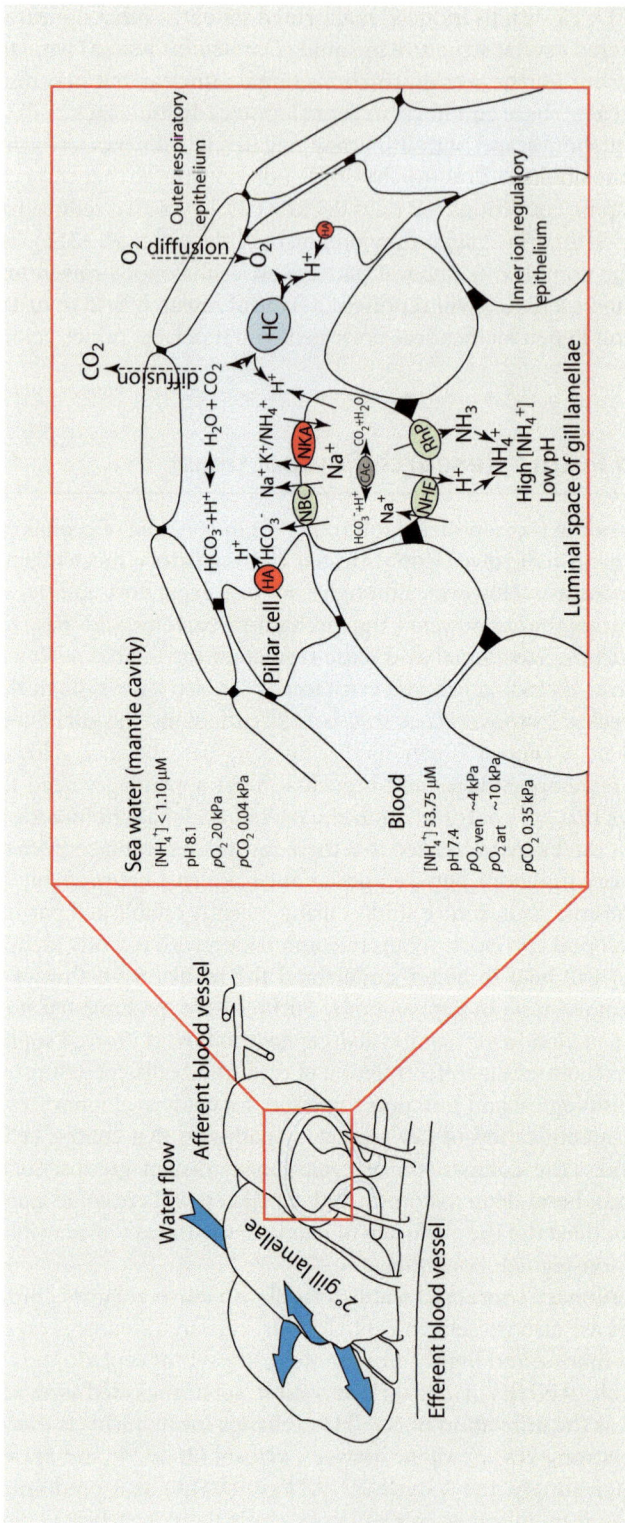

Fig. 11.3 Schematic illustration of second- and third-order lamellae of the cephalopod gill and a hypothetical model for branchial acid–base regulation and ammonia excretion. The gill lamellae consist of two epithelial layers lining a blood sinus. The third-order gill lamellae create a highly folded epithelium with concave and convex areas. While the outer convex epithelia are thin and mainly involved in gas exchange, the inner concave epithelia are rich in mitochondria and ion transporters. Accordingly the duals, respiratory and ion-regulatory function of the gill, are specially separated via the two types of epithelia. The concave parts of the third-order gill lamellae form a semitubular structure which has been hypothesized to represent the space into which H[+] and NH₄[+] are excreted via apical NHE3 and RhP. The electrochemical gradient is generated by the NKA located in basolateral membranes. The VHA has been immunohistochemically localized in pillar cells and basolateral membranes and is hypothesized to be involved in facilitating CO₂ release through local acidification (Modified after (Hu et al. 2014))

The VHA in ion-regulatory epithelia (gills and yolk epithelium) of cephalopods can be found in the basolateral membrane and is additionally expressed in pillar cells spanning through the blood space of the gill. Branchial VHA mRNA and protein expression is stimulated by acidified conditions, indicating that despite its unusual basolateral localization, it serves the animals to cope with acid–base disturbances. The exact mechanisms associated with the basolateral localization of the VHA remain unclear and represent an interesting and important future task to better understand the manifold functions of the VHA in marine animals.

Similar to the situation in crustaceans, fish, and mammals, Rhesus proteins were identified to be an important component of ion-regulatory epithelia of cephalopods. Although only one primitive form of Rhesus proteins (RhP) has been identified for invertebrates to date (e.g., crustacean (Weihrauch et al. 2004) and cephalopods (Hu et al. 2013)), their involvement in acid–base regulatory processes has been confirmed using CO_2 perturbation experiments (Hu et al. 2013, 2014). It has been suggested that apical RhP work in concert with NHEs to achieve an acid-trapping mechanism for ammonia and thus contribute to a net excretion of acid equivalents. Here acid trapping of NH_3 in branchial epithelia of squid and cuttlefish would be supported by the semitubular third-order gill lamellae which may provide an acidified compartment. In order to provide more definitive answers regarding the mechanisms and compartmentation of acid trapping in branchial epithelia, future studies need to investigate pH gradients within the microstructure of the cephalopod gill. In the context of NH_4^+ trapping, special attention should also be dedicated to excretory mechanisms in cephalopod early life stages as they typically develop in a naturally acidified and NH_4^+ enriched environment. In these early life stages, RhP was found to be upregulated upon exposure to acidified conditions and positive RhP immunoreactivity was detected in epidermal ionocytes scattered over the yolk epithelium of squid (*S. lessoniana*) (◘ Fig. 11.4a). Ammonia excretion in oviparous organisms like cephalopods is particularly interesting as like NH_3/NH_4^+ has to be excreted against a steep concentration gradient (◘ Fig. 11.4b). Finally, the stimulation of RhP during environmental hypercapnia may also be associated to the ability of these channels to facilitate CO_2 diffusion across membranes. Rh proteins were suggested to have a dual CO_2 and NH_3 transport function (Soupene et al. 2002, 2004; Kustu and Inwood 2006; Endeward et al. 2008) and thus may represent key players during acclimation to environmental hypercapnia. Functional expression experiments of RhPs in oocytes in combination with CO_2-sensitive microelectrodes (Hanstein et al. 2001) are an important task in order to functionally characterize these proteins in marine organisms.

Research addressing the effects of CO_2-driven seawater acidification on cephalopods highlights the large difference in terms of sensitivity between ontogenetic stages but also among species. While adults seem to be relatively tolerant towards CO_2-induced seawater acidification, early life stages were demonstrated to be more vulnerable. Low pH, hypercapnia, and high ammonia inside the egg capsule amplified by CO_2-driven seawater acidification are probably the major abiotic stressors that negatively affect the development of early life stages. Here it should be noted that the abiotic conditions of the PVF are controlled by environmental factors as well as the embryo itself. Thus, future studies are needed to better understand the interplay of exogenous (by the environment) *vs.* endogenous (by the embryo) processes that affect the PVF chemistry. In particular, the research on physiological responses of the embryo during exposure to hypercapnia, high-ammonia concentrations, and low pH represent important future tasks to fill gaps in current knowledge. Although the recent advances in understanding the physiological responses of *S. officinalis* and other cephalopod species towards environmental hypercapnia shed light on some major questions, even more remains to be discovered.

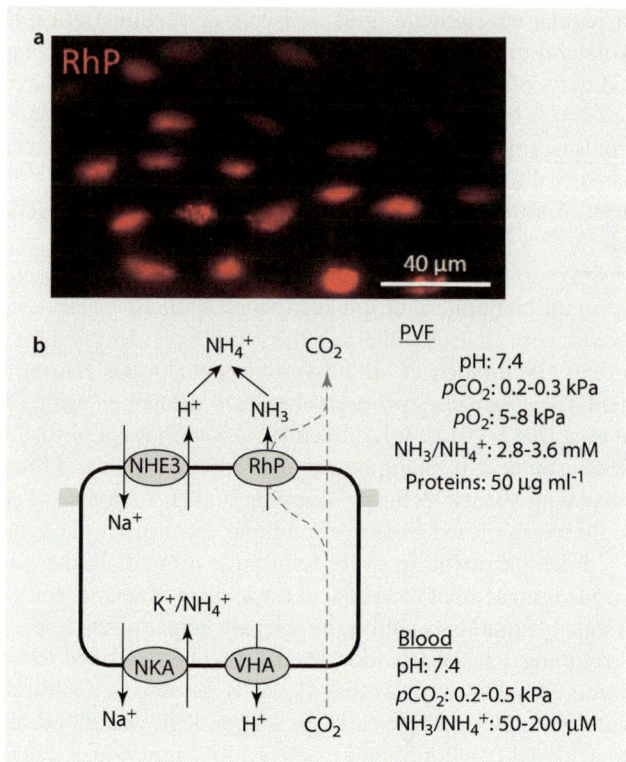

▫ Fig. 11.4 Epidermal ammonia excretion in cephalopod embryos. (**a**) Immunohistological detection of the Rhesus protein (RhP) using an antibody specifically designed against the RhP of squid (*Sepioteuthis lessoniana*). RhP-positive cells are scattered over the surface of the yolk epithelium and immunoreactivity appears in a spotted pattern associated with ionocytes. (**b**) Hypothetical model of NH_4^+ excretion in epidermal ionocytes of squid embryos. Basolateral import of NH_4^+ is achieved by the Na^+/K^+ ATPase (NKA) that can accept NH_4^+ as an alternative substrate to K^+. Apical excretion is mediated through an interplay of Na^+/H^+ exchanger 3 (NHE3) and RhP, leading to a trapping of the charged ammonium ion in the naturally acidified perivitelline fluid (PVF). Blood abiotic parameters were assumed to be similar to those found in adults (Hu et al. 2014). The role of the basolateral V-type H^+ ATPase (VHA) remains unclear and is suggested to be involved in NH_4^+ formation and/or facilitation of CO_2 formation and excretion

References

Albrecht J (2007) Ammonia toxicity in the central nervous system. In: Lajtha A, Oja SS, Schousboe A, Saransaari P (eds) Handbook of neurochemistry and molecular neurobiology. Springer, Berlin, Heidelberg

Arnold JM (1965) Normal embryonic stages of the squid, *Loligo pealii* (Lesueur). Biol Bull 128:24–32

Bonnaud L, Franko D, Vouillot L, Bouteau F (2013) A study of the electrical polarization of *Sepia officinalis* yolk envelope, a role for Na^+/K^+-ATPases in osmoregulation? Comm Integr Biol 6:26035

Boron WF (2006) Acid-base transport by the renal proximal tubule. J Am Soc Nephrol 17:2368–2382

Boucher-Rodoni R, Mangold K (1988) Comparative aspects of ammonia excretion in cephalopods. Malacologica 29:145–151

Boucher-Rodoni R, Mangold K (1989) Respiration and nitrogen excretion by the squid *Loligo forbesi*. Mar Biol 103:333–338

Boucher-Rodoni R, Mangold K (1994) Ammonia production in cephalopods, physiological and evolutionary aspects. Mar Fresh Behav Physiol 25:53–60

Brix O, Bardgard A, Cau A, Colosimo SGC, Giardina B (1989) Oxygen-binding properties of cephalopod blood with special reference to environmental temperatures and ecological distribution. J Exp Zool 252:34–42

Budelmann BU (1988) Morphological diversity of equilibrium receptor systems in aquatic invertebrates. In: Atema J, Fay RR, Popper AN, Tavolga WN (eds) Sensory biology of aquatic animals. Springer, New York, pp 757–782

Budelmann BU, Schipp R, von Boletzky S (1997) Cephalopoda. In: Harrison FW, Kohn AJ (eds) Microscopic anatomy of invertebrates. Wiley-Liss, New York

Camargo JA, Alonso Á (2006) Ecological and toxicological effects of inorganic nitrogen pollution in aquatic ecosystems: a global assessment. Environ Int 32:831–849

Cao L, Caldeira K (2008) Atmospheric CO_2 stabilization and ocean acidification. Geophys Res Lett 35:L09591. doi:10.1029/2008GL035072

Charmantier G, Charmantier-Daures M (2001) Ontogeny of osmoregulation in crustaceans: the embryonic phase. Amer Zool 41:1078–1089

Checkley J, David M, Dickson AG, Takahashi M, Radich JA, Eisenkolb N, Asch R (2009) Elevated CO_2 enhances otolith growth in young fish. Science 324:1683

Cronin ER, Seymour RS (2000) Respiration of the eggs of the giant cuttlefish *Sepia apama*. Mar Biol 136:863–870

De Leersnyder M, Lemaire J (1972) Sur la composition minérale du liquide périembryonnaire de l'oef de *Sepia officinalis* L. Cah Biol Mar 13:429–431

De Robertis EM, Kuroda H (2004) Dorsal-ventral patterning and neural induction in *Xenopus* embryos. Annu Rev Cell Dev Biol 20:285–308

Decleir W, Lemaire J, Richard A (1971) The differentiation of blood proteins during ontogeny in *Sepia officinalis* L. Comp Biochem Physiol B 40:923–928

Denton EJ, Gilpin-Brown JB (1961) The buoyancy of the cuttlefish, *Sepia officinalis* (L.). J Mar Biol Assoc UK 41:319–342

Donaubauer HH (1981) Sodium- and potassium-activated adenosine triphosphatase in the excretory organs of *Sepia officinalis* (Cephalopoda). Mar Biol 63:143–150

Dorey N, Melzner F, Martin S, Oberhänsli F, Teyssié JL, Bustamante P, Gattuso JP, Lacoue-Labarthe T (2013) Ocean acidification and temperature rise: effects on calcification during early development of the cuttlefish *Sepia officinalis*. Mar Biol 160:2007–2022

Dupont S, Ortega-Martinez O, Thorndyke MC (2010) Impact of near-future ocean acidification on echinoderms. Ecotoxicology 19:449–462

Emanuel CF, Martin AW (1956) The composition of octopus renal fluid. J Comp Physiol 39:226–234

Endeward V, Cartron JP, Ripoche P, Gros G (2008) RhAG protein of the Rhesus complex is a CO_2 channel in the human red cell membrane. FASEB J 22:64–73

Evans DH, Piermarini PM, Choe KP (2005) The multifunctional fish gill: dominant site of gas exchange, osmoregulation, acid-base regulation, and excretion of nitrogenous waste. Physiol Rev 85:97–177

Fabry VJ, Seibel BA, Feely RA, Orr JC (2008) Impacts of ocean acidification on marine fauna and ecosystem processes. J Mar Sci 65:414–432

Feely RA, Sabine CL, Hernandez-Ayon JM, Ianson D, Hales B (2008) Evidence for upwelling of corrosive "acidified" water onto the continental shelf. Science 320:1490–1492

Fehsenfeld S, Weihrauch D (2013) Differential acid-base regulation in various gills of the green crab *Carcinus maenas*: effects of elevated environmental pCO_2. Comp Biochem Physiol A 164(1):54–65. doi:10.1016/j.cbpa.2012.09.016

Gilmour KM, Perry SF (2009) Carbonic anhydrase and acid-base regulation in fish. J Exp Biol 212:1647–1661

Gruswitz F, Chaudhary S, Ho JD, Schlessinger A, Pezeshki B, Ho CM, Sali A, Westhoff CM, Stroud RM (2010) Function of human Rh based on structure of RhCG at 2.1 Å. Proc Natl Acad Sci 107:9638–9643

Guppy M, Withers P (1999) Metabolic depression in animals: physiological perspectives and biochemical generalizations. Biol Rev 74:1–40

Gutowska MA, Melzner F (2009) Abiotic conditions in cephalopod (*Sepia officinalis*) eggs: embryonic development at low pH and high pCO_2. Mar Biol 156:515–519

Gutowska MA, Pörtner HO, Melzner F (2008) Growth and calcification in the cephalopod *Sepia officinalis* under elevated seawater pCO_2. Mar Ecol Prog Ser 373:303–309

Gutowska MA, Melzner F, Langenbuch M, Bock C, Claireaux G, Pörtner HO (2010a) Acid-base regulatory ability of the cephalopod (*Sepia officinalis*) in response to environmental hypercapnia. J Comp Physiol B 180:323–335

Gutowska MA, Melzner F, Pörtner HO, Meier S (2010b) Cuttlebone calcification increases during exposure to elevated seawater pCO_2 in the cephalopod *Sepia officinalis*. Mar Biol 157:1653–1663

Hanlon RT (1990) Maintenance, rearing, and culture of teuthoid and sepioid squids. In: Gilbert DL, Adelman JW, Arnold JM (eds) Squid as experimental animals. Plenum Press, New York

Hanstein S, de Beer D, Felle HH (2001) Miniaturized carbon dioxide sensor designed for measurements within plant leaves. Sens Actuators B 81:107–114

Heisler N (1986) Acid-base regulation in animals. Elsevier Biomedical Press, Amsterdam

Henry RP (1984) The role of carbonic anhydrase in blood ion and acid-base regulation. Amer Zool 24: 241–251

Henry RP (1988) Multiple functions of carbonic anhydrase in the crustacean gill. J Exp Zool 248:19–24

Henry RP, Lucu C, Onken H, Weihrauch D (2012) Multiple functions of the crustacean gill: osmotic/ionic regulation, acid-base balance, ammonia excretion, and bioaccumulation of toxic metals. Front Physiol 3:431

Hsiao CD, You MS, Guh YJ, Ma M, Jiang YJ, Hwang PP (2007) A positive regulatory loop between *foxi3a* and *foxi3b* is essential for specification and differentiation of zebrafish epidermal ionocytes. PLoS One 2:e302

Hu MY, Yan HY, Chung WS, Shiao YC, Hwang PP (2009) Acoustically evoked potential in two cephalopods inferred using the auditory brainstem response (ABR) approach. Comp Biochem Physiol 153: 278–283

Hu MY, Sucré E, Charmantier-Daures M, Charmantier G, Lucassen M, Melzner F (2010) Localization of ion regulatory epithelia in embryos and hatchlings of two cephalopods. Cell Tiss Res 441:571–583

Hu MY, Tseng YC, Lin LY, Chen PY, Charmantier-Daures M, Hwang PP, Melzner F (2011a) New insights into ion regulation of cephalopod molluscs: a role of epidermal ionocytes in acid-base regulation during embryogenesis. Am J Physiol Regul Integr Comp Physiol 301:1700–1709

Hu MY, Tseng YC, Stumpp M, Gutowska MA, Kiko R, Lucassen M, Melzner F (2011b) Elevated seawater pCO_2 differentially affects branchial acid-base transporters over the course of development in the cephalopod *Sepia officinalis*. Am J Physiol Regul Integr Comp Physiol 300:R1100–R1114

Hu MY, Lee JR, Lin LY, Shih TH, Stumpp M, Lee MF, Hwang PP, Tseng YC (2013) Development in a naturally acidified environment: Na^+/H^+-exchanger 3-based proton secretion leads to CO_2 tolerance in cephalopod embryos. Front Zool 10:51–67

Hu MY, Guh Y-J, Stumpp M, Lee JR, Chen RD, Sung PH, Chen YC, Hwang PP, Tseng YC (2014) Branchial NH_4^+-dependent acid-base transport mechanisms and energy metabolism of squid (*Sepioteuthis lessoniana*) affected by seawater acidification. Front Zool 11:55–72

Hu MY, Hwang PP, Tseng YC (2015) Recent advances in understanding trans-epithelial acid-base regulation and excretion mechanisms in cephalopods. Tissue Barriers 3:4

Hwang PP, Lee TH (2007) New insights into fish ion regulation and mitochondrion-rich cells. Comp Biochem Physiol A 148:479–497

Jänicke M, Carney TJ, Hammerschmidt M (2007) Foxi3 transcription factors and Notch signaling control the formation of skin ionocytes from epidermal precursors of the zebrafish embryo. Dev Biol 307:258–271

Kaplan MB, Mooney TA, McCorkle DC, Cohen A (2013) Adverse effects of ocean acidification on early development of squid (*Doryteuthis pealeii*). PLos ONA 8:e63714

Kroeker KJ, Kordas RL, Crim RN, Hendriks IE, Ramajo L, Singh GS, Duarte CM, Gattuso JP (2013) Impacts of ocean acidification on marine organisms: quantifying sensitivities and interaction with warming. Glob Change Biol 19:1884–1896

Kustu S, Inwood W (2006) Biological gas channels for NH_3 and CO_2: evidence that Rh (Rhesus) proteins are CO_2 channels. Transfus Clin Biol 13:103–110

Lacoue-Labarthe T, Martin S, Oberhänsli F, Teyssié JL, Markich J, Jeffree R, Bustamante P (2009) Effects of increased pCO_2 and temperature on trace element (Ag, Cd and Zn) bioaccumulation in the eggs of the common cuttlefish, *Sepia officinalis*. Biogeosci 6:2561–2573

Langenbuch M, Pörtner HO (2002) Changes in metabolic rate and N excretion in the marine invertebrate *Sipunculus nududs* under conditions of environmental hypercapnia: identifying effective acid-base variables. J Exp Biol 205:1153–1160

Larsen EH, Deaton LE, Onken H, O'Donnell M, Grosell M, Dantzler WH, Weihrauch D (2014) Osmoregulation and excretion. Compr Physiol 4:405–573

Lee YC, Yan JJ, Cruz SA LHJ, Hwang PP (2011) Anion exchanger 1b, but not sodium-bicarbonate cotransporter 1b, plays a role in transport functions of zebrafish H^+-ATPase-rich cells. Am J Physiol Regul Integr Comp Physiol 300:C295–C307

Lin CC, Lin LY, Hsu H, Prunet P, Horng J, Hwang PP (2012) Acid secretion by mitochondrion-rich cells of medaka (*Oryzias latipes*) acclimated to acidic freshwater. Am J Physiol Integr Comp Physiol 15: R283–R291

Lindskog S, Henderson LE, Kannan KK, Liljas A, Nyman PO, Strandberg B (1971) Carbonic anhydrase. In: Boyer PD (ed) The enzymes. Academic Press, New York

Liu ST, Tsung L, Horng JL, Lin LY (2013) Proton-facilitated ammonia excretion by ionocytes of medaka (*Oryzias latipes*) acclimated to seawater. Am J Physiol Regul Integr Comp Physiol 305:242–251

Llpiński MR (2010) Cephalopod life cycles: patterns and exceptions. S Afr J Mar Sci 20:439–447

Lykkeboe G, Johansen K (1982) A cephalopod approach to rethinking about the importance of the Bohr and Haldane effects. Pacific Sci 36:305–313

Maneja RH, Piatkowski U, Melzner F (2011) Effects of ocean acidification on statolith calcification and prey capture in early life cuttlefish, *Sepia officinalis*. J Shellfish Res 30:1011

Martin M, Fehsenfeld S, Sourial MM, Weihrauch D (2011) Effects of high environmental ammonia on branchial ammonia excretion rates and tissue Rh-protein mRNA expression levels in seawater acclimated Dungeness crab *Metacarcinus magister*. Comp Biochem Physiol A 160:267–277

Meijer AJ, Lamers WH, Chamuleau RAFM (1990) Nitrogen metabolism and ornithine cycle function. Physiol Rev 70:701–748

Melzner F, Gutowska MA, Langenbuch M, Dupont S, Lucassen M, Thorndyke MC, Bleich M, Pörtner HO (2009) Physiological basis for high CO_2 tolerance in marine ectothermic animals: pre-adaptation through lifestyle and ontogeny? Biogeosci 6:2313–2331

Melzner F, Thomsen J, Koeve W, Oschlies A, Gutowska MA, Bange HW, Hansen HP, Körtzinger A (2013) Future ocean acidification will be amplified by hypoxia in coastal habitats. Mar Biol 160:1875–1888

Miller DC, Poucher S, Cardin JA, Hansen D (1990) The acute and chronic toxicity of ammonia to marine fish and mysid. Arch Environ Contam Toxicol 19:40–48

Moreau M, Leclerc C (2004) The choice between epidermal and neural fate: a matter of calcium. Int J Dev Biol 48:75–84

O'Dor RK (2002) Telemetered cephalopod energetics: swimming, soaring, and blimping. Integr Comp Biol 42:1065–1070

O'Dor RK, Webber DM (1986) The constraints on cephalopods: why squid aren't fish. Can J Zool 64: 1591–1605

Pecl GT, Steer MA, Hodgson KE (2004) The role of hatchling size in generating the intrinsic size-at-age variability of cephalopods: extending the Forsythe hypothesis. Mar Freshw Res 55:387–394

Piermarini PM, Choi I, Boron WF (2006) Cloning and characterization of an electrogenic Na/HCO_3^- cotransporter from the squid giant fiber lobe. Am J Physiol Cell Physiol 292:C2023–C2045

Portmann A (1926) Der embryonale Blutkreislauf und die Dotterresorption bei *Loligo vulgaris*. Z Morph u Ökol Tiere 5:406–423

Pörtner HO (1990) An analysis of the effects of pH on oxygen binding by squid (*Illex illecebrosus, Loligo pealei*) haemocyanin. J Exp Biol 150:407–424

Pörtner HO (1994) Coordination of metabolism, acid-base regulation and haemocyanin function in cephalopods. Mar Fresh Behav Physiol 25:131–148

Pörtner H-O, Zielinski S (1998) Environmental constraints and the physiology of performance in squids. S Afr J Mar Sci 20:207–221

Pörtner HO, Webber DM, Boutilier RG, O'Dor RK (1991) Acid-base regulation in exercising squid (*Illex illecebrosus, Loligo pealei*). Am J Physiol Regul Integr Comp Physiol 261:R239–R246

Pörtner HO, Langenbuch M, Reipschläger A (2004) Biological impact of elevated ocean CO_2 concentrations: lessons from animal physiology and earth history. J Atmos Oceanic Tech 60:705–718

Potts WTW (1965) Ammonia excretion in *Octopus dofleini*. Comp Biochem Physiol 14:339–355

Potts WTW (1994) Kinetics of sodium uptake in freshwater animals – a comparison of ion exchange and proton pump hypotheses. Am J Physiol 266:R315–R320

Randall DJ, Tsui TK (2002) Ammonia toxicity in fish. Mar Poll Bull 45:17–23

Ries JB, Cohen AI, McCorkle DC (2009) Marine calcifiers exhibit mixed responses to CO_2-induced ocean acidification. Geol Soc Am 37:1131–1134

Robertson JD (1949) Ionic regulation in some marine invertebrates. J Exp Biol 26:182–200

Romero MF, Fulton CM, Boron WF (2004) The SLC4 family of HCO_3^- transporters. Eur J Physiol 447:495–509

Rosa R, Seibel BA (2008) Synergistic effects of climate-related variables suggest future physiological impairment in a top oceanic predator. Proc Natl Acad Sci U S A 105:20776–20780

Rosa R, Trübenbach K, Repolho T, Pimentel M, Faleiro F, Boavida-Portugal J, Baptista M, Lopes VM, Dionísio G, Leal MC, Calado R, Pörtner HO (2013) Lower hypoxia thresholds of cuttlefish early life stages living in a warm acidified ocean. Proc Roy Soc B 280:20131695

Rosa R, Trübenbach K, Pimentel M, Boavida-Portugal J, Faleiro F, Baptista M, Dionísio G, Calado R, Pörtner HO, Repolho T (2014) Differential impacts of ocean acidification and warming on winter and summer progeny of a coastal squid (*Loligo vulgaris*). J Exp Biol 15:518–525

Schipp R, von Boletzky S (1976) The pancreatic appendages of dibranchiate cephalopods. Zoomorph 86:81–98

Schipp R, Mollenhauer S, Boletzky S (1979) Electron microscopical and histochemical studies of differentiation and function of the cephalopod gill (*Sepia officinalis* L.). Zoomorph 93:193–207

Seibel BA, Goffredi S, Thuesen EV, Childress J, Robison BH (2004) Ammonium content and buoyancy in midwater cephalopods. J Exp Mar Biol Ecol 313:375–387

Seymour RS, Bradford DF (1994) Gas exchange through the jelly capsule of the terrestrial eggs of the frog, *Pseudophryne bibroni*. J Comp Physiol B 157:477–481

Sherrard KM (2000) Cuttlebone morphology limits habitat depth in eleven species of *Sepia* (Cephalopoda: Sepiidae). Biol Bull 198:404–414

Shih TH, Horng JL, Hwang PP, Lin LY (2008) Ammonia excretion by the skin of zebrafish (*Danio rerio*) larvae. Am J Physiol Cell Physiol 295:C1625–C1632

Skou JC (1957) The influence of some cations on an adenosine triphosphatase from peripheral nerves. Biochim Biophys Acta 23:394–401

Smith MR, Caron J-B (2010) Primitive soft-bodied cephalopods from the Cambrian. Nature 465:469–472

Soupene E, King N, Field E, Liu P, Niyogi KK, Huang CH, Kustu S (2002) Rhesus expression in a green alga *Chlamydomonas reinhardtii* at high CO_2. Proc Natl Acad Sci U S A 99:7769–7773

Soupene E, Inwood W, Kustu S (2004) Lack of the Rhesus protein Rh1 impairs growth of the green alga *Chlamydomonas reinhardtii* at high CO_2. Proc Natl Acad Sci U S A 101:7787–7792

Strobel A, Hu MY, Gutowska MA, Lieb B, Lucassen M, Melzner F, Pörtner HO, Mark FC (2012) Influence of temperature, hypercapnia, and development on the relative expression of different hemocyanin isoforms in the common cuttlefish *Sepia officinalis*. J Exp Zool A 317:511–523

Stumpp M, Wren J, Melzner F, Thorndyke MC, Dupont ST (2011) CO2 induced seawater acidification impacts sea urchin larval development: elevated metabolic rates decrease scope for growth and induce developmental delay. Comp Biochem Physiol A 160:331–340

Stumpp M, Trübenbach K, Brennecke D, Hu MY, Melzner F (2012) Resource allocation and extracellular acid-base status in the sea urchin *Strongylocentrotus droebachiensis* in response to CO_2 induced seawater acidification. Aqua Toxicol 110–111:194–207

Virkki LV, Choi I, Davis BA, Boron WF (2002) Cloning of a Na^+-driven Cl/HCO_3^- exchanger from squid giant fiber lobe. Am J Physiol Cell Physiol 285:C771–C780

von Boletzki S (1987a) Embryonic phase. In: Boyle RP (ed) Cephalopod lifecycles. Academic Press, London, pp 5–31

von Boletzki S (1987b) Ontogenetic and phylogenetic aspects of the cephalopod circulatory system. Experientia 43:478–483

Wagner CA, Finberg KE, Breton S, Marshanski V, Brown D, Geibel JP (2003) Renal vacuolar H^+-ATPase. Physiol Rev 84:1263–1314

Wagner CA, Devuyst O, Belge H, Bourgeois S, Houillier P (2011) The rhesus protein Rhcg: a new perspective in ammonium transport and distal urinary acidification. Kidney Int 79:154–161

Watanabe S, Niida M, Maruyama T, Kaneko T (2008) Na^+/H^+ exchanger isoform 3 expressed in apical membrane of gill mitochondrion-rich cells in Mozambique tilapia *Oreochromis mossambicus*. Fish Sci 74:813–821

Weihrauch D (2006) Active ammonia absorption in the midgut of the larvae of the tobacco hornworm *Manduca sexta* L.: transport studies and mRNA expression analysis of a Rhesus-like ammonia transporter. Insect Biochem Mol Biol 36:808–821

Weihrauch D, Morris S, Towle DW (2004) Ammonia excretion in aquatic and terrestrial crabs. J Exp Biol 207:4491–4504

Weihrauch D, Wilkie MP, Walsh PJ (2009) Ammonia and urea transporters in gills of fish and aquatic crustaceans. J Exp Biol 212:1716–1730

Weiner ID, Verlander JW (2013) Renal ammonia metabolism and transport. Compr Physiol 3:201–210

Wells MJ (1990) Oxygen extraction and jet propulsion in cephalopods. Can J Zool 68:815–824

Wells JM, O'Dor RK (1991) Jet propulsion and the evolution of the cephalopods. Bull Mar Sci 49:419–432

Wells MJ, Wells J (1982) Ventilatory currents in the mantle of cephalopods. J Exp Biol 99:315–330

Wright P (1995) Nitrogen excretion: three end products, many physiological roles. J Exp Biol 198:273–281

Wright PA, Wood CM (2009) A new paradigm for ammonia excretion in aquatic animals: role of Rhesus (Rh) glycoproteins. J Exp Biol 212:2303–2312

Wu SC, Horng JL, Liu ST, Hwang PP, Wen ZH, Lin CS, Lin LY (2010) Ammonium-dependent sodium uptake in mitochondrion-rich cells of medaka (*Oryzias latipes*) larvae. Am J Physiol Cell Physiol 298:C237–C250

Yan JJ, Chou MY TK, Hwang PP (2007) Gene expression of Na^+/H^+ exchanger in zebrafish H^+-ATPase-rich cells during acclimation to low-Na^+ and acidic environments. Am J Physiol Cell Physiol 293:C1814–C1823

Service Part

© Springer International Publishing Switzerland 2017
D. Weihrauch, M. O'Donnell (eds.), *Acid-Base Balance and Nitrogen Excretion in Invertebrates*,
DOI 10.1007/978-3-319-39617-0

Index